Design for High Performance, Low Power, and Reliable 3D Integrated Circuits

Sung Kyu Lim

Design for High Performance, Low Power, and Reliable 3D Integrated Circuits

Springer

Sung Kyu Lim
School of Electrical and
Computer Engineering
Georgia Institute of Technology
777 Atlantic Drive NW
Atlanta, Georgia, USA

ISBN 978-1-4419-9541-4 ISBN 978-1-4419-9542-1 (eBook)
DOI 10.1007/978-1-4419-9542-1
Springer New York Heidelberg Dordrecht London

Library of Congress Control Number: 2012949555

Printed on acid-free paper

Springer is part of Springer Science+Business Media (www.springer.com)

To Mina, Yuna, and Jeanie

Preface

I started working on 3D IC and through-silicon-via (TSV) in 2001 when I first joined Georgia Tech. As a young faculty and a researcher, my mission was to find a topic that I can devote myself onto for many years to come and promises high risk and high return. Die stacking was not a new idea in 2001. But, the simple idea of stacking individual dies and connecting them using vias that vertically penetrate the whole die – the term TSV did not exist or widely accepted back then – made a lot of sense to me. The benefits were very obvious: shorter interconnects, shorter interconnects, and shorter interconnects (and then smaller footprint).

As many of the readers of this book are very well aware, interconnects are a huge headache in modern (and future, too) VLSI circuits and systems. Many researchers worldwide have been struggling everyday with interconnect-related issues. So, the moment you hear that interconnect lengths will reduce naturally and significantly, you begin to smile at its related benefits: higher performance, lower power consumption, fewer metal layers used, etc. And yes, extreme memory bandwidth for the architects. No, I did not forget about the cost saving for System-On-Chip (SOC) developers because they do not have to integrate all those mixed signal components into the same die!

Well, things have not exactly been the way everyone expected. Despite the huge volume of work and success stories on materials and manufacturing research and development, efforts in architecture, design, and CAD tools have been lagging behind. People have encountered electro-thermo-mechanical reliability issues associated with TSVs, and testing became highly challenging and expensive. Most of all, the killer application that will justify the huge initial investment on the manufacturing lines has not been identified until recently.[1] In the meantime, some have turned their attention to "2.5D integration," where TSVs are not used in the dies but in an interposer (silicon or glass) to integrate components mounted on it.

[1] Many agree that wide-I/O 3D DRAM for mobile applications will be the first mainstream product that commercializes TSVs.

This book came out of the effort during the last decade (2001–2012) of 3D IC design research and development at the Georgia Tech Computer-Aided Design (GTCAD) Laboratory, with more focus on the last 4 years (2009–2012). The scope of our research has expanded from physical design automation to architecture, modeling, pathfinding, and validation. We also have developed a real 3D IC that stacks one tier of 64 general-purpose cores and another tier of SRAM memory. Through this effort, we worked on the entire spectrum of design and testing for this test chip: architecture, layouts, CAD tools, package, board, and testing infrastructure. We also began to look beyond the conventional TSV-based 3D ICs and started investigating monolithic 3D IC design. The 20 chapters included in this book are organized to reflect this evolution. The first part of this book contains six chapters on design issues and solutions for high performance and low power 3D ICs.

- In Chap. 1, we study the pros and cons of two distinct ways to place through-silicon-vias (TSVs) in gate-level 3D layouts, namely, regular and irregular styles. We also study the area, wirelength, timing, and power overhead of TSVs in 3D IC layouts.
- In Chap. 2, we study how to build a Steiner tree for a given set of points in multiple dies. We also study how to relocate TSVs in a given set of Steiner trees to alleviate thermal hotspot issues.
- In Chap. 3, we study how to add buffers to a 3D net that connects gates in multiple dies in 3D IC to optimize signal delay and slew.
- In Chap. 4, we study how TSVs can be used to build a clock tree for 3D IC to reduce the total power consumption while minimizing clock skew.
- In Chap. 5, we study the issues in power delivery network design for 3D ICs and the impact of power/ground TSV usage on power supply noise.
- In Chap. 6, we study how to build a clock tree for 3D IC so that it can be used to deliver clock signal during pre-bond and post-bond testing.

The second part of this book contains 3 chapters on design-for-electrical-reliability for 3D ICs.

- In Chap. 7, we study the TSV-to-TSV coupling issues and investigate various ways to alleviate the associated problems.
- In Chap. 8, we investigate the current crowding problem at the wire-to-TSV junction in the power delivery network and its impact on IR-drop.
- In Chap. 9, we study the electro-migration failure mechanisms in TSVs caused by the current density, mechanical stress, and thermal gradient issues in 3D ICs.

The third part of this book contains 3 chapters on design-for-thermal-reliability for 3D ICs.

- In Chap. 10, we study thermal-aware architectural floorplanning for 3D IC and its impact on other metrics such as area, wirelength, and performance.
- In Chap. 11, we study gate-level placement techniques to alleviate thermal problems in 3D IC designs.

- In Chap. 12, we investigate the issues in codesign and co-analysis of thermal, power delivery, and performance targeting a 3D IC that employees micro-fluidic channels for cooling.

The fourth part of this book contains 5 chapters on design-for-mechanical-reliability for 3D ICs.

- In Chap. 13, we study the full-chip analysis of mechanical stress in 3D IC designs caused by the coefficient of thermal expansion (CTE) mismatch between TSV and silicon substrate.
- In Chap. 14, we study the impact of mechanical stress on device mobility and full-chip timing variations in 3D IC.
- In Chap. 15, we extend the full-chip study in Chap. 13 to investigate the impact of package elements on the mechanical reliability of the entire 3D chip/package system.
- In Chap. 16, we study the impact of chip/package mechanical stress on device mobility and full-chip path delay variations.
- In Chap. 17, we study the impact of TSV-induced mechanical stress on crack growth between TSV and its liner (= interfacial crack).

The last part of this book covers other topics on 3D IC design.

- In Chap. 18, we study the density, performance, and power benefit of monolithic 3D integration, where NMOS and PMOS are placed in two different tiers and connected with extremely small monolithic inter-tier vias (MIVs).
- In Chap. 19, we study the impact of TSV scaling on the area, wirelength, timing, and power quality of 3D designs done at the current and future technology nodes.
- In Chap. 20, we study the design, manufacturing, and testing of the 3D-MAPS (massively parallel processor with stacked memory), where one tier of 64 general-purpose cores and another tier of SRAM memory are bonded face to face for core-to-memory communication and utilize TSVs to communicate with the package.

These topics are mostly based on our work published at premier design and CAD conferences such as IEEE International Solid-State Circuits Conference (ISSCC), IEEE Custom Integrated Circuits Conference (CICC), ACM Design Automation Conference (DAC), IEEE International Conference on Computer-Aided Design (ICCAD), etc., during last 4 years (2009–2012).

This book is primarily intended for circuit designers and CAD tool developers from both industry and academia who are interested in learning about what the researchers at the Georgia Tech Computer-Aided Design (GTCAD) Laboratory and their colleagues have experienced from designing and validating high performance, low power, and reliable 3D ICs. However, a significant portion of this book is also based on our collaboration with people from other areas such as manufacturing, materials, testing, software applications, and computer architecture. This book discusses the needs and outcomes of such collaborations. The materials presented in this book are also based on our close collaboration with industry partners through funded research projects from Intel, IBM T. J. Watson, Samsung, Qualcomm, Mentor Graphics, and Cadence.

Despite our effort, this book may still contain errors. We will be truly grateful if you could help us correct those mistakes. Please send any report of bugs, misprints, and other errata to me at limsk@ece.gatech.edu. In the meantime, please visit our website for other resources and errata: http://users.ece.gatech.edu/limsk/3d-book.

Atlanta, GA, USA Sung Kyu Lim

Acknowledgements

I am thankful for the old and current members of the Georgia Tech Computer-Aided Design (GTCAD) Laboratory, who have done all the hard work: Mongkol Ekpanyapong, Jacob Minz, Faik Baskaya, Michael Healy, Mohit Pathak, Dae Hyun Kim, Krit Athikulwongse, Xin Zhao, Young Joon Lee, Chang Liu, Moongon Jung, Taigon Song, and Shreepad Panth. It indeed took us a long time to be where we are now, and it has been a tough but rewarding road. I will make sure to remind my future students how the seed was planted and grown by you guys.

My thanks go to my colleagues in industry, who helped our work to be practical, relevant, and forward-looking: Dr. Patrick Morrow (Intel), Dr. Clair Webb (Intel), Dr. Paul Fischer (Intel), Dr. Hong Wang (Intel), Dr. Vijay Pitchumani (Intel), Dr. Dusan Petranovic (Mentor), Dr. Kambiz Samadi (Qualcomm), Dr. Riko Radojcic (Qualcomm), Dr. Chan Seok Hwang (Samsung), Dr. Myung Soo Jang (Samsung), Dr. Michael Scheuermann (IBM), Dr. Rasit Topaloglu (IBM), Dr. Inki Hong (Cadence), Dr. Gabriel Loh (AMD), and Mr. Ho Choi (Amkor).

I have had the pleasure of working closely with the following colleagues in academia and their students: Prof. David Pan at UT Austin and his students (Jae-Seok Yang, Joydeep Mitra, Ashutosh Chakraborty, Jiwoo Pak), Prof. Hsien-Hsin Lee at Georgia Tech and his students (Mohammad Hossain, Ilya Khorosh, Dean Lewis, Tzu-Wei Lin, Guanhao Shen, Dong Hyuk Woo, Mario Vittes, Chinnakrishnan Ballapuram), Prof. Joungho Kim at KAIST and his students (Jonghyun Cho, Joohee Kim), Prof. Suresh Sitaraman at Georgia Tech and his student (Xi Liu), Prof. Saibal Mukhopadhyay and his students (Jeremy Tolbert, Minki Cho, Kwanyeob Chae, Amit Trivedi), and Prof. Paul Kohl, Prof. Rao Tummala, Prof. Yogendra Joshi, and Prof. Madhavan Swaminathan at Georgia Tech. The pioneering work on 3D IC design by Prof. James Meindl at Georgia Tech, Prof. Jason Cong at UCLA, Prof. Sachin Sapatnekar at Minnesota, and Prof. Paul Franzon at NC State are truly inspiring.

I must thank all my research sponsors: the National Science Foundation (NSF), the US Department of Defense (DOD), the Defense Advanced Research Projects Agency (DARPA), Semiconductor Research Corporation (SRC), Semiconductor Manufacturing Technology (SEMATECH), Intel, IBM, Samsung, Qualcomm, Mentor Graphics, and Cadence. Without their support and patience, this book simply does not exist.

My personal thanks go to Chuck Glaser at Springer. He was the first one to see the potential of this book and went an extra mile to convince his colleagues at Springer. He remained supportive in the midst of my numerous requests for deadline extensions. Everyone at the production team deserves my thanks for their hard work.

My sincere thanks go to my two daughters Mina and Yuna, and my wife Jeanie. Four years have passed since I wrote the first book with Springer in May 2008. Back then I felt very bad for my daughters for not being around too much when they were very young (4 and 2 years old). This time, however, they have been very nice and patient, thanks to the 4 years of maturity. Well, in fact, they were busy with their own fun and did not need their daddy that much. My wife this time did not get to design the cover page, but I still bought her flowers. Last, but not least, our parents-in-laws truly deserve all the credit.

Contents

Acronyms

1P6M	one poly six metal layers
2D IC	two-dimensional integrated circuit
3D IC	three-dimensional integrated circuit
3D-MAPS	3D massively parallel processor with stacked memory
AES	advanced encryption standard
AFD	atomic flux divergence
ALU	arithmetic and logic unit
B2B	back-to-back bonding
BCB	benzo-cyclo-butane
BEOL	back end of the line
BILP	binary integer linear programming
C4	controlled collapse chip connection
CA	thermal coupling-aware placement
CACTI	cache access and cycle time information
CAD	computer-aided design
CMOS	complementary metal oxide semiconductor
COR	cell occupancy ratio
CPU	central processing unit
CTE	co-efficient of thermal expansion
CTS	clock tree synthesis
DARPA	defense advanced research projects agency
DDR	double data rate
DEF	design exchange format
DES	data encryption standard
DME	deferred-merge and embedding
DOE	design of experiments
DP	dynamic programming
DRAM	dynamic random access memory
DRC	design rule checking
ECO	engineering change order
EDA	electronic design automation

EM	electro-migration
ERR	energy release rate
F2B	face-to-back bonding
F2F	face-to-face bonding
FEA	finite element analysis
FEM	finite element method
FF	flip-flops
FFT	fast Fourier transform
FIR	finite impulse response
FM	Fiduccia Mattheyses circuit partitioning algorithm
FO4	fan-out 4
FPGA	field-programmable gate array
FPU	floating point unit
GDSII	graphic database system version 2
GSRC	giga-scale systems research center
HFSS	high frequency structure simulator
HPWL	half perimeter wire length
I/O	input and output
ICT	interconnect technology file
ILD	inter-layer dielectric
ILP	integer linear programming
IMEC	interuniversity microelectronics center
IP	intellectual property
IPC	instructions per cycle
IR	current (I) x resistance (R)
ISA	instruction set architecture
ITRS	international technology roadmap for semiconductors
IWLS	international workshop on logic and synthesis
JPEG	joint photographic experts group
KCL	Kirchhoff's current law
KOZ	keep out zone
KVL	Kirchhoff's voltage law
LEF	library exchange format
LGA	land grid array
LP	landing pad
LP	linear programming
LPD	longest path delay
LVLS	lateral and vertical linear superposition
LVS	layout versus schematic
MCNC	microelectronics center of North Carolina
MFC	micro fluidic channel
MIPS	microprocessor without interlocked pipeline stages
MI-T	monolithic 3D integration at transistor-level
MIV	monolithic inter-tier via
MMM	method of means and medians

MOSFET	metal oxide semiconductor field-effect transistor
MST	minimum spanning tree
MTTF	mean time to failure
MUX	multiplexer
NCSU	North Carolina State University
NLP	non-linear programming
NMOS	n-type metal-oxide-semiconductor field-effect transistors
OTA	overall timing accuracy
P/G	power and ground
PCB	printed circuit board
PDE	partial differential equations
PDK	process design kit
PDN	power distribution network
PMOS	p-type metal-oxide-semiconductor field-effect transistors
PRESS	prediction error sum of squares
PTM	predictive technology model
Q3D	quasi three dimensional
QRC	Cadence parasitic resistance and capacitance extraction tool
RAT	required arrival time
RBF	radial basis function
RC	resistance and capacitance
RDL	re-distribution layer
RMSE	root mean square error
ROI	region of interest
RSM	response surface methodology
RTL	register transfer level
RUU	register update unit
SA	simulated annealing
SDF	standard delay format
SEM	scanning electron microscope
SERT	Steiner Elmore routing tree algorithm
SI	signal integrity
SIP	system-in-package
SOC	system-on-chip
SPDP	slew propagation dynamic programming
SPEC	standard performance evaluation corporation
SPEF	standard parasitic exchange format
SPICE	simulation program with integrated circuit emphasis
SRAM	static random access memory
STA	static timing analysis
STI	shallow trench isolation
TCAD	technology computer aided-design
TCG	transitive closure graph
TG	transmission gate
TLB	translation lookaside buffer

TNS	total negative timing slack
TSA	through-silicon-via spread and alignment method
TSV	through-silicon-via
T-TSV	thermal through-silicon-via
VGA	video graphics array
VGDP	van Ginneken dynamic programming
VLIW	very long instruction word
VLSI	very large scale integration
WL	wire length
WNS	worst negative timing slack

Part I
High Performance and Low Power 3D IC Designs

This part of the book covers design issues and solutions for high performance and low power 3D ICs. In Chap. 1, we study the pros/cons of two distinct ways to place through-silicon-vias (TSVs) in gate-level 3D layouts, namely, regular and irregular styles. We also study the area, wirelength, timing, and power overhead of TSVs in 3D IC layouts. In Chap. 2, we study how to build a Steiner tree for a given set of points in multiple dies. We also study how to relocate TSVs in a given set of Steiner trees to alleviate thermal hotspot issues. In Chap. 3, we study how to add buffers to a 3D net that connects gates in multiple dies in 3D IC to optimize signal delay and slew. In Chap. 4, we study how TSVs can be used to build a clock tree for 3D IC to reduce the total power consumption while minimizing clock skew. In Chap. 5, we study the issues in power delivery network design for 3D ICs and the impact of power/ground TSV usage on power supply noise. In Chap. 6, we study how to build a clock tree for 3D IC so that it can be used to deliver clock signal during pre-bond and post-bond testing.

Chapter 1
Regular Versus Irregular TSV Placement for 3D IC

Abstract Through-silicon via (TSV) is the enabling technology for fine-grained integration of multiple dies into a single 3D stack. However, TSVs occupy significant silicon area due to their sheer size, which has a great effect on the power and performance of 3D ICs. Whereas well-managed TSVs alleviate routing congestion, reduce wirelength, and improve performance, excessive or ill-managed TSVs not only increase the die area but also degrade performance and power. In this chapter, we study the impact of TSVs on the quality of 3D IC layouts. We first study two design schemes, namely TSV co-placement (irregular TSV placement) and TSV site (regular TSV placement), for the design of 3D ICs. In addition, we develop a force-directed 3D gate-level placement algorithm to find optimal locations of TSVs and gates. One key problem to solve in regular TSV placement is how to assign 3D nets to pre-placed TSVs. To solve this problem effectively, we study two TSV assignment algorithms, compare them with other TSV assignment algorithms, and analyze the impact of the quality of TSV assignment algorithms on 3D ICs. Experimental results show that the wirelength of 3D ICs is shorter than that of 2D ICs by up to 25 %. We also compare timing and power of 2D and 3D ICs.

The materials presented in this chapter are based on [19].

1.1 Introduction

Three-dimensional integrated circuits (3D ICs) are emerging as a promising way to overcome interconnect scaling problems of 2D ICs and improve performance further. In 3D ICs, gates are placed in multiple dies, and the dies are stacked vertically on top of each other as illustrated in Fig. 1.1. Since gates are distributed in multiple dies, footprint area of each die of a 3D IC becomes smaller than that of the circuit designed in 2D. Smaller footprint area results in shorter total wirelength in

S.K. Lim, *Design for High Performance, Low Power, and Reliable 3D Integrated Circuits*, DOI 10.1007/978-1-4419-9542-1_1,

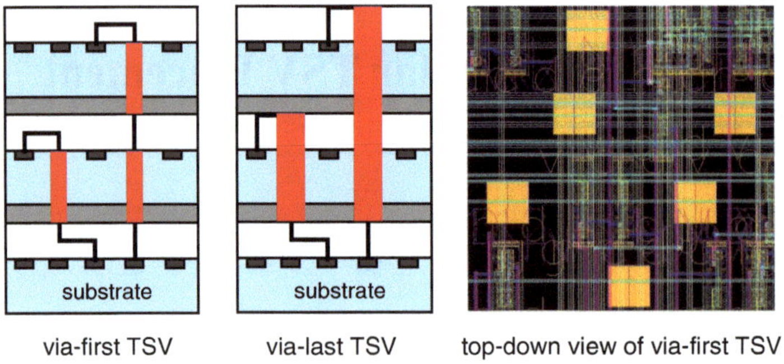

Fig. 1.1 Via-first and via-last TSVs with face-to-back bonding

3D ICs than in 2D ICs [16, 20]. Therefore, 3D ICs have a high potential to improve performance [18, 31]. Shorter wirelength can also reduce interconnect power and improve routing congestion. Less routing congestion can in turn reduce the number of metal layers used for routing in each die of 3D ICs, and the reduction of the metal layer count can contribute to cost reduction [9].

Vertical interconnects across dies in 3D ICs are enabled by through-silicon vias (TSVs). Figure 1.1 shows two types of TSVs, via-first and via-last TSVs. Via-first TSVs are fabricated during the front-end-of-line process, so they span from the backside of the bulk silicon to the bottom of the metal1 layer. On the other hand, via-last TSVs are fabricated during the back-end-of-line process, so they span from the backside of the bulk silicon to the top metal layer. After formation of TSVs, metal landing pads are attached to the TSVs. Typical size of via-first TSVs ranges from 1 to 5 μm, and that of via-last TSVs ranges from 5 to 20 μm [1, 2, 5, 22, 25].

Although TSVs play the most important role in gate-to-gate connections across dies, TSVs have negative impact on 3D IC designs. Above all, TSVs are fabricated in bulk silicon as shown in Fig. 1.1, so they consume silicon area, which otherwise can be used for gates. In addition, even empty silicon area below metal1 landing pads cannot be used for other gates. Moreover, we have to satisfy keep-out zone rules that forbid gates to be placed near TSVs. Because of these constraints and requirements, inserting excessive amount of TSVs into 3D ICs can cause serious area overhead. In addition, TSVs consume routing resources because TSVs need to be routed to gates or other TSVs through metal layers. This might cause routing congestion. Therefore, CAD tools for the design of 3D ICs should carefully take the impact of TSVs into account during placement and routing. However, most of previous works on CAD algorithms and tools for 3D ICs such as [8, 12] ignore either the sheer size of TSVs or the fact that TSVs interfere with gates and/or wires.

In this chapter, we study the design overhead of TSVs for 3D ICs. Based on DRC-clean GDSII layouts, we show a complete set of results such as wirelength, die area, timing, and power. The following specific topics are covered in this chapter:

- We study a force-directed 3D placement algorithm. This algorithm is based on the force-directed 2D placement algorithm presented in [29]. We can optimize wirelength of 3D ICs efficiently by this extended 3D placement algorithm.
- We study two 3D IC design flows, namely "TSV co-placement" and "TSV site". The TSV co-placement design scheme places TSVs and gates simultaneously, whereas the TSV site design scheme places TSVs at regular positions and then places gates.
- We study two TSV assignment algorithms for the TSV site scheme in which 3D nets are assigned to pre-placed TSVs. We also compare four TSV assignment algorithms (3D minimum spanning tree (MST)-based, 3D placement-based, minimum cost flow-based, and neighborhood search-based) and study the impact of the quality of TSV assignment algorithms on the wirelength.
- Since there exist many excellent 2D routers, we study how to use existing 2D routers to complete routing in 3D ICs.
- Since TSVs have negative effects such as occupying silicon area and having non-negligible capacitance, we study various layouts and empirically obtain and show the impact of TSVs on area, wirelength, timing, and power of 3D ICs.
- The placement and TSV assignment algorithms presented in this chapter are integrated into a commercial tool. This new tool flow generates GDSII-level 3D layouts that are fully validated. We perform various studies based on these GDSII layouts, and demonstrate the impact of TSVs on 3D IC layouts based on detailed layout data.

1.2 Existing Works

A few placement algorithms for the design of 3D ICs have been proposed in the literature. In [10], the authors randomly place standard cells within the placement area and use forces to move the cells in three dimensions to reduce cell overlap and temperature. The cells are moved from continuous space to discrete space by the legalization of the placement result. The authors sort cells in the z-direction before placing them into the nearest layer.

In [8], the authors transform a 2D placement result into 3D. The proposed transformations are based on folding and stacking a 2D design. After transformation, they use a graph-based layer assignment method to refine the 3D placement result by placing cells into multiple layers so that they reduce the number of TSVs and temperature. In [12], the authors study analytical and partitioning-based techniques for placement of 3D ICs. A recursive bisection approach is used during global placement. The authors assign a weight to each net based on its switching activity and capacitance as well as the number of TSVs. The cut direction for each bisection is selected as orthogonal to the largest of the width, height, or weighted depth of the placement area.

In [7], the authors study a multilevel non-linear programming based placement algorithm for 3D ICs. Their objective is the weighted sum of wirelength and the number of TSVs. The authors use a density penalty function to remove overlap in both the x- and y-direction as well as in the z-direction. They also use a bell-shaped density projection function to help obtain a legal placement in the z-direction. The work in [10] does not consider TSV at all in any stage. Although the works in [7, 8, 12] consider the number of TSVs, all of them do not take TSV area into account.

This work presented in this chapter is the first that considers TSV area and locations, so it is not possible to fairly compare our algorithm with previous 3D placement algorithms, which do not consider TSV area. In this chapter, therefore, we do not compare our work with any previous works on 3D placement.

1.3 Preliminaries

In this section, we introduce and explain design issues such as 3D placement and 3D design rule check (DRC). Then we speculate on TSV counts that have a huge impact on die area. Table 1.1 shows assumptions, parameters, and terminologies used in this chapter.

1.3.1 Design of 3D ICs

In 3D ICs, gates and TSVs are placed in multiple dies. Since both TSVs and gates occupy silicon area, we should avoid overlaps between them. In addition, TSVs should be routed to other TSVs or gates without violating design rules. Figure 1.2 illustrates connections to TSVs. Since a TSV is in fact connected to its M1 landing pad in the same die and Mtop landing pad in the bottom die (or backside landing

Table 1.1 Assumptions, parameters, and terminologies used in this chapter

	Value/meaning
Process technology	45 nm
Die bonding	Face-to-back
TSV shape	Square
TSV type	Via-first
TSV size ($n\times$)	Small ($1\times$), medium ($2\times$), large ($3\times$)
$n\times$ TSV width	$n \cdot 1.50\,\mu m$ ($n = 1$, 2, or 3)
$n\times$ TSV landing pad width	$n \cdot 2.07\,\mu m$ ($n = 1$, 2, or 3)
$n\times$ TSV cell width	$n \cdot 2.47\,\mu m$ ($n = 1$, 2, or 3)
Min. TSV-to-cell spacing	$n \cdot 0.4\,\mu m$ for $n\times$ TSV
Min. TSV-to-TSV spacing	$n \cdot 0.8\,\mu m$ for $n\times$ TSV
TSV resistance ($1\times$ TSV)	$100\,m\Omega$
TSV capacitance ($1\times$ TSV)	5 fF
2D net	A net whose cells exist in a single die
3D net	A net whose cells exist across multiple dies

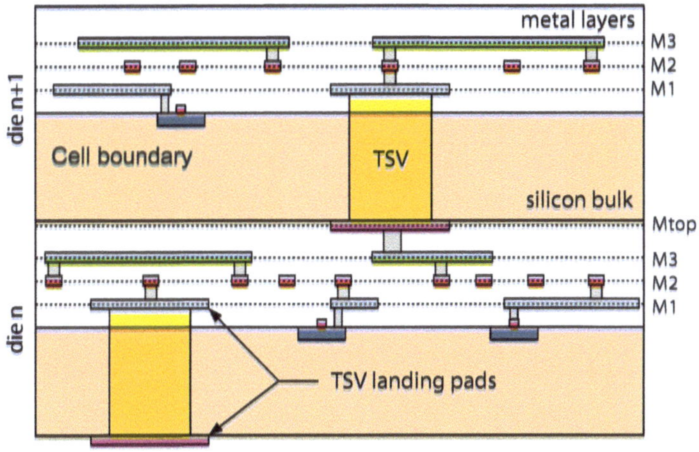

Fig. 1.2 TSVs, TSV landing pads, and connections to TSV landing pads

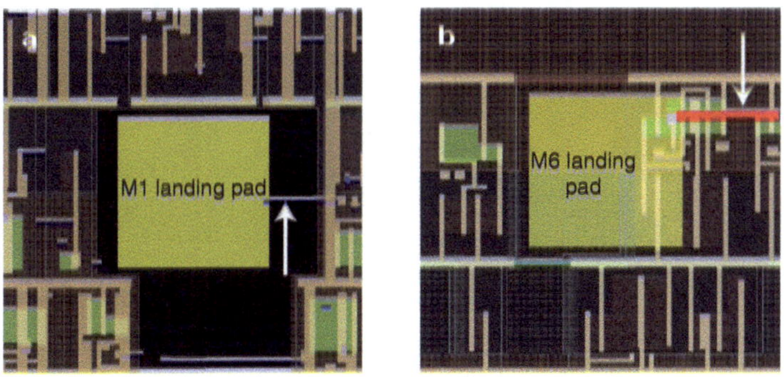

Fig. 1.3 TSV landing pads (*yellow*) and metal wires (M1 in *blue* and M6 in *red*) connected to the landing pads (Cadence Virtuoso). (**a**) A landing pad in M1. (**b**) A landing pad in M6

pad in the same die), we have to route wires to these landing pads for connections to TSVs. Figure 1.3 shows landing pads in M1 and M6 (=Mtop) and the wires connected to them in a top-down view. In case of the landing pad in Mtop in Fig. 1.3b, because it is located in Mtop, it does not interfere with gates in the same die.

3D IC layouts should also pass 3D DRC and 3D LVS as well as 2D DRC and 2D LVS. New 3D design rules include the minimum TSV-to-TSV spacing, the minimum TSV-to-cell spacing, the minimum (or maximum) TSV density, and so on. In this chapter, we apply the minimum spacing rules shown in Table 1.1 to our layouts. 3D LVS can easily be checked by existing LVS tools because LVS checks logical connections.

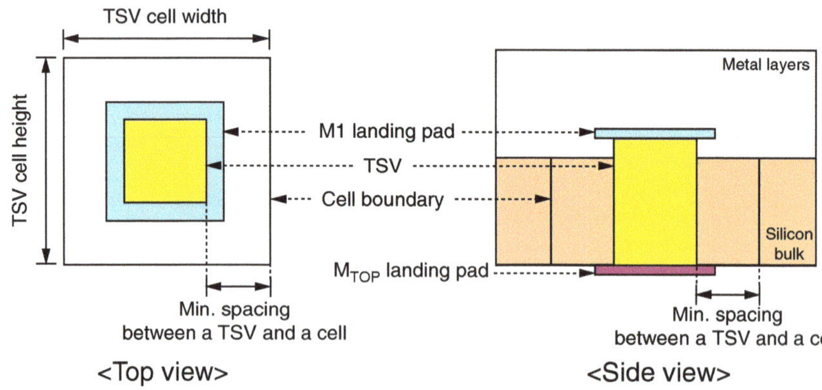

Fig. 1.4 Definition of a *TSV cell*

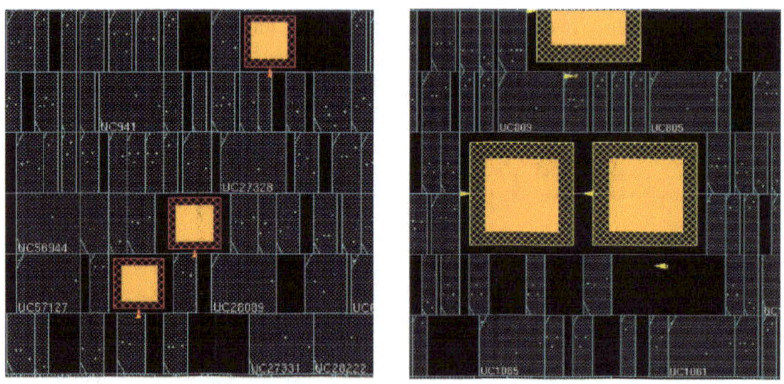

Fig. 1.5 1× TSV cell (= occupying a single standard cell row) vs. 2X TSV cell (= occupying two rows) in Cadence Encounter. *Orange squares* inside landing pads are TSVs

In our design flow, we treat TSVs as cells to automate placement and routing of TSVs while optimizing locations of cells and TSVs. In order to satisfy the minimum spacing requirement around TSVs during placement, we define a standard cell containing a TSV landing pad in M1 layer and whitespace around it. We will call this standard cell a *TSV cell* for the rest of this chapter. Figure 1.4 visualizes a TSV cell, and Fig. 1.5 shows 1× and 2× TSV cells placed in 3D IC layouts. In the 1× TSV case, a TSV cell occupies 2.47 × 2.47μm space, and a landing pad and a TSV are inside the cell.

1.3.2 *Maximum Allowable TSV Count*

Since the smallest 2D chip area is simply the total cell area, we can compute the maximum TSV count such that the chip area of a 3D IC is smaller than a predetermined number. The maximum TSV count, $N_{\mathrm{TSV}_{\max}}$, based on 2D and 3D chip areas can be calculated by the following equation:

$$N_{\mathrm{TSV}_{\max}} = (A_{\mathrm{3D}} - A_{\mathrm{2D}})/A_{\mathrm{TSV}}, \tag{1.1}$$

where A_{3D} is the sum of the area of all dies of a 3D IC, A_{2D} is the die area when the circuit is designed in 2D, and A_{TSV} is the area required by a TSV. The number of TSVs that we can use in 3D ICs is limited by $N_{\mathrm{TSV}_{\max}}$.

For example, the TSV diameter[1] of IMEC is 5um as of 2010 [14]. On the other hand, the smallest two-input NAND gate in the NCSU 45nm library [27] occupies 1.88 μm^2, and even the smallest D flip-flop occupies 4.52 μm^2 [26]. Without considering the keep-out zone, a 5um-diameter TSV occupies approximately 20 μm^2, which is $4\times$ to $10\times$ bigger than the standard cells in the 45nm libraries. If we consider the keep-out zone, this TSV occupies much larger area. Therefore, ignoring TSV area leads to a serious underestimation of TSV area overhead. For more realistic comparison, we show the average cell area of each benchmark circuit in the "Avg. cell area" column in Table 1.2. As the table shows, the area (6.1 μm^2 used in this chapter) occupied by a $1\times$ TSV cell is still larger than the average cell area.

1.3.3 Minimum TSV Count

While the maximum allowable TSV count constrains the maximum number of TSVs that can be used in 3D ICs, the minimum TSV count gives us the range of the number of available TSVs, thereby showing the adequacy of designing an IC in 3D with respect to area overhead. For example, suppose that the maximum allowable area overhead of a 3D IC is 15 % of its 2D layout area (e.g. 1 mm^2) and the TSV cell area is 6.1 μm^2. Then the maximum allowable TSV count becomes approximately 24,000. In this case, if the minimum TSV count is greater than 24,000, we cannot satisfy the area overhead constraint.

We can estimate the minimum TSV count by running min-cut partitioning because a 3D net spanning in K dies needs at least $K-1$ TSVs (one TSV between two adjacent dies). To estimate the minimum number of TSVs for each benchmark circuit used in this chapter, we use hMetis [17] for *k-way* min-cut partitioning, and Table 1.2 shows the minimum number of TSVs and their area overhead. As the table shows, some of the circuits (e.g., AL1) have huge area overhead (up to 34.15 %) caused by TSVs,[2] so they could not be designed in four dies if the maximum allowable area overhead is small (e.g., 10 %). However, this area overhead is

[1] We use diameter and width for cylindrical-shaped TSVs and square-shaped TSVs, respectively.

[2] The area overhead caused by a TSV is due to the TSV, liner around the TSV, and keep-out zone. In our work, A_{TSV} of a $1\times$ TSV is 6.1009 μm^2.

Table 1.2 Benchmark circuits and their partitioning results. We use hMETIS [17] for *k-way* min-cut partitioning. Area ratio shows the total TSV area divided by the total cell area. The area of a 1× TSVs is used. In the profile, "AL" denotes an arithmetic logic and "μP" denotes a microprocessor. Benchmark circuits marked with "*" are from IWLS 2005 benchmark suites [15], and the other benchmark circuits are from industry

					# dies						
					2		3		4		
Circuit	# gates (K)	# nets (K)	Total cell area (mm^2)	Avg. cell area (μm^2)	Min #TSV	Area ratio (%)	Min #TSV	Area ratio (%)	Min #TSV	Area ratio(%)	Profile
AL1	15	15	0.033	2.236	253	4.57	759	13.70	1,892	34.15	AL
AL2	30	30	0.082	2.754	964	7.07	1,535	11.26	2,910	21.34	AL
AL3*	77	77	0.227	2.937	733	1.94	956	2.53	2,019	5.34	IP core
AL4*	109	109	0.257	2.350	1,502	3.51	2,480	5.80	3,639	8.51	AL
AL5	324	328	1.163	3.585	452	0.23	334	0.17	1,107	0.57	AL
AL6	445	484	1.625	3.656	2,574	0.97	5,591	2.10	6,384	2.40	AL
MP1	16	16	0.039	2.460	87	1.35	465	7.22	542	8.41	μP
MP2	20	20	0.048	2.433	346	4.29	821	10.18	942	11.68	μP
MP3*	88	89	0.246	2.787	493	1.20	1,166	2.84	1,689	4.11	μP
MP4*	104	104	0.264	2.545	168	0.38	363	0.83	707	1.61	μP cores
MP5*	169	169	0.463	2.740	54	0.07	409	0.53	362	0.47	μP cores

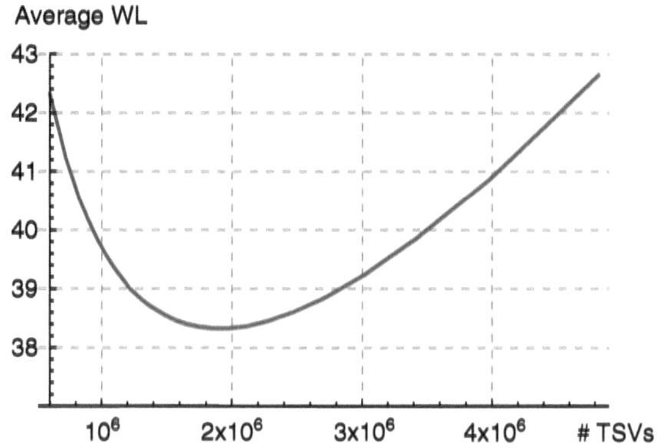

Fig. 1.6 Average wirelength (in gate socket pitches) vs. # TSVs used in a 3D IC. # cells = 40 M, # dies = 4, and TSV cell size = 6.1009 μm^2 [21]

strongly dependent on the TSV size and the average cell area. If a circuit consists of many large cells such as full adders and flip-flops or an older process such as 0.13 or 0.18 μm is used, the average cell area of the circuit will be much larger than the average cell area shown in Table 1.2. In this case, the area overhead caused by TSVs becomes smaller.

1.3.4 Tradeoff Between Wirelength and TSV Count

TSVs have two negative impact on the layout. First, they interfere with cells, thereby spreading cells out. As a result, the average distance between cells does not decrease as much as expected [20]. Second, TSVs cause routing congestion because they need to be connected to other TSVs or cells. Therefore using TSVs excessively in a 3D IC could lead to longer wirelength and worse performance than its 2D counterpart. Figure 1.6 shows a prediction result when the gate count is 40 M, the number of dies is four, and the TSV cell size is 6.1 μm^2 [21]. As the figure shows, the average wirelength goes down as the TSV count increases. However, the average wirelength goes up in the end if too many TSVs are used. This is mainly due to the area overhead caused by inserting too many TSVs. Therefore, we use the minimum number of TSVs in all of our experiments. However, in Sect. 1.7.3, we control the number of TSVs during partitioning, and observe the wirelength change as the number of TSVs varies.

1.4 3D IC Physical Design Flow

We devise two 3D IC design flows in this chapter, namely TSV co-placement and TSV site, as illustrated in Fig. 1.7. We develop these flows in such a way that we can use existing 2D routing tools while handling TSVs efficiently. By utilizing existing 2D routing tools, we can easily generate GDSII layouts of 3D ICs for in-depth analysis. Notice that our design flow is for via-first type TSVs.[3]

1.4.1 Partitioning

A way to perform 3D global placement using force-directed placement algorithms could be adding an axis along the z-direction. In this case, the quadratic wirelength function will be expressed as

$$\Gamma = \Gamma_x + \Gamma_y + \Gamma_z \,, \tag{1.2}$$

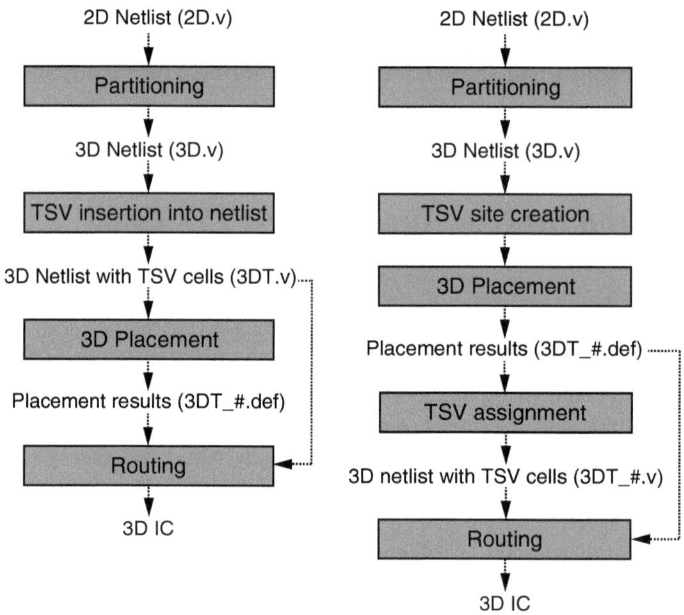

Fig. 1.7 Two 3D IC design flows developed in this chapter. (**a**) TSV co-placement, and (**b**) TSV site

[3]If we want to apply our design methodology to via-last type TSVs, we need additional steps. For the TSV co-placement flow, we need to avoid overlaps between two TSVs in adjacent dies. This can be resolved by applying another force between two TSVs in adjacent dies or by legalizing TSV locations after global placement. For the TSV-site flow, we can avoid overlaps between two TSVs in adjacent dies by using different TSV array size. If re-distribution layers exist, however, our design methodology can be directly applied to via-last type TSVs.

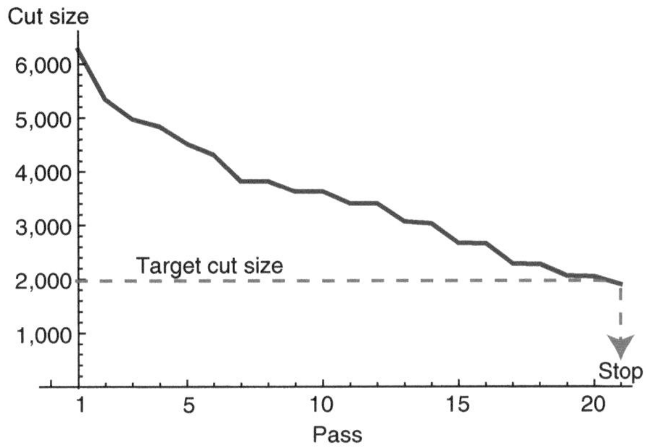

Fig. 1.8 Cut size control in partitioning. We use the AL4 circuit, and the target cut size is 2,000

where Γ_x, Γ_y, and Γ_z are wirelength along the x-, y-, and z-axis, respectively. Since Γ_x, Γ_y, and Γ_z are independent to each other, we can optimize each component of Γ separately as we do in 2D force-directed quadratic placement. However, this method cannot place cells in multiple dies unless the initial placement algorithm intentionally places cells in multiple dies. The reason is because all I/O pins are in the topmost die (die 0), so if cells are placed in the topmost die at the initial placement step, they will not be spread across multiple dies (but they are spread in the x- and y-directions). Therefore, we use partitioning as a pre-process for 3D global placement and forbid across-die movement of cells during 3D global placement.

After partitioning, we need to determine die ordering. For instance, suppose that we are designing a 3D IC using four dies. Then, a three-way partitioning produces three partitions, $p0$, $p1$, and $p2$. Assuming I/O pins are placed in die0, there exist six (= 3!) different die orderings. In this chapter, however, we do not use any particular die ordering scheme. Instead, we treat pX as die-X (e.g., $p0$ as die0). In fact, die ordering affects the minimum number of TSVs to be inserted. However, we observe that the total number of TSVs varies in a small range for different die orderings.

During partitioning, we can control the cut size to obtain the desired number of TSVs. If we want min-cut partitioning, we let the partitioner finish partitioning. However, if we want a specific cut size, we check the cut size after every pass and stop partitioning if the current cut size is less than or equal to the target cut size. Figure 1.8 shows an example of the cut size control. The min-cut size of the circuit (AL4) in the figure is approximately 1,500. However, we set the target cut size to 2,000 to increase the cut size intentionally. As the partitioning pass proceeds, the cut size decreases. At pass 20, the cut size is 2,126 and at pass 21, the cutsize becomes 1,984 which is below the target cutsize. Thus, we stop partitioning after the 21st pass.

The output of this step is a 3D netlist in which some of the nets in the original 2D netlist become 3D nets. After partitioning is completed, we compute the minimum

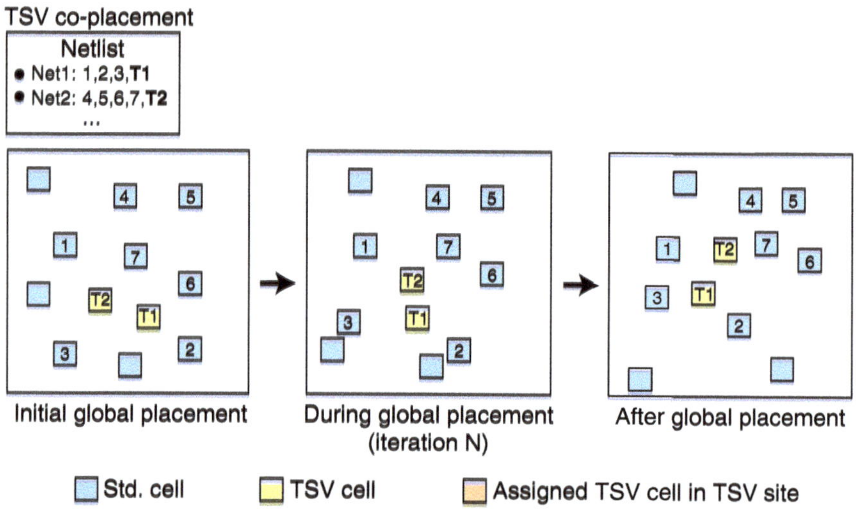

Fig. 1.9 TSV co-placement scheme: TSV insertion, 3D placement, and netlist generation

number of TSVs to be inserted. We use only one TSV for a 3D net between two adjacent dies because we want to minimize the area overhead caused by TSV insertion.

1.4.2 TSV Insertion and Placement

We study two ways to place TSVs in gate-level 3D IC designs: TSV co-placement (=irregular TSV location) and TSV site (=regular TSV location). In the TSV co-placement scheme, TSVs are added into the 3D netlist during the TSV insertion step, and then cells and TSVs are placed simultaneously during 3D placement. We explain our 3D placement algorithm in detail in Sect. 1.5. In the TSV site scheme, we pre-place TSVs uniformly on each die in the TSV site creation step, and then place cells in the 3D placement step. During 3D placement, pre-placed TSVs are treated as placement obstacles because TSVs should not overlap with any cell. After 3D placement in the TSV site scheme, we need an additional step, which we call TSV assignment, to determine which 3D nets use which pre-placed TSVs. Then, we update the 3D netlist to insert the assigned TSVs into the netlist. Figures 1.9 and 1.10 respectively illustrate the TSV co-placement and TSV site schemes. For detailed placement, we use the detailed placer of Cadence SoC Encounter [4].

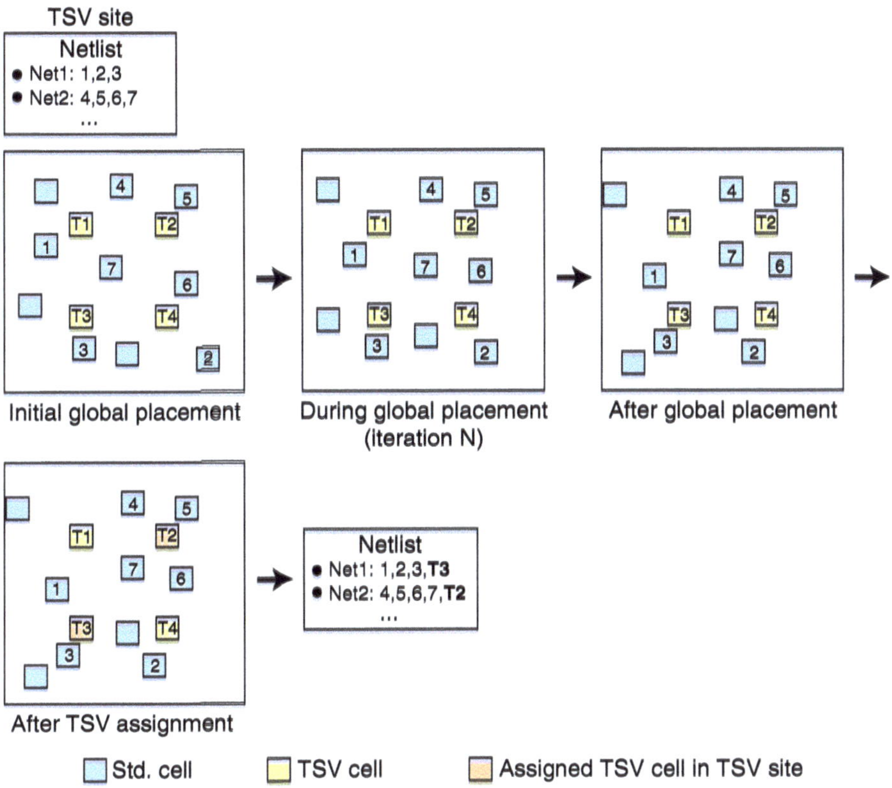

Fig. 1.10 TSV site scheme: TSV insertion, 3D placement, TSV assignment, and netlist generation

1.4.3 Routing

After 3D placement, we dump the placement result into DEF files and generate a netlist file for each die. One thing to notice is that we need to make TSV landing pads at both ends of a TSV as shown in Fig. 1.2. While we place an M1 landing pad in die(n + 1), we also have to place its corresponding Mtop landing pad in die(n) at the same location. We implement Mtop landing pads in die(n) by placing pins in the DEF file of die(n) and adding the pins into the netlist of die(n). Then, we use Cadence SoC Encounter to route each die.

1.5 3D Global Placement Algorithm

The 3D global placement algorithm used in this chapter is based on a force-directed quadratic placement algorithm [29]. We modified the algorithm to place cells and TSVs in 3D.

1.5.1 Overview of Force-Directed Placement

In quadratic placement, the optimal locations of cells are computed by minimizing the quadratic wirelength function Γ, which is expressed as

$$\Gamma = \Gamma_{\mathrm{x}} + \Gamma_{\mathrm{y}} \tag{1.3}$$

where Γ_{x} and Γ_{y} are wirelength along the x- and y-axis. Because Γ_{x} and Γ_{y} are independent to each other, they can be minimized separately to minimize Γ. The following description for the x-dimension applies similarly to the y-dimension. Here, Γ_{x} can be written in a matrix form as

$$\Gamma_{\mathrm{x}} = \frac{1}{2}\mathbf{x}^{\mathrm{T}}\mathbf{C}_{\mathrm{x}}\mathbf{x} + \mathbf{x}^{\mathrm{T}}\mathbf{d}_{\mathrm{x}} + \text{constant} \tag{1.4}$$

where $\mathbf{x} = [x_1 \cdots x_N]^{\mathrm{T}}$ is a vector representing the x-position of N cells being placed, $\mathbf{C}_{\mathrm{x}}$ is an $N \times N$ matrix representing the connectivity among the cells along the x-axis using bound-to-bound net model [29], and $\mathbf{d}_{\mathrm{x}} = [d_{\mathrm{x},1} \cdots d_{\mathrm{x},N}]^{\mathrm{T}}$ is a vector representing the connectivity from cells to pins along the x-axis. Element $c_{\mathrm{x},ij}$ of $\mathbf{C}_{\mathrm{x}}$ is the weight of connection between cell i and cell j, and element $d_{\mathrm{x},i}$ is the negative weighted position of fixed pins connected to cell i. The minimum of Γ_{x} is obtained by solving the following equation:

$$\nabla_{\mathrm{x}}\Gamma_{\mathrm{x}} = \mathbf{C}_{\mathrm{x}}\mathbf{x} + \mathbf{d}_{\mathrm{x}} = \mathbf{0} \tag{1.5}$$

Quadratic placement can be viewed as an elastic spring system when we treat Γ as the total spring energy of the system. Because the derivative of a spring energy is a force, the derivative of Γ_{x} in Eq. (1.4) can be viewed as a net force $\mathbf{f}_{\mathrm{x}}^{\mathrm{net}}$ as

$$\mathbf{f}_{\mathrm{x}}^{\mathrm{net}} = \nabla_{\mathrm{x}}\Gamma_{\mathrm{x}} = \mathbf{C}_{\mathrm{x}}\mathbf{x} + \mathbf{d}_{\mathrm{x}} \tag{1.6}$$

where $\nabla_{\mathrm{x}} = [\partial/\partial_{x_1} \cdots \partial/\partial_{x_N}]^{\mathrm{T}}$ is the vector differential operator. At equilibrium, $\mathbf{f}_{\mathrm{x}}^{\mathrm{net}}$ is zero and Γ_{x} is minimized. However, cells are crowded in few area of the chip, resulting in highly overlapped cell locations unless we apply other forces. In the force-directed quadratic placement in [29], therefore, two kinds of additional forces, move force $\mathbf{f}_{\mathrm{x}}^{\mathrm{move}}$ and hold force $\mathbf{f}_{\mathrm{x}}^{\mathrm{hold}}$, are used to remove cell overlap.

Move force is density-based force that spreads cells away from high cell density area to low cell density area to reduce cell overlap. Move force in [29] is defined for 2D ICs, thus we modify it to lower cell densities in 3D ICs. We explain the modification in Sect. 1.5.3.

Hold force is used to decouple each placement iteration from its previous iteration. It cancels out the net force that pulls cells back to the location in the previous iteration. Hold force is written as

$$\mathbf{f}_{\mathrm{x}}^{\mathrm{hold}} = -(\mathbf{C}_{\mathrm{x}}\mathbf{x}' + \mathbf{d}_{\mathrm{x}}) \tag{1.7}$$

where $\mathbf{x}' = [x'_1 \cdots x'_N]^{\mathrm{T}}$ is a vector representing the x-position of cells from the previous placement iteration. When no move force is applied, hold force holds cells at their current location. That can be shown by $\mathbf{f}_{\mathrm{x}}^{\mathrm{net}} + \mathbf{f}_{\mathrm{x}}^{\mathrm{hold}} = \mathbf{0} \Rightarrow \mathbf{C}_{\mathrm{x}}(\mathbf{x} - \mathbf{x}') = \mathbf{0} \Rightarrow \mathbf{x} = \mathbf{x}'$.

Total force $\mathbf{f}_{\mathrm{x}}$ is the summation of net force, move force, and hold force. The total force is set to zero,

$$\mathbf{f}_{\mathrm{x}} = \mathbf{f}_{\mathrm{x}}^{\mathrm{net}} + \mathbf{f}_{\mathrm{x}}^{\mathrm{move}} + \mathbf{f}_{\mathrm{x}}^{\mathrm{hold}} = \mathbf{0} \tag{1.8}$$

to minimize wirelength while removing cell overlap.

1.5.2 Overview of Our 3D Placement Algorithm

Our 3D placement algorithm is divided into three phases: initial placement, global placement, and detailed placement. In the first phase, we compute the initial cell locations by solving Eq. (1.5). In the second phase, we try to reduce cell overlaps by applying move force and hold force shown in Eq. (1.8) and solving the equation. We remove overlaps gradually because moving cells rapidly degrades the overall placement quality. Global placement continues until the amount of remaining cell overlap becomes lower than a pre-determined overlap ratio. Then we perform detailed placement by the detailed placer of Cadence Encounter.

1.5.3 Cell Placement in 3D ICs

It is not possible to extend the 2D force-directed quadratic placement algorithm to 3D placement algorithm simply by adding a z-axis variable in Eq. (1.3). The reason is that all the fixed pins in 3D ICs are on the C4-bump side, resulting in placing all the cells at the same z-position, $\mathbf{z} = \mathbf{0}$, in the initial placement [12]. In our work, therefore, we extend the force-directed quadratic placement algorithm presented in [29] by exploiting the fact that cells are already assigned into dies by a partitioner and not moving them across dies during placement. Therefore, we do not include Γ_{z} into Eq. (1.3), but let the placer focus on wirelength minimization along the x- and y-axis.

A major extension on the 2D force-directed quadratic placement algorithm in our work is to modify move force in [29] so that we remove cell overlap in each die separately. For instance, we do not apply move force between two cells at the same x and y location if they are in different dies.

The placement problem is formulated as a global electrostatic problem by treating cell area as positive charge and chip area as negative charge. The placement density D on die d can be computed by

$$D(x,y)\Big|_{z=d} = D^{\mathrm{cell}}(x,y)\Big|_{z=d} - D^{\mathrm{chip}}(x,y)\Big|_{z=d} \tag{1.9}$$

where $D^{\text{cell}}(x,y)\big|_{z=d}$ is the cell density at position (x,y) in die d, and $D^{\text{chip}}(x,y)\big|_{z=d}$ is the chip capacity at position (x,y) in die d.

After we compute D, we solve the following Poisson's equation and compute the placement potential Φ:

$$\Delta\Phi(x,y)\Big|_{z=d} = -D(x,y)\Big|_{z=d} \tag{1.10}$$

where the negative gradient of Φ indicates to which direction and how fast the cell at the position should move. Then, we model move force by connecting cell i to its target point $\mathring{x}_i$ with a spring of spring constant $\mathring{w}_i$. We compute the target point by

$$\mathring{x}_i = x'_i - \frac{\partial}{\partial x}\Phi(x,y)\Big|_{(x'_i,y'_i),z=d} \tag{1.11}$$

where x'_i is the x-position of cell i being placed on die d in the previous placement iteration. We initially define the spring constant by

$$\mathring{w}_i = \frac{A_i}{A^{\text{cell}}\big|_{z=d}}, \tag{1.12}$$

where A_i is the area of cell i, and $A^{\text{cell}}\big|_{z=d}$ is total area of cells being placed on die d. We then iteratively adjust the spring constant using quality control mechanism in [29]. Therefore, for cell i, move force is $f^{\text{move}}_{\text{x},i} = \mathring{w}_i(x_i - \mathring{x}_i)$, where x_i is the x-position of cell i. Move force $\mathbf{f}^{\text{move}}_{\text{x}}$ is finally defined for 3D ICs by

$$\mathbf{f}^{\text{move}}_{\text{x}} = \mathring{\mathbf{C}}_{\text{x}}(\mathbf{x} - \mathring{\mathbf{x}}) \tag{1.13}$$

where $\mathring{\mathbf{C}}_{\text{x}}$ is a diagonal matrix of $\mathring{w}_i$, $\mathbf{x} = [x_1 \cdots x_N]^{\text{T}}$ is a vector representing the x-position of N cells being placed, and $\mathring{\mathbf{x}} = [\mathring{x}_1 \cdots \mathring{x}_N]^{\text{T}}$ is a vector representing the target x-position of the cells. Figure 1.11 shows an illustration of the density and potential functions discussed in this section.

1.5.4 Pre-placement of TSVs in TSV Site Scheme

In the TSV site scheme, we first place TSVs evenly and then place cells. Therefore, we treat the TSVs as placement obstacles during cell placement. The number of TSVs in each row and column is computed by

$$N_{\text{TSV}_d} = N_{\text{TSV}_{d,\min}} \times K_{\text{TSV}}, \quad K_{\text{TSV}} \geq 1 \tag{1.14}$$

$$N_{\text{TSV}_{d,\text{row}}} = \lfloor\sqrt{N_{\text{TSV}_d}}\rfloor \tag{1.15}$$

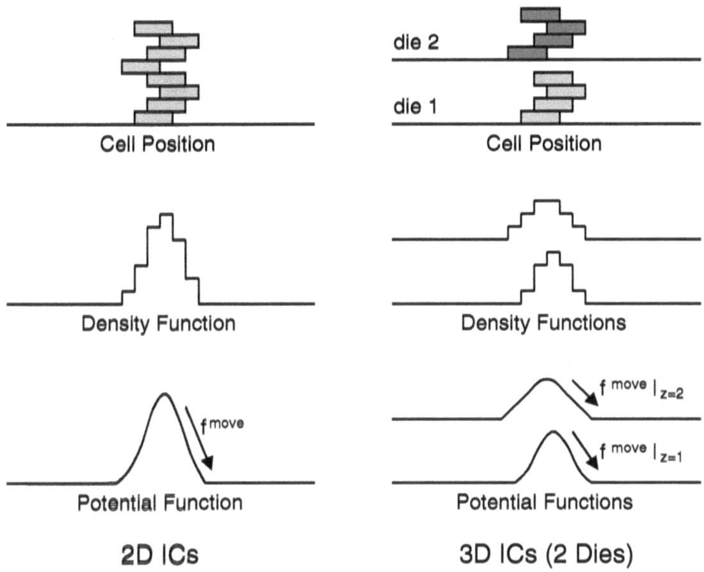

Fig. 1.11 Illustration of 2D/3D density and potential functions

$$N_{\mathrm{TSV}_{d,\mathrm{col}}} = \lceil N_{\mathrm{TSV}_d} / N_{\mathrm{TSV}_{d,\mathrm{row}}} \rceil \tag{1.16}$$

where $N_{\mathrm{TSV}_{d,\min}}$ is the minimum number of TSVs on die d, and K_{TSV} is a multiplying factor for the number of TSVs. If K_{TSV} is greater than one, we place more TSVs than the minimum TSV count so that we can have higher selectivity during TSV assignment.

Placement obstacles can be handled naturally by the means of placement density in [29]. By including the area of pre-placed TSVs when we compute placement density, we alter move force in such a way that it moves cells being placed away from pre-placed TSVs. We also include the area of pre-placed TSVs when we compute $D^{\mathrm{cell}}(x,y)\big|_{z=d}$ and $D^{\mathrm{chip}}(x,y)\big|_{z=d}$ in Eq. (1.9).

1.5.5 *Wirelength Computation for 3D Nets*

During wirelength estimation of 3D nets, we compute wirelength for each die individually as shown in Fig. 1.12. Since we use only one TSV to connect two adjacent dies for a 3D net, HPWL of the bounding box of each die is estimated.

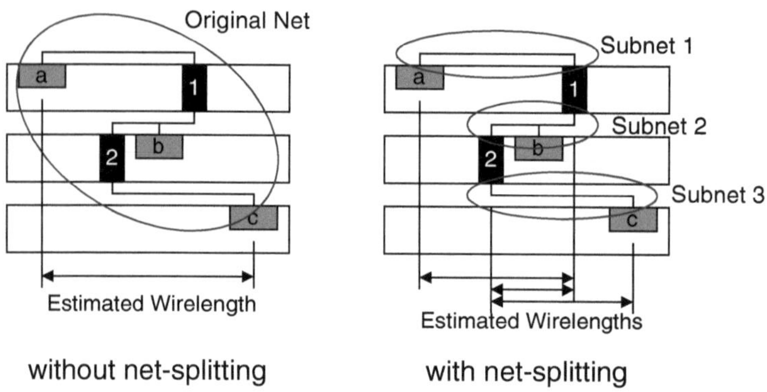

Fig. 1.12 Wirelength computation of a 3D net after subnet construction (side view)

1.6 TSV Assignment Algorithm

A TSV assignment problem in the TSV site scheme is to assign 3D nets to TSVs for given sets of dies, 3D nets, placed cells, and placed TSVs while optimizing objective functions such as total wirelength of 3D nets. The constraints in our TSV assignment problem are as follows:

- A TSV cannot be assigned to more than one 3D net.
- A 3D net should use at least one TSV.

1.6.1 Optimum Solution for TSV Assignment

The authors of [33] show the Binary Integer Linear Programming (BILP) formulation to find the optimum solution of the TSV assignment problem for two dies. Since the number of binary integer variables in the formula is too big, they also introduce and develop heuristic algorithms; an approximation method based on the Hungarian method [23] and a neighborhood search method.

If we have more than two dies and a 3D net spans in more than two dies, we have to take all the combinations of TSVs in different dies into account for the cost computation. In Fig. 1.13a, for example, the 3D net is assigned to T_1 in die 1 and T_6 in die 2, and the cost (= wirelength) is approximately $2L$. However, in Fig. 1.13b, the 3D net is assigned to T_3 in die 1 and T_6 in die 2, and the cost is approximately L. Although T_6 is used in both cases, its contribution to the cost is different. Therefore, the cost should be computed for each combination of TSVs in different dies.

The total number of combinations of TSVs is as follows:

$$N_{\text{comb}} = {}_{N_1}\mathrm{P}_{H_1} \times \cdots \times {}_{N_{\mathrm{D}-1}}\mathrm{P}_{H_{\mathrm{D}-1}} \tag{1.17}$$

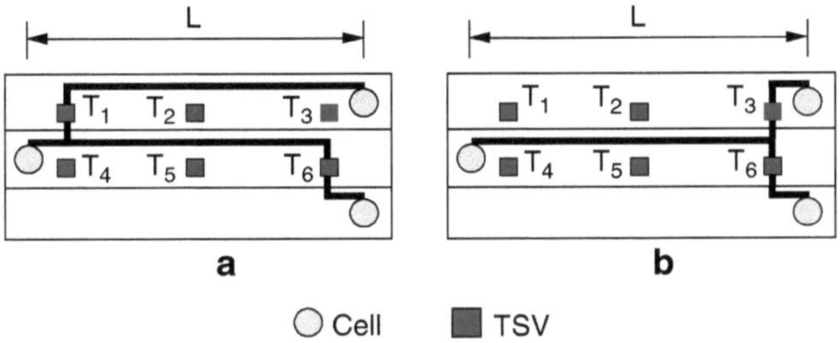

Fig. 1.13 Cost computation for different combinations of TSVs in three dies (side view). (**a**) Wirelength = $2L$ when T_1 and T_6 are selected. (**b**) Wirelength = L when T_3 and T_6 are selected

where N_{comb} is the total number of TSV combinations, D is the number of dies, N_i is the number of TSVs in die i, H_i is the number of 3D nets in die i, and P is the permutation symbol.

The optimum solution for a TSV assignment problem for more than two dies is found by the following BILP formulation:

Minimize

$$\sum_{i=1}^{N_{\text{3DNet}}} \sum_{k=1}^{CB_i} \sum_{p=1}^{N_{\text{TSV}}} d_{i,k,p} \cdot x_{i,k,p} \tag{1.18}$$

Subject to

$$\sum_{k=1}^{CB_i} \sum_{p=1}^{N_{\text{TSV}}} x_{i,k,p} = N_{\text{die}} - 1, \quad (i = 1, \cdots, N_{\text{3DNet}}) \tag{1.19}$$

$$\sum_{i=1}^{N_{\text{3DNet}}} \sum_{k=1}^{CB_i} x_{i,k,p} \leq 1 \quad (p = 1, \cdots, N_{\text{TSV}}) \tag{1.20}$$

where N_{die} is the number of dies, N_{3DNet} is the total number of 3D nets, CB_i is the total number of combinations of TSVs for the 3D net H_i, N_{TSV} is the total number of TSVs, and $d_{i,k,p}$ is the cost when the p-th TSV in the k-th combination is used for the 3D net H_i. Here, $x_{i,k,p}$ is 1 if (1) the 3D net H_i uses the k-th combination, and (2) the k-th combination uses the TSV T_p, and otherwise $x_{i,k,p}$ is 0. Equation (1.19) denotes that a 3D net uses only one combination of TSVs, and Eq. (1.20) denotes that a TSV is assigned to at most one 3D net.

The number of variables in this problem is also very big because we have to consider all the possible combinations for all 3D nets. Even if we restrict available TSVs for a 3D net to the TSVs inside a small window, the number of combinations is still big. For example, if a 3D net spans in four dies, and the window contains 20 TSVs in each die, 8,000 combinations exist for the net. Moreover, restriction on the window size may result in infeasibility of BILP. Therefore, we introduce two heuristic algorithms in the next two subsections.

1.6.2 MST-Based TSV Assignment

In this method, we use minimum spanning tree (MST) for TSV assignment as shown in Algorithm 1. We first sort 3D nets in the ascending order of their bounding box size. Since a 3D net whose bounding box is large contains more TSVs to choose inside the bounding box, we give higher priority to 3D nets having smaller bounding box. After sorting, we construct an MST using Kruskal's algorithm for each 3D net, and sort edges of the MST in the ascending order of their lengths because a short edge means short wirelength. After constructing an MST and sorting edges, we check each edge in the ascending order. If the edge spans over two adjacent dies which are not connected yet, we choose the unassigned TSV nearest to the edge and mark the TSV as "assigned to this net". We repeat this process until all dies in the 3D net are connected by TSVs. During this assignment process, the distance between an available TSV and a 3D edge is computed as follows. We first project the edge to a 2D plane so that the 3D edge becomes a 2D segment. Then, the distance is computed by calculating the Manhattan distance between the TSV and its nearest point in the segment.

```
Algorithm: MST-based TSV Assignment
Input: A set H of 3D nets
Output: Assignment result
H_sorted ← Sort (H); // in the order of bounding box size
for h ∈ H_sorted do
    E ← Construct_MST (h);
    E_sorted ← Sort (E);
    // in the ascending order of edge length
    for e ∈ E_sorted do
        d_b ← e.bot; // bottom die, e.g., die 3
        d_t ← e.top; // top die, e.g., die 0
        for i=d_t ; i¡d_b ; i++ do
            if i-th die and (i+1)-th die are not connected then
                t ← Find_TSV (e, i);
                Assign_TSV (t, h);
            end
        end
    end
end
```

Algorithm 1: MST-based TSV assignment algorithm

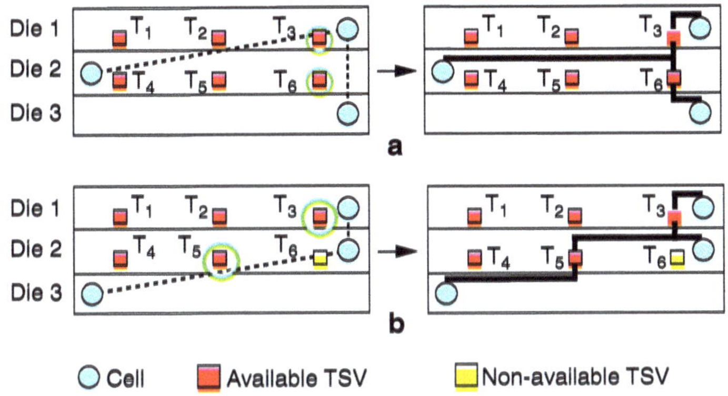

Fig. 1.14 Example: MST-based TSV assignment (side view)

Algorithm: Placement-based TSV Assignment
Input: placed gates and TSV sites
Output: each movable TSV is assigned to a TSV site
Convert all placed gates into fixed pins;
Insert movable TSVs into 3D nets in each die;
Run TSV co-placement;
Load the 3D placement result;
for *each movable TSV,* t_m **do**
    Find the nearest available TSV, t_s in TSV sites;
    Assign t_s to t_m;
end

Algorithm 2: Placement-based TSV assignment algorithm

Figure 1.14 shows two examples. In Fig. 1.14a, the shortest edge is the vertical edge. Since die 1 is not connected to die 2 yet, we find the TSV nearest to the shortest edge in die 1. In this example, T_3 is found and it is available, which means it has not been assigned to other nets, so we assign it to the 3D net. Now we need one more TSV to connect die 2 and die 3. Since the vertical edge spans from die 1 to die 3, we also find the TSV nearest to the edge in die 2. In the figure, T_6 is found and it is available, so we assign it to this 3D net. Now all dies are connected, so we stop assignment for this 3D net.

Figure 1.14b shows a different example. The shortest edge in Fig. 1.14b is the vertical edge connecting die 1 and die 2. The TSV nearest to the edge is T_3 and we assign it to the 3D net. Since this 3D net spans from die 1 to die 3, we need a TSV in die 2 to connect cells in die 2 and die 3. The TSV nearest to the next shortest edge is T_6. Since this is an unavailable TSV, which means it has already been assigned to another net, we find the next nearest TSV, T_5.

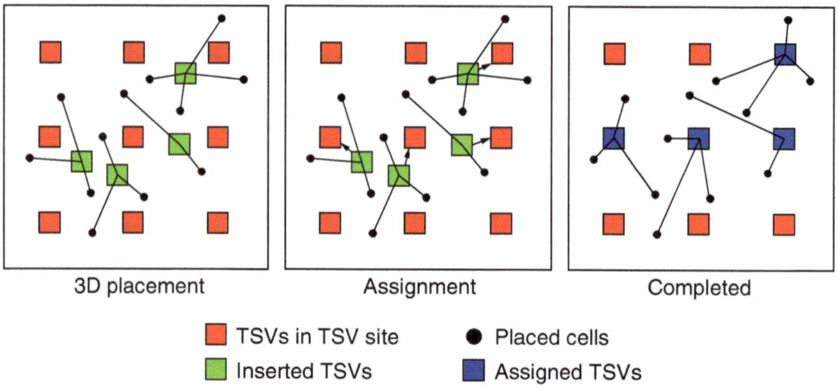

Fig. 1.15 Placement-based TSV assignment (*top view*)

1.6.3 Placement-Based TSV Assignment

The second TSV assignment method is based on 3D placement. In this method, we solve the assignment problem by a 3D placement algorithm. Figure 2 shows the placement-based TSV assignment algorithm. After placing gates in the 3D placement stage, we convert the placed gates into pins in a new 3D netlist. Therefore, there exist only pins, which are actually I/O pins and placed gates, in the netlist. Then, we insert movable TSVs into this netlist, and run TSV co-placement. After placement is finished, we load TSV locations from the 3D placement result. At this time, however, we assign TSVs inserted in the netlist to pre-placed TSVs. Figure 1.15 shows an example. In the first step, movable TSVs (green squares) are inserted into a new netlist and placed by our 3D placement algorithm. After 3D placement, we load the final locations of the movable TSVs, and assign them to their nearest pre-placed TSVs. The rightmost figure in Fig. 1.15 shows the final assignment result.

1.7 Experimental Results

We use IWLS 2005 benchmarks [15] and several industrial circuits. Table 1.2 shows the benchmark circuits and their details. We also use NCSU 45 nm technology [27] for the process technology. We implement our 3D placer and TSV assignment programs using C/C++ and Intel math kernel library 10.0 for matrix computation. We also use 64-bit Linux machines having Intel Xeon 2.5 GHz CPUs with 16 GB memory. Cadence QRC is used for RC extraction [3], and Synopsys PrimeTime is used for timing analysis [30]. For timing and power analysis, we use the typical PVT corner (supply voltage is 1.1 V and the temperature is 300 K). Figure 1.16 shows two representative layouts designed by the TSV co-placement and TSV site schemes.

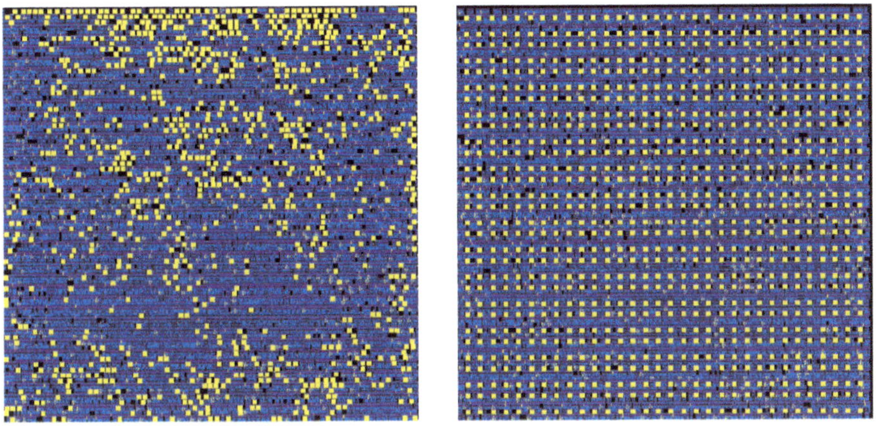

Fig. 1.16 Cadence Virtuoso snapshot of the bottommost die of AL1 designed by TSV co-placement and TSV site schemes. *Bright squares* are TSVs

1.7.1 Wirelength and Runtime Comparison

Table 1.3 shows wirelength, die area, and runtime of 2D and 3D placement results. For 2D placement, we run our placer in a 2D mode in which partitioning is not executed. For 3D placement, we use the TSV co-placement scheme, four dies, and 1× TSVs. As to wirelength, we reduce wirelength for all the circuits except MP5. In the MP5 case, the wirelength of the four-die 3D implementation is almost the same as that of its 2D implementation. Except MP5, the amount of wirelength reduction in non-microprocessor circuits (AL1–AL6) is 1–25 %, but the amount of wirelength reduction in microprocessor circuits (MP1–MP4) is 1–10 %.

In order to figure out the reason that the amount of wirelength reduction in non-microprocessor circuits is much higher than that in microprocessor circuits, we show wirelength distributions in Fig. 1.17 for AL4, which is a non-microprocessor circuit, and MP5, which is a set of microprocessors. As shown in Fig. 1.17a, long interconnections of AL4 in the 2D design become shorter in the 3D design. The longest wire in the 2D design of AL4 is about 900 μm-long, whereas the longest wire in the 3D design is about 310 μm-long. This effect obviously comes from smaller footprint area and connections in the z-direction by TSVs.

On the other hand, the wirelength distribution of the 2D design of MP5 is very similar to that of the 3D design of MP5 as shown in Fig. 1.17b. The lengths of the longest wires in the 2D and the 3D designs are also similar. Therefore, even if we stack multiple dies, the total wirelength does not change. In fact, MP5 is larger than AL4, but the longest wire of MP5 is shorter than that of AL4. This means that MP5 has few long wires. If there exist few long wires in a 2D design of a circuit, it is difficult to benefit from a 3D design of the circuit. Table 1.2 also supports this analysis. MP5 (0.463 mm^2) is bigger than AL4 (0.257 mm^2), but the longest wire

Table 1.3 Comparison of wirelength (WL), die area (Area), the minimum number of metal layers (#M) required to route all dies successfully, and runtime (Runtime) for 2D and 3D placement. For 2D placement, we run our placer without partitioning. For 3D placement, we use four-die implementation, TSV co-placement scheme, and 1× TSVs

	2D				3D				
ckt	WL (m)	Area (mm^2)	#M	Runtime (s)	WL (m)	Area (mm^2)	#M	Runtime (s)	#TSVs
AL1	0.26	0.05	4	143	0.24 (**0.95**)	0.06 (1.20)	4	51 (**0.36**)	1,305
AL2	0.63	0.12	5	266	0.50 (**0.79**)	0.14 (1.17)	4	180 (**0.68**)	2,792
AL3	1.43	0.32	4	7,758	1.26 (**0.88**)	0.34 (1.06)	4	5,840 (**0.75**)	1,906
AL4	1.94	0.37	5	18,312	1.46 (**0.75**)	0.38 (1.03)	4	3,627 (**0.20**)	4,857
AL5	9.17	1.69	5	145,466	9.07 (**0.99**)	1.96 (1.16)	5	64,023 (**0.44**)	4,368
AL6	12.96	2.22	5	159,505	11.06 (**0.85**)	2.28 (1.03)	5	102,472 (**0.64**)	10,859
MP1	0.24	0.05	4	208	0.22 (**0.90**)	0.06 (1.20)	4	69 (**0.33**)	801
MP2	0.32	0.06	4	175	0.31 (**0.97**)	0.08 (1.33)	4	100 (**0.57**)	939
MP3	1.86	0.35	4	6,742	1.72 (**0.92**)	0.36 (1.03)	4	2,510 (**0.37**)	3,084
MP4	1.72	0.37	5	4,634	1.70 (**0.99**)	0.38 (1.03)	4	1,523 (**0.33**)	1,681
MP5	2.73	0.66	5	17,655	2.73 (1.00)	0.67 (1.02)	4	3,676 (**0.21**)	659

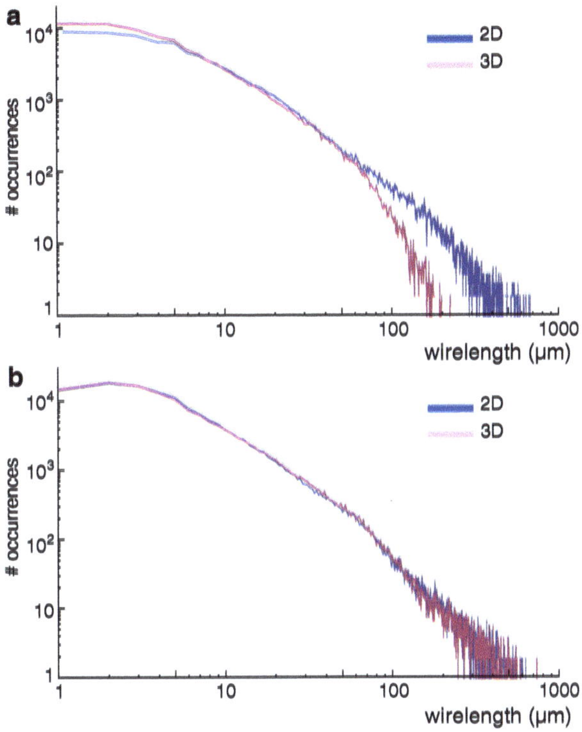

Fig. 1.17 Wirelength distribution of (**a**) AL4 whose die width is 605 μm in a 2D design and 310 μm in a 3D design (four dies). (**b**) MP5 whose die width is 812 μm in a 2D design and 410 μm in a 3D design (four dies)

of MP5 is shorter than that of AL4 (730 vs. 900 μm). Actually the die width of MP5 implemented in 3D is 410 μm, so the corner-to-corner Manhattan distance is 820 μm, which is longer than the longest wire (730 μm). However, the die width of AL4 implemented in 3D is 310 μm, so the corner-to-corner Manhattan distance is 620 μm, which is shorter than the longest wire (900 μm). Therefore, AL4 could benefit from 3D implementation, but MP5 could not.

This is also related to the min-cut partitioning result shown in Table 1.2. For example, the min-cut size of AL4 in 2-*way* partitioning is 1,502 out of 109 K nets while that of MP5 is 54 out of 169 K nets. This means that MP5 is a highly modularized circuit, so we may not be able to benefit from 3D implementation with respect to wirelength.

Regarding runtime, 3D placement in general needs shorter runtime than 2D placement. The reason is because an initial 3D placement of a circuit is likely to have less amount of overlaps than an initial 2D placement of the circuit because each die in 3D ICs has less number of cells to be placed. Since force-directed quadratic placement algorithm spends a significant portion of its runtime in overlap removal, having less number of cells in a die improves runtime. In Table 1.3, the 3D global placement is 1.3× to 5× faster than the 2D global placement.

Since the design of an IC needs routing as well as placement, we also need to compare the runtime for routing. The 3D placement generates N_{DIE} placement results. Therefore, we can run global and detailed routing for each die concurrently. Then, we can obtain the runtime for routing of a 3D IC by choosing the maximum runtime. In our simulation, the ratio between the runtime for routing of 2D ICs and the runtime for routing of 3D ICs is between 2.73 and 5.11. The reason that the runtime for routing of 3D ICs is much smaller than that for routing of 2D ICs is because the area of each die of a 3D IC is smaller than that of its 2D counterpart.

1.7.2 Metal Layers and Silicon Area Comparison

Since each die of a 3D design has less number of cells than a 2D design, the number of metal layers required for 3D designs could be less than that for 2D designs. Therefore, we find the minimum number of metal layers required to route all dies successfully. For fair comparisons, we use the same area utilization for both 2D and 3D designs. The "# ML" columns in Table 1.3 show comparisons of the minimum number of metal layers in 2D and 3D designs. Except AL5 and AL6, all the circuits are routable with four metal layers in their 3D designs, but the 2D designs of AL2, AL4, AL5, AL6, MP4, and MP5 are not routable with four metal layers because of high routing congestion.

Table 1.3 also shows area overhead of 3D IC layouts. For small circuits, area overhead is big (6–29 %). However, the area overhead in big circuits is relatively small (2–16 %). Since the area overhead is determined by the number of TSVs, if few TSVs are used for a small design, its area overhead could be negligible. Likewise, if too many TSVs are used for a large design, its area overhead could be significant.

The area of 3D designs is always larger than that of 2D designs in our experiments. However, the area of a 2D design could be larger than that of its 3D design. As seen in Table 1.3, some of the 2D designs are not routable with four metal layers. Therefore, if we have a constraint on the number of available metal layers (e.g., four metal layers), the 2D design that are not routable under the constraint should be expanded. In this case, the area of a 2D design could be larger than that of its 3D design.

1.7.3 Wirelength and TSV Count Tradeoff

Since we use partitioning as a pre-process for 3D placement, we experiment on how the TSV count affects wirelength reduction in 3D design. In this experiment, We use TSV co-placement scheme, 1× TSVs, and four dies. Figure 1.18 shows the results for AL4 and MP5. The wirelength of AL4 in 3D design monotonically increases as the TSV count increases. This indicates that the additional TSVs do not help wirelength reduction much. They rather increase die area thereby increasing the wirelength. On the other hand, the wirelength of MP5 in 3D design generally

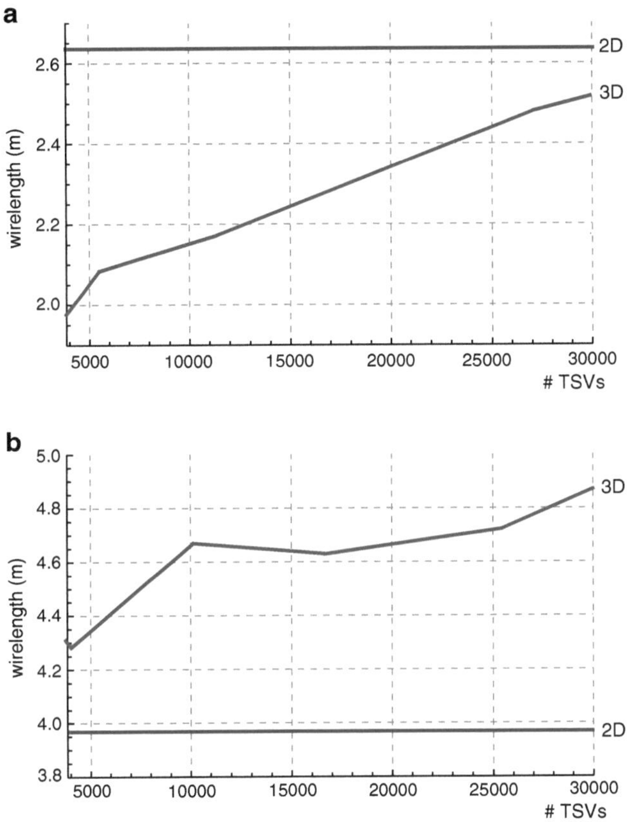

Fig. 1.18 Wirelength vs. # TSVs of (**a**) AL4, and (**b**) MP5 for 2D and 3D (four dies) designs

increases at first as the TSV count increases, but it saturates after all. Although we cannot draw a clear and obvious conclusion on the relationship between wirelength and the number of TSVs from these observations, using too many TSVs will eventually increase the die area, which will result in wirelength increase.

1.7.4 Wirelength, Die Area, and Die Count Tradeoff

As the number of dies increases, the footprint area tends to decrease,[4] so the wirelength is expected to decrease while the total die area is expected to increase.

[4]When the number of dies increases, if we ignore TSV area, the footprint area monotonically decreases. However, the number of TSVs has a great effect on the footprint area. If too many TSVs are used at a particular partitioning, the footprint area at that die count could increase.

Therefore, we observe trends of wirelength and die area when we increase the die count. In this experiment, we use the TSV co-placement scheme and 1× TSVs.

Table 1.4 shows wirelength, die area, runtime, and the number of TSVs when the die count varies from two to four. The number of TSVs in general increases as the die count goes up, thereby increasing the die area. The wirelength decreases as the die count goes up in many cases of the non-microprocessor circuits. However, a similar trend is not found in the microprocessor circuits.

For further experiment on this, we vary the number of dies (N_{die}) from 2 to 16 and observe wirelength, die area, and the number of TSVs for AL4. The wirelength of the 3D designs of AL4 dramatically decreases as N_{die} increases from two to five, then it fluctuates but in general goes up as shown in Fig. 1.19. If we increase N_{die} further, the TSV count and the die area will go up as shown in Fig. 1.20. In other words, increasing N_{die} is helpful at first, but becomes harmful in the end because the TSV count increases as N_{die} goes up and the increased TSV count in turn increases die area. We expect that similar trends will be found in other circuits. In addition, the use of numerous TSVs is definitely not helpful due to significant area overhead.

1.7.5 TSV Co-placement Versus TSV Site

Table 1.5 shows wirelength of five different placement schemes: the TSV co-placement (IR), the MST-based TSV site placement (MST), the placement-based TSV site placement (R-PL), the neighborhood search-based TSV site placement (R-NS, [33]), and the network flow-based TSV site placement (R-NF, [32]).

The TSV co-placement (IR) designs always show shorter wirelength than the TSV site placement designs. The amount of wirelength reduction of IR compared to MST, R-PL, R-NS, and R-NF is approximately 5 % on average in two-die implementations, 8 % on average in three-die implementations, and 9 % on average in four-die implementations. A reason that the TSV co-placement scheme produces shorter wirelength than the TSV site placement schemes is because the TSV co-placement scheme optimizes TSV locations and cell locations simultaneously, while pre-placed TSVs in the TSV site schemes obstruct optimal gate placement. As a result, the TSV site placement schemes have longer wirelength than the TSV co-placement scheme.

Table 1.6 shows the additional runtime required for TSV assignment in MST or R-PL. Since TSV assignment is an additional process after 3D placement, we also show runtime for 3D placement in IR and R columns. We choose TSVs for each net sequentially in MST, so the additional runtime is very small even for big circuits such as AL5 or AL6. Sorting nets in an MST takes $O(N \log N)$ where N is the total number of 3D nets, Kruskal's algorithm for MST construction takes $O(E_i \log E_i)$ where E_i is the total number of edges of net i, and selecting a TSV for an edge takes $O(T)$ where T is the total number of TSVs. Therefore, the complexity of the MST-based TSV assignment is $O(N \cdot E_i \cdot T)$. Although it is a cubic algorithm, it is very fast in reality because the number of 3D nets is not big (e.g., less than 5,000 from our

Table 1.4 Comparison of wirelength, die area, runtime, and the number of TSVs when the die count varies. We use TSV co-placement scheme and 1× TSVs. **3D-n** denotes *n*-die implementation. All the numbers except # TSVs are scaled to 2D implementation

	Wirelength			Area			Runtime			# TSVs		
Ckt	3D-2	3D-3	3D-4	3D-2	3D-3	3D-4	3D-2	3D-3	3D-4	3D-2	3D-3	3D-4
AL1	0.96	1.01	0.95	1.06	1.20	1.29	0.69	0.41	0.36	337	1,265	1,305
AL2	0.99	0.82	0.79	1.07	1.18	1.25	1.01	0.78	0.68	1,035	1,945	2,792
AL3	0.93	0.94	0.88	1.01	1.05	1.06	0.80	0.64	0.75	675	1,902	1,906
AL4	0.88	0.83	0.75	1.02	1.05	1.05	0.31	0.28	0.20	1,745	3,713	4,857
AL5	0.98	0.96	0.99	1.14	1.14	1.16	0.42	0.47	0.44	1,559	3,664	4,368
AL6	0.89	0.88	0.85	1.00	1.07	1.03	0.58	0.76	0.64	3,838	8,764	10,859
MP1	0.96	0.95	0.90	1.04	1.07	1.09	0.49	0.43	0.33	292	534	801
MP2	0.97	0.97	0.97	1.04	1.25	1.25	0.60	0.64	0.57	321	1,044	939
MP3	1.02	0.93	0.92	1.00	1.01	1.02	0.68	0.86	0.37	1,045	1,542	3,084
MP4	1.03	0.95	0.99	1.00	1.02	1.03	1.16	0.99	0.33	424	1,056	1,681
MP5	0.96	0.99	1.00	1.00	1.05	1.02	0.26	0.26	0.21	114	1,706	659

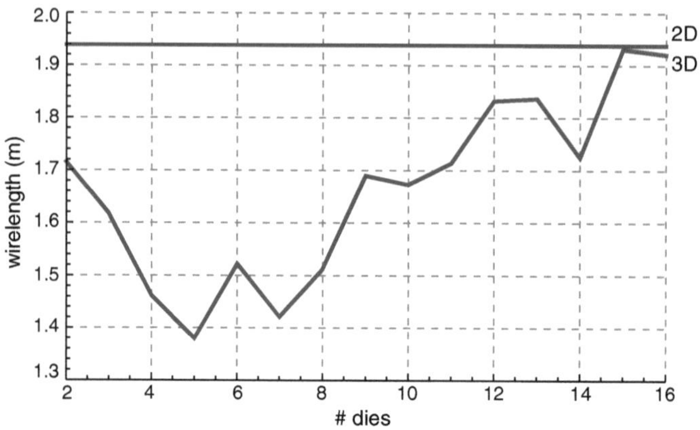

Fig. 1.19 Wirelength vs. # dies of AL4

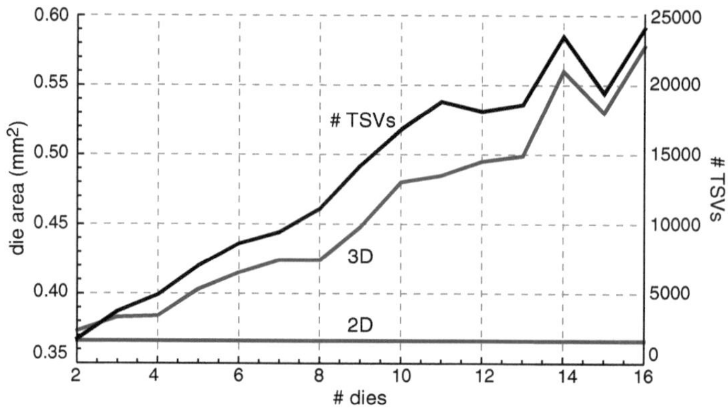

Fig. 1.20 Die area and # TSVs vs. # dies of AL4

experience) and E_i is usually quite small (e.g., less than 500 from our experience). As Table 1.6 shows, the additional runtime for TSV assignment in MST is very small compared to the runtime of 3D placement.

The additional runtime for R-PL is also very small. In the second placement phase, many pins, which consist of I/O pins and placed cells, determine the location of each TSV in the initial placement, so there is few overlap between TSVs. Therefore, the placement in R-PL needs only a few iterations of matrix computation. The numbers in parentheses in Table 1.6 show how many iterations of matrix computation we needed. Almost all of them are zero except a few cases in which only one to two iterations are necessary to remove all the overlaps among TSVs and cells.

On the other hand, the runtime for R-NS is almost negligible and the runtime for R-NF is prohibitively high. Therefore, we do not show the runtime for R-NS and R-NF in Table 1.6. Although the TSV co-placement scheme is better than the TSV

Table 1.5 Wirelength comparison for TSV placement types (scaled to the 2D placement result). **IR** denotes TSV co-placement, **R-MST** is MST-based TSV site placement, **R-PL** is placement-based TSV site placement, and **R-NS** is the TSV site placement scheme using the neighborhood search-based TSV assignment presented in [33]. We use 1× TSVs. **3D-*n*** denotes *n*-die implementation

		3D-2					3D-4				
ckt	2D	IR	R-MST	R-PL	R-NS	R-NF	IR	R-MST	R-PL	R-NS	R-NF
AL1	0.26*m* (1.00)	0.96	1.02	1.04	1.05	1.00	0.95	1.05	1.05	1.07	1.04
AL2	0.63*m* (1.00)	0.99	1.10	1.12	1.14	1.07	0.79	0.93	0.92	0.92	–
AL3	1.43*m* (1.00)	0.93	1.00	1.00	0.99	0.98	0.88	0.95	0.96	0.96	0.94
AL4	1.94*m* (1.00)	0.88	0.93	0.94	0.94	0.93	0.75	0.82	0.85	0.83	–
AL5	9.17*m* (1.00)	0.98	1.01	1.02	1.03	1.00	0.99	1.03	1.03	1.03	–
AL6	12.96*m* (1.00)	0.89	0.93	0.96	0.95	–	0.85	0.95	0.99	0.98	–
MP1	0.24*m* (1.00)	0.96	1.05	1.05	1.06	1.04	0.90	0.98	0.98	0.97	0.97
MP2	0.32*m* (1.00)	0.97	1.00	1.00	1.00	1.00	0.97	1.02	1.03	1.03	1.03
MP3	1.86*m* (1.00)	1.02	1.04	1.04	1.04	1.04	0.92	1.01	1.02	1.01	–
MP4	1.72*m* (1.00)	1.03	1.07	1.08	1.08	1.07	0.99	1.06	1.06	1.06	1.04
MP5	2.73*m* (1.00)	0.96	1.01	1.01	1.02	1.00	1.00	1.05	1.05	1.07	1.05
Geomean	1.00	0.96	1.01	1.02	1.03		0.90	0.98	0.99	0.99	

Table 1.6 Comparison of additional runtime for TSV assignment for the two TSV site placement schemes shown in Table 1.5. **IR** and **R** denote runtime of TSV co-placement and that of TSV site placement respectively. **A(R-MST)** (or **A(R-PL)**) denotes the additional runtime for TSV assignment of R-MST (or R-PL). The unit is second. (The numbers in parentheses denote total iteration counts of matrix computation during placement)

	3D-2				3D-4			
	Placement		Assignment		Placement		Assignment	
ckt	IR	R	A(R-MST)	A(R-PL)	IR	R	A(R-MST)	A(R-PL)
AL1	98	77	0.14	0.18 (0)	51	70	0.25	0.77 (0)
AL2	267	260	0.30	0.51 (0)	180	195	0.70	2.11 (0)
AL3	6,229	5,691	0.69	0.81 (0)	5,840	4,950	0.99	0.70 (0)
AL4	5,622	5,287	0.34	2.77 (0)	3,627	3,911	1.37	4.76 (0)
AL5	61,113	63,034	5.02	19.25 (0)	64,023	47,863	10.53	63.21 (2)
AL6	92,865	137,784	23.81	8.59 (0)	102,472	101,031	40.32	13.51 (0)
MP1	102	119	0.97	0.89 (1)	69	75	0.97	0.61 (0)
MP2	105	114	0.25	0.21 (0)	100	82	0.30	0.44 (0)
MP3	4,613	5,932	0.84	1.32 (0)	2,510	4,925	1.37	2.35 (0)
MP4	5,354	5,130	0.10	0.93 (0)	1,523	2,201	0.41	1.78 (0)
MP5	4,551	6,312	0.15	1.04 (0)	3,676	3,563	0.22	2.00 (0)

Table 1.7 Comparison of wirelength, die area, and runtime when the TSV size varies. We use the TSV co-placement scheme with four dies (Scaled to 1× TSV case)

	Wirelength			Area			Runtime		
ckt	1×	2×	3×	1×	2×	3×	1×	2×	3×
AL1	1.00	1.04	1.68	1.00	1.46	2.20	1.00	1.87	2.21
AL2	1.00	1.14	1.37	1.00	1.68	2.74	1.00	1.63	1.95
AL3	1.00	1.08	1.32	1.00	1.16	1.72	1.00	1.55	1.80
AL4	1.00	1.18	1.36	1.00	1.31	2.05	1.00	1.32	1.30
AL5	1.00	1.07	1.12	1.00	1.12	1.30	1.00	1.44	1.62
AL6	1.00	1.11	1.14	1.00	1.35	1.58	1.00	1.34	1.77
MP1	1.00	1.20	1.75	1.00	1.38	2.01	1.00	1.38	2.11
MP2	1.00	1.06	1.34	1.00	1.37	1.92	1.00	1.50	1.86
MP3	1.00	1.07	1.21	1.00	1.20	1.70	1.00	2.11	3.59
MP4	1.00	1.03	1.09	1.00	1.11	1.40	1.00	1.52	2.30
MP5	1.00	1.01	1.05	1.00	1.08	1.36	1.00	1.03	1.14
Geomean	1.00	1.09	1.29	1.00	1.28	1.77	1.00	1.49	1.89

site placement schemes with respect to wirelength, the TSV site placement scheme has its own advantages such as better heat dissipation and stronger package bonding according to [11, 13, 33].

1.7.6 *Impact of TSV Size*

Using big TSVs results in large die area overhead, thereby degrading total wirelength. It also causes more serious overlap among TSVs and cells, thereby increasing runtime for 3D placement. Therefore, we study the impact of TSV size on wirelength, die area, and runtime. Table 1.7 shows the results for 1×, 2×, and 3× TSV size. Wirelength always increases as the TSV size increases, and so does the die area. When the TSV becomes 2× bigger, the wirelength increases by 9 % on average and the die area increases by 29 % on average. However, the wirelength becomes 29 % longer and the die area becomes 77 % larger on average when the TSV becomes 3× bigger. The runtime increase is also not negligible. As the TSV size goes up from 1× to 2× and 3×, the runtime increases by 49 and 89 % respectively. Therefore, we observe that the use of bigger TSVs causes serious wirelength and runtime overhead as well as area overhead.

1.7.7 *Timing and Power Comparison*

One of the major goals to build 3D ICs is to improve performance. Therefore, we present timing analysis results in this section though our placer is not a timing-

Fig. 1.21 TSV RC model

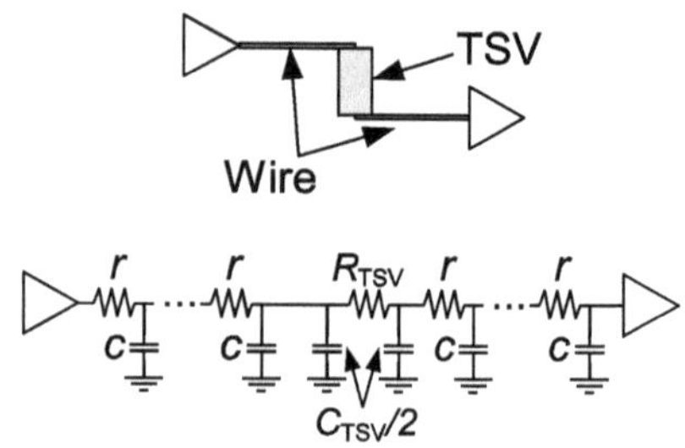

driven placer. For timing optimization, authors of [24] present timing scaling-based and timing budgeting-based timing optimization methods. In our work, we use the timing scaling-based optimization method. The optimization process is as follows. First, parasitic resistance and capacitance of each die are extracted. After we obtain SPEF files of each die, we create a top-level SPEF file in which TSVs are represented by TSV resistance and capacitance. Figure 1.21 shows the TSV RC model used in this chapter. Then, timing analysis is performed by Synopsys PrimeTime. This timing analysis generates delays of 3D paths as well as 2D paths. Then, input and output timing constraints of each die are generated by the tool in [24]. With these new constraints, we optimize each die separately in Cadence Encounter. To fully optimize a given circuit, we repeat this whole timing optimization process four times. We assume that the clock distribution is ideal.

Table 1.8 shows timing analysis results after timing optimization. The 2D design of AL1 has smaller WNS and TNS than the 3D designs, and the 2D design of AL2 has smaller WNS and TNS than the 3D designs except the case where the TSV co-placement is used for two-die implementation. On the other hand, some of the 3D designs outperform the 2D designs in the AL3 and AL4 cases. The reason that the 3D designs of the small circuits (AL1 and AL2) do not outperform 2D designs is because AL1 and AL2 are too small, so the TSV capacitance results in delay overhead. In the AL3 case, most of the 3D designs satisfy the timing constraints, but the 2D design has still non-negligible WNS and TNS. In the AL4 case, the worst negative slacks of all the 2D and 3D designs are similar, but the 3D designs have much smaller TNS.

Table 1.9 shows power analysis results in which we report the total power reported by PrimeTime. Although the wirelength of 3D designs are shorter than 2D designs, power consumptions of the 3D designs are similar to those of the 2D designs. The reason that shorter wirelength does not necessarily result in less power is due to non-negligible TSV capacitance (5 fF in our simulation[5]).

[5]The TSV capacitance used in [21] is 37 fF for a square-shaped TSV whose width and height are 5 and 50 μm, respectively. In our case, TSV width is 1.50 μm and TSV height is 20 μm, so we actually obtain 4.43 fF for our TSV capacitance by linear scaling because TSV capacitance is almost linearly proportional to TSV width and TSV height [28].

Table 1.8 Worst negative slack (WNS) and total negative slack (TNS) **after timing optimization**. We use 1× TSVs, and TSV capacitance is 5 fF. **3D-n** denotes *n*-die implementation, **IR** denotes the TSV co-placement, and **MST** is the MST-based TSV site placement. The unit of all the numbers is nanosecond

				3D-2				3D-4			
		2D		IR		MST		IR		MST	
ckt	Target freq (MHz)	WNS	TNS	WNS	TNS	WNS	TNS	WNS	TNS	WNS	TNS
AL1	200	−1.85	−61	−2.02	−73	−2.19	−68	−2.74	−125	−2.25	−78
AL2	100	−3.22	−441	−3.15	−484	−3.40	−517	−3.62	−502	−3.45	−460
AL3	200	−0.61	−1,578	0.02	0.00	0.09	0.00	0.04	0.00	−0.42	−1,197
AL4	140	−0.28	−2,974	−0.13	−183	−0.39	−304	−0.21	−1,506	−0.50	−316

Table 1.9 Power analysis results **after timing optimization**. We use 1× TSVs, and TSV capacitance is 5 fF. **3D-n** denotes *n*-die implementation, **IR** denotes the TSV co-placement, and **MST** is the MST-based TSV site placement. All numbers are scaled to 2D designs

		3D-2		3D-4	
ckt	2D	IR	MST	IR	MST
AL1	1.00	0.98	0.99	1.02	1.02
AL2	1.00	0.97	0.99	1.04	1.05
AL3	1.00	0.99	1.00	0.98	0.98
AL4	1.00	1.02	1.08	1.05	1.07

1.8 Conclusions

In this chapter, we studied the design impact of TSVs on the 3D stacked IC layout. We first discussed the design issues newly introduced in 3D ICs and studied two 3D IC design flows, TSV co-placement and TSV site. In the TSV co-placement scheme, we place gates and TSVs simultaneously whereas we place TSVs uniformly, place gates while treating pre-placed TSVs as obstacles, and then assign 3D nets to pre-placed TSVs in the TSV site scheme. For 3D placement, we extended an existing force-directed 2D placement algorithm to 3D. We also studied two TSV assignment algorithms for the TSV site scheme. The experimental results showed that our 3D placer outperforms its 2D version with respect to wirelength, the number of metal layers used in routing, and runtime for placement. However, die area increases by 2–29 % due to TSV insertion. We also observed differences between non-microprocessor circuits and microprocessor circuits from diverse angles such as die area, wirelength distribution, and partitioning results. Finally we showed timing and power analysis results. For small circuits, 3D designs are inferior to 2D designs. However, for relatively large circuits (e.g., # gates is bigger than 50 K), 3D designs show better timing than 2D designs after timing optimization.

References

1. K. Bernstein, P. Andry, J. Cann, P. Emma, D. Greenberg, W. Haensch, M. Ignatowski, S. Koester, J. Magerlein, R. Puri, A. Young, Interconnects in the third dimension: design challenges for 3D ICs, in *Proceedings of ACM Design Automation Conference* (IEEE, Piscataway, 2007), pp. 562–567
2. E. Beyne, P.D. Moor, W. Ruythooren, R. Labie, A. Jourdain, H. Tilmans, D.S. Tezcan, P. Soussan, B. Swinnen, R. Cartuyvels, Through-silicon via and die stacking technologies for microsystems-integration, in *Proceedings of IEEE International Electron Devices Meeting* (IEEE, Piscataway, 2008), pp. 1–4
3. Cadence Design Systems, QRC Extraction Users Manual 8.1.2
4. Cadence Design Systems, Soc Encounter, 2009. http://www.cadence.com

5. H. Chaabouni, M. Rousseau, P. Leduc, A. Farcy, R.E. Farhane, A. Thuaire, G. Haury, A. Valentian, G. Billiot, M. Assous, F.D. Crecy, J. Cluzel, A. Toffoli, D. Bouchu, L. Cadix, T. Lacrevaz, P. Ancey, N. Sillon, B. Flechet, Investigation on TSV impact on 65 nm CMOS devices and circuits, in *Proceedings of IEEE International Electron Devices Meeting* (IEEE, Piscataway, 2010)
6. J. Cong, S.K. Lim, Edge separability based circuit clustering with application to multi-level circuit partitioning. IEEE Trans. Comput. Aided Des. Integr. Circuits Syst. **23**(3), 346–357 (2004)
7. J. Cong, G. Luo, A Multilevel analytical placement for 3D ICs, in *Proceedings of Asia and South Pacific Design Automation Conference* (IEEE, Piscataway, 2009)
8. J. Cong, G. Luo, J. Wei, Y. Zhang, Thermal-aware 3D IC placement via transformation, in *Proceedings of Asia and South Pacific Design Automation Conference* (IEEE, Piscataway, 2007), pp. 780–785
9. X. Dong, Y. Xie, System-level cost analysis and design exploration for three-dimensional integrated circuits (3D ICs), in *Proceedings of Asia and South Pacific Design Automation Conference* (IEEE, Piscataway, 2009), pp. 234–241
10. B. Goplen, S. Sapatnekar, Efficient thermal placement of standard cells in 3D ICs using a force directed approach, in *Proceedings of IEEE International Conference on Computer-Aided Design* (ACM, New York; IEEE, Piscataway, 2003)
11. B. Goplen, S. Sapatnekar, Thermal via placement in 3D ICs, in *Proceedings of International Symposium on Physical Design* (ACM, New York, 2005), pp. 167–174
12. B. Goplen, S. Sapatnekar, Placement of 3D ICs with thermal and interlayer via considerations, in *Proceedings of ACM Design Automation Conference* (IEEE, Piscataway, 2007), pp. 626–631
13. H. Hua, C. Mineo, K. Schoenfliess, A. Sule, S. Melamed, R. Jenkal, W.R. Davis, Exploring compromises among timing, power and temperature in three-dimensional integrated circuits, in *Proceedings of ACM Design Automation Conference* (IEEE, Piscataway, 2006), pp. 997–1002
14. IMEC, 3D stacked IC (3D-SIC), 2008. http://www.imec.be
15. IWLS, IWLS 2005 benchmarks, 2005. http://www.iwls.org/iwls2005
16. J.W. Joyner, P. Zarkesh-Ha, J.A. Davis, J.D. Meindl, A three-dimensional stochastic wire-length distribution for variable separation of strata, in *Proceedings of IEEE International Interconnect Technology Conference* (IEEE, Piscataway, 2000), pp. 126–128
17. G. Karypis, V. Kumar, hMETIS, a hypergraph partitioning package version 1.5.3, 2007. http://glaros.dtc.umn.edu/gkhome/metis/hmetis/download
18. D.H. Kim, S.K. Lim, Through-silicon-via-aware delay and power prediction model for buffered interconnects in 3D ICs, in *Proceedings of ACM/IEEE International Workshop on System Level Interconnect Prediction* (ACM, New York, 2010), pp. 25–32
19. D.H. Kim, K. Athikulwongse, S.K. Lim, A study of through-silicon-via impact on the 3D stacked IC layout, in *Proceedings of IEEE International Conference on Computer-Aided Design* (ACM, New York, 2009)
20. D.H. Kim, S. Mukhopadhyay, S.K. Lim, Through-silicon-via aware interconnect prediction and optimization for 3D stacked ICs, in *Proceedings of ACM/IEEE International Workshop on System Level Interconnect Prediction* (ACM, New York, 2009), pp. 85–92
21. D.H. Kim, S. Mukhopadhyay, S.K. Lim, TSV-aware interconnect length and power prediction for 3D stacked ICs, in *Proceedings of IEEE International Interconnect Technology Conference* (IEEE, Piscataway, 2009), pp. 26–28
22. M. Koyanagi, T. Fukushima, T. Tanaka, High-deisnty through silicon vias for 3-D LSIs, in *Proceedings of the IEEE* (IEEE, Piscataway 2009), pp. 49–59
23. H.W. Kuhn. The hungarian method for the assignment problem. Nav. Res. Logist. **2**, 83–97 (1955)
24. Y.-J. Lee, S.K. Lim, Timing analysis and optimization for 3D stacked multi-core microprocessors, in *Proceedings of International 3D System Integration Conference* (IEEE, Piscataway, 2010)

25. H.Y. Li, E. Liao, X.F. Pang, H. Yu, X.X. Yu, J.Y. Sun, Fast electroplating TSV process development for the via-last approach, in *IEEE Electronic Components and Technology Conference* (IEEE, Piscataway, 2010), pp. 777–780
26. Nangate, Nangate 45 nm open cell library, 2009. http://www.nangate.com
27. NCSU, FreePDK45, 2009. http://www.eda.ncsu.edu/wiki/FreePDK
28. I. Savidis, E.G. Friedman, Closed-form expressions of 3-D via resistance, inductance, and capacitance. IEEE Trans. Electron Devices **56**(9), 1873–1881 (2009)
29. P. Spindler, U. Schlichtmann, F.M. Johannes, Kraftwerk2 – a fast force-directed quadratic placement approach using an accurate net model. IEEE Trans Comput. Aided Des. Integr. Circuits Syst. **27**(8), 1398–1411 (2008)
30. Synopsys, PrimeTime, 2008. http://www.synopsys.com
31. T. Thorolfsson, K. Gonsalves, P.D. Franzon, Design automation for a 3DIC FFT processor for synthetic aperture radar: a case study, in *Proceedings of ACM Design Automation Conference* (ACM, New York, 2009), pp. 51–56
32. M.-C. Tsai, T.-C. Wang, T. Hwang, Through-silicon via planning in 3-D floorplanning, in *IEEE Transactions on VLSI Systems* (IEEE, New York, 2010)
33. H. Yan, Z. Li, Q. Zhou, X. Hong, Via assignment algorithm for hierarchical 3-D placement, in *Proceedings of IEEE International Conference on Communications, Circuits and Systems* (IEEE, Piscataway, 2005), pp. 1225–1229

Chapter 2
Steiner Routing for 3D IC

Abstract In this chapter, we study a performance and thermal-aware Steiner routing algorithm for 3D stacked ICs. Our algorithm consists of two steps: tree construction and tree refinement. Our tree construction algorithm builds a delay-oriented Steiner tree under a given thermal profile. We show that our 3D tree construction involves minimization of two-variable Elmore delay function. In our tree refinement algorithm, we reposition the through-silicon-vias (TSVs) used in existing Steiner trees while preserving the original routing topology for further thermal optimization under performance constraint. We employ a novel scheme to relax the initial NLP formulation to ILP and consider all TSV from all nets simultaneously. Our tree construction algorithm outperforms the popular 3D-maze routing by 52% in terms of performance at the cost of 15% wirelength and 6% TSV count increase for four-die stacking. In addition, our TSV relocation results in 9% maximum temperature reduction at no additional area cost. We also provide extensive experimental results including (i) the wirelength and delay distribution of various types of 3D interconnects, (ii) the impact of TSV RC parasitics on routing and TSV relocation, and (iii) the impact of various bonding styles on routing and TSV relocation. Lastly, we provide results on two-die stacking.

The materials presented in this chapter are based on [13].

2.1 Introduction

Technology feature sizes continue to shrink to meet performance demands on integrated circuits. This, coupled with growing overall chip dimensions, leads to greater consumption of the available delay and power budgets by the interconnect structures on these chips. As global and semi-global wires become increasingly expensive and clock frequencies become higher and higher, designers seek new

S.K. Lim, *Design for High Performance, Low Power, and Reliable 3D Integrated Circuits*, DOI 10.1007/978-1-4419-9542-1_2,

architectures and technologies that rely less on sending signals across the chip. However, few scalable solutions have been proposed. One such solution is three-dimensional (3D) integration.

In a 3D integrated circuit, transistors may be fabricated on top of other transistors, resulting in multiple layers of active components. These transistors may then be wired to other transistors on the same device layer, to transistors on different device layers, or both, depending on the process technology. In the wafer-bonding approach [8], discrete wafers are "glued" together using vertical copper interconnects, permits multiple wafers and multiple third-dimension interconnects, overcoming the above limitations. 3D integration offers tremendous potential for keeping Moore's Law on track. It provides a means to continue to increase device density by stacking more transistors in the same footprint. 3D integration also addresses the wire delay problem by enabling the replacement of long and slow global interconnects with short and fast vertical routes.

The conventional 3D integration, so called System-In-Package (SIP), involves stacking packaged chips with wire-bonding based communication. Our target technology is to stack bare dies, not packaged chips, and utilize Through-Silicon-Via (TSV) to establish interconnect among the dies. With this approach, the saving on the total amount of wiring, delay, and its power consumption easily outnumbers that of stacked packages. In addition, the absence of off-chip communication naturally translates into smaller delay and low power. Compared with System-On-Chip (SOC) implementation, TSV-based 3D die-stacking helps reduce noise and interference among mixed signal components since these are separated into different dies. The advancement of TSV technology has matured enough to shrink the via size to a few micron dimensions, thereby contributing little to the area, delay, and power consumption of the overall system.

One of the major concerns of 3D ICs is thermal dissipation. Stacking of different device layers combined with the low thermal conductivity of the bonding material may result in excessively high on-chip temperature. The location of TSVs in a Steiner tree has high impact on the overall topology as well as the delay at the sink nodes of the tree, since it determines the amount of wiring done at all intermediate dies that the tree spans. Moreover, TSVs play a significant role in lowering the temperature of the chip. The reason is that they establish thermal paths to the heat sink when placed in the middle of a hotspot. Thus, several existing works utilize TSVs to lower on-chip temperature of 3D ICs [4, 5, 10, 11, 16, 17].

In this chapter, we formulate and solve the new *Performance and Thermal-aware 3D IC Steiner Routing* problem for multi-pin net routing in 3D stacked ICs. We emphasize that this problem is different from the conventional *2D Steiner Routing with Multiple Routing Layers*. The main reason is that the pins in 3D ICs are located in *multiple* device layers, whereas the pins in the 2D ICs are located in a single device layer as shown in Fig. 2.1. The following specific topics are covered in this chapter:

- We build efficient algorithms to construct Steiner trees for 3D ICs. Unlike the existing works on 3D Steiner tree construction [4, 7, 12, 17] that focus on

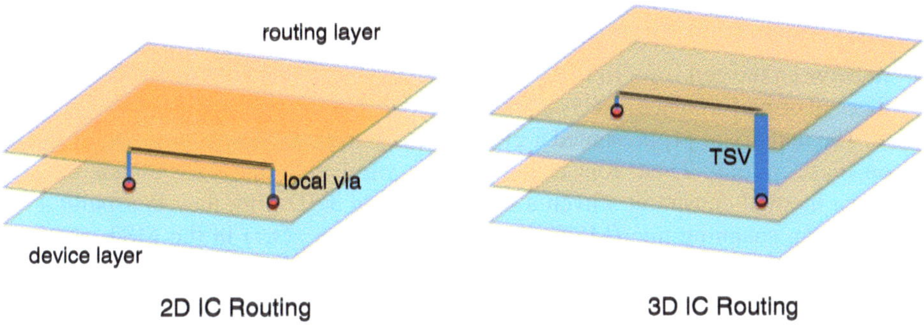

Fig. 2.1 Difference between 2D and 3D IC routing. In 3D IC, pins are located in multiple device layers, whereas in 2D IC, all pins lie in the same device layer

wirelength and thermal optimization, our tree construction algorithm optimizes thermal-aware performance. We believe that performance still remains as an important design metric, and thermal-aware delay model is important in Steiner tree construction. Unlike the existing works that decompose a 3D interconnect into a set of 2D interconnects, our router considers all pins in all dies and their bonding styles simultaneously and constructs performance-oriented Steiner trees while determining the optimal location for TSVs.

- We formulate and solve the new *TSV Relocation* problem for thermal optimization in 3D stacked ICs. Unlike the existing works on TSV-related thermal optimization that insert *additional* dummy TSVs for thermal optimization, our thermal optimization is based on relocating *existing* TSVs while maintaining the original routing topology. Thus, our thermal optimization does not require valuable routing resource nor rip-up-and-reroute. We employ a novel scheme to relax the initial NLP formulation to ILP and consider all TSV from all nets simultaneously.
- We study related experimental results. Our tree construction algorithm outperforms the popular 3D-maze routing by 52% in terms of performance at the cost of 15% wirelength and 6% TSV count increase for four-die stacking. In addition, our TSV relocation results in 9% maximum temperature reduction at no additional area cost.

2.2 Existing Works

The history of routing algorithm development for 3D ICs is relatively short. The authors of [7] presented a set of standard cell-based physical design tools for 3D ICs. Their 3D global routing algorithm is based on a 3D extension of [3], where the routing region is recursively partitioned into a series of x-, y-, and z-direction cuts in a top-down fashion. Routing topologies are then gradually optimized and refined as more and more subregions are generated by the partitioning. The z-direction cuts introduce vertical connection (= TSVs), and the authors utilize the routing channels in between the cell rows to insert TSVs. The objective is to reduce wirelength, and thermal or performance effects are not considered.

The authors of [4] presented a global routing algorithm for 3D ICs that is based on maze routing. Their goal is to minimize wirelength and TSV counts under a given thermal constraint. Given a set of points in a 3D grid, they first build a 3D minimum spanning tree (MST). Then for each edge $e(s,t)$ in the MST, shortest-path based maze search from s to t is performed.[1] Once the routing is completed, dummy TSVs are inserted in the layout whitespace for thermal optimization. Then, the whole process of routing and TSV insertion is performed in a multi-level routing framework. They later presented a follow-up work [5], where the TSV insertion and refinement is performed based on non-linear programming formulation.

The authors of [17] performed thermal-aware global routing for 3D ICs while utilizing "thermal vias" and "thermal wires". The authors first decompose each multi-pin net in the given netlist into a set of two-pin nets based on minimum spanning tree (MST) construction. Next, routing congestion map is obtained based on L/Z shape topology assumption. The location of TSVs are then determined for all inter-die two-pin nets based on the congestion map. The authors complete the routing for all nets by performing 2D maze routing in each die separately. Based on this initial tree construction, thermal analysis is performed to identify hotpots. The authors then perform thermal via/wire insertion and rip-up-and-reroute alternatively to remove temperature and congestion violations iteratively.

Note that the authors of [4] and [17] insert *additional* dummy TSVs for thermal optimization. On the other hand, our thermal optimization is based on relocating *existing* TSVs while maintaining the original routing topology. Thus, our thermal optimization does not require valuable routing resource nor rip-up-and-reroute. In addition, our routing trees are built under performance and thermal constraints.[2]

The authors of [14] presented an analytical delay model for inter-die 3D interconnects. Their delay model is a function of TSV location/height as well as the related wires. Using this model, they determine the delay-optimal TSV location along a 3D interconnect that connects gates in different dies. However, this model is based on two-pin connections, and they do not perform routing. The authors of [12] perform block-level global routing for 3D System-In-Package, where the fundamental problem is similar: utilize routing layers in between multiple device layers as well as the routing channels around the modules in each device

[1] We provide comparison between our 3D router to this so called 3D maze router in the experimental section.

[2] In addition to the thermal via insertion during routing, the authors of [10] insert thermal vias after placement, while the authors of [11] try to redistribute whitespace in a given floorplan to allocate space for thermal vias. The authors of [16] insert both dummy thermal TSVs and power/ground TSVs simultaneously to reduce thermal and power supply noise.

layer to complete routing. The goal is to minimize wirelength, layer, congestion, and crosstalk. They formulate and design heuristics for 3D pin redistribution, net distribution, and channel assignment problems.

2.3 Preliminaries

2.3.1 Problem Formulation

We assume the following are given: (1) a set of m nets $\{n_0, n_1, \cdots, n_{m-1}\}$, where each net is represented by a list of pins $n_i = \{p_0, p_1, \cdots, p_{k-1}\}$ with p_0 as the driver, (2) a 3D routing grid G that represents the routing resource in a given 3D stacked IC, where each grid node represents a routing region and each edge denotes the adjacency among the regions, (3) each x/y grid edge is associated with horizontal/vertical wire capacity and z with TSV capacity, (4) the location of each pin $p(x,y,z)$ in G, and (5) a 3D thermal grid Z with thermal resistance on all edges and power consumption on all nodes.[3] A *3D Steiner Tree* is defined to be a set of 2D (=planar) Steiner trees connected by TSVs.

The goal of the *Performance and Thermal-aware 3D Steiner Routing* problem is to generate a 3D Steiner tree for each net while satisfying the capacity constraints specified in the underlying G.[4] The objective is to minimize (1) the maximum temperature among all nodes in the thermal grid, and (2) the maximum Elmore delay among all pins in each tree, where the delay is computed based on the current thermal distribution. Note that the thermal resistance values for some edges in Z changes based on the number of TSVs assigned, which changes during routing and TSV relocation.

This chapter uses the temperature dependent interconnect delay model presented in [1]. The line resistance per unit length can be calculated as: $r(x) = r_0(1 + \beta \cdot T(x))$, where r_0 is the resistance at 0°C, β is the temperature co-efficient of resistance, and $T(x)$ denotes the temperature at location x.

Depending on the number and the location of pins in each net, there exist the following four types of nets to be routed:

- *Single-die-two-pin (SD2P) nets*: a net in this group connects two pins that are located in the same die.
- *Single-die-multi-pin (SDMP) nets*: a net in this group connects more than two pins that are located in the same die.

---

[3]Our thermal grid Z dimension is an integer multiple of our routing grid G dimension. This ensures that all nets in a routing grid can be assigned to a single thermal grid.

[4]Note that we do not explicitly minimize routing/via congestion but implicitly address it by satisfying the capacity constraints. Each edge capacity can be set independently to reflect the availability of the routing resource. For example, we reduce the capacity of vertical routing grid edges that go through a high placement density region.

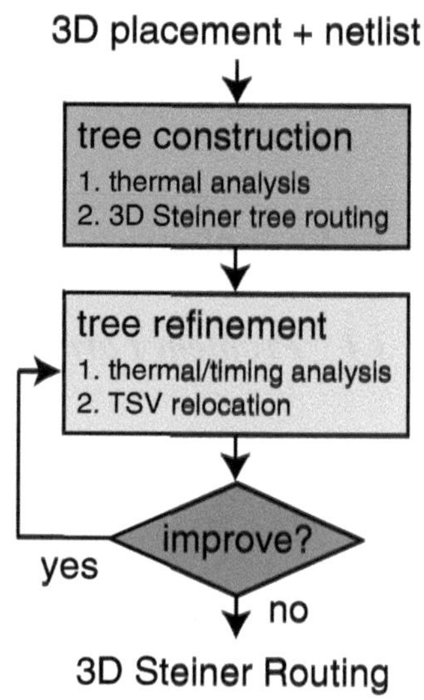

Fig. 2.2 Overview of our performance and thermal-aware 3D Steiner routing

- *Multi-die-two-pin (MD2P) nets*: a net in this group connects two pins that are located in two different dies.
- *Multi-die-multi-pin (MDMP) nets*: a net in this group connects more than two pins that are located in multiple different dies.

Note that the Steiner routers for the conventional 2D ICs deal with the first two types, whereas 3D Steiner routers need to route all four types. The last two types require TSVs for inter-die connections.

2.3.2 *Overview of the Approach*

Optimizing performance and thermal objectives simultaneously during 3D Steiner routing is a challenging task. We solve this problem in two phases, namely, tree construction and tree refinement. The main goal during tree construction is to obtain initial trees that optimize performance under the given thermal profile. The main goal of tree refinement is to relocate the TSVs used in the initial trees for thermal optimization under the given timing constraint.

Figure 2.2 shows an overview of our performance and thermal-aware 3D Steiner routing process. Given a netlist and its 3D placement, we first perform thermal analysis to obtain the thermal distribution to be used during tree construction. Next, we construct a performance-oriented routing tree for each net one by one under the non-uniform thermal profile. The temperature values are updated periodically to reflect the thermal resistance and temperature changes from TSV usage.[5] In our tree refinement phase, we first perform timing analysis and obtain timing slack for each pin. We also perform thermal analysis to identify thermal hotspots. We then minimize the 3D on-chip temperature by relocating TSVs used in each tree under the given timing constraints. The goal is to move TSVs closer to the hotspots so that the thermal resistance values are reduced in those regions. Note that our TSV relocation preserves the original tree topology while optimizing the thermal objective. We recompute the timing slacks and temperature values to reflect the relocation at every iteration. Lastly, we repeat the whole tree refinement phase until no more thermal improvement is possible.

2.4 3D Steiner Tree Construction

2.4.1 Overview of the Algorithm

The basic approach of our 3D Steiner tree construction algorithm is similar to SERT [2], where an existing tree is incrementally grown by connecting a new sink pin to it. SERT starts with the driver pin and selects the next sink pin that minimizes Elmore delay when connected to the driver. This process continues until all sink pins are connected to the tree that is growing. The goal is to minimize the maximum Elmore delay among all sink pins of the tree. Here the biggest challenge is to compute the point on the tree where the new pin connects to. There are three major differences between SERT and our work: (1) all the pins in SERT are located in the same die, whereas our 3D algorithm handles the pins located in multiple dies. This 3D case requires the usage of TSVs, and the location of these TSVs has huge impact on the topology of the tree as well as the sink pin delay. (2) The delay optimization in SERT is based on single variable, whereas our algorithm deals with two-variable function optimization, (3) our interconnect delay is computed based on the given thermal profile.

A pseudocode of our algorithm is shown in Algorithm 3. Our routing algorithm consists of two phases: construction (line 1–14) and rip-up-and-reroute (15–16). We construct 3D Steiner trees during the construction phase while ignoring congestion, and then fix the capacity violation by rip-up-and-reroute. Given a net n, our 3D Steiner tree T_n initially contains the driver pin (line 2). We store the remaining pins of n in a set Q_n (line 3). We then examine all pin-edge pairs (line 5–6) and compute the impact of connecting the pin to the edge on Elmore delay under the given thermal profile Z, where the pin is chosen from Q_n and the edge is from T_n. Specifically, the

[5]The thermal resistance change from a single TSV insertion is very small. In addition, updating thermal map after every net is computationally prohibitive for a large design. Thus, we choose to update thermal distribution periodically. We use the full 3D thermal grid model for our temperature analysis [15].

Algorithm: 3D Steiner Tree Construction
input : netlist NL, routing graph G, thermal profile Z
output: 3D Steiner tree for each net
for *each net* $n \in NL$ **do**
    $T_n = p_0(n)$;
    $Q_n =$ set of pins of n except p_0;
    while $Q_n \neq \emptyset$ **do**
        for *each pin* $a \in Q_n$ **do**
            for *each edge* $e \in T_n$ **do**
                $x =$ connection point for $a \rightarrow e$;
                $y =$ TSV location on $e(x,a)$;
                update $dly(p)$ for all $p \in T_n \cup a$;
                $X(a,e) = \max\{dly(p)\}$;
            end
        end
        $(a_{min}, e_{min}) =$ pin+edge pair with min X;
        $T_n = T_n \cup e_{min}$;
        remove a_{min} from Q_n;
    end
    update Z periodically;
end
for *each non-timing critical* T_n *violating capacity* **do**
    rip-up-and-reroute T_n under Z;
end

Algorithm 3: Pseudocode of performance and thermal-aware 3D Steiner tree construction algorithm. In case e and a are located in different planes, $e(r,a)$ will utilize TSVs

delay impact is calculated based on the increase in temperature-dependent Elmore delay among all pins currently in T_n (line 9–10), where $dly(p)$ is the Elmore delay at pin p. This requires the computation of connection point x and TSV location y (line 7–8) (to be discussed in Sect. 2.4.2). Next, we select the pin-edge pair that results in the minimum max-delay increase (line 11) and add the pin to T_n (line 12–13). Since TSV insertion affects the thermal resistance of the related area, we perform thermal analysis periodically (not after every net routing) (line 14).

Our rip-up-and-reroute is done on non-timing critical nets, i.e., the nets with smaller max-delay values (line 15–16). Specifically, we first sort the nets that utilize routing edges that violated the capacity constraint based on their timing slack values. We then rip up the nets one-by-one in the sorted order and reroute it until the violation is completely removed. We use a maze router that minimizes weighted path length to reroute a net, where the weight considers the remaining routing capacity and temperature. In this case, our cost function penalizes routing edges that are more congested and/or located nearby hotspots. For a two-pin net, our maze router tries to find the source-to-sink shortest weighted path such that the routing capacity is not violated. For a multi-pin net, we first decompose the net into a set of two-pin nets based on their MST (Minimum Spanning Tree). We then route each two-pin sub-net using our maze router described above.

2.4.2 Computing Connection Point and TSV Location

For a given multi-pin net and a partial tree, our goal is to find the next pin (and its connection point and TSV location) so that adding this pin, compared with other pins, minimizes the delay increase in the overall tree. A final Steiner tree is obtained once all the pins are added. This section discusses how to compute the connection point and TSV location. Our discussion is based on two-die case for the simplicity of the discussion, but our algorithm is applicable to multiple die stacking without any modification. Let r_1 and c_1 denote the unit length resistance and capacitance values for die 1. r_2 and c_2 are similarly defined for die 2. The capacitance and resistance of a TSV connecting the two die are denoted C_{via} and R_{via}.

Given a pin p and an edge $e \in T$, the *connection point* is defined as the point on e to which p is connected. The connection point computation for 2D case has been presented in [2], where the Elmore delay change on an entire tree caused by adding a new pin to the tree is a function of a single variable x, the location of connection point. We extend this work by introducing a second variable y that represents the location of TSV. We then optimize the two-variable delay function and determine the location of connection point ($=x$) and TSV ($=y$) for 3D case.

Referring to Fig. 2.3, $e(p,c)$ and $e(q,b)$ are edges on T. p is the parent node of $e(p,c)$, and q is the parent node of $e(q,b)$. a is the new pin that needs to connect to $e(p,c)$. Edge $e(p,c)$ lies on die 1 with interconnect parasitics r_1 and c_1, whereas a lies on die 2 with interconnect parasitics r_2 and c_2. d is the point on $e(p,c)$ that is of the shortest distance to a. x is the connection point, and y is the location of TSV.

Our first goal is to derive Elmore delay equations that are the functions of x and y. In what follows, we let δx denote the distance between node p and node x, δq, δa, δb, δc, and δd are defined similarly. δy is the distance between x and y, and δz is the distance between y and a. Let T_b denote the subtree rooted at node b. In order to compute the Elmore delay change on all sink pins in T caused by adding a to T, we consider the following four cases:

1. Delay at the node to be added ($=$ node a)
2. Delay at the subtree located after the connection point ($=$ node c)
3. Delay at the subtree that could be located either before or after the connection point ($=$ node b)
4. Delay of the nodes not in T_p.

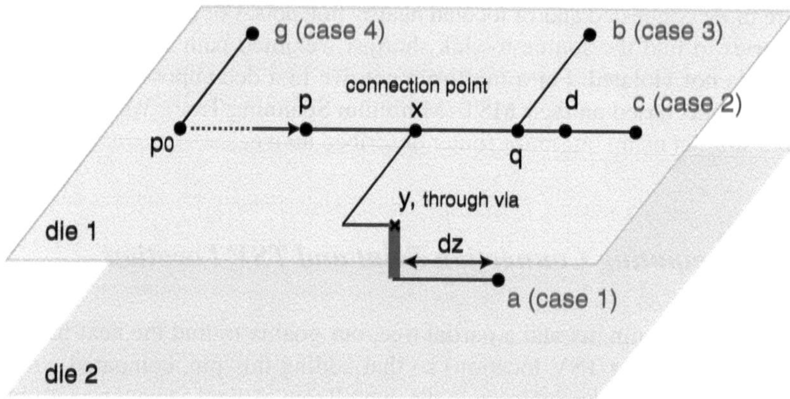

Fig. 2.3 Illustration of how pin a connects to $e(p,c) \in T$. $e(y,a)$ is routed in the *bottom die*, where as all other edges are routed in the *top die*. x is the location of connection point on $e(p,c)$. y is the location of the TSV inserted on $e(x,a)$. $e(q,b)$ is another branch in T. g is another sink that is not a part of the subtree rooted at p. d is the shortest distance point on $e(p,c)$ from a, and δz is the distance between the TSV and a. The Elmore delay of $T \cup a$ is a function of both x and y

Figure 2.3 illustrates these four cases:

2.4.2.1 Case 1

We handle the delay at node a. In this case, $d(a)$ is a sum of four functions:

$$d(a) = f_1 + f_2 + f_3 + f_4$$

f_1 is the delay from node p_0 to p. The delay from node p to a can be further divided into (1) the delay from node p to x ($= f_2 + f_3$), and (2) the delay from node x to a ($= f_4$). In addition, the delay from node p to x depends on edge $e(q,b)$. Thus, we consider the delay from p to x as a summation of two terms f_2 and f_3, where f_2 is the delay from p to x without considering $e(q,b)$ and T_b. f_3 is the additional delay from p to x when considering $e(q,b)$ and T_b. Thus,

$$f_1 = K_0 + K_1\{c_1\delta y + C_{via} + c_2\delta z + c_1\delta c + C_c + C_b + c_1(\delta b - \delta q)\}$$

$$f_2 = r_1\delta x\left(c_1\frac{\delta x}{2} + c_1\delta y + C_{via} + c_2\delta z + c_1(\delta c - \delta x) + C_c\right)$$

$$f_3 = \begin{cases} r_1\delta x(c_1(\delta b - \delta q) + C_b), & \text{if } \delta x \leq \delta q \\ r_1\delta q(c_1(\delta b - \delta q) + C_b), & \text{if } \delta x \geq \delta q \end{cases}$$

$$f_4 = r_1\delta y\left(c_1\frac{\delta y}{2} + C_{via} + c_2\delta z\right) + R_{via}\left(\frac{C_{via}}{2} + c_2\delta z\right) + r_2c_2\frac{\delta z^2}{2}$$

where $\delta z = \delta a - (\delta x + \delta y)$, K_0 is the sum of resistance and capacitance products along $p_0 \rightarrow p$ path, K_1 is the sum of resistance along $p_0 \rightarrow p$ path, and C_i is the capacitance of the sub-tree rooted at node i.

2.4.2.2 Case 2

The new delay at node c is given by

$$d(c) = f_1 + f_2 + f_3' + f_4'$$

where

$$f_3' = r_1 \delta q (c_1(\delta b - \delta q) + C_b)$$

$$f_4' = r_1(\delta c - \delta x)\left\{\frac{c_1(\delta c - \delta x)}{2} + C_c\right\}$$

f_2' is the delay seen at node c due to the branch $e(q,b)$, and f_4' is the delay from node x to node c without considering the branch $e(q,b,)$.

2.4.2.3 Case 3

The new delay at node b is given by

$$d(b) = f_1 + f_2'' + f_3''$$

where

$$f_2'' = \begin{cases} r_1 \delta x(c_1 \delta y + C_{via} + c_2 \delta z), & \text{if } \delta x \leq \delta q \\ r_1 \delta q(c_1 \delta y + C_{via} + c_2 \delta z), & \text{if } \delta x \geq \delta q \end{cases}$$

$$f_3'' = r_1 \delta q\left\{c_1 \frac{\delta q}{2} + c_1(\delta b - \delta q) + C_b + C_c\right\}$$

f_2'' is the added delay at node b from adding a new pin a. f_3'' is the delay from node p to q without considering the effect of the new pin a.

2.4.2.4 Case 4

For all other nodes not in T_p, the added delay is a function of the added capacitance, which is linear in terms of x and y and given by

$$\Delta C = c_1(\delta x + \delta y) + C_{via} + c_2 \delta z$$

These cases identify the four possible ways by which delay at the new node a and the other existing nodes of the tree may change due to the addition of a. The objective is to find the location of connection point x and the location of TSV y for the new node

a such that the total increase in delay is minimal under given thermal profile. As discussed earlier, we use this connection point computation to identify the pin-edge pair that results in the minimum increase of maximum Elmore delay under given thermal distribution.[6]

2.4.3 Optimization of Delay Equations

We discuss how the delay equations derived in the previous section can be used to generate a small set of possible optimum location points. We first consider the conditions needed to determine the minimum of a general quadratic function of two variables. We later show how the delay equations derived in the previous section can be optimized using these conditions.

In general, for a quadratic function of two variables $F(\delta x, \delta y)$, the maximum or the minimum of the function depends upon the values of $\frac{\partial^2 F}{\partial \delta x^2}$ and the determinant of the Hessian matrix H_1:

$$\begin{bmatrix} \frac{\partial^2 F}{\partial \delta x^2} & \frac{\partial^2 F}{\partial \delta x \partial \delta y} \\ \frac{\partial^2 F}{\partial \delta x \partial \delta y} & \frac{\partial^2 F}{\partial \delta y^2} \end{bmatrix}$$

where F is the delay function under consideration. The above values for a quadratic function of two variables are always constant.

We have $0 \leq \delta x \leq \delta d$ and $0 \leq \delta y \leq \delta a$, so we consider the following cases:

- Case A: If $\frac{\partial^2 F}{\partial \delta x^2} \leq 0$ and $H_1 \geq 0$, the minimum can be found at the boundary points, i.e., $\delta x = 0$ or $\delta x = \delta d$ and $\delta y = 0$ or $\delta y = \delta a$. Thus, we have four points to look for the minimum.
- Case B: If $\frac{\partial^2 F}{\partial \delta x^2} \leq 0$ $\frac{\partial^2 F}{\partial \delta y^2} \leq 0$ and $H_1 = 0$, we have a concave function, and the minimum lies on the boundary points.
- Case C: If $\frac{\partial^2 F}{\partial \delta x^2} = 0$, $\frac{\partial^2 F}{\partial \delta y^2} = 0$, and $H_1 = 0$, then $F(\delta x, \delta y)$ is a linear function of δx and δy, and the minimum lies at the boundary points.
- Case D: If $H_1 < 0$, the critical point found is a saddle point, and the minimum lies at the boundary. The set of boundary points may be found by setting $\delta x = 0$ or $\delta x = \delta d$ and minimizing $F(\delta x, \delta y)$ as a function of δy, or setting $\delta y = 0$ or $\delta y = \delta a$ and minimizing $F(\delta x, \delta y)$ as a function of δx.

We show that the Elmore delay at each sink node in T can be optimized by considering any of the four cases shown above. Thus, there is only a fixed number of points (x, y) for which the Elmore delay values are minimized. Details are included in appendix section "Optimization of Two-Variable Delay Equations".

---

[6] In case of connecting two pins located in non-adjacent dies, we use a stacked TSV so that no routing in the intermediate layers is used. For example, a pin in die 1 and die 3 requires two TSV that are vertically aligned so that there is no routing necessary in die 2.

2.5 3D Tree Refinement with TSV Relocation

2.5.1 *Overview of the Algorithm*

The motivation behind our TSV relocation is to move as many TSVs into thermal hotspots as possible while preserving the original tree topology we obtain during our construction step. The TSVs in hotspots reduce the thermal resistance in these areas and establish heat conducting paths to the heat sink. The objective is to remove hotspots while not violating the timing and routing capacity constraints.

TSVs are usually etched or drilled through device layers by special techniques and are costly to fabricate [5]. Thus, large number of TSVs will degrade the yield and reliability of the 3D chip. This is the drawback of the existing works that add *additional* dummy TSV to reduce temperature [4, 5, 10, 11, 16, 17]. During our TSV relocation, however, no new dummy TSVs are added but the existing TSVs are relocated. The objective is to reduce the maximum on-chip temperature as much as possible using TSV relocation so that additional TSVs needed may be kept at minimum.[7]

In general, thermal optimization with TSVs is non-linear in nature due to the well-known relation $T = PR$, where T is the temperature matrix, P is the power vector, and R is the thermal resistance matrix. We have $R \propto \frac{1}{a}$, where a is the number of TSVs. In addition, general solutions available for solving non-linear problems cannot be applied directly to large size problems. In this chapter, we study a novel solution that helps us effectively overcome the non-linear nature of this problem. We study a relaxed ILP based formulation in which the number of integer variables are kept at minimum. Our ILP-based method optimizes TSVs on *all* nets simultaneously, which is more rigorous than a sequential approach that optimizes the nets one by one. In addition, we target all hotspots simultaneously instead of iteratively targeting one by one. Our experimental results in Sect. 2.6 demonstrate the advantage of this approach.

2.5.2 *Movable Range*

We start our TSV relocation phase with the set of 3D Steiner trees we obtain from the construction step. All pins in each tree T_i are associated with timing constraint that denotes the required arrival time in terms of Elmore delay. Each TSV $v \in T_i$ is associated with the *movable range* that denotes the range of new location along its route to the connection point so that the timing constraints are not violated. We perform thermal-aware static timing analysis to compute timing slack for all points.

[7] An approach which combines both TSV relocation and dummy TSV insertion should provide best results and is outside the scope of this chapter.

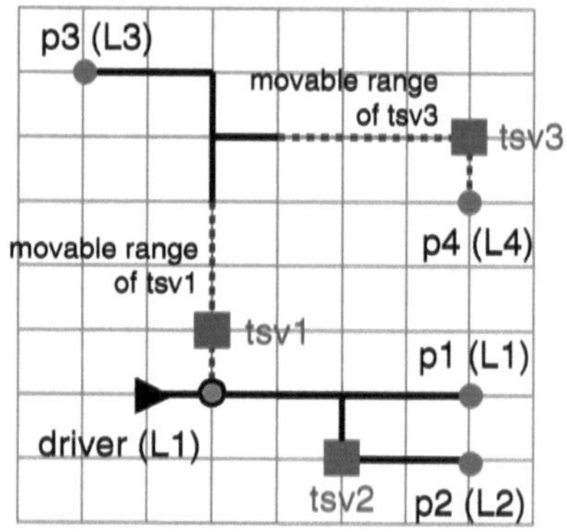

Fig. 2.4 Illustration of movable range. This figure is a *top-down* view of a multi-die stack. The driver is represented by a *triangle*, the sinks are represented by *dots*, and the TSVs are represented by *square boxes*. The driver is located in die 1 (=L1). The movable range of each TSV is represented by the *dotted lines*

This static timing analysis gives us the available slack on each pin. The TSVs are then moved along the Steiner tree edge to determine the movable range of each TSV. Note that new TSV location translates to new delay at the sinks. The range ensures that no timing constraints are violated during the relocation. An illustration is shown in Fig. 2.4. In case the movable range of a TSV v is a single point, v is non-movable; otherwise it is movable. Our goal is then to find a new location for each movable TSV in each Steiner tree so that the maximum temperature among all nodes in the thermal grid is minimized while the timing and routing resource capacity constraints are not violated. Note that we preserve the original topology of the Steiner trees. All that is changing is the location of TSVs for thermal optimization, where the movable TSVs are moved into thermal hotspots under timing constraint to reduce the thermal resistivity.

2.5.3 *Compact Thermal Analysis*

Figure 2.5 shows the fast thermal model used in our TSV relocation method, which we adopted from [5]. In this model each heat source is considered as a current source, and the temperature as voltage level. The 3D structure is divided into smaller region, which is represented by its thermal resistance. In this model a tile structure is imposed on the surface, where each tile is approximated as a resistive chain as shown in Fig. 2.5. Temperature equations are then constructed based on the voltage equation $V = I \cdot R$. For example, the temperature at node 4 is given by $T_4 = T_3 + (P_5 + P_4) \cdot R_5$.

In 3D ICs, the heat sinks are attached to the bottom or top side of the 3D IC stack, with other boundaries being adiabatic. Thus, the dominant heat flow is in the vertical direction. For the purpose of optimization, we view each tile stack in Fig. 2.5 as an independent thermal resistive chain. In this case, we do not consider effects of lateral thermal dissipation, which can be justified by the fact the thermal conductivity of epoxy material used for bonding is much lower than that

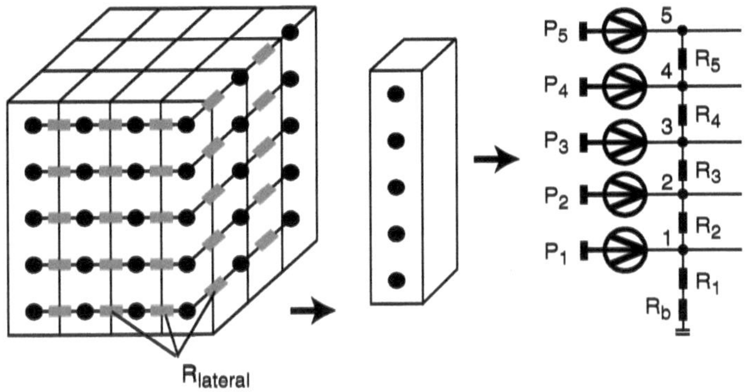

Fig. 2.5 Thermal model used, where R_b denotes the thermal resistance to the heat sink. The convention used in our ILP formulation is that adding TSVs at point (= tile) i reduces the value of R_i

of silicon itself. This essentially means that it is difficult for heat to dissipate in the vertical direction as compared with horizontal direction. To accurately verify the temperature reduction, a full (= vertical and lateral) resistive thermal model [15] (considering lateral resistances) is run twice, once before and once after our TSV relocation phase. The final temperature values reported in our experiments are based on this full resistive model.

Another important reason for our adoption of this vertical heat flow model is that we can formulate our simultaneous TSV relocation using ILP and solve it efficiently as discussed in the subsequent sections. To solve for the temperature values at all nodes, all temperature equations are constructed and reduced to the form $T = P \cdot R$, where T, P and R are all vectors. These equations can be solved directly by using the values of power and the thermal resistance at each tile.

2.5.4 Non-linear Programming Formulation

In the following sections, we first show how the TSV relocation problem can be formulated as a NLP (non-linear programming) formulation. We then show how the NLP is converted to an ILP (integer linear programming) formulation. The ILP formulation adds a large number of integer variables in the problem, thus making it difficult to solve. Lastly, we present our fast ILP problem formulation that reduces the number of integer variables significantly.

The NLP based formulation is defined as follows (Table 2.1 explains the notations we use in the formulation):

Table 2.1 Variables and constants used in our NLP/ILP formulations

$T_{i,j,k}$	Temperature at tile (i,j,k)
$\alpha_{i,j,k}$	Temperature-related weight for tile (i,j,k), which is computed by $T^{org}_{i,j,k}/T^{org}_{max}$, where $T^{org}_{i,j,k}$ denotes the original temperature for (i,j,k) before the optimization, and T^{org}_{max} is the maximum value among all $T^{org}_{i,j,k}$ values. *Constant*
$vmax$	Maximum number of TSVs each tile can accommodate. *Constant*
$\beta^m_{i,j,k}$	Becomes 1 when a TSV is moved to tile (i,j,k) so that the total number of movable TSVs at (i,j,k) changes from $m-1$ to m
$V^{org}_{i,j,k}$	Original number of TSVs (= movable + non-movable) in tile (i,j,k) before the optimization. *Constant*
$V^m_{i,j,k}$	Number of TSVs in tile (i,j,k), which is just m
$V^{opt}_{i,j,k}$	Number of TSVs (= movable + non-movable) in tile (i,j,k) after the optimization
$M^{x,y,k}_{i,j,k}(n)$	Becomes 1 if a TSV in net n is moved from tile (i,j,k) to (x,y,k); 0 otherwise
$G^{cur}_{i,j,k}$	Current usage of wires and TSVs for a grid (i,j,k) in 3D routing grid G
$G^{max}_{i,j,k}$	Capacity constraint of wires and TSVs for a grid (i,j,k) in 3D routing grid G. *Constant*
$R^m_{i,j,k}$	Thermal resistance of tile (i,j,k) having m TSVs. *Constant*
$R^{no}_{i,j,k}$	Thermal resistance of tile (i,j,k) with no TSVs. *Constant*
α	Thermal resistance of one TSV. *Constant*
$\gamma^m_{i,j,k}$	Becomes 1 if the number of TSVs in the tile (i,j,k) is m
$P_{i,j,k}$	Power at tile (i,j,k)
$\Delta G_{i,j,k}$	Change in the routing resource usage at tile (i,j,k)
$\delta^m_{i,j,k}$	It is the temperature difference between tile i,j,k and $i,j,k-1$ if the number of TSVs in tile i,j,k is m. We have $\delta^m_{i,j,k} = (P_{i,j,n} + \cdots + P_{i,j,k}) \cdot R^m_{i,j,k}$. *Constant*
N	Entire netlist
N_{mov}	Set of nets whose vias are movable
K	Total number of dies in the stack
$N^{in}_{i,j,k}$	Set of nets that contain TSVs that are moved into tile (i,j,k)
$N^{out}_{i,j,k}$	Set of nets that contain TSVs that are moved out of tile (i,j,k)

Minimize

$$\sum_{(i,j,k)\in Z} \alpha_{i,j,k} \cdot T_{i,j,k} \tag{2.1}$$

Subject to:

$$T_{i,j,k} = T_{i,j,k-1} + (P_{i,j,K} + \cdots + P_{i,j,k}) \times R^{opt}_{i,j,k} \tag{2.2}$$

$$R^{opt}_{i,j,k} = \frac{\alpha}{\alpha/R^{no}_{i,j,k} + V^{opt}_{i,j,k}} \tag{2.3}$$

$$V^{opt}_{i,j,k} = V^{org}_{i,j,k} + \Delta V_{i,j,k} \tag{2.4}$$

$$\Delta V_{i,j,k} = \sum_{n\in N^{in}_{i,j,k}} M^{i,j,k}_{x,y,k}(n) - \sum_{n\in N^{out}_{i,j,k}} M^{v,w,k}_{i,j,k}(n) \tag{2.5}$$

$$G^{cur}_{i,j,k} \leq G^{max}_{i,j,k}, \ \forall (i,j,k) \tag{2.6}$$

$$G^{cur}_{i,j,k} = G^{org}_{i,j,k} + \sum_{n \in N} M^{v,w,k}_{x,y,k}(n) \cdot \Delta G_{i,j,k}, (i,j,k) \text{ on path } (x,y,k) \to (v,w,k) \tag{2.7}$$

$$\sum_{n \in N_{mov}} M^{x,y,k}_{i,j,k}(n) = 1 \tag{2.8}$$

$$M^{x,y,k}_{i,j,k}(n) \in \{0,1\} \tag{2.9}$$

Equation (2.1) is our objective function, where we minimize the weighted sum of temperature values at all thermal tiles.[8] The weights $\alpha_{i,j,k}$ are computed based on the initial temperature measured before the TSV relocation. In this case, the higher the $\alpha_{i,j,k}$ is, the lower the $T_{i,j,k}$ we desire.[9]

Equation (2.2) gives the temperature at each tile based on our fast thermal model illustrated in Fig. 2.5, where K is the maximum height of our thermal tile (= total number of dies in the 3D stack). Equation (2.3) shows the variation of thermal resistance based on the number of TSVs in a tile. This is obtained from solving the following parallel resistance relation:

$$\frac{1}{R^{opt}_{i,j,k}} = \frac{1}{R^{no}_{i,j,k}} + \frac{V^{opt}_{i,j,k}}{\alpha}$$

Equation (2.4) is the definition of $V^{opt}_{i,j,k}$. Equation (2.5) states that the total change in the number of TSVs for tile (i,j,k) is the total number of TSVs moved into tile (i,j,k) minus the total number of TSVs moved out of tile (i,j,k). Equation (2.6) ensures that the routing resource (= wires and TSVs) capacity constraints are satisfied.

Equation (2.7) shows how the routing resource usage is updated after a TSV is moved from tile (x,y,k) to (v,w,k). Note that the usage of *all* tiles along the path from (x,y,k) to (v,w,k) is affected. Let $G^{org}_{i,j,k}$ denote the original usage at tile (i,j,k) before the move. Then, the new usage, denote by $G^{cur}_{i,j,k}$, is computed by adding the total amount of change made on (i,j,k), where the total amount of change is computed by summing the various $\Delta G_{i,j,k}$ changes based on whether the corresponding vias are moved or not. We note that $\Delta G_{i,j,k}$ can take both positive or

[8]Thermal tiles and thermal grid nodes are used interchangeably in this section.

[9]Note that this objective is not directly minimizing the maximum temperature among all tiles. This is necessary in our formulation; otherwise, we can no longer relax the $\beta^{i}_{i,j,k}$ variables to be continuous as explained in Appendix Explanation of Eq. (2.15), which in turn means that our ILP formulation will contain excessive amount of integer variables. We note, however, that giving larger weights to the higher temperature tiles in the objective function helps us ensure that hotspots are given a preference.

negative values based on whether the corresponding via move increases or decreases the routing capacity at the given location. Lastly, the whole process is done for all nets that contain relocated TSVs.

Equation (2.8) ensures that only one TSV per net is moved. Note that this restriction is unavoidable since the movable range of a TSV is computed independent of other TSVs. Once a TSV is moved, it affects the timing constraint, movability, and the range of *all* other TSVs in the same net. The ultimate way to perform TSV relocation is to consider all TSVs from all nets simultaneously, which is computationally expensive. However, our method that considers one TSV from all nets simultaneously is better than a sequential approach that considers all TSVs from a single net. Equation (2.9) states that $M_{i,j,k}^{x,y,k}(n)$ are binary integer variables.

We note that this original TSV relocation problem is non-linear due to the inverse relation between thermal resistance and number of TSVs in a tile ($= R_{i,j,k}^{opt}$ *vs* $V_{i,j,k}^{opt}$) in Eq. (2.3). In the next section, we study a simplified integer linear programming formulation that overcomes this non-linear problem formulation.

2.5.5 Integer Linear Programming Formulation

From the NLP formulation, we see that the number of TSVs in each tile is an integer variable. Our ILP based formulation differs from the NLP in the following way: we replace Eqs. (2.2) and (2.3) with the following:

$$T_{i,j,k} = T_{i,j,k-1} + \gamma_{i,j,k}^{0} \times \delta_{i,j,k}^{0} + \cdots + \gamma_{i,j,k}^{vmax} \times \delta_{i,j,k}^{vmax} \tag{2.10}$$

$$1 \cdot \gamma_{i,j,k}^{1} + 2 \cdot \gamma_{i,j,k}^{2} + \cdots + vmax \cdot \gamma_{i,j,k}^{vmax} = V_{i,j,k}^{opt} \tag{2.11}$$

$$\sum_{m=0}^{vmax} \gamma_{i,j,k}^{m} = 1, \ \forall (i,j,k) \tag{2.12}$$

$$\gamma_{i,j,k}^{m} \in \{0,1\} \tag{2.13}$$

Equation (2.10) is a new way to calculate the temperature at each tile (refer to Table 2.1 for the definition of the related variables and constants). In this equation, $\gamma_{i,j,k}^{m}$ are the new integer variables, whereas $\delta_{i,j,k}$ are the constants that are calculated for each possible value of the number of TSVs in a tile. Equation (2.11) equates the $\gamma_{i,j,k}^{m}$ variable with the optimum number of TSVs in a tile. Compared with the non-linear Eq. (2.3), this Eq. (2.11) shows linear relationship between $V_{i,j,k}^{opt}$ vs. $\gamma_{i,j,k}$. Equation (2.12) ensures that for each tile, only one $\gamma_{i,j,k}^{m}$ takes a value 1. Lastly, Eq. (2.13) ensures that $\gamma_{i,j,k}^{m}$ is either 0 or 1. All other equations in our NLP formulation remain same in our ILP formulation.

The number of new integer variables $\gamma_{i,j,k}^{m}$ is proportional to the number of tiles in our thermal grid plus the number of TSVs movable in each grid. Adding such a large number of integer variables makes the problem harder to solve. In our next section, we study our fast ILP formulation, which removes the need of integer $\gamma^m_{i,j,k}$ variables, thus reducing the number of integer variables required significantly.

2.5.6 Fast Integer Linear Programming Formulation

Our fast ILP-based TSV relocation is formulated as follows (Table 2.1 explains the notations we use in the formulation):

Minimize

$$\sum_{(i,j,k)\in Z} \alpha_{i,j,k} \cdot T_{i,j,k} \tag{2.14}$$

Subject to:

$$T_{i,j,k} = T_{i,j,k-1} + \delta^0_{i,j,k} - \beta^1_{i,j,k} \cdot (\delta^0_{i,j,k} - \delta^1_{i,j,k}) - \cdots - \beta^{vmax}_{i,j,k} \cdot (\delta^{vmax-1}_{i,j,k} - \delta^{vmax}_{i,j,k}) \tag{2.15}$$

$$\beta^1_{i,j,k} + \beta^2_{i,j,k} + \cdots + \beta^{vmax}_{i,j,k} = V^{opt}_{i,j,k} \tag{2.16}$$

$$V^{opt}_{i,j,k} = V^{org}_{i,j,k} + \Delta V_{i,j,k} \tag{2.17}$$

$$\Delta V_{i,j,k} = \sum_{n\in N^{in}_{i,j,k}} M^{i,j,k}_{x,y,k}(n) - \sum_{n\in N^{out}_{i,j,k}} M^{v,w,k}_{i,j,k}(n) \tag{2.18}$$

$$G^{cur}_{i,j,k} \le G^{max}_{i,j,k},\ \forall(i,j,k) \tag{2.19}$$

$$G^{cur}_{i,j,k} = G^{org}_{i,j,k} + \sum_{n\in N} M^{v,w,k}_{x,y,k}(n) \cdot \Delta G_{i,j,k}, (i,j,k) \text{ on path } (x,y,k) \to (v,w,k) \tag{2.20}$$

$$0 \le \beta^m_{i,j,k} \le 1 \tag{2.21}$$

$$M^{x,y,k}_{i,j,k}(n) \in \{0,1\} \tag{2.22}$$

$$\sum_{n\in N_{mov}} M^{x,y,k}_{i,j,k}(n) = 1 \tag{2.23}$$

Equation (2.15) is a new way to compute the temperature at tile (i,j,k), which is different from Eq. (2.10). A detailed explanation of this equation is included in appendix section "Explanation of Eq. (2.15)". Equation (2.16) states that the total number of TSVs in a tile (i, j, k), which is specified by the $\beta^i_{i,j,k}$ values $(1 \leq i \leq vmax)$, should be equal to $V^{opt}_{i,j,k}$. Equation (2.21) restricts the range of values $\beta^n_{i,j,k}$ can take. All other equations are the same as discussed in NLP formulation.

A few points are worth mentioning. First, in order to overcome the restriction of moving just one TSV per net (=Eq. (2.23)), we repeat the entire relaxed ILP multiple times so that multiple TSVs from a single net are given a chance to relocate in an iterative fashion. We stop the iteration if the improvement on both the maximum and average temperature is minimal. Our related experiment shown in Table 2.5 suggests that the most of the temperature saving is obtained during the first iteration and that the overall algorithm converges within a small number of iterations. Second, the number of integer variables ($= M^{x,y,k}_{i,j,k}(n)$) can be huge if the number of nets is larger, or the thermal grid is finer. This makes our fast ILP formulation less desirable for a large problem instance. However, we overcome this limitation by relaxing these integer M variables and solving our fast ILP problem. We round the continuous variables based on a threshold value $\lambda = 0.5$. All variables above λ are converted into 1 provided that they do not violate the routing capacity constraint, whereas all other variables are converted to zero. Table 2.6 shows the impact of this relaxation on the solution quality and runtime.

2.6 Experimental Results

2.6.1 Experimental Setting

We implemented our router named 3D Elmore Router in C++/STL and ran our experiments on a Linux server running at 2.5 GHz and having 16 GB of memory. We tested our algorithms with three sets of benchmark: ISCAS89, ITC99, and ISPD98. We report the total wirelength, total number of TSVs used, the maximum thermal-aware Elmore delay among all sinks, the maximum temperature among all nodes in the thermal grid, and runtime in seconds for each circuit. We obtained 3D placement using an algorithm that is similar to [9]. The following are the details of our experimental setting:

1. We use four-die stacking for our 3D IC, where the top two and bottom two dies are bonded in face-to-face and the middle two in back-to-back unless specified otherwise.
2. We assume all four dies have different unit-length resistance and capacitance values [14] as follows unless specified otherwise: $r_1 = 86\,\Omega/\text{mm}$ and $c_1 = 396\,\text{fF/mm}$, $r_2 = 175\,\Omega/\text{mm}$ and $c_2 = 100\,\text{fF/mm}$, $r_3 = 74\,\Omega/\text{mm}$ and $c_3 = 279\,\text{fF/mm}$, and $r_4 = 154\,\Omega/\text{mm}$ and $c_4 = 120\,\text{fF/mm}$.
3. The dimension of our face-to-face TSVs is $1 \times 1 \times 10\,\mu\text{m}$. The parasitics are $R_{via} = 17.2\,\Omega/\text{mm}$ and $C_{via} = 371.8\,\text{fF/mm}$. The dimension of our back-to-back TSVs is $10 \times 10 \times 40\,\mu\text{m}$. The parasitics are $R_{via} = 0.172\,\Omega/\text{mm}$ and $C_{via} = 1943.8\,\text{fF/mm}$.

Table 2.2 3D routing grid dimensions and edge capacities for four die stack

				XYZ capacities		
ckt	#nets	v-bound	Grid dim	X & Y	Z (F2F)	Z (B2B)
s9234	5,844	5,563	20 × 20	8	3	1
b14_opt	5,646	5,546	40 × 40	8	3	2
s13207	8,727	8,298	40 × 40	10	4	2
s15850	10,397	9,915	40 × 40	10	5	4
b20_opt	12,501	12,174	40 × 40	16	6	4
b21_opt	12,678	12,288	60 × 60	16	6	4
b22_opt	18,086	17,911	60 × 60	20	7	5
ibm09	52,989	53,483	80 × 80	60	20	10
ibm10	68,004	67,651	100 × 100	75	20	10
ibm11	70,028	70,524	120 × 120	85	25	12
ibm13	84,191	82,989	140 × 140	95	25	12
ibm17	184,227	180,001	140 × 140	300	40	20

4. The "v-bound" column in Table 2.2 shows the lower bound of the TSV usage for each circuit. For MD2P nets, the lower bound is the number of dies in between the two pins plus 1. For MDMP nets, we use the fewest possible TSVs to connect all pins in the dies.
5. The routing grid dimensions used for four die stack are shown in Table 2.2. The dimensions were increased based on the circuit size (= number of nets), and the routing capacities were chosen so that about 10% of the nets need to be re-routed after initial tree generation.
6. The thermal grid dimension is $20 \times 20 \times 4$ for four-die stacked 3D IC. For thermal analysis, we use the following thermal conductivity values: silicon is 150 W/mK, copper is 285 W/mK, and epoxy (= bonding) layer is 0.05 W/mK. The power generated in each thermal grid is proportional to the number of cells placed in it, multiplied by a random value ranging from 1 to $10^7\,\mathrm{W/m^2}$ to account for the gate-level switching activity factor. We use our compact thermal model discussed in Sect. 2.5.3 for TSV relocation and [15] for all other purposes.

2.6.2 Tree Construction Results

We implement two existing 3D routers for comparison. The first is 3D maze router by [4] discussed in Sect. 2.2. The second is 3D A-tree router, where we extend the original 2D version [6] into 3D. More specifically, we first converted the 3D problem to a 2D problem by mapping the pin locations to the 2D plane. Then we perform 2D A-tree routing [6]. Lastly, our 2D solution is translated back to 3D, where the connection to the pins not located on the same die as the driver are connected using TSVs.

Table 2.3 shows a comparison between 3D maze [4], 3D A-tree [6], and our 3D Elmore routing. Our baseline is 3D maze router. We observe that our 3D Elmore

Table 2.3 Comparison between 3D maze [4], 3D A-tree [6], and our 3D Elmore routers. The "v-bound" column shows the lower bound on TSV count

	3D maze router [4]				3D A-tree router [6]				3D elmore router			
ckt	Wire	Delay	TSV	cpu	Wire	Delay	TSV	cpu	Wire	Delay	TSV	cpu
s9234	0.167	0.072	5,700	8	0.169	0.047	5,734	1	0.18	0.023	5,613	2
b14_opt	0.19	0.086	7,366	10	0.21	0.071	7,818	5	0.23	0.072	6,937	5
s13207	0.39	0.091	9,707	15	0.41	0.077	10,342	7	0.43	0.064	9,595	9
s15850	0.49	0.11	11,763	46	0.51	0.11	12,924	10	0.56	0.092	11,516	15
b20_opt	0.947	0.31	18,423	108	1.04	0.25	21,730	46	1.19	0.22	18,703	57
b21_opt	0.97	0.29	18,627	128	1.06	0.138	21,688	44	1.2	0.145	19,314	58
b22_opt	2.15	0.48	27,116	186	2.33	0.32	31,255	87	2.64	0.24	29,090	104
ibm09	29.4	222.2	104,481	639	33.01	124.7	122,465	245	36.9	103.6	118,013	353
ibm10	51.4	600.9	136,071	1,263	57.4	337.5	157,983	349	61.9	273.3	153,483	601
ibm11	53.3	457.9	131,815	1,859	60.5	264.6	152,422	456	61.2	218.4	141,922	711
ibm13	83.5	953.3	167,311	4,000	90.7	509.4	195,459	1,113	94.7	438.7	183,813	1,411
ibm17	157.4	1,723.4	398,145	7,754	169.8	1,124.6	408,861	2,122	172.4	899.8	405,400	2,456
RATIO	1.00	1.00	1.00	1.00	1.09	0.59	1.1	0.28	1.15	0.48	1.06	0.36

Table 2.4 Delay and wirelength distribution among the four different types of nets. We report the average max-sink delay and wirelength values among all nets in each type

	SD2P		SDMP		MD2P		MDMP	
ckt	dly	Wire	dly	Wire	dly	Wire	dly	Wire
s9234	0.004	0.014	0.005	0.029	0.005	0.013	0.128	0.107
b14_opt	0.005	0.012	0.015	0.043	0.003	0.011	0.32	0.116
s13207	0.020	0.028	0.06	0.08	0.018	0.029	0.6	0.2
s15850	0.023	0.031	0.12	0.075	0.02	0.03	0.63	0.21
b20_opt	0.021	0.03	0.11	0.08	0.019	0.028	0.68	0.246
b21_opt	0.022	0.03	0.13	0.085	0.018	0.029	0.65	0.24
b22_opt	0.047	0.042	0.16	0.153	0.041	0.046	1.1	0.36
ibm09	0.71	0.189	2.3	0.6	0.64	0.19	8.5	1.64
ibm10	1.24	0.25	3.59	0.79	1.09	0.22	12.6	1.95
ibm11	1.24	0.248	4.28	0.81	1.21	0.21	14.3	1.98
ibm13	2.15	0.34	6.4	1.04	2.01	0.33	28.4	2.85
ibm17	2.3	0.37	7.8	1.2	2.1	0.36	29.2	2.9
RATIO	1.0	1.0	3.2	3.51	0.92	0.94	12.4	8.1

router achieves 52% average delay improvement over 3D maze routing and 11% improvement over 3D A-tree. The delay reported is the maximum path delay of the final layout, which we obtain using a static timing analyzer. The TSV count is comparable between 3D Elmore router and 3D A-tree, while 3D maze uses 6% less TSVs. In case of wirelength, 3D maze and 3D A-tree obtained comparable results, but our 3D Elmore router uses 15% higher wirelength. Lastly, our 3D Elmore router runs 3× faster than 3D maze router and about 40% slower than 3D A-tree router.

We see that the number of TSVs needed by the 3D maze or 3D Elmore router is about twice as many as the minimum required. The number of TSVs used in 3D A-tree algorithm is the highest.[10] We observe from circuit `b21_opt` that 3D A-tree performs better than our 3D Elmore in terms of performance. We observe that this was due to congestion that caused larger number of nets to be ripped and re-routed in our 3D Elmore algorithm. In some cases, we observe that 3D A-tree caused lower congestion thus required less re-routing.

2.6.3 *Delay and Wirelength Distribution*

Our first goal is to collect the wirelength and delay statistics among the four types of nets in 3D Steiner routing mentioned in Sect. 2.3.1: single-die-two-pin (SD2P), single-die-multi-pin (SDMP), multi-die-two-pin (MD2P), and multi-die-multi-pin (MDMP) nets. Table 2.4 shows the statistics, where we report the average max-sink

[10]Our 3D A-tree is a performance-oriented router to be used for delay comparison, and TSV minimization is not considered.

delay and wirelength values among all nets in each type. We observe that MDMP nets have the largest delay and wirelength on average, which suggests that MDMP nets are the hardest to route in general (12.4× delay and 8.1× wirelength compared with SD2P). This is reasonable since they contain multiple pins in multiple dies and thus require multiple TSVs. We also observe that multi-pin nets incur larger delay and wirelength compared with two-pin nets (SDMP vs. SD2P and MDMP vs. MD2P). We also observe that MD2P nets have 8% smaller delay compared with SD2P on average. This is primarily due to the benefit of 3D connection, where TSV-based 3D connections tend to have smaller delay.

2.6.4 TSV Relocation Results

To evaluate the effectiveness of our TSV relocation algorithm, we implemented a fast greedy algorithm that tries to move TSVs into thermal hotspots in an iterative fashion: we choose a single hotspot and relocate movable TSVs into it. We then repeat this process for the next hotspot until no more temperature improvement can be obtained. In addition, we developed two TSV relocation methods based on our ILP formulation introduced in Sect. 2.5.6: single ILP and multiple ILP. Under the single ILP method, we perform our ILP-based TSV relocation once. Under the multiple ILP method, we repeat the ILP-based TSV relocation until there is no more gain on temperature reduction. In this case, we report the number of iterations taken. Note that our ILP-based methods target *all* hotspots simultaneously.

Table 2.5 shows the maximum temperature, average temperature and standard deviation obtained by the greedy method and our ILP-based methods (single iteration vs. multiple iteration). We observe that our ILP-based simultaneous methods achieve consistent improvement over the greedy approach at the expense of additional runtime. We obtained 9% maximum temperature and 16% average temperature reduction with our ILP-based methods, whereas the greedy method improves the max/ave temperature by only 1%. Note that this free saving does not require any additional area for dummy TSV insertion. The runtime for our biggest circuit `ibm17` that contains 184 K nets is around 3,045 s for single ILP. This shows that our fast ILP method scales well with the complexity of the circuit while maintaining high quality solutions. In Table 2.5 we also show the impact of multiple iterations on our fast ILP. We observe that the temperature saving between single vs. multiple ILP is comparable for maximum temperature. In case of average temperature and standard deviation, however, our multiple ILP outperformed single ILP by 6%. We also observe that our multiple ILP method converges quickly to a high quality solution within a few iterations.

Table 2.6 shows the impact of the M-variable relaxation in our fast-ILP method. Due to the excessive runtime of our original/slow ILP, we used the four smallest circuits. In case of the two bigger circuits, we gave our slow ILP one full day and have it report the best solution discovered so far. We observe that our slow ILP obtained 6% better results on the maximum temperature, but the runtime

Table 2.5 TSV relocation results. T_{max}, T_{ave}, T_{std} respectively denotes the maximum temperature, average temperature, and standard deviation among all thermal tiles. The runtime is in seconds

	Initial temp			Greedy				Single ILP				Multiple ILP				
ckt	T_{max}	T_{avg}	T_{std}	T_{max}	T_{avg}	T_{std}	cpu	T_{max}	T_{avg}	T_{std}	cpu	T_{max}	T_{avg}	T_{std}	cpu	itr
s9234	92.6	52.3	23.7	91.9	51.6	23.7	2	84.5	45.1	20.7	67	83.1	40.1	18.3	246	5
b14_opt	114.5	46.1	32.3	114.1	45.8	32.2	2	107.6	38.7	27.8	89	105.8	33.6	25.1	315	6
s13207	112.1	66.2	27.6	111.3	66.1	27.6	2	101.2	53.2	24.3	82	101.1	52.8	24.0	196	3
s15850	115.1	44.3	34.3	114.0	44.0	34.1	2	103.1	37.5	27.6	175	102.6	35.3	26.9	324	3
b20_opt	108.3	38.9	37.3	108.1	38.8	37.0	3	98.7	30.3	32.3	209	96.4	27.6	29.8	813	7
b21_opt	114.1	45.7	24.2	113.5	45.5	24.0	3	109.4	37.8	20.1	206	108.4	35.7	19.4	612	3
b22_opt	114.5	54.6	18.4	114.2	54.6	18.4	4	105.1	49.6	15.2	228	104.5	47.8	14.7	388	2
ibm09	94.2	47.8	23.1	93.8	47.3	22.9	11	87.1	43.6	21.2	702	84.8	39.5	18.7	2,113	6
ibm10	113.4	54.2	21.0	112.9	54.0	20.9	13	105.6	45.2	16.4	452	104.0	32.3	18.3	1,566	5
ibm11	108.4	45.3	27.8	107.7	45.1	27.7	19	99.7	35.2	21.0	832	97.9	32.3	18.3	4,211	8
ibm13	95.1	52.4	27.9	94.5	52.1	27.7	24	86.6	43.8	22.3	1,387	86.1	41.8	21.8	2,346	2
ibm17	111.2	48.6	19.2	110.8	48.2	19.0	38	101.5	41.7	15.4	3,045	101.0	39.6	14.1	6,836	3
RATIO	1.00	1.00	1.00	0.99	0.99	0.99	1.00	0.91	0.84	0.83	64.3	0.90	0.78	0.77	160.1	–

Table 2.6 Impact of M-variable relaxation (fast-ILP vs slow-ILP) in terms of maximum temperature and runtime

	Fast-ILP		Slow-ILP	
ckt	T_{max}	cpu	T_{max}	cpu
s9234	84.5	1.1 min	80.1	234 min
b14_opt	107.6	1.4 min	99.5	342 min
s13207	101.2	1.3 min	94.2	>1-day
s15850	103.1	2.9 min	98.4	>1-day
RATIO	1.0	1.0	0.94	–

Table 2.7 Various TSV sizes and their parasitic resistance (Ω/mm) and capacitance (fF/mm). Dimensions are in μm

	Face-to-face TSV				
	Width	Height	Depth	R	C
Size I	0.5	0.5	5	68.8	288.5
Size II	1	1	10	17.2	371.8
Size III	2	2	15	4.3	554.2
	Back-to-back TSV				
	Width	Height	Depth	R	C
Size I	5	5	20	0.688	1,229.1
Size II	10	10	40	0.172	1,943.8
Size III	20	20	60	0.043	2,798.2
	Face-to-back TSV				
	Width	Height	Depth	R	C
Size I	4	4	15	1.07	982.8
Size II	8	8	35	0.267	1,519.1
Size III	15	15	55	0.076	2,604.3

required is prohibitive. Based on this runtime trend, one can expect that our slow ILP will not be able to handle bigger circuits, which in turn shows that our M-variable relaxation is the key to making our ILP method scalable.

2.6.5 Impact of TSV Dimension and Parasitics

Next, we study the impact of TSV dimension and parasitics on delay, wirelength, TSV count, and temperature. Note that TSVs play an important role in determining the overall routing topology as well as the underlying thermal profile. Table 2.7 shows three different TSV sizes and their RC parasitics we use in our experiment. Size II is the default set that is used in all of the previous experiments as discussed in Sect. 2.6.1. Table 2.8 shows our 3D Elmore routing results using these sizes, where Size I, the smallest, is our baseline. We first observe that the delay increases as the TSV dimension grows. This is mainly due to the wirelength increase, which is caused by the routing congestion from using larger TSVs. We observe that these large TSVs have detrimental impact on MDMP (multi-die-multi-pin) nets, which are more likely to become critical nets. Another factor is the higher parasitic capacitance values for larger TSVs. Since Elmore delay model penalizes heavily

Table 2.8 Impact of TSV dimension on delay, wirelength, and TSV counts. We use the three TSV sizes shown in Table 2.7

	Size I			Size II			Size III		
ckt	Delay	Wire	TSV	Delay	Wire	TSV	Delay	Wire	TSV
s9234	0.02	0.171	5,694	0.023	0.18	5,613	0.032	0.183	5,558
b14_opt	0.048	0.213	7,062	0.072	0.23	6,937	0.122	0.217	6,911
s13207	0.047	0.43	9,749	0.064	0.43	9,595	0.077	0.44	9,657
s15850	0.066	0.53	11,703	0.092	0.56	11,516	0.163	0.57	11,351
b20_opt	0.16	1.18	19,274	0.22	1.19	18,703	0.31	1.21	18,187
b21_opt	0.125	1.2	19,818	0.145	1.2	19,314	0.228	1.24	18,810
b22_opt	0.229	2.61	29,818	0.24	2.64	29,090	0.48	2.69	28,847
ibm09	96.0	36.6	119,125	103.6	36.9	118,013	127.8	37.3	116,724
ibm10	252.1	60.6	155,349	273.3	61.9	153,483	390.2	65.7	151,519
ibm11	198.6	60.2	142,990	218.4	61.2	141,922	268.7	67.6	139,202
ibm13	419.5	92.5	184,428	438.7	94.7	183,813	563.4	109.5	181,456
ibm17	824.8	171.1	416,458	899.8	172.4	405,400	987.6	187.4	401,345
RATIO	1.0	1.0	1.0	1.07	1.01	0.98	1.34	1.1	0.97

Table 2.9 Impact of TSV dimension on TSV relocation and the maximum temperature reduction. We use the three TSV sizes shown in Table 2.7. T_{ini} and T_{max} respectively denote the maximum temperature before and after TSV relocation

		Size I	Size II	Size III
ckt	T_{ini}	T_{max}	T_{max}	T_{max}
s9234	92.6	88.61	84.5	82.38
b14_opt	114.5	109.3	107.6	105.1
s13207	112.1	104.2	101.2	99.7
s15850	115.1	104.1	103.1	100.4
b20_opt	108.3	101.8	98.7	95.4
b21_opt	114.1	111.5	109.4	106.2
b22_opt	114.5	108.2	105.1	105.5
ibm09	94.2	91.3	87.1	84.5
ibm10	113.4	109.8	105.6	101.4
ibm11	108.4	105.2	99.7	94.5
ibm13	95.1	89.2	86.6	82.3
ibm17	111.2	107.8	101.5	99.2
RATIO	1.0	0.95	0.91	0.89

on capacitance increase, having larger TSVs results in more delay increase. Next, the actual TSV count decreases as the TSV dimension increases. This is because our delay-driven router may avoid using TSVs, especially for short interconnects, to minimize the overall delay. However, the TSV count reduction is only 3%, indicating that our 3D Elmore router still makes a heavy use of TSVs (up to 400 K for `ibm17`).

We also conducted experiments to observe the impact of TSV dimension and parasitics on temperature saving obtained by our TSV relocation algorithm. The results are shown in Table 2.9. We observe that larger TSVs result in more temperature saving (5, 9, and 11%). This is mainly due to the smaller thermal resistivity for larger TSVs. We note that circuit `b22_opt` does not follow this trend. This occurs since different TSV dimensions may result in different routing solutions, thereby influencing the thermal profile and temperature saving opportunity.

Table 2.10 Impact of bonding style, where we use face-to-back bonding only in our four-die stack

	3D maze			3D A-tree			3D elmore		
ckt	Delay	Wire	TSV	Delay	Wire	TSV	Delay	Wire	TSV
s9234	0.018	0.146	2,195	0.019	0.155	2,402	0.015	0.161	2,113
b14_opt	0.078	0.194	3,680	0.18	0.042	4,459	0.039	0.219	3,914
s13207	0.074	0.388	3,497	0.07	0.41	4,262	0.07	0.43	3,676
s15850	0.103	0.504	4,298	0.093	0.53	5,561	0.083	0.57	4,680
b20_opt	0.354	0.975	8,210	0.228	1.06	12,207	0.186	1.15	10,084
b21_opt	0.38	0.966	7,977	0.163	1.05	11,367	0.156	1.13	9,834
b22_opt	0.555	2.18	11,408	0.466	2.38	16,824	0.281	2.71	15,185
ibm09	228.1	30.3	44,667	135.7	33.9	65,692	116.1	35.8	59,955
ibm10	578.7	52.3	49,855	355.2	58.5	77,334	312.1	59.1	74,329
ibm11	409.6	52.6	67,076	254.1	59.7	75,281	215.5	60.2	61,353
ibm13	906.1	81.5	72,546	491.9	87.5	98,964	451.7	90.3	79,688
ibm17	2,234.6	174.5	160,435	1,224.5	187.8	192,349	978.8	206.9	178,013
RATIO	1.0	1.0	1.0	0.56	1.09	1.30	0.47	1.15	1.14

2.6.6 *Impact of Bonding Style*

So far the top two and bottom two dies are bonded in face-to-face, and the middle two in back-to-back in our four-die stack as discussed in Sect. 2.6.1. We now study the impact of bonding style on 3D routing results. In our new four-die stack, all dies are bonded face-to-back. We use $R_{via} = 0.267\,\Omega/\text{mm}$ and $C_{via} = 1{,}519.1\,\text{fF/mm}$ for the face-to-back TSVs. One important difference between the "F2F+B2B" stack and "F2B-only" stack is that the TSV upper bound is different: face-to-face bonding allows more TSVs than face-to-back or back-to-back, primarily due to the sizes. Therefore, we re-ran our 3D placer to obtain a new placement that minimizes inter-die connections for all face-to-back bonding. The routing results are shown in Table 2.10. We first observe that the TSV count is considerably lower compared with Table 2.3. This is mainly due to the absence of face-to-face bonding in the new F2B-only stack. On the other hand, the delay values in Table 2.10 are larger than those in Table 2.3. This is mainly due to the larger wirelength and parasitic capacitance for the face-to-back TSVs. The reason for this wirelength increase with F2B-only stacking is because the number of inter-die connection is minimized, resulting in less opportunity for wirelength reduction. Lastly, both Tables 2.3 and 2.10 show the same trend in terms of delay, wirelength, and TSV count among 3D maze, 3D A-tree, and 3D Elmore routers.

2.6.7 *Two-die versus four-die Stacking*

So far our 3D stack contained four dies as discussed in Sect. 2.6.1. We now use two-die stack in our experiment, where the two dies are bonded face-to-face. Our wire and TSV parasitics are as follows: $r_1 = 86\,\Omega/\text{mm}$, $c_1 = 396\,\text{fF/mm}$, $r_2 = 175\,\Omega/\text{mm}$,

$c_2 = 100\,\text{fF/mm}$, $R_{via} = 17.2\,\Omega/\text{mm}$, and $C_{via} = 371.8\,\text{fF/mm}$. Table 2.11 shows a comparison between 3D maze, 3D A-tree, and our 3D Elmore router. Our baseline is 3D maze router. We observe that our 3D Elmore router achieves 54% average delay improvement over 3D maze routing and 11% improvement over 3D A-tree. This significant delay reduction came at the cost of 13% wirelength and 10% TSV count increase compared with 3D maze router. This trend is almost the same as what we saw in Table 2.3 for the four-die stack case.

TSV relocation results are shown in Table 2.12. We again observe a similar trend as in Table 2.5: greedy method does not produce good results, and our single ILP method produces comparable results to the multiple ILP method in a shorter runtime. The difference between two-die and four-die stack is twofold. First, the temperature reduction is smaller with two-die (6%) compared with four-die (9%). This is because there are fewer hotspots and fewer TSVs in two-die stack. Second, the ILP runtime is also smaller in two-die stack due to the smaller problem size.

2.7 Conclusions

We studied two new problems that are important for 3D stacked IC technology: 3D Steiner tree construction and TSV relocation. Our routing algorithm is based on a constructive method, where a 3D Steiner tree is grown by connecting a new pin to the existing tree. We derived two-variable delay equations and optimized them to compute the location of the TSVs under given thermal profile. For TSV relocation, we devised an innovative technique which helps avoid the non-linear optimization required for temperature optimization. Our formulation can handle large number of TSVs simultaneously for an effective temperature optimization.

Appendix

Optimization of Two-Variable Delay Equations

Assuming $a_0 = r_2/r_1$ and $b_0 = c_1/c_2$, the optimization of two-variable delay functions shown in Sect. 2.4.3 allow the computation of x (= connection point) and y (= TSV location) as follows:

- For $d(a)$ we have

$$\frac{\partial^2 F}{\partial \delta x^2} = r_1 c_2 (a_0 - b_0 - 2)$$

$$\frac{\partial^2 F}{\partial \delta y^2} = r_1 c_2 (a_0 + b_0 - 2)$$

$$H_1 = -(r_1 c_2)^2 \{(a_0 + b_0 - 2) 2 b_0\}$$

Table 2.11 Two-die stack results: comparison between 3D maze [4], 3D A-tree [6], and our 3D Elmore routers. The "v-bound" column shows the lower bound on TSV count

	3D maze routing				3D A-tree				3D elmore routing			
ckt	Delay	Wire	TSV	cpu	Delay	Wire	TSV	cpu	Delay	Wire	TSV	cpu
s9234	0.04	0.3	5,137	13	0.04	0.33	5,250	2	0.017	0.3	4,976	4
b14_opt	0.083	0.38	6,102	12	0.04	0.42	6,560	4	0.04	0.42	6,056	6
s13207	0.13	0.71	7,960	15	0.11	0.75	8,338	4	0.99	0.79	8,216	7
s15850	0.17	0.92	9,616	26	0.82	0.98	10,049	7	0.068	1.04	9,890	10
b20_opt	0.41	1.76	14,205	35	0.17	1.96	15,724	12	0.15	2.1	15,445	15
b21_opt	0.35	1.75	14,253	35	0.16	1.96	16,004	12	0.19	2.07	15,650	16
b22_opt	0.99	4.1	20,557	86	0.38	4.54	22,545	40	0.34	4.8	23,571	56
ibm09	973.5	58.5	79,677	411	461.9	66.7	89,134	234	388.7	68.7	93,184	287
ibm10	2,245.8	98.2	109,685	530	1,156.6	106.7	119,065	241	1,023.5	113.4	128,630	284
ibm11	1,617.6	102.4	100,321	771	804.8	116.1	112,061	303	657.1	125.2	113,573	344
ibm13	2,578.6	174.5	139,355	987	1,667.6	186.6	140,774	331	1,473.9	195.6	142,872	413
ibm17	6,208.3	338.9	338,854	1,526	3,177.2	357.8	359,394	411	2,790.1	375.3	370,599	791
RATIO	1.0	1.0	1.0	1.0	0.53	1.08	1.07	0.36	0.46	1.13	1.1	0.5

Table 2.12 Two-die stack results: TSV relocation results. T_{lnl} and T_{max} respectively denote the maximum temperature before and after TSV relocation

		Greedy		Single ILP		Multiple ILP		
ckt	T_{lnl}	T_{max}	cpu	T_{max}	cpu	T_{max}	cpu	iter
s9234	64.5	64.2	1.04	61.4	7.04	61.4	11.25	2
b14_opt	79.3	78.1	3.12	74.5	20.56	74.2	36.8	2
s13207	66.79	66.13	2.96	62.74	22.1	61.93	48.9	2
s15850	88.23	87.6	3.12	84.3	24.5	83.2	66.9	3
b20_opt	84.6	83.9	3.67	80.2	33.1	79.8	60.4	2
b21_opt	72.8	72.1	3.5	67.7	32.6	67.1	67.8	2
b22_opt	84.4	83.5	5.12	79.2	38.9	78.6	81.3	2
ibm09	77.9	76.8	12.3	73.5	148.9	73.1	405.6	3
ibm10	81.5	80.2	13.6	77.4	156.7	77.1	280.7	2
ibm11	81.9	81.1	15.3	77.2	221.3	76.5	515.7	3
ibm13	83.5	82.4	14.8	78.1	356.9	77.5	678.4	2
ibm17	86.2	85.4	19.4	81.3	623.8	80.5	1,356.7	2
RATIO	1.0	0.989	1.0	0.943	17.22	0.936	36.86	–

Thus, we see that when $H_1 = 0$, $\frac{\partial^2 F}{\partial \delta x^2} \leq 0$ and $\frac{\partial^2 F}{\partial \delta y^2} = 0$, the optimal delay is found at points according to the Case B. If $H_1 < 0$, optimal delay is found at points according to the Case D. If $H_1 > 0$, we have $\frac{\partial^2 F}{\partial \delta x^2} \leq 0$, so the optimal delay is found at points according to the Case A.

- For $d(b)$ we need to evaluate two cases: (1) when $x \geq b$, we have $\frac{\partial^2 F}{\partial \delta x^2} = 0$, $\frac{\partial^2 F}{\partial \delta y^2} = 0$, and $H_1 = 0$. Thus, the optimal delay is found at points according to the Case C. (2) when $x \leq b$, we have $\frac{\partial^2 F}{\partial \delta x^2} = -2r_1c_2$, $\frac{\partial^2 F}{\partial \delta y^2} = 0$, and $H_1 = -(r_1c_2)^2(b_0 - 1)^2$. Thus, if $H_1 = 0$, the optimal delay is found at points according to the Case B. Otherwise, they are found at points according to the Case D.
- For $d(c)$ we have $\frac{\partial^2 F}{\partial \delta x^2} = -2r_1c_2$, $\frac{\partial^2 F}{\partial \delta y^2} = 0$, and $H_1 = -(r_1c_2)^2(b_0 - 1)^2$. If $H_1 = 0$, the optimal delay is found at points according to the Case B. Otherwise, they are found at points according to the Case D.
- For all other nodes not in T_p, we have $\frac{\partial^2 F}{\partial \delta x^2} = 0$, $\frac{\partial^2 F}{\partial \delta y^2} = 0$, and $H_1 = 0$ since the delay is a linear function of δx and δy. Thus, the optimal delay is found at points according to the Case C.

Since a_0 and b_0 values are dependent on the interconnect parameters at each die, we see that the number of points (x,y) is a fixed constant.

Explanation of Eq. (2.15)

From Fig. 2.5 we note that the temperature at tile (i, j, k) having n TSVs is computed as follows:

$$T_{i,j,k} = T_{i,j,k-1} + (P_{i,j,K} + \cdots + P_{i,j,k}) \times R^{m}_{i,j,k}$$

We can write $\delta^m_{i,j,k}$ (refer to Table 2.1 for definition) as follows:

$$\delta^m_{i,j,k} = (P_{i,j,K} + \cdots + P_{i,j,k}) \times R^m_{i,j,k}$$

We see that $\delta^m_{i,j,k} \propto R^m_{i,j,k} \propto 1/V^m_{i,j,k}$, thus $\delta^m_{i,j,k}$ is strictly decreasing for increasing values of m and $m > 0$. It can be seen easily that the temperature of a given tile having m TSVs can be rewritten as

$$T_{i,j,k} = T_{i,j,k-1} + \delta^0_{i,j,k} - (\delta^0_{i,j,k} - \delta^1_{i,j,k}) - \cdots\cdots - (\delta^{m-1}_{i,j,k} - \delta^m_{i,j,k})$$

We define $\Delta T^m_{i,j,k}$ as follows:

$$\Delta T^m_{i,j,k} = \delta^{m-1}_{i,j,k} - \delta^m_{i,j,k}$$

which is equal to the coefficient of the variable $\beta^m_{i,j,k}$. Note that $\Delta T^m_{i,j,k}$ is strictly decreasing when m is increasing. This enables us to use non-integer values for the variable $\beta^m_{i,j,k}$ (refer to Table 2.1 for its definition). The reason is that for any value of $V^m_{i,j,k}$, $\beta^m_{i,j,k}$ will always reach its maximum allowed value of 1 before $\beta^{m+1}_{i,j,k}$ starts having a non-zero value. This is due to the fact that $\Delta T^m_{i,j,k} > \Delta T^{m+1}_{i,j,k}$, which corresponds to a greater decrease in objective function per unit change of $V^m_{i,j,k}$. In other words, if $\beta^m_{i,j,k} < 1$ and $\beta^{m+1}_{i,j,k} > 0$, then we can always find a solution with a lower cost by (1) adding γ to $\beta^m_{i,j,k}$ so that $\beta^m_{i,j,k} = 1$, and (2) adjusting $\beta^{m+1}_{i,j,k}$ with $\beta^{m+1}_{i,j,k} - \gamma$.

Thus, we see in our new reduced ILP formulation that the extra variables $\beta_{i,j,k}$ are not constrained to be integers and that the only integer variables we need are the $M^{x,y,k}_{i,j,k}(n)$ variables. Note that this assumption is no longer valid if we try to minimize the maximum temperature of the tiles, since $T_{i,j,k}$ is no longer present in the objective function. There is no constraint on the $\beta^m_{i,j,k}$ values, so they cannot be relaxed as continuous variables. Thus, the resulting ILP will have an extremely large number of integer variables.

References

1. A. Ajami, K. Banerjee, M. Pedram, Effects of non-uniform substrate temperature on the clock signal integrity in high performance designs, in *Proceedings of IEEE Custom Integrated Circuits Conference*, San Diego, 2001, pp. 233–236
2. K. Boese, A. Kahng, B. McCoyy, G. Robins, Near-optimal critical sink routing tree constructions, in *IEEE Transactions on Computer-Aided Design of Integrated Circuits and Systems*, **14**(12), 1417–1436 (1995)
3. M. Burnstein, R. Pelavin, Hierarchical wire routing, in *IEEE Transactions on Computer-Aided Design of Integrated Circuits and Systems*, **2**(4), 223–234 (1983)
4. J. Cong, Y. Zhang, Thermal-driven multilevel routing for 3-D ICs, in *Proceedings of Asia and South Pacific Design Automation Conference*, Shanghai, 2005, pp. 121–126

5. J. Cong, Y. Zhang, Thermal via planning for 3-D ICs, in *Proceedings of IEEE International Conference on Computer-Aided Design*, San Jose, 2005
6. J. Cong, K.-S. Leung, D. Zhou, Performance driven interconnect design based on distributed RC delay model, in *Proceedings of ACM Design Automation Conference*, Dallas, 1993
7. S. Das, A. Chandrakasan, R. Reif, Design tools for 3-D integrated circuits, in *Proceedings of Asia and South Pacific Design Automation Conference*, Kitakyushu, 2003, pp. 53–56
8. A. Fan, A. Rahman, R. Reif, Copper wafer bonding. Electrochem. Solid-State Lett. **2**, 534–536 (1999)
9. B. Goplen, S. Sapatnekar, Efficient thermal placement of standard cells in 3D ICs using a force directed approach, in *Proceedings of IEEE International Conference on Computer-Aided Design*, San Jose, 2003
10. B. Goplen, S. Sapatnekar, Placement of thermal vias in 3-D ICs using various thermal objectives. IEEE Trans. Comput.-Aided Des. Integr. Circuits Syst. **25**(4), 692–709 (2006)
11. X. Li, Y. Ma, X. Hong, S. Dong, J. Cong, LP based white space redistribution for thermal via planning and performance optimization in 3D ICs, in *Proceedings of Asia and South Pacific Design Automation Conference*, Seoul, 2008, pp. 209–212
12. J. Minz, S.K. Lim, Block-level 3D global routing with an application to 3D packaging. IEEE Trans. Comput.-Aided Des. Integr. Circuits Syst. **25**(10), 2248–2257 (2006)
13. M. Pathak, S.K. Lim, Thermal-aware steiner routing for 3D stacked ICs, in *Proceedings of IEEE International Conference on Computer-Aided Design*, San Jose, 2007
14. V. Pavlidis, E. Friedman, Interconnect delay minimization through interlayer via placement in 3-D ICs, in *Proceedings of Great Lakes Symposum on VLSI*, Chicago, 2005
15. T.-Y. Wang, C.C.-P. Chen, 3-D thermal-ADI: a linear-time chip level transient thermal simulator, in *IEEE Transactions Computer-Aided Design of Integrated Circuits and Systems*, **21**(12), 1434–1445 (2002)
16. H. Yu, J. Ho, L. He, Simultaneous power and thermal integrity driven via stapling in 3D ICs, in *Proceedings of IEEE International Conference on Computer-Aided Design*, San Jose, 2006
17. T. Zhang, Y. Zhan, S.S. Sapatnekar, Temperature-aware routing in 3D ICs, in *Proceedings of Asia and South Pacific Design Automation Conference*, Yokohama, 2006, pp. 309–314

Chapter 3
Buffer Insertion for 3D IC

Abstract Despite the momentum 3D IC technology has gained recently, there has been little progress on timing optimization for 3D ICs. In this chapter, we first study the fact that Through-Silicon-Vias (TSVs) have large parasitic capacitances that increase signal slew. Next, we develop a buffer insertion algorithm that improves the delay of both 3D and 2D nets in a 3D IC with explicit consideration of signal slew. The effectiveness of this technique is demonstrated with various nets and full-chip results. Compared with the well-known van Ginneken algorithm and the timing-constraint-based 2D optimization by a commercial software, our algorithm finds buffering solutions with lower slew-aware delay and buffer usage with tolerable runtime overhead.

The materials presented in this chapter are based on [8].

3.1 Introduction

For high performance 3D ICs, it is crucial to perform thorough timing optimization, especially when the 3D nets are on timing critical paths. Among timing optimization techniques, buffer insertion is known to be the most effective way. However, currently there is no commercial design software that performs buffer insertion on multiple die designs simultaneously. The through-silicon-vias (TSVs) have a large parasitic capacitance that increases the signal slew and the delay on the downstream. Even for 2D ICs, today's advanced technology nodes (e.g., 28 nm node) experience high slew degradation along nets, which in turn increases gate delay.

L.P.P.P. van Ginneken [15] adopted dynamic programming (VGDP). VGDP has been used in slew buffering [5], which fixes slew design rule violations but does not optimize timing. A non-VGDP buffer insertion algorithm with slew consideration was presented in [13]. However, their delay models could not adopt effective capac-

S.K. Lim, *Design for High Performance, Low Power, and Reliable 3D Integrated Circuits*, DOI 10.1007/978-1-4419-9542-1_3,

itance [14], and hence considering the fact that TSVs affect effective capacitance much, this algorithm is not suitable for 3D ICs. Also, their framework relies on nonlinear optimization which would incur runtime issues for large net instances, which was not reported. In contrast, the VGDP framework is known for efficiency and flexibility, yet there has been no work that considers realistic signal slew in VGDP framework. In [9] the authors considered slew in VGDP framework, however their slew model is not realistic and the implementation is complicated due to piecewise linear functions. In this chapter we study the bottom-up slew propagation DP (SPDP), which is a modified version of VGDP, to perform delay optimization with the consideration of slew for TSV-based 3D ICs. By considering slew in DP framework, we achieve lower buffered delay compared with the original VGDP.

There is a common belief in 3D IC community that timing optimization can be handled with existing 2D electronic design automation (EDA) tools, with a little modifications for TSV handling. However, since a 2D EDA tool handles each die separately, it cannot consider the whole 3D path and timing optimization quality is worse compared with true 3D buffer insertion methods. The following specific topics are covered in this chapter:

- We study the impact of slew caused by TSVs in 3D nets on gate and net delays. With a buffered 3D net, we discuss how severe TSV-induced slew degradation is, and how the buffer solution could be improved.
- We incorporate a reasonably accurate slew model into the van Ginneken DP framework for delay minimization. We study a slew binning idea to explicitly and efficiently consider slew-aware delay during solution search. In addition, using the slew information we study several efficient pruning rules, limiting search space and reducing runtime.
- We compare and analyze the buffer insertion solutions of our SPDP with VGDP and the timing-constraint-based 2D buffer insertion using commercial EDA softwares. With full-chip 3D IC designs, we demonstrate how much timing could be improved if we apply 3D buffer insertion instead of timing-constraint-based 2D optimization.

3.2 Problem Definition

We define our buffer insertion problem as follows: After placement and routing stage, on a given routed net topology with placed TSV pin pads and landing pads, we insert buffers (from a given buffer library) at candidate locations to maximize the required arrival time (RAT) at the source gate. This is equivalent to minimizing the delay from the source gate to the critical sink gate. We assume that the input slew at the source gate, the loading capacitance at the sink output, and the RAT at the sink output are given. As discussed in Sect. 3.3, since the input slew (S_i) of the sink gate affects the sink gate delay, we include the sink gate delay in RAT calculation during DP. Thus, we minimize the delay from the input of the source gate to the output of the critical sink gate.

This problem is different from the delay-constrained buffer insertion problem [9, 13], where we try to find a buffer insertion solution that minimizes resource usage (e.g., area, power) under a delay target. For these net instances, we just need a fast buffer insertion algorithm with reasonable quality in delay. In contrast, for nets on critical paths, we should find a buffer insertion solution that provides the lowest delay to the critical sink. Our target nets for buffer insertion are these 'hard' instances, for which finding a better solution is more important than finding a reasonably good solution quickly. Also note that our problem differs from slew minimization, which does not produce the lowest delay to critical sinks. Our goal is to minimize delay, and incorporating slew in delay calculation provides more accurate delay calculation during DP, however minimizing slew does not produce the lowest delay to critical sinks.

Also, we are allowed to spend a reasonable time for buffer insertion, because in the post-route physical design optimization, the most time-consuming part is the RC extraction and the timing update. It is better to use a little more time to find a better solution, because it will lead to less number of optimization loops and earlier timing closure.

3.3 Motivational Example

In our 3D IC design experience, we observed that the slew degradation due to the TSVs is quite severe, even after buffer insertion. As a motivational example, Fig. 3.1a shows a 3D net with two TSVs. The source gate is NAND2_X1 and the sink1/2 gates are AOI21_X1. We perform buffer insertion using VGDP, SPDP, and timing-constraint-based 2D optimization by Cadence Encounter. For simplicity of demonstration, we assume that the buffer library consists of a single buffer (BUF_X4) and an inverter (INV_X4), the input slew (S_i) at the source gate is 40 ps, and the load capacitance at sink1/2 is 20 fF. The delay, slew, and arrival time (AT) values in Fig. 3.1 are obtained by layout and PrimeTime 3D static timing analysis (STA). The VGDP places a buffer right after source (for boosting) and another at 750 μm. Due to the large TSV capacitance and the long wirelength driven by buf1, the slew degradation from buf1 to buf2 is large, thus the S_i at the buf2 is quite high which increases the delay of buf2. Also, the S_i at the critical sink gate is large. Since our SPDP considers slew during DP, inv2 is much closer to TSV1 than buf1 in VGDP solution is, and inv3 is also closer to TSV1 than buf2 in VGDP solution is. As a result, the S_i at inv3 is only 35 ps, which reduces the delay of inv3. Also, the S_i at the critical sink is lower with SPDP which reduces sink gate delay. This lower slew is especially helpful because the sink gate delay is sensitive to S_i. Comparing AT values at sink1 output, we see that SPDP achieves 4.3% delay reduction compared with VGDP. The timing-constraint-based 2D buffer insertion with a commercial design software does not produce a good result. It inserted buffers that usually have higher intrinsic delay than inverters. Moreover, during Die 1 optimization the tool

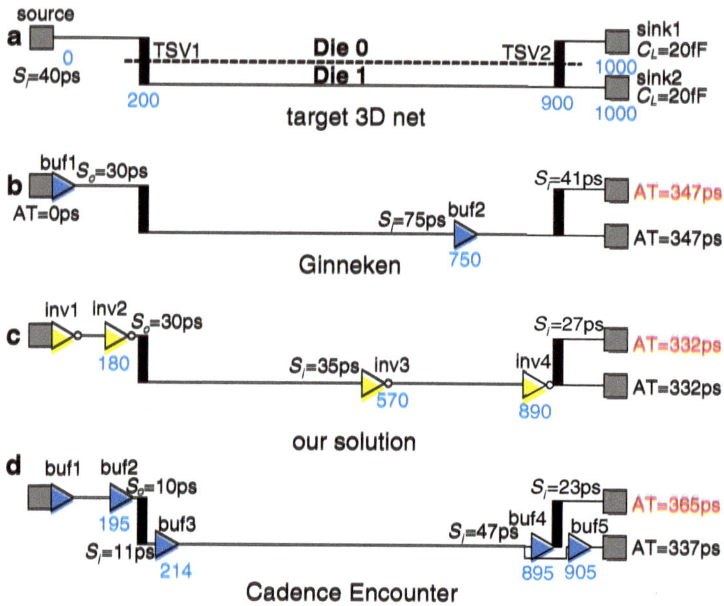

Fig. 3.1 A motivational example. Numbers shown in *blue* represent the distance from source gate in µm. (**a**) Target net and buffer insertion solutions with (**b**) VGDP, (**c**) SPDP, and (**d**) timing-constraint-based 2D optimization by Cadence Encounter, S_o means the output slew of a gate

does not know exactly where buf2 is, thus the location of buf3 is needlessly too close to TSV1. Even though S_i at the critical sink is the lowest, the AT is the worst among the three buffer insertion solutions. This clearly demonstrates why timing-constraint-based 2D buffer insertion is not thorough enough for timing-critical nets.

3.4 Delay and Slew Models

3.4.1 Target 3D IC and TSV Structure

In this chapter, as shown in Fig. 3.2a, we assume four dies are stacked, and TSVs go through Die 0, 1, and 2. Our TSV macro occupies six standard cell rows as shown in Fig. 3.2b. Due to the reliability issues and performance variation, gates and buffers should be placed outside TSV keep-out zone. The TSVs have large parasitics that affect timing. Each TSV has a parasitic capacitance (C_{TSV}) and a resistance (R_{TSV}), and is represented by a π-model with two capacitors and a resistor as shown in Fig. 3.2c. Based on our physical assumptions such as TSV liner thickness and doping concentration, using the formula in [7] we calculated C_{TSV} and R_{TSV}, which are shown in Table 3.1. The inductance of TSV is ignored because it is not dominant

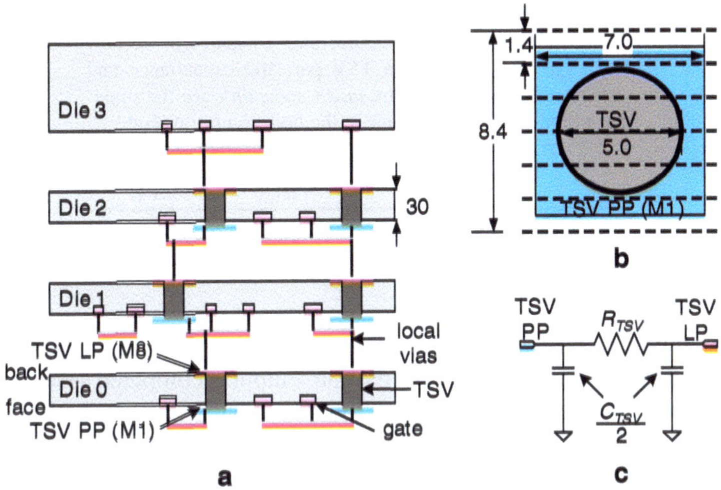

Fig. 3.2 (**a**) *Side view* of the 3D IC, (**b**) *top view* of a TSV, and (**c**) TSV RC model. TSV PP (M1) and TSV LP (M8) represent TSV pin pad on metal1 and TSV landing pad on metal8, respectively. *Dashed lines* in (**b**) denote standard cell row boundaries. Dimensions are in μm

under a few GHz signal frequency. We assume that the unit length capacitance and resistance of net wires are C_m and R_m. Due to the TSVs, 3D nets no longer have uniform RC characteristics, which needs to be reflected in the buffer insertion DP framework.

3.4.2 Delay and Slew Models for Gates

Linear gate delay model has been extensively used in timing optimization works [1,5,9,15]. Given the lumped load capacitance (C_L) at the output pin of the gate g, the linear gate delay ($D_{g,lin}$) and the output slew ($S_{o,lin}$) are expressed as follows:

$$D_{g,lin} = K_g + R_g \cdot C_L, \quad S_{o,lin} = SK_g + SR_g \cdot C_L \tag{3.1}$$

where K_g and R_g are intrinsic delay and output resistance of gate g, and SK_g and SR_g are intrinsic slew and slew resistance.

As discussed in [2], the linear gate delay model is inaccurate because (1) due to the resistive shielding [12], the lumped load capacitance is an overestimate of the effective capacitance [14] seen at the gate output, and (2) gate delay is not a linear function of load capacitance. The first problem can be solved by adopting the effective capacitance model, while the second one is dealt by the k-factor equation based model [14]. In the effective capacitance calculation, the RC network is

Table 3.1 Parameters used in this chapter. The C_m and R_m mean unit length capacitance and resistance of *metal5*. The C_{TSV} and R_{TSV} mean TSV parasitic capacitance and resistance, respectively. The *maxS* and *minS* are the maximum/minimum allowed slew in the bottom-up traversal

C_m	0.102 fF/μm	R_m	1.5 Ω/μm
C_{TSV}	59 fF	R_{TSV}	0.1 Ω
Bin size	2.0 ps	dS	0 bin
$maxS$	70 ps	$minS$	1.0 ps

reduced to a π-model (C_n, R_π, C_f) in which R_π models the resistive shielding effect. Then the effective capacitance (C_{ef}) at the gate output is computed as in [14]. Using effective capacitance model is essential for 3D IC buffer insertion because TSVs have high capacitance that causes the lumped capacitance (C_L) to overestimate gate delay much, which would discourage buffer usage. Given a two pin net with a fixed wirelength, when we change the TSV location from the source gate towards the sink gate, the source gate delay decreases because the TSV capacitance is shielded by the increased upstream resistance and effective capacitance decreases. The lumped capacitance model fails to capture this trend, thus using it for 3D IC buffer insertion is discouraged.

The original k-factor equations for gate delay and output slew are:

$$D_{g,orgk} = (kd_1 + kd_2 C_{ef})S_i + kd_3 C_{ef}^3 + kd_4 C_{ef} + kd_5 \tag{3.2}$$

$$S_{o,orgk} = (ks_1 + ks_2 C_L)S_i + ks_3 C_L^2 + ks_4 C_L + ks_5 \tag{3.3}$$

where S_i is the input slew at the gate g and $kd_1 - kd_5$ and $ks_1 - ks_5$ are curve-fitting parameters. Note that the parameter values differ for signal sense (rise/fall). In addition, C_g is the input pin capacitance of the gate. Also note that C_L is used for S_o calculation, because C_{ef} tends to underestimate S_o [14]. The library defines the maximum allowed C_L and S_i per each gate.

The problem with the original k-factor equations is that the models are linearly dependent on S_i. However, most of the gates require higher order polynomial equations for accuracy. Thus, we adopt a new k-factor equations for gate delay and output slew:

$$\begin{aligned} D_{g,newk} = {} & kd_1 + kd_2 C_{ef} + kd_3 S_i + kd_4 C_{ef}^2 + kd_5 C_{ef} S_i \\ & + kd_6 S_i^2 + kd_7 C_{ef}^3 + kd_8 C_{ef}^2 S_i + kd_9 C_{ef} S_i^2 + kd_{10} S_i^3 \end{aligned} \tag{3.4}$$

$$\begin{aligned} S_{o,newk} = {} & ks_1 + ks_2 C_L + ks_3 S_i + ks_4 C_L^2 + ks_5 C_L S_i \\ & + ks_6 S_i^2 + ks_7 C_L^3 + ks_8 C_L^2 S_i + ks_9 C_L S_i^2 \end{aligned} \tag{3.5}$$

The $D_{g,newk}$ is a third order polynomial in both C_{ef} and S_i, and the $S_{o,newk}$ is a third order polynomial in C_L and second order in S_i. As discussed in Sect. 3.6, during the bottom-up slew propagation, we need to compute S_i, given S_o and C_L. The computation is straightforward; by solving the above second order polynomial equation we obtain S_i.

From our library characterization experiments, we found that the new k-factor equations fit the library data better than the original one. Thus we use the new k-factor equation based delay and slew models.

3.4.3 Delay and Slew Models for Nets

Our net delay calculator uses Elmore delay model. It is easy to compute and the delay is additive [2], which helps pruning during DP traversal. The shortcoming of Elmore delay is that it may deviate from the actual delay by orders of magnitude [2]. For higher accuracy, a moment-matching based delay metric such as WED [10] or D2M [3] can be used. The model assumes step input signals, yet in real circuits input signals have finite slews, thus the model tends to underestimate the actual delay. The PERI method [6] converts the delay from the delay metrics for step inputs to the delay with ramp inputs. From layout simulations for various 3D nets, we observed that the WED model combined with PERI is quite accurate compared with PrimeTime results.

However, with the moment-based net delay models the optimality of DP framework solution is not guaranteed, because dominance relation cannot be defined as discussed in [2]. For two solutions, $a1$:$(q1, C_{ef}1, m1)$ and $a2$:$(q2, C_{ef}2, m2)$, even if $q1 \geq q2$ and $C_{ef}1 \leq C_{ef}2$, depending on the upstream solutions, the seemingly inferior $a2$ may give a better solution on the upstream side. No correct pruning scheme for VGDP with moment-based delay models currently exists. We observed that buffer insertion with moment-based net delay model quite often produced worse solutions than buffer insertion with Elmore net delay model did. Thus, we do not employ a moment-based net delay model in our SPDP.

As shown in Fig. 3.3, for slew degradation (S_d) on nets, we use the Bakoglu's metric [4]. Applying the PERI method for slew [6], we obtain the slew propagated through wires and TSVs.

3.5 Ginneken-3D Algorithm

We first extend the original Ginneken algorithm designed for 2D ICs [15] into 3D and name it Ginneken-3D. Our Ginneken-3D algorithm is similar to VGG in [2], with extensions for 3D IC handling. From the layout of all dies, we build a binary tree $T = (V, E)$ per each target 2D/3D net, where V is a set of nodes and E is a set of edges. A TSV is represented by an edge connecting nodes in different dies.

The net wires are segmented by 20 μm to generate internal nodes for candidate buffer locations [1]. The TSV-related information, such as keep-out-zone, should be considered in generating these candidate locations. In addition, we are given a set of buffers, B. The VGDP comprises two steps: a bottom-up then a top-down traversal. During the bottom-up traversal, we generate candidate solutions at the leaf vertices and propagate them bottom-up. A *candidate solution* (or a *solution*) a is a data tuple (q, C, b, a_l, a_r) associated with a node $v \in V$, where q is RAT, C is load capacitance, b is an inserted buffer if any, and a_l and a_r are the left and the right child solutions from which a is generated. With effective capacitance model [14], the C of the solution is replaced by a tuple (C_n, R_π, C_f), which represents the π-model. Thus, a solution becomes $(q, (C_n, R_\pi, C_f), b, a_l, a_r)$. Each node has its own solution list, and the solutions are propagated bottom-up. The VGDP assumes a default input slew for delay computation of gates, including buffers and inverters [2].

The efficiency of the VGDP comes from the pruning of solutions at each node during the bottom-up traversal. The pruning scheme presented in the original work [15] is simple yet accurate because dominance relationship can be defined clearly for linear gate delay model with lumped capacitance and Elmore net delay. For example, for two solutions, $a1$:$(q1, C1)$ and $a2$:$(q2, C2)$, if $q1 \geq q2$ and $C1 \leq C2$, then $a1$ always produces a better solution than $a2$ on the upstream side. However, with slew consideration the pruning should be performed more intelligently, because the dominance relation no longer holds, as we will discuss in Sect. 3.6.

After bottom-up traversal is finished, only one solution survives at the root node, because the C at the root node is the same (= input capacitance of source gate) for all solutions, and after pruning only the solution with the largest q survives. From the best solution, the top-down traversal is performed. The best solutions at child nodes are obtained by following the child solution pointers (a_l and a_r) stored during the bottom-up traversal. In this top-down traversal, the best buffer insertion solution is obtained by checking if the best solution of a node includes an inserted buffer.

3.6 Bottom-Up Slew Propagation DP

It is well known that slew affects delay. Physically, slew is determined top-down, as shown in Fig. 3.3a. The function F is our third order polynomial slew model obtained from timing library. Note that propagated slew is not additive. During the bottom-up traversal of VGDP, the slew at the current node is unknown until the buffer (or the gate) on the upstream is determined. This is why slew consideration in DP framework is hard. To overcome this hurdle, we *guess* the slew at each node during the bottom-up traversal. Each solution has an additional entry for propagated slew, $S = (S_b, S_d)$, which consists of slew base (S_b) and slew degradation (S_d). The top-down slew equations in Fig. 3.3a can be solved to get the bottom-up slew equations in Fig. 3.3b. Since slew is not additive, we need to calculate slew at a node from slew base and slew degradation. The function G can be found by solving F for S_l. Note that due to the direction of slew calculation, S_b in slew calculation is

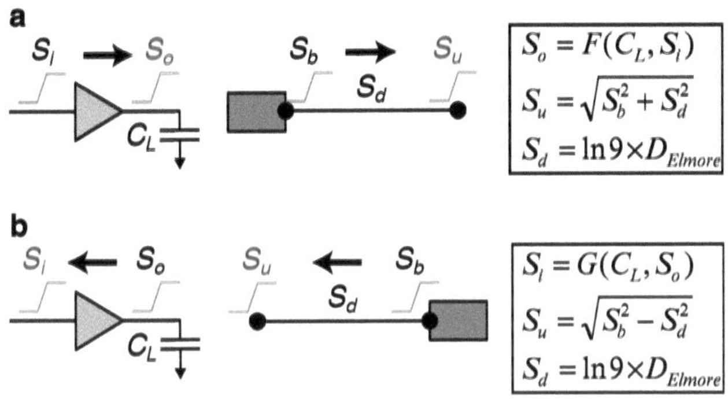

Fig. 3.3 Gate and net slew calculations in (**a**) *top-down* and (**b**) *bottom-up* traversal

different for the top-down and the bottom-up traversal; in the top-down traversal we define S_b as the output slew of the gate on the upstream, whereas in the bottom-up traversal S_b is the input slew of the gate on the downstream. A solution now becomes $(q, (C_n, R_\pi, C_f), (S_b, S_d), b, a_l, a_r)$.

The bottom-up traversal and the top-down traversal of our SPDP are outlined in Algorithms 4 and 5. Compared with VGDP, our new or modified ideas are highlighted in blue. We now explain the algorithm in detail.

3.6.1 *Generation of Sink Solutions*

Per each sink node, we create a set of solutions, each with the S set to a different trial S_l (tS_l) as shown in Line 3, Algorithm 4. Since we observe that in good buffer insertion solutions the S_l at the sink is in [10, 50] ps range, we generate the solutions for this range. One may think that we may start from a single solution with a single tS_l, perform the buffer insertion, then vary tS_l until we find the best result. However, finding the best solution by scanning tS_l is not efficient because the buffer insertion results change unpredictably with different tS_l mainly due to the discreteness of buffer candidate locations and buffer strengths. Furthermore, this approach cannot handle multi-pin nets efficiently because of numerous possible tS_l combinations at the sinks. Thus, we study the slew binning to find the best solution more efficiently.

3.6.2 *Slew Binning and Slew-Based Pruning*

Our slew binning is different from [5]. We divide the allowed slew range by a predefined *bin size* to create multiple slew bins, each with its own ID (*bin*). A slew value is associated with a corresponding slew bin. Now a solution is represented

Algorithm: Bottom-up Traversal of Slew Propagation Dynamic Programming
Input: a graph G=(V,E) with topologically sorted node list $Vlist$, a buffer library B
Output: list of solutions for each node v

```
foreach node v of Vlist in reverse order do
    if v has no child then
        make sink solutions with varied tS_i values for different slew bins and
        add them to v;
    end
    else if v has one child v_c then
        propagate the solutions of v_c to v;
    end
    else if v has two children v_cl, v_cr then
        merge solutions of v_cl and v_cr with slew consideration and add it to v;
    end
    if v is a feasible buffer location then
        for each solution, make a buffered solution if possible and add it to v;
    end
    for all solutions, calculate net delay of the parent wire or TSV of v and
    update q;
    for all solutions, calculate slew degradation along the parent wire or TSV
    and update S;
    prune solutions of v with slew consideration;
end
```

Algorithm 4: Bottom-up traversal of SPDP

as (q, (C_n, R_π, C_f), (S_b, S_d), bin, b, a_l, a_r). If bin size is small, the difference of slew among solutions in the same bin is small. This property provides a good pruning opportunity (Line 12, Algorithm 4); for the solutions in the same bin, we can compare q and C as in VGDP to check dominance relation. Note that this is an approximation; even if two solutions at a node have the same slew value, depending on their S_b and S_d, the slew on the upstream may differ because slew is not additive. However, we observe that this pruning works well in practice; pruning only when two solutions have similar S_b and S_d produces a buffering solution of almost the same quality with more than 20% runtime overhead.

At sink nodes, when bin size is 2 ps, total 20 solutions are generated in [10, 50] ps range. Compared with a single solution generation at sinks in VGDP, this multiple solution generation at sinks increases the run time of SPDP. Thus, it is crucial that the pruning scheme is efficient. During the bottom-up traversal we limit the maximum of the propagated slew, $maxS$, so that the search space is limited. We may set $maxS$ larger than the maximum S_i at sink gates because buffers can recover degraded slews very well and sometimes wires towards non-critical sinks may have

Algorithm: Top-down Traversal of Slew Propagation Dynamic Programming

Input: a graph $G=(V,E)$ with topologically sorted node list $Vlist$ with solutions from bottom-up traversal

Output: list *Blist* of buffer locations and types

```
foreach solution a of the root node do
    compute the gate delay and q at the source gate input with propagated
    slew;
end
find the top N_best solutions with the highest q's;
foreach top N_best solution at the root node do
    mark the current solution at the root node;
    foreach node v of Vlist do
        if marked solution at v has an inserted buffer b then
            calculate gate delay and output slew;
        end
        if v has a left child v_cl then
            mark the solution a_l of a at v_cl;
            propagate top-down delay and slew to v_cl;
        end
        if v has a right child v_cr then
            mark the solution a_r of a at v_cr;
            propagate top-down delay and slew to v_cl;
        end
    end
end
mark the best solution at the root node with lowest D_top-down to critical sink;
foreach node v of Vlist do
    if v has the best solution a with an inserted buffer b then
        add the location and type of b in Blist;
    end
    if v has a left child v_cl then
        mark the best solution a_l of a at v_cl;
    end
    if v has a right child v_cr then
        mark the best solution a_r of a at v_cr;
    end
end
```

Algorithm 5: Top-down traversal of SPDP

larger slew between buffers. The *maxS* effectively limits the maximum number of slew bins at each node, and the runtime complexity depends on the number of allowed slew bins. As shown in Table 3.2, *maxS* effectively limits search space and

Table 3.2 Delay and runtime of SPDP with varied *maxS* for critical nets in a 3D IC design

maxS (ps)	60	70	80	90
Maximum $D_{top-down}$(ps)	440.54	440.54	440.54	440.54
Average $D_{top-down}$(ps)	155.94	155.88	155.85	155.85
Total runtime (s)	7.81	10.25	13.93	17.75

Table 3.3 Delay and runtime with varied bin sizes for critical nets in a 3D IC design

Bin size (ps)	1.0	2.0	3.0
Maximum $D_{top-down}$ (ps)	433.96	440.54	456.43
Average $D_{top-down}$ (ps)	155.44	155.88	156.28
Total runtime (s)	32.27	10.25	5.44

runtime. Considering the delay and runtime tradeoff, we determine $maxS = 70$ ps. The minimum slew during DP, *minS*, is 1 ps.

In our slew binning scheme, we propagate a single slew value in each solution. Propagating bins which have ranges of slew (i.e. [min, max]) as in [5] cannot be applied to our SPDP because: (1) the delay and the slew calculation using [min, max] slew complicates the pruning; (2) slew range expands quickly as we perform S_o-to-S_i conversion for buffers. Usually, buffers have very low slopes in $S_i - S_o$ graphs; a narrow range on the S_o side corresponds to a wide range on the S_i side. After going through three buffers, the propagated slew range usually covers all good slew range (1–70 ps), rendering the propagation pointless.

The buffered delay and the runtime with varied bin sizes are shown in Table 3.3. With a larger bin size, runtime decreases because fewer solutions are generated at the sink gates and pruning applies to more solutions. Yet too large bin size degrades solution quality. We set the bin size as 2 ps for the delay and runtime tradeoff.

3.6.3 Merging Solutions

For multi-pin nets, during bottom-up traversal we merge children solutions at the merging node, as shown in Fig. 3.4. The $q3$ and $C3$ of the merged solution $a3$ are calculated as in VGDP. Yet we need to be careful in merging the slew values. Physically, $S1$ and $S2$ should be the same as $S3$. In the previous slew buffering work [5] the authors used the *max* operation, $S3 = max(S1, S2)$, because they propagated a maximum slew constraint. If we adopt this *max* operation in SPDP, it may propagate wrong slew values, which in turn makes delay calculations on the downstream inaccurate and pruning on the upstream incorrect. Thus, we only merge the child solutions when $S1$ and $S2$ are very close to each other. However, S_b1 and S_d1 may be different from S_b2 and S_d2. Since slew is not additive, depending on the upstream slew degradation, the propagated slew calculated from $S1$ and $S2$ may differ, which incurs inevitable slew calculation error on the upstream. Since the delay to the

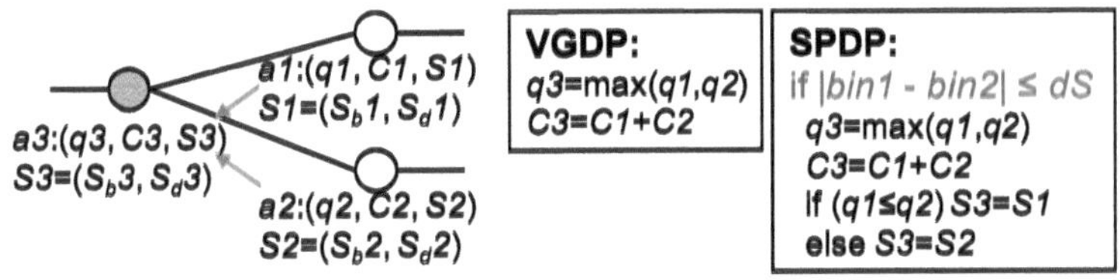

Fig. 3.4 Solution merge rule for VGDP and SPDP

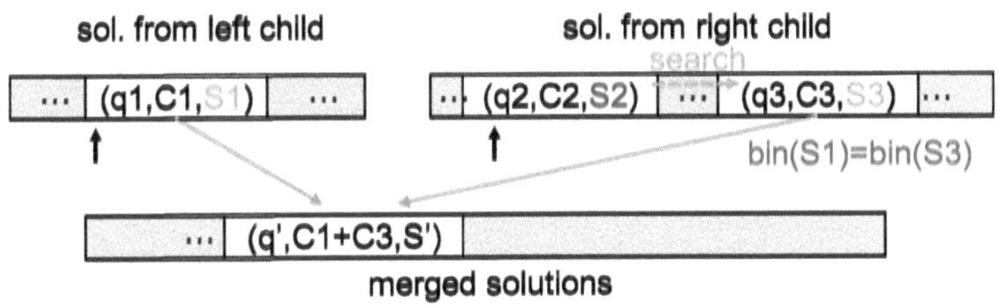

Fig. 3.5 Slew matching technique. The q′ and S′ are determined as in Fig. 3.4

critical sink has to be more accurate than those to other sinks, we take the S of the timing-critical child (with lower q) as the merged slew.

As shown in Fig. 3.5, our data structure for solutions at a node is a list, in which the solutions are sorted in ascending order of both q and C, but not in S. In merging process of VGDP, a left/right pointer is pointing to the left/right child solution to be merged (refer to [9]). After creating a solution by merging, only the timing-critical side pointer moves towards right (= larger q and C). However, since SPDP merges solutions when the slew of left and right solutions match, if solutions on left side are all timing critical and all these solutions do not belong to the slew bin that matches the slew bin of the current right solution, no further solution is merged (i.e. right pointer is stuck). To avoid this, on the non-timing-critical side, we actively search for the solution that matches slew of the solution on timing-critical side, using a separate pointer. Since each list usually contains lots of solutions with different slews, finding a solution with the matching slew bin does not take many steps from the current pointer. This technique improved solution quality without much runtime overhead.

In our implementation, we merge the solutions when the slew bin ID of $S1$ and $S2$ differ within a threshold, dS (Line 7, Algorithm 4). We investigate on allowing this small difference to check if slew calculation error during bottom-up might cause wrong pruning at merge. In Table 3.4, the percentage of merged solutions, buffered delay, and runtime with various dS are shown. With a larger dS, more child solutions are merged and the runtime increases. The percentage of merged solution is high, because of the above slew matching technique. From the results, it is obvious that $dS = 0$ produces the best delay with the lowest runtime. Thus, we set the $dS = 0$, meaning that the solutions are merged when they have the same slew bin ID.

Table 3.4 Percentage of merged solutions, delay, and runtime with varied dS for critical multi-pin nets in a 3D IC design

dS (bins)	0	1	2	3
Merged sols (%)	82.3	91.9	93.5	94.4
Maximum $D_{top-down}$ (ps)	413.00	440.54	486.45	543.38
Average $D_{top-down}$ (ps)	209.55	216.33	221.24	226.88
Total runtime (s)	9.55	10.25	11.71	12.09

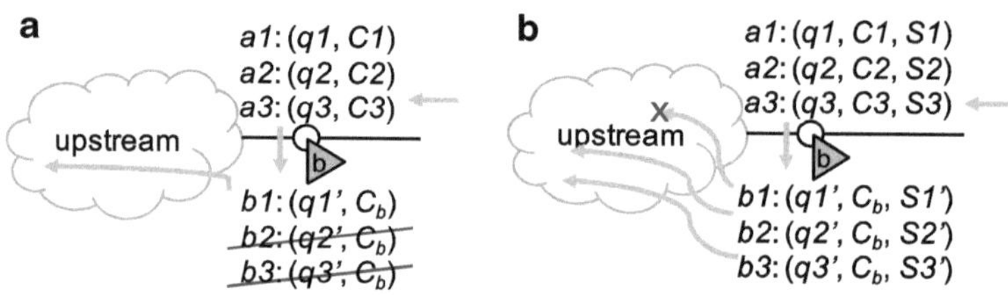

Fig. 3.6 Different buffer insertion scheme for (**a**) VGDP and (**b**) SPDP

3.6.4 Buffer Insertion

The propagated slew provides a very efficient pruning mechanism in the buffer insertion step (Line 9, Algorithm 4). For any solution at a node during the bottom-up traversal, the C is known, and for a buffer b with the given C, we can convert $S_o(g)$ to $S_i(g)$ as explained in Sect. 3.6. If the converted S_i is out of the $[minS, maxS]$ range, we prune the solution. Instead of pruning out the solution, if we keep the solution with a default S_i value, say 40 ps, the delay calculations on the downstream of the solution become incorrect because the slew at the current node has been changed. This delay calculation error may cause a better solution pruned by the default-slew solution, leading to worse buffer insertion results.

Another problem in the buffer insertion step is the dominance relation. As shown in Fig. 3.6a, in VGDP when the buffered solutions $a1' - a3'$ are generated from the propagated solutions $a1 - a3$, the Cs of $a1' - a3'$ are the same, C_b. Thus the buffered solution with the highest q dominates all the other buffered solutions for the buffer b. Thus, only *one* new buffered solution is added per each buffer. However, in SPDP, since $S1'$ and $S2'$ also affect the upstream solutions, it is not correct to just compare q for pruning. For example, even if $q1' \geq q2'$ and $S1' \leq S2'$, it is possible that solution $a1'$ gets pruned out on the upstream side due to slew-based pruning, while solution $a2'$ can survive up to the root node and possibly become the best solution. This means that per each buffer type we should make buffered solutions from *all* candidate solutions (Line 9, Algorithm 4), increasing the algorithm complexity from $O(|B|^2|V|^2)$ to $O(|B|^{|V|+1}|V|^2)$, where $|B|$ is the size of buffer library and $|V|$ is the number of nodes. However, thanks to the slew-based pruning rules discussed in this section, the runtime of SPDP is acceptable.

3.6.5 Multiple Solution Tracking

After bottom-up, we have multiple solutions at the root node. After $So - Si$ conversion, the Si of a solution at the root node may not match the input slew assumption. We intentionally allow it because delay and slew of solutions will be more accurately evaluated in the top-down traversal. In the top-down traversal, we propagate the slew top-down and calculate the gate and net delays with the propagated slew. The top-down delay calculation ($D_{top-down}$, calculated by our internal delay/slew models) may have small difference from the bottom-up delay calculation ($D_{bottom-up}$, also calculated internally) because the top-down slew values may differ from the bottom-up ones thus effective capacitance and gate/net delays change. Due to the slew calculations at solution merge and the aforementioned input slew condition at the root node, $D_{bottom-up}$ has an inherent error. Thus, we sort the solutions based on q and pick the top N_{best} solutions. Then, we perform the top-down solution tracking from each best solution (Lines 4–14, Algorithm 5), and check $D_{top-down}$. The one with the lowest $D_{top-down}$ is chosen as the final buffer insertion solution. We observe that the buffer insertion quality generally improves with higher N_{best} and after 30 it saturates. Since the top-down delay calculation is straightforward, the runtime overhead due to multiple solution tracking is negligible.

3.7 3D IC Design Methodologies

We develop design methodologies for the following buffer insertion methods in this chapter: (1) Encounter-3D: The timing-constraint-based 2D buffer insertion for 3D ICs with Cadence Encounter, (2) Ginneken-3D: The original VGDP with extensions for handling 3D ICs, with a fixed input slew of 40 ps for all gates, (3) SPDP: The proposed SPDP algorithm, with parameters in Table 3.1. Figure 3.7 shows the overall full-chip design flow for the three buffer insertion methods. Starting from partitioned and placed design, in Cadence Encounter we run a preliminary optimization for 2D nets without timing constraints on die boundaries (TSV ports) to fix DRVs within dies. Then with netlist and RC parasitic files extracted by Cadence QRC for all dies plus the top level netlist and the RC parasitic file that models TSVs, we run Synopsys PrimeTime to perform true 3D STA and generate timing constraints on die boundaries as we normally do for hierarchical designs. With the timing constraints, in Encounter we run die-by-die 2D optimization in pre-route mode first. For fair comparisons, we only allow buffer/inverter insertion in the optimization. Then die-by-die routing is performed, followed by RC extraction. A 3D STA is performed to obtain updated timing constraints on die boundaries. With the timing constraints, in Encounter we run post-route 2D optimization, which is the final design for Encounter-3D. Finally, a 3D STA is performed to obtain timing results such as worst negative slack (WNS) and total negative slack (TNS).

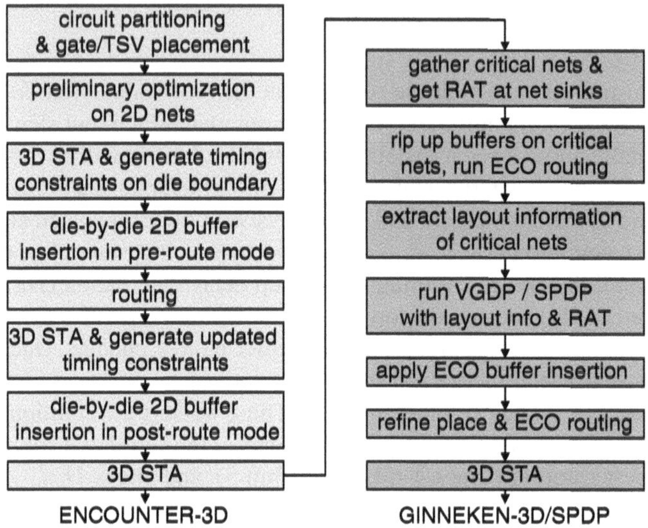

Fig. 3.7 Overall full-chip design flow for the buffer insertion methods. The ECO means engineering change order

Starting from the final design of Encounter-3D, in PrimeTime we gather the nets on top 5% critical paths and the RAT of sink gates in these nets. In Encounter, we rip up buffers on the critical nets and run ECO routing to repair broken routing due to buffer deletions. Then we extract the layout information of critical nets from Encounter. The Ginneken-3D or SPDP are performed to find buffer insertion solutions, which are fed back to Encounter using ECO buffer insertion commands. Then we run placement legalization and ECO routing, which is the final design for Ginneken-3D and SPDP. Finally, a 3D STA is performed.

Table 3.1 summarizes the parameters we used in the experiments. The wire parasitics from a moderately congested layout (extracted by Cadence QRC) matches the C_m and R_m with less than 5 and 1% error. The SPDP parameters discussed in Sect. 3.6 are also summarized in Table 3.1.

3.8 Experimental Results

To demonstrate the effectiveness of the proposed buffer insertion algorithm, we perform buffer insertion on various nets and full-chip designs. We run the experiments on a Linux server with Intel Xeon processors running at 2.5 GHz and 48 GB main memory. In this study, we use Nangate 45 nm standard cell library [11]. The buffer set in Nangate 45 nm standard cell library consists of six non-inverting buffers (BUF_X1/2/4/8/16/32) and six inverting buffers (INV_X1/2/4/8/16/32). Each has its own parameters such as C_b, $kd_1 - kd_{10}$, etc. Maximum C_L allowed at the buffer

Table 3.5 Summary of target design information. The '#nets(critical)' means the number of nets in the whole design and the critical nets selected for buffer insertion. Die size is in μm, and the 'clock' means target clock period in ns

Name	#gates	#nets (critical)	Die size	#TSVs	Clock
*ckt*1	12,924	13,256 (455)	350 × 350	1,203	1.00
*ckt*2	46,677	48,426 (3,408)	500 × 500	3,102	1.00
*ckt*3	50,375	55,454 (1,607)	700 × 700	8,596	1.00
*ckt*4	253,554	331,177 (7,405)	1,300 × 1,300	22,303	1.50
*ckt*5	546,460	714,782 (14,102)	1,900 × 1,900	42,325	2.00

output is defined by library. We assume four dies are stacked in our 3D IC. Our TSV size is 5 um diameter and 30 um height, and the TSV macro occupies six standard cell rows.

3.8.1 *Buffer Insertion Results*

Our five target designs are summarized in Table 3.5, and the buffer insertion results are shown in Table 3.6. Note that the runtime of Encounter-3D is not reported, because Encounter performs many internal steps during buffer insertion thus runtime for buffer insertion alone cannot be measured. Compared with Encounter-3D, Ginneken-3D improves WNS and TNS by 31.4 and 41.0% on average, which means applying 3D buffer insertion is advantageous over timing-constraint-based 2D optimization. Compared with Ginneken-3D, our SPDP further improves WNS and TNS by 8.7 and 10.9%, and the maximum achievable clock frequency is 3.2% higher, while using 4% less number of buffers. The reason why Ginneken-3D used more buffers than SPDP is that Ginneken-3D inserted offloading buffers whenever possible to reduce little bits of delay to the critical sink, while SPDP avoided it thanks to our slew-aware merging.

The cumulative runtime of SPDP for all five designs is about 3.1 times that of Ginneken-3D, which is acceptable because (1) number of critical nets to be buffered are small compared with the total net count, and (2) buffering is one of many optimization steps and it usually consumes around 10% of the total optimization time. Thus, for these "hard" net instances, we are allowed to spend more time on buffering to improve timing. Compared with Encounter-3D, our SPDP produces 37.4 and 47.4% better WNS and TNS, and 19.6% higher max clock frequency, yet uses 7.4% more buffers because our algorithm does not minimize number of buffers. Note that it is possible to run buffer usage reduction (area reclamation) on non-timing-critical side paths of the target nets as a post-step, which is outside our scope. These full-chip results clearly demonstrate that our SPDP algorithm is superior to Ginneken-3D and Encounter-3D in WNS, TNS and maximum achievable clock frequency.

Table 3.6 Comparison of buffer insertion results. The '#bufs' means the number of buffers in the design, and the f_{max} stands for maximum achievable clock frequency. Runtime values of Ginneken-3D and SPDP include bottom-up and top-down traversals in DP. The WNS, TNS, f_{max}, and runtime are in ps, ns, MHz, and s respectively

		*ckt*1	*ckt*2	*ckt*3	*ckt*4	*ckt*5	Ratio
Encounter-3D	#bufs	5,134	12,587	30,812	79,510	188,014	0.894
	WNS	−528.54	−1,466.49	−1,367.94	−1,213.83	−604.58	1.458
	TNS	−69.42	−973.07	−4,576.33	−1,941.03	−631.20	1.694
	f_{max}	654.22	405.43	422.31	368.48	383.94	0.863
Ginneken-3D	#bufs	5,338	15,596	31,459	88,070	213,212	1.000
	WNS	−392.62	−832.77	−1,156.31	−665.62	−507.43	1.000
	TNS	−56.41	−695.42	−3,885.07	−151.64	−47.62	1.000
	f_{max}	718.07	545.62	463.76	461.76	398.82	1.000
	Runtime	1.307	11.934	23.332	88.033	294.522	1.000
SPDP	#bufs	5,183	13,297	31,177	84,312	205,420	0.960
	WNS	−353.711	−740.974	−1,106.94	−620.617	−423.778	0.913
	TNS	−53.60	−607.34	−3,483.79	−130.73	−35.54	0.891
	f_{max}	738.71	574.39	474.62	471.56	412.58	1.032
	Runtime	2.441	24.186	97.339	233.674	929.970	3.072

Table 3.7 Comparison of buffer insertion results with different TSV capacitance for *ckt*3 in Sect. 3.8. The '#bufs' stands for the number of buffers in the design, the f_{max} for maximum achievable clock frequency

TSV capacitance		59 fF	30 fF	Ratio
Encounter-3D	#bufs	30,812	29,145	0.946
	WNS	−1,367.94	−1,216.80	0.890
	TNS	−4,576.33	−3,921.68	0.857
	f_{max}	422.31	451.10	1.068
Ginneken-3D	#bufs	31,459	29,822	0.948
	WNS	−1,156.31	−1,062.62	0.919
	TNS	−3,885.07	−3,842.19	0.989
	f_{max}	463.76	484.82	1.045
	Runtime	23.332	22.107	0.947
SPDP	#bufs	31,177	29,478	0.946
	WNS	−1,106.94	−1,048.15	0.947
	TNS	−3,483.79	−2,949.59	0.847
	f_{max}	474.62	488.25	1.029
	Runtime	97.339	86.868	0.892

3.8.2 Endpoint Slack Histograms

To visualize the timing quality of buffer insertion results, in Fig. 3.8 we compare the timing endpoint slack histograms for design *ckt*2 with Encounter-3D, Ginneken-3D, and SPDP. In Encounter-3D result, the long tail towards left (slack < -1.2ns) is because Encounter-3D could not optimize several critical 3D nets effectively. Compared with those of Ginneken-3D or SPDP, the overall histogram bars of Encounter-3D lie on the left side, meaning that the overall buffer insertion quality is

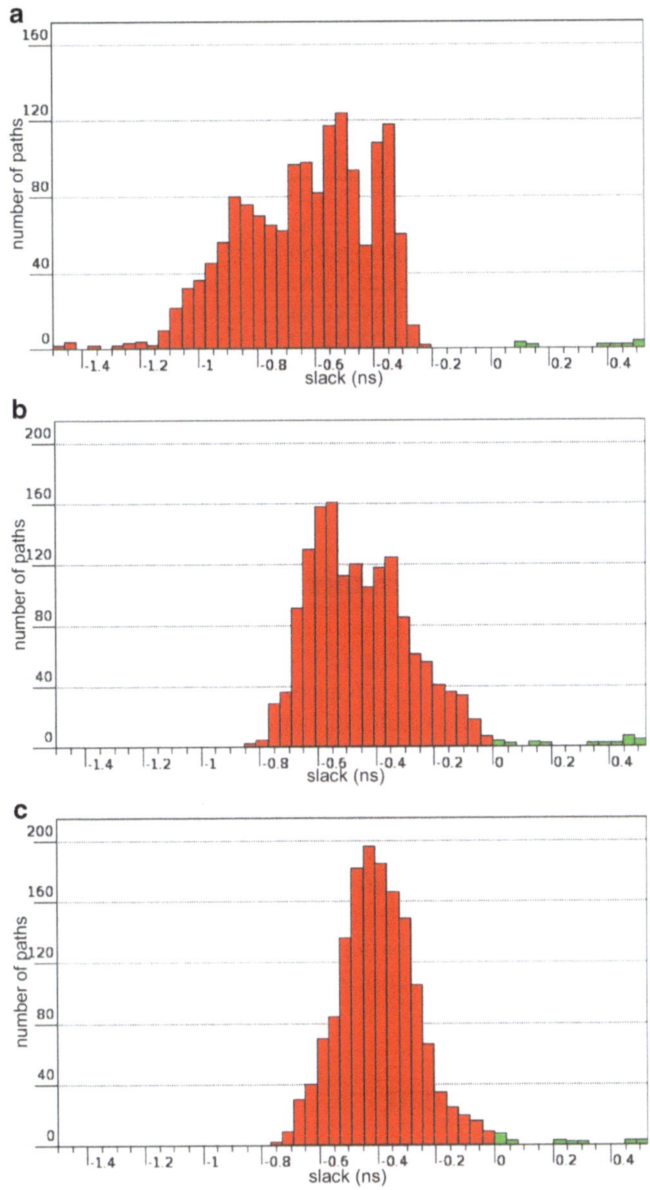

Fig. 3.8 Endpoint slack histograms for *ckt*2 with (**a**) Encounter-3D, (**b**) Ginneken-3D, and (**c**) our SPDP

the worst among the three methods. Compared with the Ginneken-3D graph, in the SPDP graph the leftmost bar (= WNS) as well as the overall distribution are on the right side, meaning better timing results.

Table 3.8 Information of the nets on the critical path with Encounter-3D for design *ckt*3. The 'WL', '#TSVs' and '#cand. buf loc' stand for the wirelength, the number of TSVs, and the number of candidate buffer locations in the net, respectively

Net	From cell/pin	To cell/pin	Fanout/WL/#TSV	#cand. buf loc
*n*1	DFF_X1/CK	NAND2_X2/A1	8 / 595.30 / 1	39
*n*2	NAND2_X2/A1	AND2_X2/A1	4 / 95.68 / 0	8
*n*3	AND2_X2/A1	AND2_X4/A2	16 / 1818.35 / 1	108
*n*4	AND2_X4/A2	OAI22_X1/A2	24 / 4527.09 / 6	262
*n*5	OAI22_X1/A2	NOR4_X1/A4	1 / 16.04 / 0	1
*n*6	NOR4_X1/A4	AND4_X4/A2	1 / 465.78 / 2	28
*n*7	AND4_X4/A2	AND4_X4/A1	1 / 272.89 / 0	14
*n*8	AND4_X4/A1	NAND4_X4/A2	1 / 333.68 / 2	21
*n*9	NAND4_X4/A2	NAND2_X4/A2	2 / 1343.20 / 0	69
*n*10	NAND2_X4/A2	NAND4_X4/A4	1 / 119.70 / 0	6
*n*11	NAND4_X4/A4	DFF_X1/D	1 / 658.53 / 0	33

3.8.3 *Impact of TSV Capacitance*

We now study how much impact the TSV capacitance has on the design quality. As TSV manufacturing technology advances, the unwanted TSV capacitance may be reduced. In Table 3.7, the results of buffer insertion with the three methods in Sect. 3.8 are compared. When the TSV capacitance is reduced from 59 to 30 fF, the three methods use less number of buffers, because of the reduced TSV capacitance. In addition, the WNS, TNS, and maximum achievable clock frequency are improved. For Encounter-3D, WNS improves by 11%, which suggests that timing-constraint-based 2D optimization may not work well especially when TSV capacitance is large. Ginneken-3D did not show much improvement in TNS, possibly because of the reduced buffer usage. The runtime reduced because less number of nets were identified as critical.

3.8.4 *Critical Path Analysis*

We analyze the buffer insertion results for the nets on the critical path from the design *ckt*3 in Sect. 3.8. The target net information and the buffer insertion results are summarized in Tables 3.8 and 3.9. The column 2 and 3 in Table 3.8 show the instance, pin, and cell names of the source gate and the critical sink gate of the net. The second last row in Table 3.9 shows the setup time at the timing endpoint F/F. Compared with Ginneken-3D, SPDP reduces the path delay by 6.8%, using 33% less number of buffers.

The biggest difference between SPDP and Ginneken-3D is observed for net *n*4. Our SPDP inserted fewer buffers than Ginneken-3D yet produced lower delay and

Table 3.9 Comparison of buffer insertion results for design *ckt*3. The '#bufs' means the number of buffers/inverters inserted on the net. The 'delay' is measured from the source input to the critical sink input of the net, and 'slew' is the S_i of the critical sink. Delay and slew are in ps

	Encounter-3D			Ginneken-3D			SPDP		
Net	#bufs	Delay	Slew	#bufs	Delay	Slew	#bufs	Delay	Slew
*n*1	4	199.04	6.48	7	140.86	40.33	2	155.66	17.43
*n*2	2	41.10	8.75	4	53.40	6.58	2	48.98	8.97
*n*3	8	75.81	15.49	9	66.27	33.56	8	79.98	18.45
*n*4	21	962.32	37.87	24	557.38	51.51	13	466.27	36.90
*n*5	0	31.91	17.54	0	35.16	20.55	0	32.05	17.59
*n*6	7	338.13	16.60	1	234.19	57.92	2	223.08	35.80
*n*7	1	87.07	16.07	0	89.73	30.86	0	80.79	28.15
*n*8	6	266.00	7.84	1	111.09	28.64	2	105.31	18.27
*n*9	5	210.79	27.11	3	153.87	43.29	2	167.29	54.96
*n*10	0	33.71	18.78	0	42.97	23.66	0	49.95	28.05
*n*11	2	109.93	17.76	0	126.19	90.18	2	106.82	29.06
Setup		43.96			68.52			48.55	
Sum	56	2,399.76	190.28	49	1,679.60	427.07	33	1,564.73	293.63

slew at the sink gate input, which is helpful because the delay of sink gate OAI22_X1 is sensitive to the S_i. Although Encounter-3D inserted similar number of buffers as Ginneken-3D, it inserted too many buffers on the critical path, which increases path delay due to the buffer intrinsic delay. Ginneken-3D does not consider slew, thus the slew of Ginneken-3D varies in a wide range. On the other hand, SPDP produces low slew values in most cases to reduce sink gate delay, except for *n*9 where the delay of sink gate (NAND2_X4) is not very sensitive to the S_i. With Ginneken-3D, the S_i to the timing endpoint (sink of *n*11) is quite high, which increases the setup time. The Encounter-3D produced the minimum S_i among the three methods, however it did not lead to the minimum delay.

Figure 3.9 shows layout snapshots for the critical path of design *ckt*3 with SPDP buffer insertion. It is the path shown in Table 3.9. The gates, buffers or inverters, TSV pin pads, and TSV landing pads are highlighted in different colors. It starts from Die 2 (#1), goes down to Die 3 then comes back to Die 2 (#5), goes up to Die 1 then Die 0 (#9), comes back to Die 1 then Die 2 (#13), goes down to Die 3 and comes back to Die 2 (#17), goes up to Die 1 and comes back to Die 2 (#21), and finally goes down to Die 3 on which the timing endpoint (#24) exists. The reason why there are many TSVs on the path is that (1) we use a multi-way partitioner with cut-size control for wirelength vs. area tradeoff, and (2) the nets on the path involves a lot of sinks on different dies.

On the path, we observe that there are several booster buffers for weak cells such as the one around #1 on Die 2 and #3 on Die 3. We also observe that some buffers, such as the one before #8 on Die 1 or #2 on Die 2 or #4 on Die 3, are placed close to TSV pin pad/laning pad to reduce slew degradation through TSVs. When there are many sinks clustered together (around #10 on Die 0, #3 on Die 3, or #16 on die Die 3), our SPDP inserted several buffers in the area to perform proper offloading

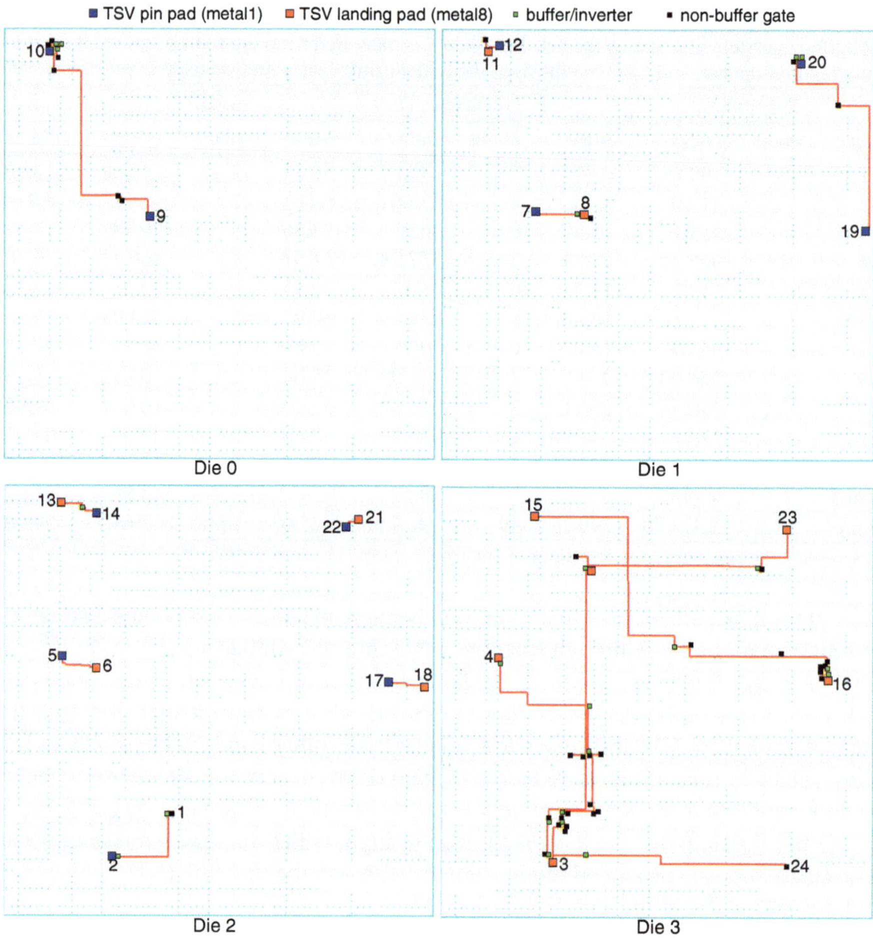

Fig. 3.9 Layout snapshots for the critical path of *ckt*3 with SPDP. The numbers in the Figure represent the order in the path. *Cyan dots* denote the location of TSV pin pads and landing pads outside the critical path

to gates on non-timing-critical side paths thus reduce slew degradation along the critical path. Depending on how much load exists on a die, the delay could be lower by not inserting a buffer for feed-through nets such as #5–#6, #17–#18, or #21–#22 on Die 2. The buffer right before #16 on Die 3 is a very strong one, thus it drives two TSVs to another buffer around #20 on Die 1.

3.9 Conclusions

We studied buffer insertion for 3D ICs with slew considerations in van Ginneken DP framework. Compared with the fixed-slew van Ginneken algorithm, our slew-aware SPDP algorithm reduced the overall delay with a reasonable runtime overhead. In addition, it outperformed the timing-constraint-based 2D buffer insertion with a commercial tool.

References

1. C. Alpert, A. Devgan, Wire segmenting for improved buffer insertion, in *Proceedings of ACM Design Automation Conference*, Anaheim, 1997, pp. 588–593
2. C.J. Alpert, A. Devgan, S.T. Quay, Buffer insertion with accurate gate and interconnect delay computation, in *Proceedings of ACM Design Automation Conference*, New Orleans, 1999, pp. 479–484
3. C.J. Alpert, A. Devgan, C. Kashyap, A two moment RC delay metric for performance optimization, in *Proceedings of International Symposium on Physical Design*, San Diego, 2000, pp. 73–78
4. H.B. Bakoglu, *Circuits, Interconnects, and Packaging for VLSI* (Addison-Wesley, Reading, MA 1990)
5. S. Hu, C.J. Alpert, J. Hu, S.K. Karandikar, Z. Li, W. Shi, C.Z. Sze, Fast algorithm for slew-constrained minimum cost buffering. IEEE Trans. Comput.-Aided Des. Integr. Circuits Syst. **26**(11), 2009–2022 (2007)
6. C.V. Kashyap, C.J. Alpert, F. Liu, A. Devgan, Closed-form expressions for extending step delay and slew metrics to ramp inputs for RC trees. IEEE Trans. Comput.-Aided Des. Integr. Circuits Syst. **23**(4), 509–516 (2004)
7. G. Katti, M. Stucchi, K.D. Meyer, W. Dehaene, Electrical modeling and characterization of through silicon via for three-dimensional ICs. IEEE Trans. Electron Devices **57**(1), 256–262 (2010)
8. Y.-J. Lee, I. Hong, S.K. Lim, Slew-aware buffer insertion for through-silicon-via-based 3D ICs, in *Proceedings of IEEE Custom Integrated Circuits Conference*, San Jose, 2012
9. J. Lillis, C.-K. Cheng, T.-T. Y. Lin, Optimal wire sizing and buffer insertion for low power and a generalized delay model. IEEE J. Solid-State Circuits **31**(3), 437–447 (1996)
10. F. Liu, C. Kashyap, C.J. Alpert, A delay metric for RC circuits based on the weibull distribution. IEEE Trans. Comput.-Aided Des. Integr. Circuits Syst. **23**(3), 443–447 (2004)
11. Nangate Inc., Nangate 45nm open cell library, available online at http://www.nangate.com/?page_id=22
12. P.R. O'Brien, T.L. Savarino, Modeling the driving-point characteristic of resistive interconnect for accurate delay estimation, in *Proceedings of IEEE International Conference on Computer-Aided Design*, Santa Clara, 1989, pp. 512–515
13. Y. Peng X. Liu, Low-power repeater insertion with both delay and slew rate constraints, in *Proceedings of ACM Design Automation Conference*, San Francisco, 2006, pp. 302–307
14. J. Qian, S. Pullela, L. Pillage, Modeling the effective capacitance for the RC interconnect of CMOS gates. IEEE Trans. Comput.-Aided Des. Integr. Circuits Syst. **13**(12), 1526–1535 (1994)
15. L.P.P.P. van Ginneken, Buffer placement in distributed RC-tree networks for minimal elmore delay, in *Proceedings of IEEE International Symposium on Circuits and Systems*, New Orleans, 1990, pp. 865–868

Chapter 4
Low Power Clock Routing for 3D IC

Abstract This chapter focuses on low-power and low-slew clock network design and analysis for through-silicon-via (TSV)-based three-dimensional stacked ICs (3D ICs). First, we study the impact of the TSV count and the TSV RC parasitics on clock power consumption. Several techniques are introduced to reduce the clock power consumption and slew of the 3D clock distribution network. We analyze how these design factors affect the overall wirelength, clock power, slew, and skew in 3D clock network design. Second, we study a two-step 3D clock tree synthesis method: (1) 3D abstract tree generation based on the three-dimensional method of means and medians (3D-MMM) algorithm; (2) buffering and embedding based on the slew-aware deferred-merge buffering and embedding (sDMBE) algorithm. We also extend the 3D-MMM method (3D-MMM-ext) to determine the optimal number of TSVs to be used in the 3D clock tree so that the overall power consumption is minimized. Related SPICE simulation indicates that: (1) a 3D clock network that uses multiple TSVs significantly reduces the clock power compared with the single-TSV case; (2) as the TSV capacitance increases, the power savings of a multiple-TSV clock network decreases; and (3) our 3D-MMM-ext method finds a close-to-optimal design point in the TSV count vs. power consumption tradeoff curve very efficiently.

The materials presented in this chapter are based on [27].

4.1 Introduction

In three-dimensional integrated circuits (3D ICs), the clock distribution network spreads over the entire stack to distribute the clock signal to all the sequential elements. Clock skew, defined as the maximum difference in the clock signal arrival times from the clock source to all sinks, is required to be less than 3 or 4 % of the clock period in an aggressive clock network design according to the International

S.K. Lim, *Design for High Performance, Low Power, and Reliable 3D Integrated Circuits*, DOI 10.1007/978-1-4419-9542-1_4,

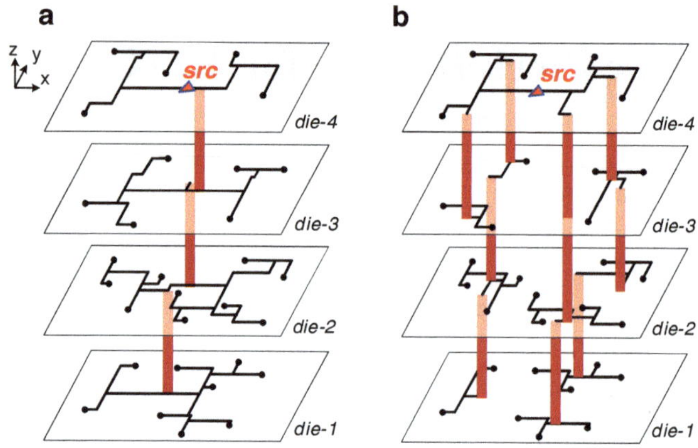

Fig. 4.1 Four-die stack 3D clock networks with two different TSV counts. (**a**) Uses single TSV between adjacent dies; (**b**) uses ten TSVs. The overall wirelength is shorter in (**b**)

Technology Roadmap for Semiconductors (ITRS) projection [9]. Thus, clock skew control, which was well studied in 2D ICs [18], is still a primary objective in the 3D clock network design. However, the clock signal in 3D ICs is distributed not only along the X and Y directions, but also along the Z direction using through-silicon vias (TSVs). The clock distribution network drives large capacitive loads and switches at a high frequency. This leads to an increasingly large proportion of the total power dissipated in the clock distribution network. In some applications, the clock network itself is responsible for 25 % [6] and even up to 50 % [28] of the total chip power consumption. Moreover, clock slew must also be taken into consideration when designing a 3D clock network, because a large clock slew may cause a setup/hold time violation. Thus, low power, skew, and slew remain important design goals in 3D clock networks.

TSVs provide the vertical interconnections to deliver the clock signal to all dies in the 3D stack. The TSV count is an important factor that characterizes the physical and electrical properties of the clock network. The 3D integration with TSVs has been intensively studied in both chip-to-chip and chip-to-wafer communications [14]. The fabrication and characterization of TSVs are being explored in many companies and institutions [21]. TSV reliability issues are also studied [23].

The low-power 3D clock network design demands a thorough investigation on how the TSV count and the TSV parasitic affect the clock performance. Existing work has demonstrated that the total wirelength of a 3D clock network decreases significantly if more TSVs are used [12, 15, 25, 26]. According to the observations made in [15], the die that contains the clock source includes a complete tree, while other dies can have subtrees as illustrated in Fig. 4.1. A 3D clock tree that utilizes multiple TSVs tends to reduce the overall wirelength as more and more TSVs are used. However, the analysis of TSV *RC* parasitics on the clock network has not been addressed in the literature. If a 3D clock tree utilizes many TSVs that have large TSV *RC* parasitics, the clock delay and power consumption contributed by the TSVs may increase significantly. Using more TSVs helps to reduce the wirelength and thus power consumption, but the TSV capacitance increases the clock power consumed at the same time. Our experiments indicate that the larger the TSV capacitance is, the faster the clock power consumption increases when more and more TSVs are used.

In this chapter, we study the impact of various design parameters on the wirelength, clock power, slew, and skew of the 3D clock network. These parameters include the TSV count,[1] the TSV parasitics, the maximum loading capacitance of the clock buffers, and the supply voltage. We also develop clock network synthesis algorithms for low-power 3D clock network design. The following specific topics are covered in this chapter:

- We study the impact of the TSV count and the TSV parasitics on the clock power. We show the "TSV count vs. power dissipation" tradeoff curves for various TSV parasitic values and discuss how the TSV count and the TSV capacitance together determine the overall clock power consumption.
- We discuss the impact of the TSV count and clock buffer insertion on clock slew control. Our study shows that using multiple TSVs helps to reduce the maximum and average slew as compared with the single-TSV case. In addition, specifying an upper bound for the load capacitance of each clock buffer remains an efficient way to control the maximum slew of the 3D clock network design.
- We study an effective way to determine the optimal number of TSVs for the 3D clock tree so that the overall power consumption is minimized. Our method predicts the impact of adding a new TSV into the current clock topology on the overall power consumption during the top-down abstract tree generation. This prediction helps to decide whether pairing of two clock nodes in different dies and using a TSV for this pair is useful for power reduction or not. Related experiments indicate that our method finds a close-to-optimal design point in the TSV count vs. power consumption tradeoff efficiently as compared with a straightforward exhaustive search method.

4.2 Existing Works

In 3D clock tree design and optimization, TSV planning plays an important role in constructing a low-power 3D clock network. Minz et al. [15] proposed the first work on 3D clock routing algorithms. They discovered that the total wirelength decreases significantly when more TSVs are used in the 3D clock network. They also studied the thermal impact on the 3D clock network and proposed thermal-aware 3D clock tree synthesis method to balance the clock skew caused by the thermal variations. Kim and Kim [12] presented a 3D embedding method to reduce wirelength. However, they do not consider power consumption or slew rate and do not provide any SPICE simulation results. Zhao et al. [26] developed a clock

[1]In this chapter, we use "TSV count" to refer to the total number of TSVs used in a 3D clock tree.

design method to support the pre-bond testing for 3D ICs. They also discussed the impact of the TSV counts on the pre-bond testable clock tree. They observed that using multiple TSVs helps a pre-bond testable 3D clock network achieve low power consumption. However, this work did not take into account the impact of the TSV capacitance on the clock power.

Pavlidis et al. [16] presented measurement data on a fabricated 3D clock distribution network. Arunachalam and Burleson [2] proposed the use of a separate layer for the clock distribution network to reduce power. Their simulations show 15–20 % power reduction over the same 2D chip clock network. However, they focused on a simple H-tree and did not perform any design-level optimization.

Due to the significant dimension of the TSVs occupying the layout space [13], the impact of TSV parasitics, especially the capacitance, should be taken into consideration for a low-power 3D clock network design. Existing work mainly focuses on 15 fF TSVs. But the TSV capacitance can vary from tens of femto-farads up to a few hundred femto-farads, depending on the material, TSV diameter, oxide thickness, and TSV height [19, 22]. We observe that large TSV capacitance values significantly affect the existing discussions on the TSV count vs. power tradeoff. In this case, the multiple-TSV insertion reduces the total wirelength and power, but the large TSV parasitic capacitance increases the power consumption. As a result, the total clock power may increase in the multiple-TSV case. Therefore, a thorough study on the impact of both the TSV capacitance and the TSV count on the overall 3D clock power is required. Given the TSV parasitic impedance, a straightforward approach to find the optimal TSV usage for low-power design is to exhaustively search the entire range of the TSV count. This approach, however, requires prohibitive design time and is thus not practical.

4.3 Preliminaries

4.3.1 *Electrical and Physical Model of 3D Clock Network*

In this chapter, a 3D clock network is modeled as a distributed RC network. The sink nodes, which represent flip-flops and clock input pins of IP or memory blocks, are modeled as capacitive loads. Wire segments and TSVs are modeled with a π model,[2] which is a classical way to represent the parasitics of a clock network. Each buffer or driver is constructed with two inverters. Note that prior work has focused on the electrical modeling of TSVs [3, 11, 19, 22]. Our 3D clock routing algorithm is flexible to handle a more complicated TSV parasitic model if desired.

---

[2] In this chapter, wire segments denote the edges of the abstract tree, and are not uniformly distributed. Depending on the TSV insertion and buffer insertion on the abstract tree, a src-to-sink path usually contains tens of wire segments, with each segment length varies from tens of micro-meters to a few hundreds of micro-meters.

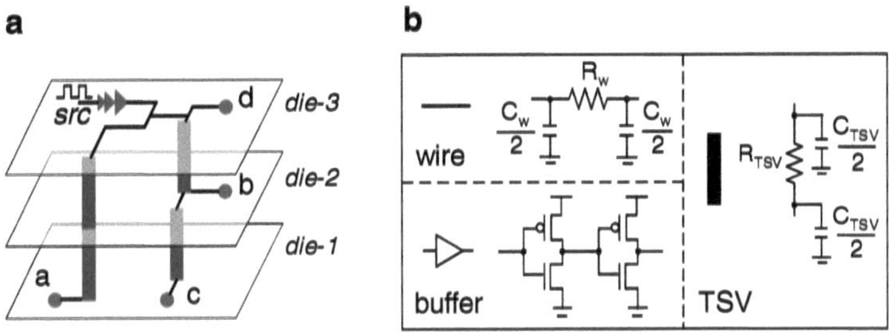

Fig. 4.2 A sample clock tree and its electrical model. (**a**) A sample three-die clock network using four TSVs, where the clock source is in die-3. Sink *a* in die-1 uses two TSVs that are vertically aligned, and Sink *b* in die-2 uses one TSV to connect to the clock source. (**b**) Electrical models of the clock wire segments, TSVs, and buffers/drivers

The *TSV bound* is defined as a user specified constraint on the maximum allowed TSV number per die. The TSV bound is usually decided before clock synthesis and is based upon the process technology. Different from the TSV bound, the *TSV count* (*#TSVs*) is the total number of TSVs utilized in the 3D clock tree. For an n-die 3D stack, #TSVs is usually less than or equal to $(n-1)\times$ TSV bound.

A three-die clock interconnect using four TSVs is shown in Fig. 4.2: the clock source is located in die-3; Sink *a* in die-1 connects to the source using two TSVs that are vertically aligned; Sink *c* in die-1 connects to the source by two TSVs; and Sink *b* in die-2 uses one TSV.

4.3.2 Problem Formulation

The *3D Clock Tree Synthesis* problem can be formulated as following: Given a set of sinks in all dies, a TSV bound, a pre-determined clock source location, and the parasitics of the wires, buffers, and TSVs, the 3D clock tree synthesis constructs a fully-connected 3D clock network such that (1) the clock sinks in all dies are connected by a single tree; (2) the TSV count in each die is under the TSV bound; (3) the clock skew is minimized (and zero under the Elmore delay model [5]); (4) the clock slew is below the constraint; and (5) the wirelength and the clock power are minimized.

The clock skew is the maximum difference among the arrival times at the clock sinks. In the existing clock tree synthesis tools, the Elmore delay model is a popular measure of the *RC* delay and skew. The primary goal of our 3D clock tree synthesis is to guarantee a zero Elmore-skew clock network. In order to achieve more accurate timing information and to evaluate our clock synthesis performance, we use SPICE simulation on our 3D clock trees. The simulated clock skew is constrained to less than 3 % of the clock period. The clock slew is defined as the transition time from 10 to 90 % of the clock signal at each sink.

The TSV bound constraint plays an important role in achieving low-power 3D clock networks. This bound reflects the impact of the TSV usage on routing congestion, capacitive coupling, stress-induced manufacturing issues, and so on. By varying the TSV bounds, we obtain different 3D clock networks with different qualities. Note that the TSV bound and the actual TSV usage in each die may be different because the bound only puts the limit on the maximum TSV usage for each die.

4.4 3D Clock Tree Synthesis

4.4.1 Overview

Our 3D clock tree synthesis algorithm consists of two major steps: (1) 3D abstract tree generation and (2) slew-aware buffering and embedding. First, we generate a 3D abstract tree based on our *3D Method of Means and Medians* (3D-MMM) algorithm. The 3D-MMM algorithm basically determines which pair of nodes (sink nodes or merging points) to connect together and utilizes TSVs if necessary, while building a binary tree in a top-down fashion. Note that our 3D-MMM algorithm works in such a way that there is always one die that contains a *single tree* which connects all sinks in the die, whereas the sinks in other dies are connected with *multiple trees*. In this case, the clock source is located in the die that contains the single tree.

Once a 3D abstract tree is obtained, we determine the routing topology and exact geometric locations for all the nodes, TSVs, and buffers. Our *slew-aware deferred-merge buffering and embedding* (sDMBE) method is a two-phase approach, which is based on the classic deferred-merge and embedding (DME) algorithm [4] in 2D clock routing. sDMBE first visits each node in a bottom-up fashion, determines the merging type for a pair of subtrees, inserts buffers if necessary, and calculates the merging distances based on the zero-Elmore-skew equations. The outcomes of sDMBE during the first phase are the merging segments, which store a collection of feasible locations of the internal nodes in the 3D abstract tree. During the second phase, sDMBE visits the whole abstract tree in a top-down manner while deciding the exact merging locations of the internal nodes, buffers, TSVs, and exact routing topology until all sinks are connected via a single tree.

4.4.2 3D Abstract Tree Generation

The first step of our 3D clock tree synthesis is the 3D abstract tree generation using the 3D-MMM algorithm. A 3D abstract tree indicates the hierarchical connection information among the sink nodes, internal nodes, TSVs, and the root node. The 3D abstract tree of an n-die stack clock network is an n-colored binary tree, which is used to identify the die index for all the nodes.

We develop the 3D-MMM algorithm to generate a 3D abstract tree for the given clock sinks in a top-down manner, which is an extension of the Method of Means

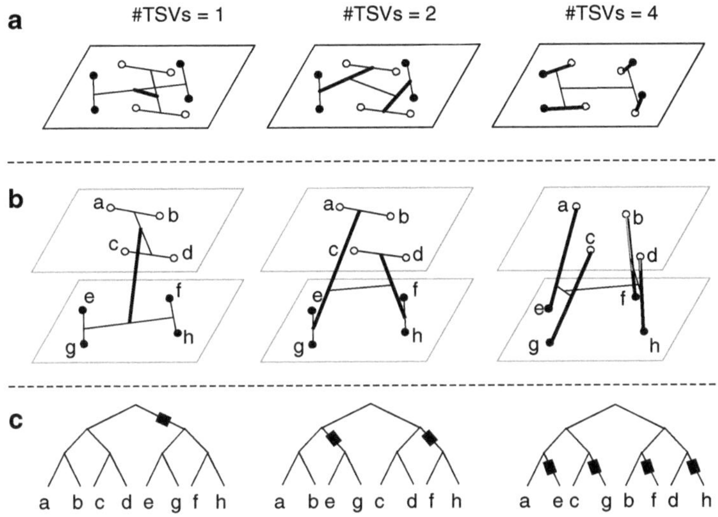

Fig. 4.3 The 3D abstract trees generated by our 3D-MMM algorithm under various TSV bounds. (**a**) 2D view, where *thick lines* denote TSV connection. (**b**) 3D view. (**c**) Binary abstract trees, where the *squares* denote TSVs

and Medians (MMM) algorithm [10]. The 3D abstract trees generated by the 3D-MMM algorithm with various TSV bounds are shown in Fig. 4.3. Note that a larger TSV bound tends to move TSVs closer to the sink nodes and causes more vertical clock connections than horizontal connections. However, the overall wirelength is reduced due to the short horizontal connection length. The basic idea of our 3D-MMM algorithm is to recursively divide the given sink set into two subsets until each sink belongs to its own set. A TSV is used if we decide to merge a pair of nodes in different dies. In this case, our goal is to evenly distribute the TSVs across the die area and to satisfy the given TSV bound, which is shown to improve manufacturability [24].

Let $S = \{s_1, s_2, .., s_k\}$ denote a set of sinks, where the locations of the sinks have been decided before the 3D clock tree synthesis. We assume that the maximum allowed TSV count for each die in S (TSV bound) is also given. Each s_i is a triplet of (x_i, y_i, z_i), where z_i is the die index of s_i, and x_i and y_i are the X and Y coordinates of s_i. Let $stack(S)$ denote the number of dies the sinks in Set S are located in. In each recursive partitioning, we divide Set S into two Subsets S_1 and S_2 based on the following two cases:

- *Z-cut*: if the TSV bound is one, the given Sink Set S is partitioned such that the sinks from the same die belong to the same subset. The connection between S_1 and S_2 needs one TSV in-between adjacent dies. Note that 3D-MMM is a bi-partitioning process. When the sinks of S distribute to more-than-two dies (i.e., $stack(S)>2$), we need $stack(S) - 1$ iterations of Z-direction partitions to split the

input : Sink set $S = \{s_1, \cdots, s_k\}$, source die index Z_s
output: Subsets S_T and S_B
$Z_{\min} = min(z_1,..,z_i,..,z_k)$, $s_i = (x_i, y_i, z_i) \in S$
$Z_{\max} = max(z_1,..,z_i,..,z_k)$, $s_i = (x_i, y_i, z_i) \in S$
if $Z_s \le Z_{min}$ **then**
  $S_T = \{s_1,..,s_i,..,s_{kI}\}$, $z_i \in [Z_{\min}+1, Z_{\max}]$
  $S_B = \{s_{kI+1},..,s_j,..,s_k\}$, $z_j = Z_{\min}$
else if $Z_s \ge Z_{\max}$ **then**
  $S_T = \{s_1,..,s_i,..,s_{kI}\}$, $z_i = Z_{\max}$
  $S_B = \{s_{kI+1},..,s_j,..,s_k\}$, $z_j \in [Z_{\min}, Z_{\max}-1]$
else
  $S_T = \{s_1,..,s_i,..,s_{kI}\}$, $z_i = Z_s$
  $S_B = \{s_{kI+1},..,s_j,..,s_k\}$, $z_j \neq Z_s$
end

Algorithm 6: Pseudo code of the Z-cut procedure, Z-cut(S, S_T, S_B), which corresponds to Line 6 in the 3D-MMM algorithm, Algorithm 7.

sink set into subsets, so that the sinks belonging to the same die are in the same subset. Furthermore, the order of the Z-cut also depends on the source die index. Algorithm 6 shows the details of the Z-cut procedure.

- *X/Y-cut*: if the TSV bound is larger than one, or if the sinks in Set S belong to the same die, Set S is partitioned geometrically by a horizontal line (X-cut or Y-cut), and Z-dimension is ignored. If the subsets contain sinks from different dies, we potentially need multiple TSVs to connect those sinks.

At the end of each partitioning, we propagate the TSV bound constraint by assigning a TSV bound for each new subset.

The 3D abstract tree generation using the 3D-MMM algorithm is shown in Algorithm 7. The recursive method takes as inputs a set of 3D clock sinks and a TSV bound. If the size of the given sink set (i.e., $|S|$) is one, then we reach the bottom level of the abstract tree (Lines 3–4). If the TSV bound is one, Z-cut is applied to partition Sink Set S into two Subsets S_1 and S_2 (Lines 6–7). As previously discussed, once the TSV bound is one, our 3D-MMM performs $stack(S) - 1$ Z-direction partitions, so that the sinks belong to the same die are in the same subset. In order to guarantee that only one TSV is used between adjacent dies, the order of die-wise Z-cut depends on the source-die index and the die indices in Sink Set S as illustrated in Algorithm 6. In the case that the above conditions are not satisfied, Set S is partitioned geometrically by a horizontal line (X-cut or Y-cut), so called X/Y-cut (line 9). And the Z-dimension of each sink is ignored. The cut line is drawn at the median of the X or Y coordinates of the sinks. The TSV bound is divided for the two subsets (line 10). The bound for each subset is calculated by (1) estimating the number of TSVs required by each subset and (2) dividing the given Bound B according to the ratio of the estimated TSVs. For each subset, we assume the minimum sink size in each die as the

```
input  : 3D Sink set S and a TSV bound B
output: Root of Sink Set S
S1 and S2 ⟵ subsets of S
if |S| = 1 then
  | return root(S)
else if B = 1 and stack(S) > 1 then
  | Z-cut(S, S1, S2)
  | B1 = B2 = 1
else
  | Geometrically divide S into S1, S2
  | Find B1, B2 such that B1 + B2 = B
end
root(S1) ⟵ AbsTreeGen3D(S1, B1)
root(S2) ⟵ AbsTreeGen3D(S2, B2)
leftChild(root(S)) ⟵ root(S1)
rightChild(root(S)) ⟵ root(S2)
return root(S)
```

Algorithm 7: Pseudo code of AbsTreeGen3D(S, B) in the 3D-MMM algorithm.

estimation of the number of TSVs. The procedure is called recursively for each of Subsets S_1 and S_2 with different TSV bounds (lines 11–12). The roots of the subtrees are connected by the root of the higher-level tree (lines 13–15). The complexity of the algorithm is $O(n \cdot logn)$, where n is the number of nodes.

Corresponding to the n-die stack clock sinks, the 3D abstract tree is an n-colored binary tree, where each node (i.e., sinks, internal nodes, and the root) is assigned a color to represent which die it belongs to. The dies are numbered from 1 to n from the bottom to the top. Let $c(p)$ be the color index for Node p, where $c(p) \in \{1,2,..,n\}$. For example, $c(p) = 1$ means that Node p is located in die-1. Let $c(src)$ denote the die index, where the clock source is located. During the top-down 3D abstract tree generation, we color the nodes corresponding to the sink sets. Considering Node p with Sink Set S, let Z_{max} and Z_{min} be the maximum and minimum die indices of the sinks within Set S. The color of p is determined as follows:

$$c(p) = \begin{cases} c(src), & \text{if } p \text{ is the root} \\ Z_{min}, & \text{else if } Z_{min} > c(src) \\ Z_{max}, & \text{else if } Z_{max} < c(src) \\ c(src), & \text{otherwise.} \end{cases} \quad (4.1)$$

Considering Edge e with two Terminal Nodes n_1 and n_2. The following are true: (1) if $c(n_1) = c(n_2)$, Edge e will be routed in the same die as Nodes n_1 and n_2; (2) if $c(n_1) \neq c(n_2)$, then $|c(n_1) - c(n_2)|$ number of TSVs will be inserted along Edge e.

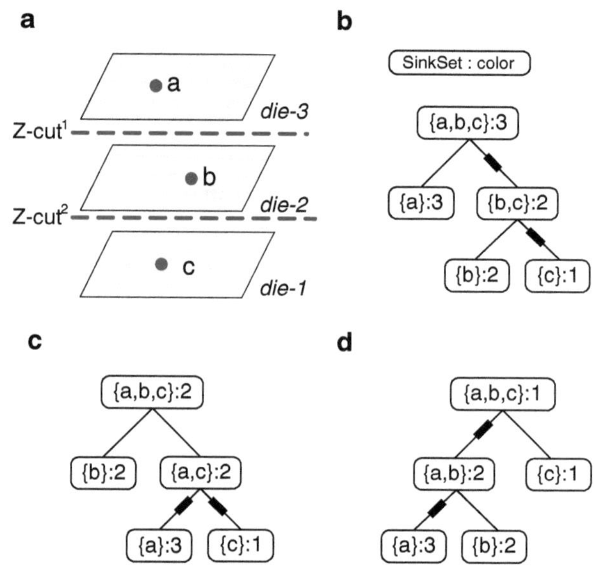

Fig. 4.4 Three-colored 3D abstract trees after applying Z-cut twice on the three-die stack sink set $\{a,b,c\}$, when the clock source is located in (**b**) die-3, (**c**) die-2, and (**d**) die-1. Each node in the abstract tree contains the corresponding sink set and a color index. (**b**) First applies Z-cut^1 and then Z-cut^2, whereas (**d**) applies Z-cut^2 first and then Z-cut^1

Figure 4.4 shows an illustration, where 3D abstract trees for a sink set $\{a,b,c\}$ are shown after applying Z-cut twice. Figure 4.4b–d are the three abstract trees, where the clock source is located in die-3, die-2 and die-1, respectively. Each node in the abstract tree contains the sink set and color information. The abstract trees in Fig. 4.4b is obtained by Z-cut^1 first and then Z-cut^2. Whereas, Fig. 4.4d applies Z-cut^2 first and then Z-cut^1. Figure 4.4c first extracts the sinks of the clock source die and then applies a Z-cut. The primary goal of using a different Z-cut sequence is to guarantee that only one TSV is necessary between adjacent dies after $stack(S) - 1$ Z-cuts.

4.4.3 *Slew-Aware Buffering and Embedding*

The second step of our 3D clock tree synthesis is slew-aware buffering and embedding: Given a 3D abstract tree, the goal is to determine the exact geometric locations of all the nodes, TSVs, and buffers, such that the wirelength of the embedded and buffered clock tree is minimized, the load capacitance of each buffer does not exceed the pre-defined maximum value (CMAX), and the clock skew is zero under the Elmore delay model. We develop the slew-aware deferred-merge buffering and embedding (sDMBE) algorithm to geometrically embed (route) the abstract tree.

sDMBE is a two-phase algorithm and is based on the deferred-merge embedding (DME) algorithm [4], which has been widely used in 2D clock synthesis. The first phase in sDMBE is to determine the merging types and to construct the merging

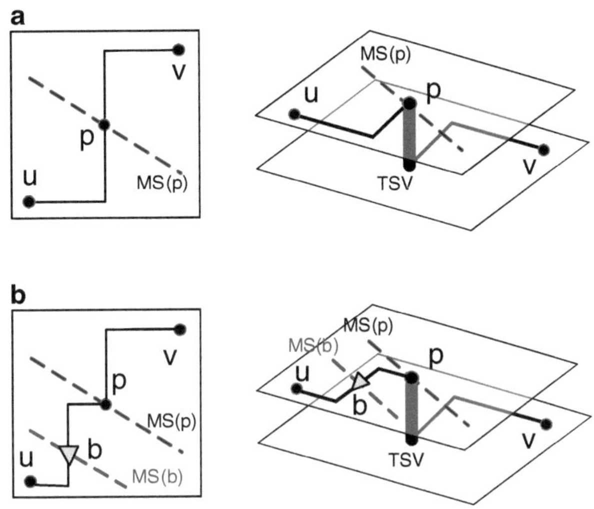

Fig. 4.5 Samples of 3D merging segments for (**a**) an unbuffered tree, and (**b**) a buffered tree

segments for each pair of subsets in a bottom-up traversal. Different from the existing 2D synthesis [1, 7, 20], which focused on slew-aware buffer insertion after clock routing, sDMBE performs buffer insertion during the bottom-up procedure. The goal of slew-aware buffering in sDMBE is to locate buffers while merging subsets so that the load capacitances of buffers are within the given bound (CMAX). The impact of CMAX on the 3D clock slew is discussed in Sect. 4.6.5. Merging segments are obtained based on the merging distances, which are computed under the zero-skew equations in the Elmore delay model and the wirelength minimization goals. The second phase of sDMBE is to decide the exact locations of internal nodes, buffers, and TSVs in a top-down fashion and determine the routing topology of the overall clock nets. The complexity of our approach is $O(n)$, which makes it feasible for incremental clock routing or inclusion in a solution search framework.

Two samples of merging segments for unbuffered and buffered 3D clock trees are shown in Fig. 4.5. When merging Child Nodes u and v to Parent Node p, sDMBE first decides the merging type based on the given 3D abstract tree and the CMAX constraint. Corresponding to the merging type among clock wires, buffers, and TSVs, we obtain the merging distances between Nodes p and u, p and v in Fig. 4.5a, the distances between Node p and Buffer b, Buffer b and Node u, Nodes p and v in Fig. 4.5b.

4.5 Extension of 3D-MMM Algorithm

As illustrated earlier in Fig. 4.1, the overall wirelength of the 3D clock tree reduces as more and more TSVs are used. Figure 4.6 provides another demonstration that higher usage of TSVs leads to shorter wirelength. This raises an important question:

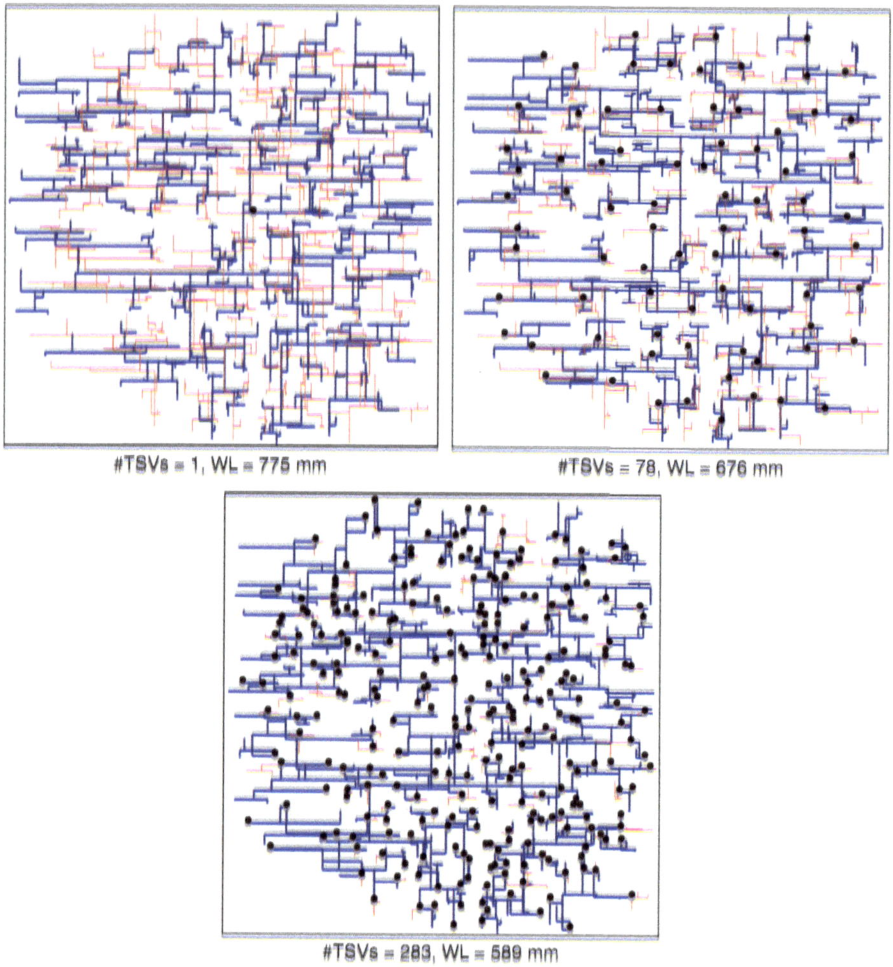

Fig. 4.6 3D clock trees for the two-die stack r_3 with varying TSV bounds. The *black dots* are the TSV location candidates. And the *bold* and *thin lines* illustrate the clock nets in die-1 and die-2, respectively

what is the optimal number of TSVs for a 3D clock tree that leads to the minimum possible power consumption? One obvious way to answer this question is by trying *all* possible TSV counts and choosing the best power result (an exhaustive search). This method, however, is very time consuming and requires prohibitive runtime as shown in Table 4.2. Thus, our goal is to find this TSV count that leads to the minimum (or close-to-minimum) power result in much shorter runtime. This calls for careful attentions to the impact of the TSV count not only on the overall wirelength but also the total number of buffers and the total TSV capacitance as these factors equally affect the overall power consumption.

We develop our new low-power 3D clock tree synthesis method, named 3D-MMM-ext, by extending our 3D-MMM algorithm presented in Sect. 4.4.2. The goal of the 3D-MMM-ext is to construct a low-power clock network by wisely assigning clock TSVs during the 3D abstract tree generation. In each top-down partition, let S be the current sink set. Let $Z(S)$ denote the vertical distance Set S spans, which can be expressed as:

$$Z(S) = Z_{\max} - Z_{\min} \tag{4.2}$$

where $Z_{\max}$ and $Z_{\min}$ are the maximum and minimum die indices of the sinks within Set S. Note that $Z(S)$ also indicates the minimum number of TSVs required by the clock network connecting all the sinks in S. Different from the 3D-MMM algorithm, which decides the cut direction (Z-cut or X/Y-cut) based on the TSV bound (lines 5 and 8 in Algorithm 7), the key technique of 3D-MMM-ext is to determine the cutting orientation of the current iteration (i.e., Z-cut or X/Y-cut) by looking ahead to the next cutting iteration while estimating and comparing the costs of the following two cases:

- Case-1: apply Z-cut at the current iteration and then apply X/Y-cut on each die once in the following iterations;
- Case-2: apply X/Y-cut at the current iteration and postpone Z-cut to the next iteration.

Note that for the n-die stack case, Z-cut means applying die-wise partitions in multiple iterations until the sinks having the same die index are partitioned into the same subset. In the case-1 style partition, Sink Set S has $stack(S) - 1$ times Z-cuts and $stack(S)$ times X/Y-cuts. S in the case-2 has one X/Y-cut and $2 \times (stack(S) - 1)$ Z-cuts. Let S_i^z and S_i^{xy} represent the subsets after case-1 and case-2 style partitions, respectively. The sinks within the set S_i^z (or S_i^{xy}) are in the same die. Figure 4.7 shows an example of determining the current cut direction using the 3D-MMM-ext on the sink set S. Figure 4.7a shows the case-1 style partition, where Z-cut is applied during the current iteration and then X/Y-cut^1 and X/Y-cut^2 are applied on die-1 and die-2, respectively. Figure 4.7b illustrates the case-2 partition results. We also show a part of the 3D abstract tree corresponding to case-1 and case-2 partitions, respectively. We have the following relation:

$$S = \bigcup_{i=1}^{4} S_i^z = \bigcup_{i=1}^{4} S_i^{xy} \tag{4.3}$$

By comparing the cost of case-1 (P_z) and the cost of case-2 (P_{xy}), the cut direction of the current iteration is determined as follows:

$$\text{Current Cut} = \begin{cases} \text{X/Y-cut}, & \text{if } P_z > P_{xy} \\ \text{Z-cut}, & \text{otherwise.} \end{cases} \tag{4.4}$$

This means that if selecting Z-cut during the current iteration helps reduce power, then we choose Z-cut; otherwise, we choose X/Y-cut. Cost P_z is defined as follows:

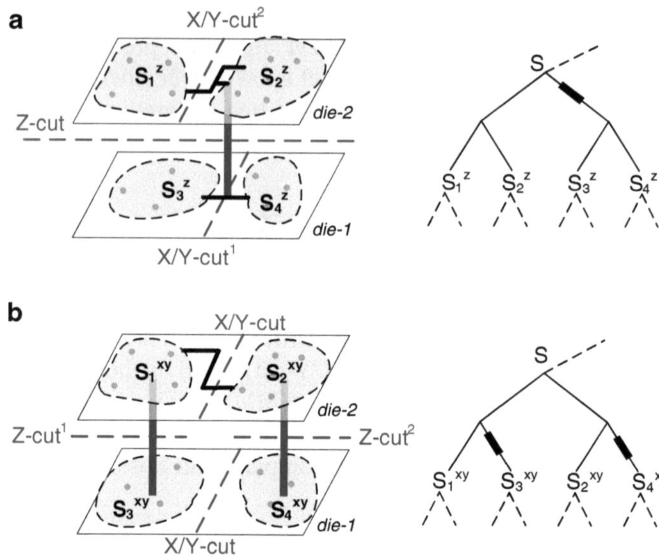

Fig. 4.7 The 3D-MMM-ext algorithm performed on a two-die stack with Sink Set S. We show the 3D abstract trees, cut orders, and the subsets from case-1 and case-2 style partitions. (**a**) Case-1, where we apply Z-cut at the current iteration, and then X/Y-cut^1 and X/Y-cut^2 in die-1 and die-2, respectively and (**b**) case-2, where we apply X/Y-cut at the current iteration, and then Z-cut^1 and Z-cut^2. P_z and P_{xy} are the cost of merging S_i^z and S_i^{xy} in (**a**) and in (**b**), respectively

$$P_z = \sum_{i \in cond1} P(S_i^z) + \sum_{j,k \in cond2} P(S_j^z, S_k^z) \tag{4.5}$$

Similarly,

$$P_{xy} = \sum_{i \in cond1} P(S_i^{xy}) + \sum_{j,k \in cond2} P(S_j^{xy}, S_k^{xy}) \tag{4.6}$$

Let S_i represent either S_i^{xy} or S_i^z. The first item $P(S_i)$ in the cost function is the cost of Subset S_i, where $cond1$ covers the final subsets after the look-ahead partitions. The second item $P(S_j, S_k)$ in the cost function is the cost of connecting Subsets S_j and S_k. $P(S_j, S_k)$ mainly comes from TSVs, global wires, and buffers. Therefore, $cond2$ covers all pairs of subtrees in the 3D abstract tree, where we merge those final subsets to their parent Sink Set S during the bottom-up traversal.

Considering the two-die stack examples in Fig. 4.7, P_z and P_{xy} can be expressed as follows:

$$\begin{aligned} P_z = &\sum_{i=1}^{4} P(S_i^z) + P(S_1^z, S_2^z) + P(S_3^z, S_4^z) \\ &+ P(S_1^z \cup S_2^z, S_3^z \cup S_4^z) \end{aligned} \tag{4.7}$$

$$P_{xy} = \sum_{i=1}^{4} P(S_i^{xy}) + P(S_1^{xy}, S_3^{xy}) + P(S_2^{xy}, S_4^{xy})$$
$$+P(S_1^{xy} \cup S_3^{xy}, S_2^{xy} \cup S_4^{xy}) \quad (4.8)$$

To estimate the cost for each sink set, we use the half-parameter wirelength model for $P(S_i^z)$ and $P(S_i^{xy})$. Then, $P(S_j, S_k)$ is estimated as follows:

- If no TSV is required to connect S_j and S_k:

$$P(S_j, S_k) \approx \mathrm{CD}(S_j, S_k) \quad (4.9)$$

  where $\mathrm{CD}(S_j, S_k)$ is the distance between the centers of Subsets S_j and S_k. In Fig. 4.7, $P(S_1^z, S_2^z)$, $P(S_3^z, S_4^z)$ and $P(S_1^{xy} \cup S_3^{xy}, S_2^{xy} \cup S_4^{xy})$ belong to this case.
- If TSVs are needed to provide inter-die connection between S_j and S_k:

$$P(S_j, S_k) \approx \mathrm{CD}(S_j, S_k) + \alpha \times C_{\mathrm{TSV}}/c \quad (4.10)$$

  where C_{TSV} is the TSV capacitance, c is the unit-length capacitance of the clock line, and α is an estimator representing the cost of TSV insertion. We use the following empirical equation to calculate α:

$$\alpha = (2 \times |Z(S_j) - Z(S_k)| + 3) \times \beta \quad (4.11)$$

  where $\beta = 0.05$, 0.05 and 0.1 if the TSV capacitance is 15, 50, and 100 fF, respectively. In Fig. 4.7, $P(S_1^z \cup S_2^z, S_3^z \cup S_4^z)$, $P(S_1^{xy}, S_3^{xy})$ and $P(S_2^{xy}, S_4^{xy})$ belong to this case.

4.6 Experimental Results

We first examine a two-die stack to investigate the impact of the TSV count and the TSV parasitics on clock power consumption. Next, we show the efficiency of the 3D-MMM-ext algorithm in finding the optimal number of TSVs to be used for minimum power consumption. We then present the results of our clock slew control method. Lastly, we show the impact of scaling the supply voltage on 3D clock power consumption. We validate our claims with SPICE simulation results.

4.6.1 Simulation Settings

We construct a zero-Elmore-skew 3D clock network by using the 3D clock tree synthesis methods developed in Sects. 4.4 and 4.5. We then extract the netlist of

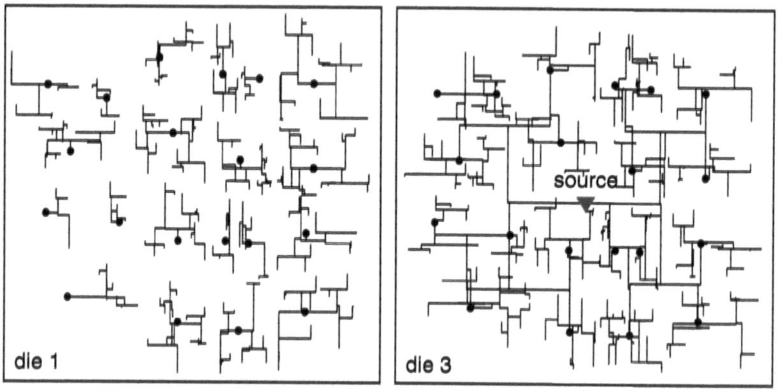

Fig. 4.8 Clock trees in die-1 and die-3 of a sample six-die 3D clock network, where the clock source is located in die-3. *Black dots* denote TSVs. The TSV bound is set to 20. Die-1 contains many local trees, whereas die-3 contains a single global tree

the entire 3D clock network for SPICE simulation. After the simulation, we obtain highly accurate power consumption and timing information of the entire clock network. Note that our 3D clock tree has zero skew under the Elmore delay model but may have nonzero clock skew from SPICE simulation. Thus, we constrain the SPICE clock skew to be less than 3 % of the clock period at a frequency of 1 GHz. The slew is constrained within 10 % of the clock period. The clock power mainly comes from the switching capacitance of the interconnect, sink nodes, TSVs, and clock buffers.

The technology parameters are based on the 45 nm Predictive Technology Model [17]: per unit-length wire resistance is 0.1 Ω/um, and per unit-length wire capacitance is 0.2 fF/um. The buffer parameters are: driving resistance is 122 Ω, input capacitance is 24 fF, and intrinsic delay is 17 ps. The TSV resistance is 35 mΩ. In order to study the impact of the TSV *RC* parasitics on the 3D clock network, we vary the linear oxide thickness and choose three typical TSV capacitance values (i.e., 15, 50, 100 fF). The supply voltage is set to 1.2 V unless otherwise specified. The maximum load capacitance of each clock buffer, denoted CMAX, is set to 300 fF for slew control unless otherwise specified.

Our analysis focuses on two-die and six-die 3D clock networks. In the six-die case, the clock source is located in the middle die (die-3) as suggested in [25], unless otherwise specified. As a result, die-3 in a six-die clock network contains a complete tree. The IBM benchmarks r_1 to r_5 [8] are used. Since r_1 to r_5 are originally designed for 2D ICs, we randomly distribute the sinks into two or six dies. We then scale the footprint area by $\sqrt{N}$ to reflect the area reduction in the 3D design.

Sample clock trees in die-1 and die-3 of a six-die 3D clock network are shown in Fig. 4.8. The triangle denotes the clock source in die-3. Each die contains up to 20 TSVs. Note that die-3 has a single global tree that connects all the sinks, and die-1 contains multiple local trees that are connected to the clock source using TSVs.

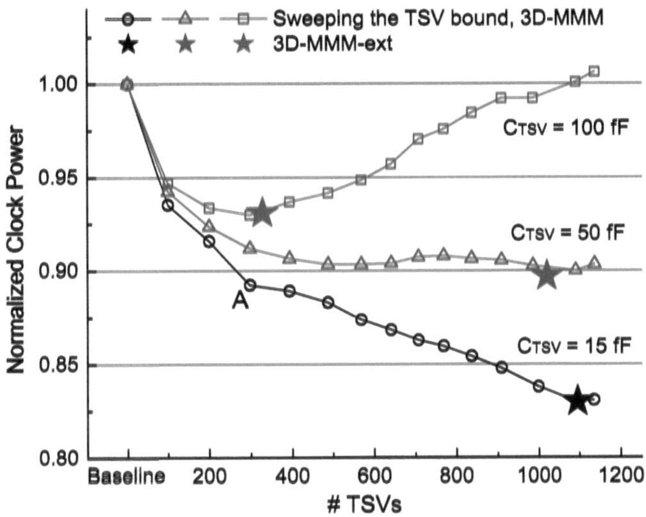

Fig. 4.9 Impact of the TSV capacitance and count on clock power for the two-die *r*5. The TSV capacitance (C_{TSV}) is set to 15, 50, and 100 fF. Our baseline is the clock tree that uses one TSV between adjacent dies. For each C_{TSV}, we show the 3D-MMM results by sweeping the TSV count. We also highlight the 3D-MMM-ext results for each C_{TSV}, which are marked as stars near to the trends

4.6.2 Impact of TSV Count and Parasitic Capacitance

To investigate the impact of the TSVs on clock power consumption, we use a two-die stack implementation of the biggest benchmark r_5, which has 3,101 sink nodes with input capacitances varying from 30 to 80 fF. Figure 4.9 shows three clock power trend curves for a TSV capacitance (C_{TSV}) of 15, 50, 100 fF, respectively. On the x-axis we show the total number of TSVs used in each entire 3D clock tree, which is obtained by imposing a different TSV bound. Our baseline 3D clock network contains only one TSV between adjacent dies.

The clock power is affected by both the TSV count and the TSV capacitance as shown in Fig. 4.9. First, using 15 fF TSVs in the clock network construction, the clock power decreases significantly when more TSVs are used. We are able to obtain a low-power clock network design by relaxing the TSV bound. We can achieve up to 17.0 % power reduction as compared with the single-TSV case. The power savings mostly comes from wirelength reduction because the clock wire capacitance significantly affects the overall power consumed by the clock network. When more TSVs are used, the number of local trees in the non-source dies increases while their size decreases. This means that the multiple-TSV case encourages more local clock distribution in 3D designs while reducing the overall wirelength. Second, if the TSV has a large capacitance (e.g., 50, 100 fF), the contribution of the TSV capacitance to the overall power consumption is non-negligible. As a result, when the TSV count increases, the overall clock power reduction becomes slower. Particularly, if the

TSV capacitance is 100 fF, clock power does not decrease when the TSV count exceeds a certain amount and eventually starts increasing. In this case, the clock power from the TSV capacitance increases faster than the power decreases from wirelength reduction.

From this trend study, we conclude that given a TSV parasitic capacitance, there exists an optimum number of TSVs that results in the minimum 3D clock power. This in turn allows us to choose the right TSV bound for a given power budget. If a power savings of 10 % is required for using the 15 fF TSVs, the TSV bound of 300 can be used based on Point *A* in Fig. 4.9.

4.6.3 Exhaustive Search Results

A straightforward way to find the "min-power TSV count", i.e., the number of TSVs used in a 3D clock tree that leads to the minimum overall clock power consumption, is to exhaustively sweep the TSV bound from one to infinity,[3] constructing and simulating the entire 3D clock network corresponding to each TSV bound. By plotting the TSV count vs. power trend curve, we are then able to find the optimum solution. Figure 4.10 shows the clock power trend based on 1,137 3D clock trees we generated and simulated for the two-die stack r_5. We assume the TSV parasitic capacitance is 100 fF.

We observe that the lowest power comes from the clock network that uses 250 TSVs with 1.190 W clock power and 2,004,250 μm wirelength. In addition, we observe that the exhaustive search result agrees with the TSV count vs. power trend we presented in the previous section, although power fluctuates locally in a small range of the TSV count. If the TSV count exceeds 600, the clock power is much more sensitive to the TSV count increase. Using one more TSV may lead to the clock power increasing or decreasing by 1 %. This is because, when using a large amount of TSVs, the clock network has a large number of smaller local trees. This means that the TSV capacitance itself is comparable to or even larger than that of a single local clock tree. In this case, using a few more TSVs leads to a large fluctuation in clock power.

The proposed exhaustive search method does allow us to find the min-power TSV count, but it is too costly in terms of runtime. The smaller step size we use for the TSV count in the search, the lower power of a 3D clock network we find, but more simulations as well as runtime are required. Note that the typical SPICE simulation time of a two-die r_5 clock network is around 200 s. Repeating this 1,137 times is prohibitive.

---

[3]Note that the TSV bound of infinity means that we do not impose any restriction on the maximum number of TSVs used in each die. This usually results in a high usage of TSVs that mainly targets at wirelength minimization.

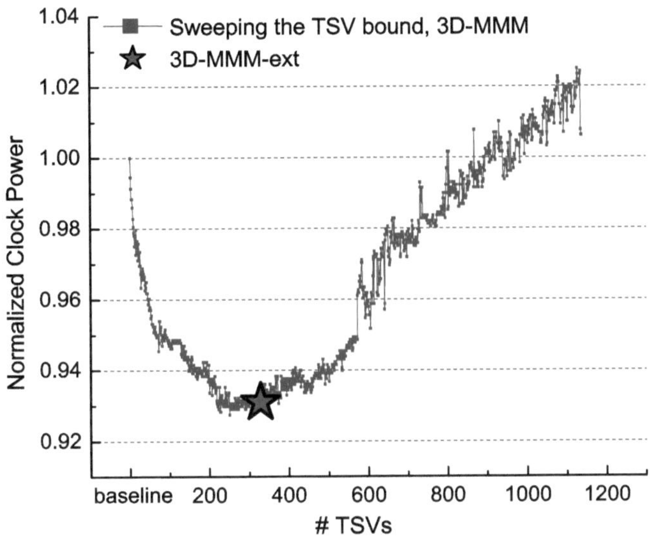

Fig. 4.10 Clock power trends for the two-die stack r_5 based on exhaustive search within the TSV count range [1, 1, 137]. The TSV capacitance is 100 fF. We also plot the 3D-MMM-ext algorithm result. The exhaustive search covers 1,137 simulations on various clock trees. The runtime for each simulation is around 200 s

4.6.4 3D-MMM-ext Algorithm Results

In Fig. 4.10, the star indicates the solution obtained by our 3D-MMM-ext algorithm. Our algorithm does not involve any exhaustive search on the TSV count, but relies on our look-ahead-based method to control the TSV usage and to minimize the overall power consumption. We observe that our 3D-MMM-ext generates a 3D clock tree that has a similar quality as the one obtained by the exhaustive research but at a fraction of runtime. The runtime required for 3D-MMM-ext is comparable to that of generating a single 3D clock tree. The solution quality obtained by our 3D-MMM-ext algorithm can also be seen in Fig. 4.9, where the stars indicate the 3D trees produced by 3D-MMM-ext. The power consumption at these points is comparable to the minimum power solutions found in each curve.

Tables 4.1 and 4.2 present more detailed comparisons of wirelength (μm), buffer count (#Bufs), clock power (W), clock skew (ps), number of simulations (#sims), and the total simulation runtime (s) between the exhaustive search and the 3D-MMM-ext algorithm. We use two-die 3D stacks. We also show the wirelength and power reduction of 3D-MMM-ext with respect to the exhaustive search. First, the clock power of 3D-MMM-ext is comparable to that of the exhaustive search. In most cases, 3D-MMM-ext has less than 1 % power difference. In some cases, 3D-MMM-ext achieves even lower power (i.e., positive reduction) than the exhaustive search. This is mainly because the low-power design obtained by the exhaustive search depends on the sweeping granularity and simulation times.

Table 4.1 Exhaustive search vs. our 3D-MMM-ext algorithm. We report wirelength in μm, power in mW, and skew in ps

Two-die stack results

	Exhaustive search					3D-MMM-ext					% imprv	
ckt	#TSV	WL	#buf	pwr	Skew	#TSV	WL	#buf	pwr	Skew	WL	pwr
TSV capacitance = 15 fF												
r_1	91	220,362	275	0.122	14.6	93	221,443	282	0.125	9.3	−0.5	−2.5
r_2	222	433,639	573	0.250	14.1	211	445,647	588	0.255	14.2	−2.8	−2.0
r_3	320	582,035	778	0.342	12.1	297	583,274	779	0.342	13.5	−0.2	0.0
r_4	715	1,157,160	1,587	0.696	16.1	660	1,165,529	1,594	0.698	16.8	−0.7	−0.3
r_5	1,129	1,728,660	2,496	1.062	20.2	1,096	1,737,100	2,509	1.065	19.8	−0.5	−0.3
TSV capacitance = 50 fF												
r_1	95	218,257	292	0.129	11.9	85	221,719	293	0.130	11.9	−1.6	−0.8
r_2	222	438,370	602	0.267	14.4	205	448,195	618	0.271	13.6	−2.2	−1.5
r_3	253	605,079	848	0.368	14.4	288	589,654	845	0.366	15.7	2.5	0.5
r_4	660	1,171,810	1,723	0.748	17.0	639	1,165,253	1,727	0.745	15.0	0.6	0.4
r_5	1,091	1,753,390	2,726	1.155	18.3	1,020	1,749,543	2,684	1.151	17.8	0.2	0.3
TSV capacitance = 100 fF												
r_1	56	230,940	301	0.135	10.1	45	238,242	303	0.137	12.6	−3.2	−1.5
r_2	76	493,957	654	0.284	13.3	87	492,966	661	0.287	13.0	0.2	−1.1
r_3	60	674,674	883	0.383	12.9	112	645,062	897	0.383	13.4	4.4	0.0
r_4	254	1,293,830	1,926	0.793	19.4	247	1,286,784	1,891	0.787	18.2	0.5	0.8
r_5	250	2,004,250	2,799	1.190	14.0	328	1,953,453	2,798	1.194	19.0	2.5	−0.3

Table 4.2 Runtime comparison between exhaustive search vs. our 3D-MMM-ext algorithm. Runtime is reported in seconds

Two-die stack results				
	Exhaustive search		3D-MMM-ext	
ckt	# simulations	Runtime	# simulations	Runtime
TSV capacitance = 15 fF				
r_1	37	602.5	1	16.8
r_2	29	1,059.5	1	32.5
r_3	31	1,712.3	1	50.5
r_4	41	4,981.5	1	107.1
r_5	41	9,104.3	1	187.7
TSV capacitance = 50 fF				
r_1	37	623.5	1	17.6
r_2	29	1,087.8	1	36.5
r_3	31	1,508.0	1	48.1
r_4	41	5,391.1	1	114.6
r_5	41	9,152.9	1	186.3
TSV capacitance = 100 fF				
r_1	37	618.7	1	16.0
r_2	29	1,156.0	1	33.5
r_3	31	1,733.6	1	55.1
r_4	41	5,798.7	1	125.2
r_5	41	9,323.5	1	179.8

Second, the simulation runtime comparison reveals the effectiveness of our 3D-MMM-ext algorithm. The 3D-MMM-ext requires only a single simulation, whereas the exhaustive search requires 29–41 simulations.

The comparisons between using a single TSV and using multiple TSVs (obtained with the 3D-MMM-ext algorithm) are presented in Tables 4.3 and 4.4. We use a two-die and a six-die implementation of our benchmark designs. First, the 3D-MMM-ext is able to find the low-power 3D clock trees. For the two-die stacks in Table 4.3, the 3D-MMM-ext reduces the clock power by around 16.1–18.8, 10.3–13.7, and 6.6–8.3 % as compared with the single-TSV cases and achieves wirelength savings around 24.0–26.5, 23.9–26.6, and 16.6–18.9 %, when the TSV capacitance is 15, 50, and 100 fF, respectively. In the case of six-die stacks shown in Table 4.4, our 3D-MMM-ext reduces power by up to 36.1, 26.4, and 9.1 %, and reduces wirelength by up to 50.7, 47.4, and 17.3 %.

The comparisons between placing the clock source in die-1 and in die-3 are presented in Table 4.5 lists, where the 3D clock designs are for the six-die stacks using the 3D-MMM-ext algorithm. When moving the clock source to the middle die (die-3), the 3D-MMM-ext achieves further power savings, especially in the case when the TSV capacitance is 100 fF. In addition, in most of the cases, e.g., the six-die stacks with 15 and 50 fF TSVs, the middle-die 3D-MMM-ext uses fewer TSVs and achieves lower power than the cases when the src is in die-1.

In most cases, the simulated clock skew is less than 20 ps, which is less than the 30 ps constraint. In the case of the six-die 3D stack of r_5, Fig. 4.11 shows the spatial distribution of the propagation delay for the die containing the clock source.

Table 4.3 Impact of TSV usage. We compare single vs. multiple TSVs (= source in die-3) for two die stacks. We report wirelength in μm, power in mW, skew in ps, and runtime (= cpu) in seconds

Two die stack results													
	Single TSV					Multiple TSVs						% imprv	
ckt	WL	#buf	pwr	Skew	cpu	#TSV	WL	#buf	pwr	Skew	cpu	WL	pwr
TSV capacitance = 15 fF													
r_1	291,421	327	0.149	10.5	17.6	93	221,443	282	0.125	9.3	16.8	24.0	16.1
r_2	602,484	706	0.314	15.4	43.2	211	445,647	588	0.255	14.2	32.5	26.0	18.8
r_3	775,194	930	0.410	17.4	55.2	297	583,274	779	0.342	13.5	50.5	24.8	16.6
r_4	1,586,630	1,990	0.855	18.2	122.8	660	1,165,529	1,594	0.698	16.8	107.1	26.5	18.4
r_5	2,341,420	2,897	1.283	17.0	188.0	1,096	1,737,100	2,509	1.065	19.8	187.7	25.8	17.0
TSV capacitance = 50 fF													
r_1	291,498	327	0.149	12.4	18.1	85	221,719	293	0.130	11.9	17.6	23.9	12.8
r_2	602,485	706	0.314	15.2	38.4	205	448,195	618	0.271	13.6	36.5	25.6	13.7
r_3	775,056	930	0.410	17.2	53.2	288	589,654	845	0.366	15.7	48.1	23.9	10.7
r_4	1,586,880	1,991	0.855	14.8	121.5	639	1,165,253	1,727	0.745	15.0	114.6	26.6	12.9
r_5	2,341,360	2,897	1.283	16.8	220.1	1,020	1,749,543	2,684	1.151	17.8	186.3	25.3	10.3
TSV capacitance = 100 fF													
r_1	291,421	328	0.149	9.9	17.5	45	238,242	303	0.137	12.6	16.0	18.2	8.1
r_2	601,929	707	0.313	13.5	40.0	87	492,966	661	0.287	13.0	33.5	18.1	8.3
r_3	775,029	930	0.410	17.3	54.2	112	645,062	897	0.383	13.4	55.1	16.8	6.6
r_4	1,586,630	1,992	0.855	15.7	131.3	247	1,286,784	1,891	0.787	18.2	125.2	18.9	8.0
r_5	2,341,460	2,897	1.283	17.1	187.6	328	1,953,453	2,798	1.194	19.0	179.8	16.6	6.9

Table 4.4 Impact of TSV usage. We compare single vs. multiple TSVs (= source in die-3) for six-die stacks. We report wirelength in μm, power in mW, skew in ps, and runtime (= cpu) in seconds

Six-die stack results													
	Single TSV					Multiple TSVs						% imprv	
ckt	WL	#buf	pwr	Skew	cpu	#TSV	WL	#buf	pwr	Skew	cpu	WL	pwr
TSV capacitance = 15 fF													
r_1	272,109	332	0.144	19.4	19.0	297	138,223	214	0.092	12.8	10.5	49.2	36.1
r_2	566,944	684	0.298	16.1	45.0	668	280,901	445	0.191	18.2	29.7	50.5	35.9
r_3	717,479	887	0.388	15.0	57.0	965	376,634	626	0.264	17.1	45.8	47.5	32.0
r_4	1,496,180	1,870	0.816	18.5	119.8	2,195	752,370	1,316	0.551	17.6	84.0	49.7	32.5
r_5	2,299,220	2,935	1.265	19.6	205.3	3,497	1,133,262	2,070	0.854	21.4	154.0	50.7	32.5
TSV capacitance = 50 fF													
r_1	272,849	332	0.144	17.4	17.7	275	143,626	257	0.106	18.5	11.5	47.4	26.4
r_2	567,686	684	0.299	15.0	46.6	631	302,068	562	0.230	20.3	35.2	46.8	23.1
r_3	719,610	891	0.389	14.3	66.1	918	403,235	775	0.316	18.5	50.2	44.0	18.8
r_4	1,493,990	1,870	0.815	15.0	123.0	2,045	810,708	1,680	0.670	27.0	95.1	45.7	17.8
r_5	2,299,590	2,935	1.266	19.3	217.8	3,270	1,250,269	2,644	1.051	23.4	189.8	45.6	17.0
TSV capacitance = 100 fF													
r_1	273,951	332	0.145	16.6	16.8	30	234,821	309	0.133	29.0	17.1	14.3	8.3
r_2	566,803	685	0.298	11.1	45.1	80	468,805	638	0.271	28.9	41.2	17.3	9.1
r_3	720,705	893	0.390	14.2	61.6	75	651,298	873	0.374	23.1	60.3	9.6	4.1
r_4	1,497,240	1,873	0.817	14.0	126.5	115	1,333,034	1,804	0.769	23.8	118.8	11.0	5.9
r_5	2,300,620	2,935	1.266	19.2	183.6	180	2,014,167	2,780	1.179	28.3	186.7	12.5	6.9

Table 4.5 Impact of clock source location. We compare clock source located in die-1 vs. die-3 (both using 3D-MMM-ext) for six-die stacks. We report wirelength in μm, power in mW, skew in ps, and runtime (= cpu) in seconds

Six-die stack results														
	Source in die-1						Source in die-3						% imprv	
ckt	#TSV	WL	#buf	pwr	Skew	cpu	#TSV	WL	#buf	pwr	Skew	cpu	WL	pwr
TSV capacitance = 15 fF														
r_1	375	141,353	227	0.095	15.6	11	297	138,223	214	0.092	12.8	10	2.2	3.2
r_2	798	287,536	479	0.197	20.0	32	668	280,901	445	0.191	18.2	29	2.3	3.0
r_3	1,196	376,081	665	0.268	14.5	49	965	376,634	626	0.264	17.1	45	−0.1	1.5
r_4	2,594	766,596	1,371	0.561	17.5	95	2,195	752,370	1,316	0.551	17.6	84	1.9	1.8
r_5	4,133	1,167,350	2,174	0.876	20.2	163	3,497	1,133,262	2,070	0.854	21.4	154	2.9	2.5
TSV capacitance = 50 fF														
r_1	345	147,503	284	0.113	16.0	3	275	143,626	257	0.106	18.5	11	2.6	6.2
r_2	742	309,985	647	0.243	25.3	42	631	302,068	562	0.230	20.3	35	2.6	5.3
r_3	1,063	423,253	899	0.339	20.3	62	918	403,235	775	0.316	18.5	50	4.7	6.8
r_4	2,335	856,880	1,931	0.719	24.4	113	2,045	810,708	1,680	0.670	27.0	95	5.4	6.8
r_5	3,688	1,349,599	3,086	1.143	25.5	195	3,270	1,250,269	2,644	1.051	23.4	189	7.4	8.0
TSV capacitance = 100 fF														
r_1	20	261,396	322	0.143	15.4	17	30	234,821	309	0.133	29.0	17	10.2	7.0
r_2	40	537,705	661	0.294	13.8	39	80	468,805	638	0.271	28.9	41	12.8	7.8
r_3	45	709,790	902	0.393	14.1	54	75	651,298	873	0.374	23.1	60	8.2	4.8
r_4	90	1,409,870	1,824	0.798	16.0	109	115	1,333,034	1,804	0.769	23.8	118	5.4	3.6
r_5	100	2,154,326	2,827	1.225	17.1	180	180	2,014,167	2,780	1.179	28.3	186	6.5	3.8

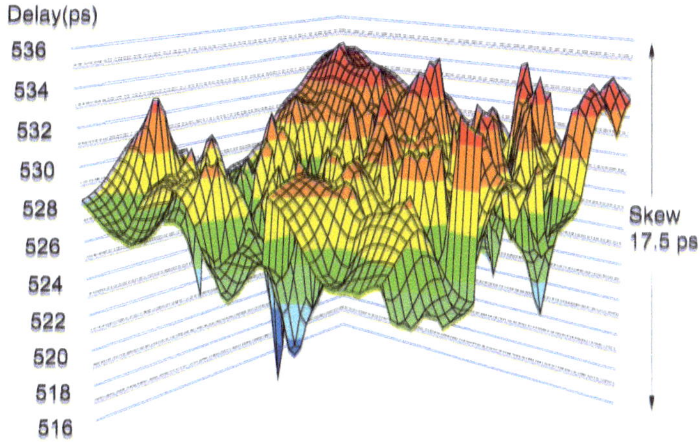

Fig. 4.11 Spatial distribution of propagation delay (ps) and clock skew (ps) of the clock source die for the six-die stack r_5. The TSV count is 3,497

The TSV count is 3,497. We observe that the clock skew among the six dies varies within [17.5 ps, 21.4 ps]. The skew of the entire 3D clock network is 21.4 ps. Referring to the TSV *RC* parasitics and the 300 fF CMAX constraint, the delay along each TSV is in the order of 0.01 ps. Compared with the >500 ps src-to-sink delay, this means that the TSV itself contributes a negligible portion of delay to the entire src-to-sink delay. Note that our 3D clock tree synthesis algorithm builds a zero-skew tree under the Elmore delay model, which in practice shows discrepancy between SPICE simulation results.

4.6.5 Low-Slew 3D Clock Routing

Our goal in this experiment is to show that the TSV count also affects the clock slew distribution. Figure 4.12 shows the slew distribution of the six-die 3D clock tree for r_5 among all sinks. The clock slew constraint is set to 100 ps, which is 10 % of the clock period. The slew distribution of the single-TSV clock tree is shown in Fig. 4.12a, whereas Fig. 4.12b shows the slew distribution of the multiple TSV clock tree using the 3D-MMM-ext. In the single-TSV clock tree, slew varies within [34.2 ps, 82.7 ps] with an average slew of 53.9 ps. The slew distribution of the multiple-TSV case is in the range of [29.1 ps, 80.3 ps] with an average slew of 46.8 ps. Compared with the single-TSV case, the multiple-TSV case reduces the maximum slew and average slew by 2.4 and 7.1 ps, respectively. The main reason for the improved slew distribution of the multiple-TSV 3D tree is the shorter wirelength, which in turn reduces the capacitive load. Thus, we conclude that multiple TSVs are effective in improving the slew distribution.

The impact of CMAX, the maximum clock buffer load capacitance, on slew variations (min, average, max) and power consumption in the single-TSV and

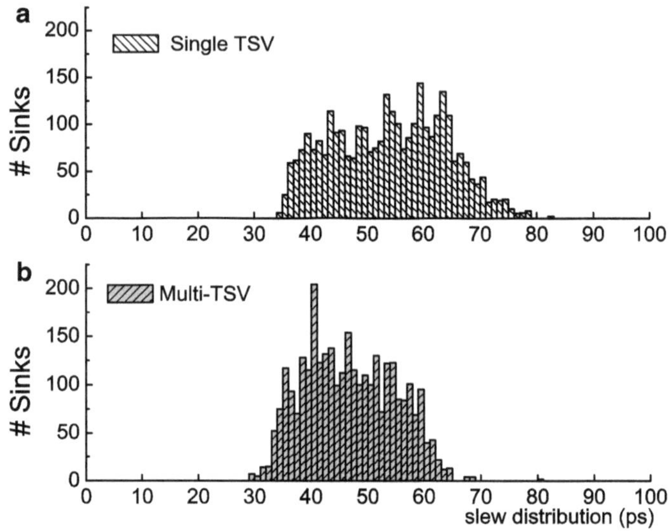

Fig. 4.12 Slew distribution of six-die 3D clock network among all sinks. Slew constraint is set to 10 % of the clock period, and CMAX is 300 fF. (**a**) Slew distribution in the single-TSV clock tree and (**b**) in the multiple-TSV clock tree

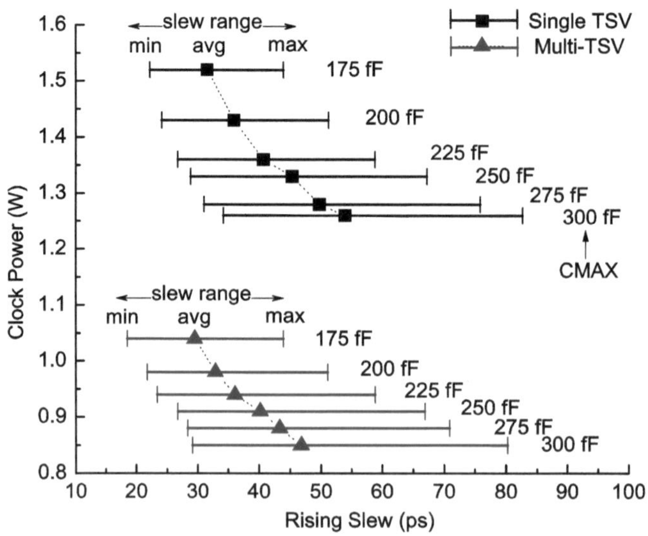

Fig. 4.13 Slew variations and power comparisons between single-TSV and multiple-TSV clock trees. CMAX varies from 175 to 300 fF

multiple-TSV clock trees is shown in Fig. 4.13. First, CMAX remains an efficient means to control the maximum slew in 3D clock network design. Both the single-TSV and multiple-TSV cases have similar trends as CMAX varies from 300 to

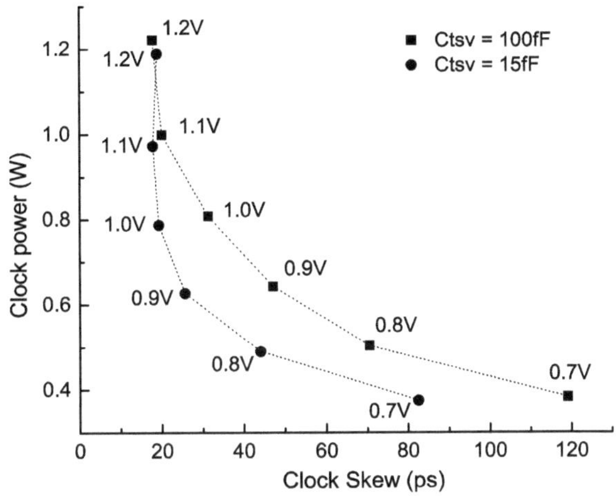

Fig. 4.14 Impact of scaling the supply voltage on clock power and clock skew. The supply voltage decreases from 1.2 to 0.7 V. We compare two clock networks using 15 and 100 fF TSVs. Each network uses 125 TSVs

175 fF: a smaller CMAX reduces the maximum slew, but increases the clock power. This is because each buffer stage is allowed to drive a smaller capacitance with smaller CMAX, which in turn requires more buffers and thus consumes more power. Second, given a certain CMAX, multiple-TSV clock trees always have reduced maximum slew and less average slew as compared with the single-TSV cases. Third, we note that the multiple-TSV case always consumes less power than the single-TSV case. Therefore, we conclude that the multiple-TSV case achieves both low power and better slew results.

4.6.6 Scaling the Supply Voltage

In this section, we study the impact of supply voltage scaling on 3D clock power, clock skew and slew. The clock skew and power changes when the supply voltage is scaled down from 1.2 to 0.7 V. These changes are shown in Fig. 4.14, for a clock frequency of 1 GHz. We first compare the two clock networks based on 15 and 100 fF TSV capacitance. Both of the clock networks use 125 TSVs. We first observe that both clock networks have a similar trend when the supply voltage is scaled down: The clock power is reduced from around 1.2 to 0.4 W, which is more than a 65 % power reduction. Second, the clock skew increases from 20 to 80 ps if the TSV capacitance is 15 fF, and from 20 to 120 ps for 100 fF TSVs. Moreover, the clock skew for a 100 fF TSV capacitance increases faster than that for a 15 fF TSV capacitance. This is mainly because the former uses 2,830 clock buffers, whereas the latter uses 2,789 clock buffers. The more buffers a 3D clock tree contains, the faster the clock skew degrades with the supply voltage scaling down. Thus, if the maximum simulated clock skew is set to 40 ps, the clock network can normally operate above 0.8 and 0.9 V, using 15 and 100 fF TSVs, respectively.

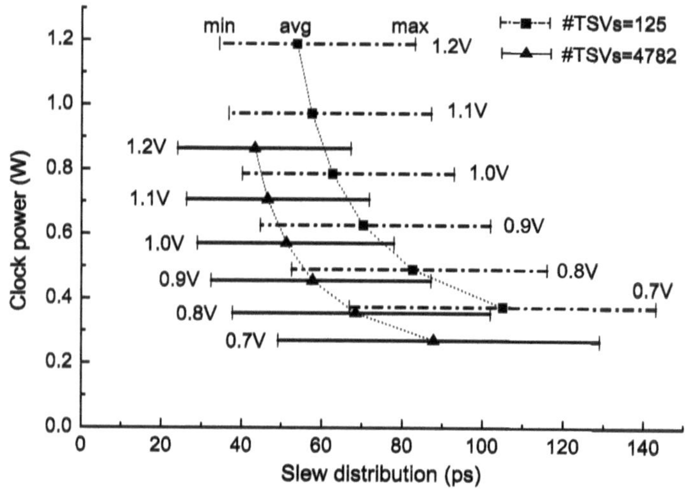

Fig. 4.15 Impact of supply voltage scaling on the clock slew distribution and clock power. Supply voltage decreases from 1.2 to 0.7 V. The TSV capacitance is 15 fF. We compare two clock networks using 125 and 4,782 TSVs

The impact of scaling the supply voltage on the clock slew distribution and power changes is shown in Fig. 4.15. The supply voltage is scaled down from 1.2 to 0.7 V, and the clock frequency is kept at 1 GHz. We compare two clock networks: the first uses 125 TSVs, and the second uses 4,782 TSVs. Both clock networks are based on 15 fF TSVs. We find that the clock network using 4,782 TSVs always has better control on slew distribution regardless of the supply voltage value. In addition, the clock tree using 4,782 TSVs consumes lower power than the tree using 125 TSVs for all the voltage levels. As discussed earlier, this is due to the faster reduction in capacitance from shorter wirelength and fewer buffers than the TSV capacitance increase by using more TSVs. When the supply voltage scales down, the power difference between these two clock networks is reduced.

4.6.7 *Comparison with Existing Work*

We show the comparison of our work with [12] in Table 4.6. Note that [12] does not support buffer insertion or provide any SPICE simulation results. However, we attempted a comparison with [12] by disabling our buffer insertion. We use the same benchmark settings and report the skew and delay values in the Elmore delay model. We observe that our method uses 21.3–33.7 % fewer TSVs than [12] while using 5.2–5.8 % more wirelength. Note that in our work we can control the TSV count versus wirelength tradeoff by tweaking the TSV bound. In addition, these results come from unbuffered clock trees. Our sDMBE algorithm supports buffer insertion, which helps to properly control wire snaking and therefore better minimizes the wirelength.

Table 4.6 Comparisons with [12]

	MMM-3D + ZCTE-3D [12]			Ours		
ckt	#TSVs	WL	Delay	#TSVs	WL	Delay
r_1	83	1,441,849	1.64	55	1,521,459	1.68
r_2	197	2,831,346	4.34	155	2,978,537	4.33
r_3	276	3,725,294	6.37	214	3,918,503	6.51
r_4	653	7,424,886	19.28	510	7,856,725	19.43
r_5	1,052	10,940,984	35.20	811	11,528,598	35.94

4.7 Conclusions

In this chapter, we studied design optimization techniques for reliable low-power and low-slew 3D clock network design. We studied the impact of the TSV count and the TSV capacitance on clock power trends. We observed that using more TSVs helps reduce the wirelength and power consumption and shows better control over clock slew variations. However, in the case of a large TSV parasitic capacitance, clock power could increase if too many TSVs are used. We also observed that a smaller maximum loading capacitance on the clock buffers efficiently lowers the 3D clock slew. Furthermore, we developed a low-power 3D clock tree synthesis algorithm called 3D-MMM-ext. Experimental results show that our 3D-MMM-ext algorithm constructs low-power 3D clock designs that have comparable power and reliability to an exhaustive search with a few orders of magnitude shorter runtime.

References

1. C. Albrecht, A.B. Kahng, B. Liu, I.I. Mandoiu, A.Z. Zelikovsky, On the skew-bounded minimum-buffer routing tree problem. IEEE Trans. Comput. Aided Design Integr. Circuits Syst. **22**(7), 937–945 (2003)
2. V. Arunachalam, W. Burleson, Low-power clock distribution in a multilayer core 3D microprocessor, in *Proceedings of the 18th ACM Great Lakes Symposium on VLSI*, Orlando, 2008, pp. 429–434
3. T. Bandyopadhyay, R. Chatterjee, D. Chung, M. Swaminathan, R. Tummala, Electrical modeling of through silicon and package vias, in *IEEE International Conference on 3D System Integration, 2009. 3DIC 2009*, San Francisco, pp. 1–8
4. K.D. Boese, A.B. Kahng, Zero-skew clock routing trees with minimum wirelength, in *Proceedings of Fifth Annual IEEE International ASIC Conference and Exhibit, 1992*, Rochester, 1992, pp. 17–21
5. W.C. Elmore, The transient analysis of damped linear networks with particular regard to wideband amplifiers. J. Appl. Phys. **19**(1), 55–63 (1948)
6. E.G. Friedman, Clock distribution networks in synchronous digital integratedcircuits. Proc. IEEE **89**(5), 665–692 (2001)
7. S. Hu, C.J. Alpert, J. Hu, S.K. Karandikar, Z. Li, W. Shi, C.N. Sze, Fast algorithms for slew-constrained minimum cost buffering. IEEE Trans. Comput. Aided Design Integr. Circuits Syst. **26**(11), 2009–2022 (2007)
8. GSRC Benchmark, http://vlsicad.ucsd.edu/GSRC/bookshelf/Slots/BST

9. International technology roadmap for semiconductors (ITRS), http://www.itrs.net/
10. M. Jackson, A. Srinivasan, E. Kuh, Clock routing for high-performance ICs, in *Proceedings of ACM Design Automation Conference*, Orlando, 1990, pp. 573–579
11. G. Katti, M. Stucchi, K. De Meyer, W. Dehaene, Electrical modeling and characterization of through silicon via for three-dimensional ICs. Electron Devices IEEE Trans. **57**(1), 256–262 (2010)
12. T.-Y. Kim, T. Kim, Clock tree embedding for 3D ICs, in *Proceedings of Asia and South Pacific Design Automation Conference*, Taipei, 2010, pp. 486–491
13. D.H. Kim, K. Athikulwongse, S.K. Lim, A study of through-silicon-via impact on the 3D stacked IC layout, in *Proceedings IEEE International Conference on Computer-Aided Design*, San Jose, 2009, pp. 674–680
14. J.U. Knickerbocker, P.S. Andry, B. Dang, R.R. Horton, M.J. Interrante, C.S. Patel, R.J. Polastre, K. Sakuma, R. Sirdeshmukh, E.J. Sprogis, S.M. Sri-Jayantha, A.M. Stephens, A.W. Topol, C.K. Tsang, B.C. Webb, S.L. Wright, Three-dimensional silicon integration. IBM J. Res. Dev. **52**(6), 553–569 (2008)
15. J. Minz, X. Zhao, S.K. Lim, Buffered clock tree synthesis for 3D ICs under thermal variations, in *Proceedings of Asia and South Pacific Design Automation Conference*, Seoul, 2008, pp. 504–509
16. V.F. Pavlidis, I. Savidis, E.G. Friedman. Clock distribution networks for 3-D integrated circuits, in *Custom Integrated Circuits Conference, 2008. CICC 2008. IEEE*, San Jose, 2008, pp. 651–654
17. Predictive Technology Model, http://ptm.asu.edu/
18. P.J. Restle, T.G. McNamara, D.A. Webber, P.J. Camporese, K.F. Eng, K.A. Jenkins, D.H. Allen, M.J. Rohn, M.P. Quaranta, D.W. Boerstler, C.J. Alpert, C.A. Carter, R.N. Bailey, J.G. Petrovick, B.L. Krauter, B.D. McCredie, A clock distribution network for microprocessors. IEEE J. Solid-State Circuits **36**(5), 792–799 (2001)
19. I. Savidis, E.G. Friedman, Closed-form expressions of 3-D via resistance, inductance, and capacitance. IEEE Trans. Electron Devices **56**(9), 1873–1881 (2009)
20. G.E. Tellez, M. Sarrafzadeh, Minimal buffer insertion in clock trees with skew and slew rate constraints. IEEE Trans. Comput. Aided Design Integr. Circuits Syst. **16**(4), 333–342 (1997)
21. J. Vardaman, 3-D Through-silicon vias become a reality (2007). http://www.highbeam.com/doc/1G1-164627024.html
22. R. Weerasekera, M. Grange, D. Pamunuwa, H. Tenhunen, L.-R. Zheng, Compact modeling of through-silicon vias (TSVs) in three-dimensional (3-D) integrated circuits, in *IEEE International Conference on 3D System Integration, 2009. 3DIC 2009*, San Francisco, 2009, pp. 1–8
23. S.L. Wright, P.S. Andry, E. Sprogis, B. Dang, R.J. Polastre, Reliability testing of through-silicon vias for high-current 3D applications, in *Proceedings of the 58th Electronic Components and Technology Conference, 2008. ECTC 2008*, Orlando, 2008, pp. 879–883
24. J.-S. Yang, K. Athikulwongse, Y.-J. Lee, S.K. Lim, D.Z. Pan, TSV stress aware timing analysis with applications to 3D-IC layout optimization, in *Proceedings of ACM Design Automation Conference*, Anaheim, 2010, pp. 803–806
25. X. Zhao, S.K. Lim, Power and slew-aware clock network design for through-silicon-via (TSV) based 3D ICs, in *Proceedings of Asia and South Pacific Design Automation Conference*, Taipei, 2010, pp. 175–180
26. X. Zhao, D.L. Lewis, H.H.S. Lee, S.K. Lim, Pre-bond testable low-power clock tree design for 3D stacked ICs, in *Proceedings of IEEE International Conference on Computer-Aided Design*, San Jose, 2009, pp. 184–190
27. X. Zhao, J. Minz, S.K. Lim, Low-power and reliable clock network design for through silicon via based 3D ICs. IEEE Trans. Compon. Packag. Manuf. Technol. **1**(2), 247–259 (2011)
28. Q.K. Zhu, *High-Speed Clock Network Design* (Springer, New York, 2003)

Chapter 5
Power Delivery Network Design for 3D IC

Abstract In this chapter, we first study the impact of P/G TSVs on the power supply noise as well as 3D IC layouts. We perform sign-off static IR-drop analysis on GDSII layouts of 2D and 3D IC designs using commercial-grade tools. We also explore the impact of 3D P/G network topology on IR-drop by varying P/G TSV pitch. Next, we study a non-regular P/G TSV placement algorithm to further reduce the number of P/G TSVs used, while satisfying the given IR-drop noise requirement. Compared with the conventional regular structure, our non-regular P/G TSV placement algorithm reduces the P/G TSV count, wirelength, and footprint area by 59.3, 3.4, and 3.5 % on average, respectively. Next, we study the TSV RC variation impact on 3D power delivery network (PDN). First, we model TSV RC variation due to process variation. Then, we perform sign-off power supply noise analysis of 3D PDN in GDSII layouts which contain power/ground (P/G) TSV RC variation model. We explore the effect of TSV RC variation range, the number of variation sources (P/G TSV count), the number of C4 bumps, and TSV size on the robustness of PDN under TSV RC variation. Our results show that TSV RC variations cause negligible influence on 3D PDN due to much smaller parasitic values of TSVs compared with that of entire PDN.

The materials presented in this chapter are based on [5, 6].

5.1 Introduction

Three-dimensional (3D) system integration has emerged as a promising technology to continue Moore's law beyond the conventional 2D ICs. Through-silicon-via (TSV) is a key enabler for 3D integration, which provides vertical interconnections between stacked dies. Using 3D integration with TSVs, the average and the maximum wirelength between different dies as well as within a same die can

S.K. Lim, *Design for High Performance, Low Power, and Reliable 3D Integrated Circuits*, DOI 10.1007/978-1-4419-9542-1_5,

be significantly reduced compared to 2D ICs. The shorter wirelength implies improvement in performance, power, and footprint area. Although TSVs can improve wirelength, performance and area, they occupy non-negligible silicon area. Excessive or ill-placed TSVs not only increase die area, but also have negative impacts on these objectives in 3D ICs [9].

Power delivery is believed to be one of the biggest challenges in 3D stacked ICs. With the rapid advance of fabrication technology and increase in number of gates in unit chip area, power consumption of a chip increases. As multiple dies are stacked together into a smaller footprint, delivering current to all parts of the 3D stack while meeting the power noise constraints becomes highly challenging. This is mainly because the number of TSVs available for power/ground (P/G) nets is limited, causing severe routing congestion if many 3D connections are desired [10]. In this chapter, we explore IR-drop noise problems in 3D ICs with TSV based on GDSII layouts.

Even though there are many works on 3D power delivery network (PDN), there is none on TSV RC variation impact on power supply noise in 3D PDN, to the best of our knowledge. Process variation is one of the critical issues of semiconductor fabrication, which affects yield, performance, and power consumption. These process variations change the TSV parasitic characteristics, hence affect the quality of PDN. In this chapter, we explore the impact of TSV RC parasitic variation on the robustness of 3D PDN. The following specific topics are covered in this chapter:

- We study the impact of P/G TSVs on the IR-drop noise as well as 3D IC layouts, and traditional layout-related metrics such as wirelength and footprint area. We perform IR-drop noise analysis on 2D and 3D IC GDSII layouts using existing 2D commercial grade tools. We show how to extend 2D commercial tools to handle TSVs and 3D die stacking during layout generation.
- We study the impact of 3D P/G network topology on IR-drop. P/G TSVs are inserted in a regular fashion (conventional 2D-array style) for flip-chip bonding and the pitch of both P/G TSVs and P/G bumps are assumed to be same. We vary P/G TSV pitch from 150 um down to 50 um to examine its impact on IR-drop as well as layouts.
- We study a non-regular P/G TSV placement algorithm to further reduce the number of P/G TSVs used while achieving the given IR-drop noise requirement. The results from both conventional regular and non-regular P/G TSV placement are compared in terms of IR-drop noise, P/G TSV count, and other metrics such as wirelength and footprint area.
- We study TSV RC variation modeling due to process variation. We perform both static (IR-drop) and dynamic noise (voltage droop) analysis on GDSII level 3D IC layouts with TSV RC variation model using existing 2D commercial sign-off level analysis tools.
- We study the impact of number of variation sources (P/G TSV count), number of P/G bumps, and TSV RC variation range on the power supply noise. Also we study the impact of P/G TSV size and its variation on the 3D PDN quality.

5.2 Existing Works

In general, the objective of P/G TSV optimization is to minimize power noise with minimum number of P/G TSVs. Previous works on 3D power delivery networks employed regular P/G TSV placement or optimized the density of P/G TSVs in each P/G tile to meet power noise requirement.

A physical model of 3D power distribution network is presented in [4]. Their model assumed that power is fed from the package through power I/O bumps distributed over the bottom-most die and travels to the upper dies using TSVs and solders. Therefore, P/G TSV locations are predetermined by regularly placed power I/O bumps. Three different TSV topologies for 3D P/G network have been explored in terms of power integrity in [3]: (1) a large single TSV aligned to a C4 bump, (2) multiple TSVs around a C4 bump, (3) and evenly distributed TSVs throughout a die. Again, density and location of P/G TSV were predetermined.

It has been shown that 3D die stacking has a higher impact on IR-drop than Ldi/dt noise [8]. 3D stacking inherently increases the resistance of a 3D P/G network due to P/G TSVs, which directly impacts IR-drop. On the other hand, Ldi/dt noise due to time varying activities in the modules is caused by dominant off-chip inductive components. They also examined the effects of P/G TSV spacing as well as C4 bump spacing on power noise. However, these approaches also assumed regularly placed P/G TSVs with predefined density.

A simultaneous TSV optimization scheme for both power and thermal integrity is proposed in [14]. They first divide each die into N tiles and define possible TSV placement density. Depending upon the power and thermal noise level in each tile, a minimum TSV density pattern is selected. However, this work only considers noise in P/G planes in the package without considering on-chip power supply routing.

5.3 P/G TSV Impact on 3D IC Layouts

Our target 3D structure is illustrated in Fig. 5.1. We assumed adjacent dies are bonded in a face-to-back (F2B) fashion. Depending on their type, via-first TSVs interfere with a device layer, whereas via-last TSVs, which pierce through all stacked dies, interfere with both device and metal layers. In our work, both P/G TSVs and signal TSVs are via-first type assuming only one type of TSV is preferable in a manufacturing process. Thus signal TSVs affect a device layer and top-most and bottom-most metal layer, i.e. M1 and M6 in 130 nm technology using six metal layers. However, P/G TSVs are routed through stacked local vias in each die as shown in Fig. 5.1, hence affect metal layers as well as device layer in a similar way as via-last TSVs. Thus, P/G TSVs cause severe routing congestion if many 3D connections are required. Note that power C4 bumps are connected to P/G TSV landing pads at M6 using redistribution layer (RDL).

Figure 5.2 shows a part of signal net routing result of M5 and M6 for a FFT circuit (256 point and 8-bit precision) using commercial grade tool, i.e. Cadence

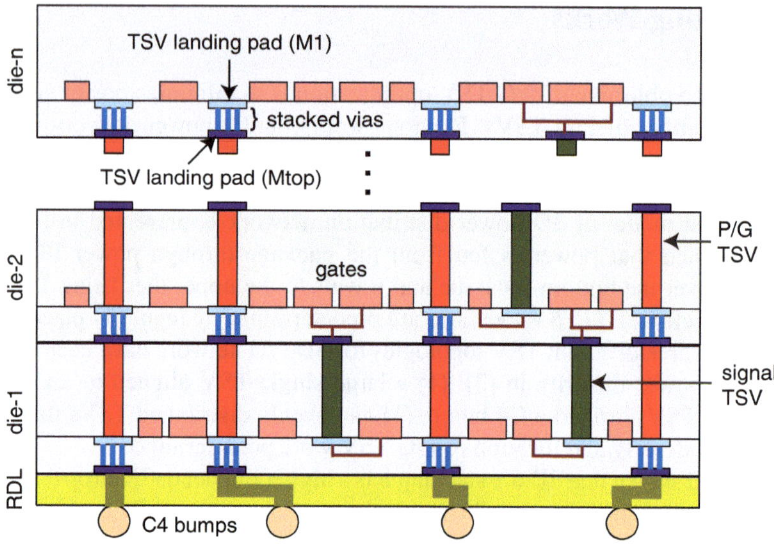

Fig. 5.1 Target 3D structure with via-first TSVs. P/G TSVs are vertically connected with stacked local via arrays

SoC Encounter. We placed P/G TSVs regularly with 50 μm pitch on top of P/G bumps in this design. This figure clearly shows that not only the space that P/G TSVs occupy, but space between P/G TSVs are not fully exploited for signal net routing, hence causes more severe routing congestions than expected. In the case of M6 (yellow), wires are routed in vertical direction, and space between P/G TSVs in vertical direction is not used well since horizontal space between P/G TSVs limits the routing capacity for vertical M6 wires. This phenomenon mostly occurs in higher metal layers which are typically used for long signal net connections. It is possible that regularly placed P/G TSVs make a bottleneck for long wires to route. This might cause wirelength increase and performance degradation. If routing space is not enough, foot print area needs to increase to mitigate routing problems.

P/G net routing for standard cell rows also becomes challenging. Figure 5.3 shows a part of P/G net routing result for the same FFT circuit. As P/G TSV size becomes larger than a standard cell height, it is inevitable that single power (ground) TSV also covers the region that ground (power) nets are supposed to be routed. Therefore, power (ground) nets should detour around ground (power) TSVs to avoid short between power and ground, which is an additional source of routing congestion. Moreover, P/G nets should also avoid short with an M1 landing pad of a signal TSV. Thus, P/G net routing in 3D ICs consumes more routing resources compared to 2D ICs, hence reduces available routing resources for signal net routing.

These observations call for P/G TSV count reduction. In a chip design phase, we can estimate power consumption profiles based on simulations and power library for standard cells and macro blocks. Thus, it is possible to identify the region that consumes more power and demands more current than other regions. These high

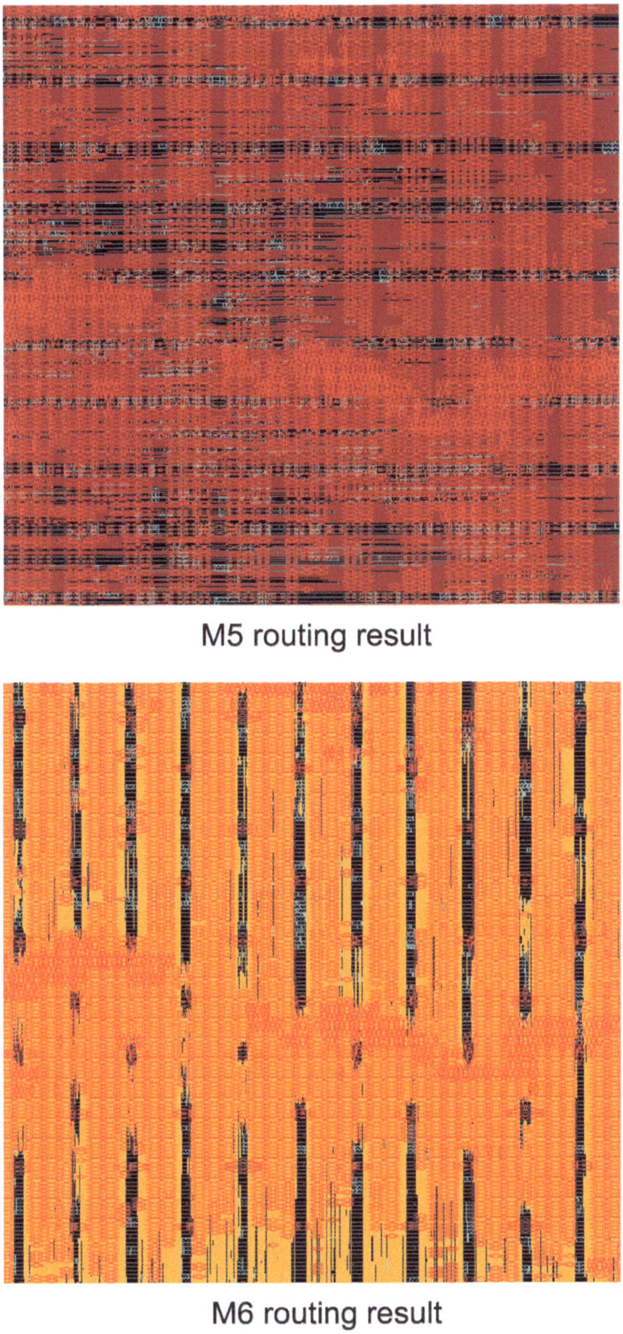

Fig. 5.2 Routing congestion in M5 and M6 due to regularly placed P/G TSVs

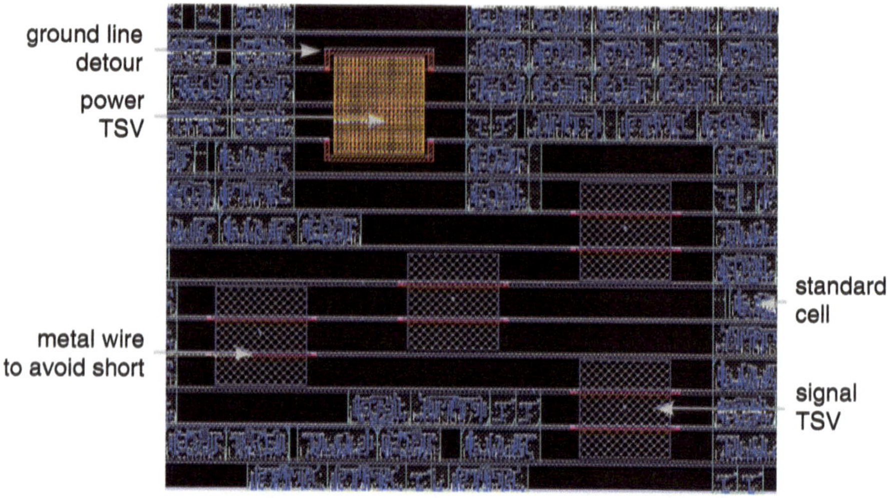

Fig. 5.3 Power distribution network layout. Additional routing resources are needed for P/G net to detour P/G TSVs and signal TSVs

power consuming locations are susceptible to IR-drop noise violation. If IR-drop noise is estimated accurately based on the given power profile, we can identify the power noisy spots. By placing P/G TSVs more in these power noisy spots than other regions, we can efficiently reduce IR-drop noise while using minimum number of P/G TSVs as well as saving routing resources compared to conventional regularly placed P/G TSVs.

5.4 Non-regular Power/Ground TSV Placement Algorithm

A standard cell based design is used in this chapter. P/G rings are routed on the periphery of circuits and P/G stripes which provide power and ground for each standard cell are routed horizontally. Therefore, we can build a series resistor chain along P/G stripes with current sources, which represents standard cells, based on a given circuit layout and a power profile. IR-drop noise estimation is performed on this resistive circuit. To handle large circuits with millions of nodes, we adopt an equivalent circuit modeling method [11, 12]. Our P/G TSV placement algorithm is applied to this simplified P/G network to obtain the optimal P/G TSV locations.

5.4.1 Equivalent Circuit for Series Resistors

Consider a series resistor chain in the P/G network in Fig. 5.4. There will be some voltage V_s, between the two series ends, N_1 and N_n. Conceptually, a voltage source

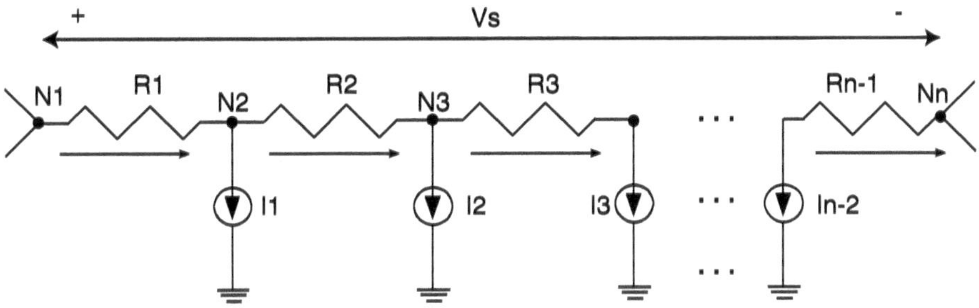

Fig. 5.4 Series resistor chain in P/G network

Fig. 5.5 Series resistor equivalent circuit

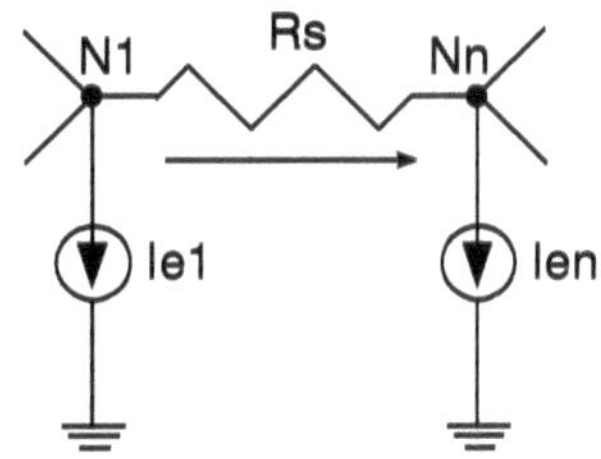

with V_s can be added between the nodes N_1 and N_n without disturbing the network. Suppose the positive current direction for resistive branch R_i is from N_i to N_{i+1}. Superposition can be applied to this network to produce equivalent circuit shown in Fig. 5.5, where the positive current direction of R_s is from N_1 to N_n. The equivalent resistor R_s is just the sum of all the resistors in series.

$$R_s = \sum_{i=1}^{n-1} R_i \tag{5.1}$$

Superposition can be used to determine how the current from each current source divides between the two ends. All current sources except the one in question are replaced by open circuits, while the voltage source between nodes N_1 and N_n is replaced by a short circuit. The resulting system is a simple current divider, and the additional current at N_1 and N_n is sum of all the divided currents. The equivalent current I_{e_1} and I_{e_n} can be calculated as follows [11, 12]:

$$I_{e_1} = \sum_{i=1}^{n-2} \frac{\sum_{j=i+1}^{n-1} R_j}{R_s} I_i \tag{5.2}$$

$$I_{e_n} = \sum_{i=1}^{n-2} \frac{\sum_{j=1}^{i} R_j}{R_s} I_i \tag{5.3}$$

Once the network has been solved with the equivalent series circuit and the voltages at the end nodes are known, the intermediate node voltages and currents are calculated based on superposition as follows:

$$V_{i+1} = V_i - \frac{R_i}{R_s} V_s - R_i I_{e_i} \tag{5.4}$$

$$I_{e_i+1} = I_{e_i} - I_i \tag{5.5}$$

5.4.2 *Equivalent Circuit Modeling for P/G TSV Placement*

The equivalent circuit modeling method was originally developed for fast P/G network simulation [12]. To apply this method for our P/G TSV placement algorithm, we have to take into account the following differences:

1. In a series resistor chain circuit, not all the node voltage information need to be considered for possible P/G TSV placement location. We only need to examine the nodes whose voltages are the local minimum (worst IR-drop) or the local maximum (worst ground bounce). This means that these local maximum and local minimum voltage nodes cannot be suppressed, since these nodes are required to maintain the node voltage information. Thus, other nodes except these local minimum and local maximum can be simplified with two current sources and one resistor as shown in Fig. 5.5. Since each power (ground) stripe in a design will have a single local minimum (maximum) node, each P/G stripe will be divided into two sub-chain circuits. Currents will flow in the same direction in each sub-chain and we only need to consider the voltages at two ends.
2. When we insert P/G TSVs in the worst IR-drop node, the current flow direction is changed in affected P/G stripes. Even though the total current demands for these P/G nets are unchanged, current direction is altered with additional current flow from P/G TSVs. Therefore, the local minimum or the local maximum nodes in these P/G stripes are changed; hence node voltages in these stripes should be updated.
3. Inserting P/G TSVs in some P/G stripes not only changes node voltages at these P/G stripes, but also alters the boundary voltages and currents at the ends of these P/G stripes, which will affect the boundary voltages and currents for other P/G stripes. Figure 5.6 shows a simple example. Before inserting a power TSV, currents are flowing inwards to the worst IR-drop node. After inserting the power TSV, the amount of current flowing from both ends will be reduced depending upon the amount of current that power TSV provides. This will change the amount that outer power ring supplies. Hence IR-drop through the power ring and the boundary voltages and currents in adjacent power stripes will be changed as well.

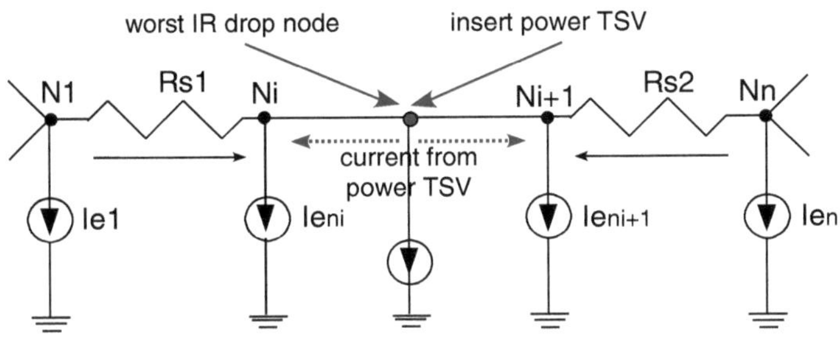

Fig. 5.6 Boundary voltage and current change due to P/G TSV insertion

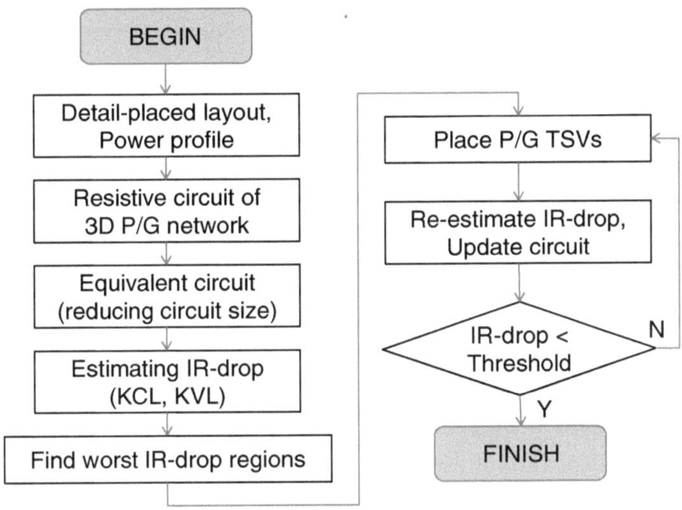

Fig. 5.7 Flow chart of non-regular P/G TSV placement algorithm

5.4.3 Non-regular P/G TSV Placement Algorithm

We construct P/G resistive network based on the detailed cell placement results along with a power profile for each stacked die. Then, we calculate IR-drop noise based on Kirchhoff's voltage law (KVL) and current law (KCL), and identify the local minimum and the local maximum node voltages for power and ground net respectively. With this information, we build the simplified circuit to handle large size circuits efficiently using the equivalent circuit model. Next, P/G TSVs are inserted where IR-drop constraint is violated, and IR-drop is re-evaluated using equivalent circuit model. If the current design with P/G TSVs inserted meets the target IR-drop threshold, the P/G TSV placement algorithms finishes. If not, we insert additional P/G TSVs to the current IR-drop violating regions. The flow chart of P/G TSV placement algorithm is shown in Fig. 5.7.

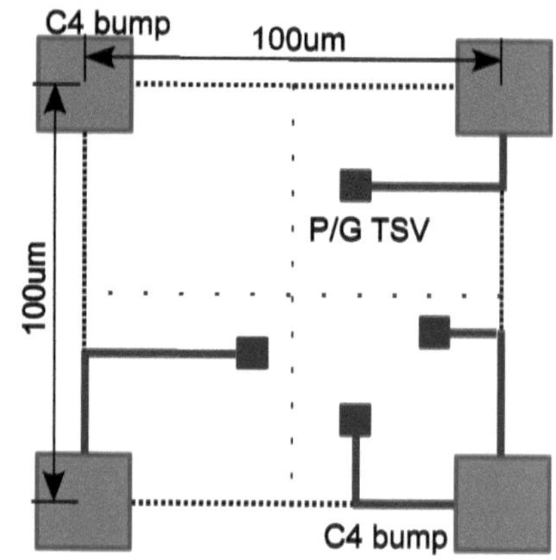

Fig. 5.8 P/G TSVs are routed to the nearest C4 bump using RDL

The algorithm first constructs 2D P/G network for each die separately. Based on a given power profile, the amount of current that each gate requires is computed, in our case power consumption values of each gate is divided by the nominal power supply voltage 1.5 V, and then every gate is replaced by a corresponding current source. Wire resistance between adjacent gates is calculated by their relative distance and the resistivity of that wire segment. We first assume that P/G bumps are available above P/G rings on the periphery with a predetermined pitch. P/G TSVs are inserted for these locations for 3D P/G connections, which do not affect routing resources in a core region. Then, we calculate the total current demand for each P/G net, and compute boundary voltages and currents at both ends of P/G nets. When we compute the vertical current flow between dies and IR-drop through P/G TSVs, it is assumed that P/G TSVs supply current to the nearest P/G nets. Based on these boundary conditions, each node voltage in 3D P/G nets is estimated by using KVL and KCL. Then, we can identify the worst IR-drop nodes and simplify circuits using equivalent circuit model.

Based on the IR-drop estimation, we insert P/G TSVs where IR-drop exceeds the constraint. Since we perform our algorithm based on a detailed placement result, it is desirable to minimize the change to a design layout. Since signal TSVs are already placed at the locations to improve wirelength and timing objective, we avoid overlaps between signal TSVs and P/G TSVs. If an optimal P/G TSV location overlaps with gates, we move these gates to nearest white spaces using a Cadence SoC Encounter's place refinement. We also consider IR-drop from a C4 bump to a P/G TSV. We assume P/G redistribution layer (RDL), where C4 bumps and P/G TSVs are routed, exists using the top-most metal layer. We further assume that a P/G TSV is routed to the nearest C4 bump only, and that each C4 bump is an ideal voltage source shown in Fig. 5.8.

With the newly inserted P/G TSVs, node voltages that are attached to P/G TSVs are evaluated again and the simplified circuit is updated. Then, worst IR-drop nodes are re-computed based on new boundary conditions. At this phase, we use the simplified circuit to reduce the computational time. Then, we check whether

IR-drop constraint is met for all stacked dies. If the first run is unsuccessful, we insert P/G TSVs at the current worst IR-drop nodes that violate IR-drop constraint, and recalculate IR-drop. We perform iterative P/G TSV insertions until IR-drop constraint is achieved for all dies.

5.4.4 *Validation Methodology*

To validate our IR-drop estimation algorithm, we compare the results for both 2D and 3D IC GDSII layouts using existing 2D commercial-grade tools. Our 3D IR-drop analysis tool is based on Cadence VoltageStorm, which is designed for 2D ICs. We take the following steps to handle 3D designs using VoltageStorm:

1. We modify the interconnect technology file (ICT), which contains information on all layers (device layer, dielectric layer, metal layer, vias, and TSVs) and their relative position and resistance values, to model our two die–stacked configuration.
2. We create a 3D technology file (TCH), which contains resistive and capacitive information for all metal and via layers using Cadence Techgen.
3. We generate a 3D library exchange format (LEF) file so that layers and gates in different dies can be distinguished by the tools. For example, M1 in top-most die and M1 in bottom-most die should be differentiated so that 2D tools distinguish these M1 layers.
4. Power consumption data and layer mapping files, which maps design to appropriate LEF and GDS layers, are modified as well to be used for different dies accordingly.
5. Finally, we create a 3D design exchange format (DEF) file from the final layout of each die to form a single 3D design. After all these preparations are ready, we can run 3D IR-drop analysis on this combined design using Cadence VoltageStorm.

Using our IR-drop estimation method, we were able to match both 2D and the two die-stacked 3D IR-drop results from the VoltageStorm within 7 % error. Our computed resistance value based on the ICT file for each P/G wire segment overestimated by 6 % compared to VoltageStorm. Since this is deterministic error, resistivity value is tuned to match the results. Due to the tools limitation on number of layers it can process, we validated our algorithm up to two die-stacked 3D ICs.

5.5 Power/Ground TSV Placement Results

The proposed non-regular P/G TSV placement algorithm has been implemented in C programming language. The experiments were performed on a 64-bit Linux server with two quadcore Intel Xeon 2.5 GHz CPUs and 16 GB main memory. We use four FFT circuits for our analysis. All circuits are synthesized using Synopsis Design

Table 5.1 Benchmark circuits

Circuit	# gates (K)	# signal TSVs	TSV area (%)	Placement density (%)	Clock freq (MHz)	Profile
FFT1	200	794	8.9	77.4	200	256 point 8 bit precision
FFT2	405	1,397	8.2	77.3	142	256 point 16 bit precision
FFT3	910	7,089	16.3	77.2	111	512 point 16 bit precision

Table 5.2 TSV specifications

Item	Value
TSV Diameter (μm)	6
TSV Landing pad (μm)	10
TSV Keep out zone (μm)	6.225
TSV Cell size (μm × μm)	18.45 × 18.45
TSV Height (μm)	30
TSV Resistance (mΩ)	30

Compiler with the physical library for the target 130 nm technology, and designed using Cadence SoC Encounter to 2D and two die-stacked 3D ICs, which are listed in Table 5.1. The number of signal TSVs was chosen for signal TSVs to cover around 10 % of the chip area, and the overall placement density including both standard cells and TSVs is targeted to 80 %. Experimental settings for the TSV used in our simulations are shown in Table 5.2, which are similar to the data of manufactured TSVs in [1]. Note that the TSV size is large, which occupies five standard cell rows in this setting (a standard cell row height is 3.69 μm).

5.5.1 *IR-Drop Analysis Results for 2D and 3D Designs*

We first compare IR-drop analysis results between our 2D and 3D designs. We assume that I/O cells are located on the periphery of a chip, and I/O cell pitch is 100 μm. We further assume that 50 % of all I/O cells are dedicated to power and ground connections. In this section, P/G TSVs are inserted only on the periphery of a chip where I/O cells locate. In our 3D designs, all I/O cells are placed in a bottom die nearest to C4 bumps. P/G C4 bumps are connected to a bottom die through RDL, then P/G TSVs are used to deliver power from a bottom die to a top die.

I/O cell density is kept the same for both 2D and 3D designs for fair comparison, and each P/G I/O cell contains 5 P/G TSVs, respectively. The clock frequency for power simulation is set by the slower clock frequency between 2D and 3D designs. Then, statistical power analysis is performed with 0.2 toggle probability for all nets excluding clock nets.

Tables 5.3 and 5.4 show IR-drop analysis results of our 2D and 3D designs. We make the following observations. First, the footprint area of 3D is larger than 50 % of 2D footprint area mainly because of the large TSV. Second, the total wirelength in 3D design is larger than 2D, again mainly because of the large

Table 5.3 IR-drop, footprint area, wirelength, and power consumption for 2D designs

Ckt	Area (μm × μm)	WL (mm)	Power (mW)	IR-drop (mV)	# P/G IO cell
FFT1	2,271 × 2,271	13,018	845	178	42
FFT2	3,243 × 3,243	32,197	1,240	217	62
FFT3	4,754 × 4,754	67,495	2,120	304	94

Table 5.4 IR-drop, footprint area, wirelength, and power consumption for 3D designs. The numbers in parentheses are ratios to 2D

Ckt	Area (μm × μm)	WL (mm)	Power (mW)	IR-drop(mV) (top/bot)	# P/G IO cell / P/G TSV
FFT1	1,744 × 1,744 (0.59)	14,780 (1.14)	839 (0.99)	175/157	34/170
FFT2	2,412 × 2,412 (0.55)	35,772 (1.11)	1,246 (1.00)	226/200	46/230
FFT3	3,851 × 3,851 (0.65)	85,405 (1.27)	2,162 (1.02)	314/310	74/370

Table 5.5 Number of P/G TSVs used with different pitch. The number of P/G TSVs in core/periphery is shown separately

Ckt	Peri	150 μm	125 μm	100 μm	75 μm	50 μm
FFT1	0/170	221/170	313/170	481/170	1,013/170	2,113/190
FFT2	0/230	481/230	685/230	1,013/230	2,113/230	5,725/270
FFT3	0/370	1,201/370	1,741/370	2,665/370	5,613/390	14,965/430

TSV size [9]. In addition, we still use Cadence 2D placement and routing tools for each die separately, which gave us sub-optimal 3D layout quality. Third, the power consumption is comparable in both 2D and 3D designs. The combined effect of smaller footprint area and larger total wirelength of 3D leads to this comparable power consumption in both 2D and 3D designs. Lastly, the IR-drop is worse in 3D designs compared to 2D designs as circuit size becomes larger. This IR-drop increase in large 3D designs is mainly because the on-chip power delivery resource (P/G I/O cells and P/G TSVs) is less in 3D designs.

5.5.2 *Impact of 3D P/G Network Topology on IR-Drop*

The results shown in the Sect. 5.5.1 indicate that inserting P/G TSVs only on the periphery of a chip is not sufficient to alleviate IR-drop noise problems in 3D ICs. In this section, P/G TSVs are inserted in a conventional 2D-array fashion for flip-chip bonding. We also vary the P/G TSV pitch from 150 μm down to 50 μm to examine its impact on IR-drop. Most commercial products today have C4 bump pitches around 100–200 μm, however, researchers have demonstrated micro-bumps with pitches below 10 μm [3]. In this experiment, the pitch of both P/G TSVs and P/G bumps are kept same, which is an ideal case for IR-drop noise reduction. The number of P/G TSVs used for different TSV pitch is shown in Table 5.5.

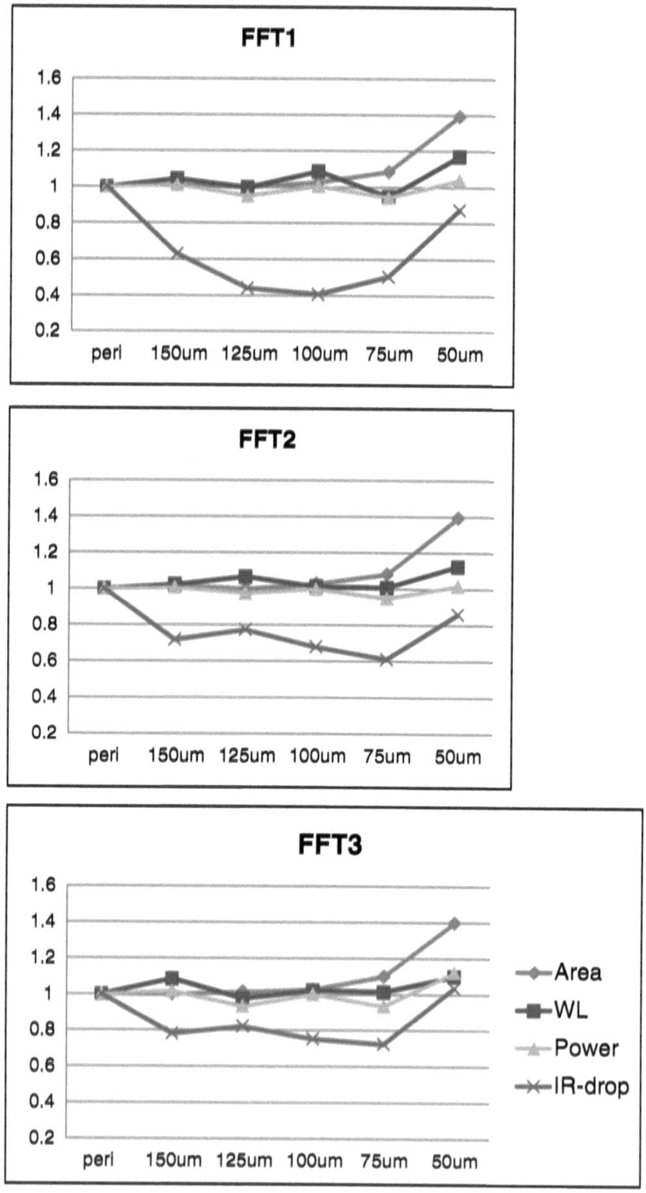

Fig. 5.9 Impact of P/G TSV pitch on IR-drop, footprint area, wirelength, and power

Figure 5.9 shows the impact of P/G TSV pitch on IR-drop, power, wirelength, and footprint area. The results are normalized to the baseline 3D design that has P/G TSVs only on the periphery. As the P/G TSV pitch decreases and the available

Table 5.6 Regular P/G TSV (100 μm pitch) placement results

Ckt	# P/G TSV (core/peri)	Area (μm × μm)	WL (mm)	IR-drop (mV)
FFT1	481/170	1,776 × 1,776	16,056	71
FFT2	1,013/230	2,444 × 2,444	36,240	153
FFT3	2,665/370	3,902 × 3,902	87,289	236

Table 5.7 Our non-regular P/G TSV placement results

Ckt	# P/G TSV (core/peri)	Area (μm × μm)	WL (mm)	IR-drop (mV)	IR-drop % error	Runtime (s)
FFT1	160/170 (49.4 % ↓)	1,776 × 1,776 (2.6 % ↓)	15,028 (6.4 % ↓)	75	5.6	2.63
FFT2	266/230 (60.0 % ↓)	2,444 × 2,444 (3.5 % ↓)	35,152 (3.1 % ↓)	148	3.3	6.76
FFT3	592/370 (68.4 % ↓)	3,902 × 3,902 (4.2 % ↓)	86,507 (1.0 % ↓)	251	6.4	16.44

P/G TSVs increase, IR-drop significantly improves up to 100 or 75 μm pitch depending on benchmark circuits. With 100 μm P/G TSV pitch, IR-drop improves 60 % compared to the baseline design for FFT1.

However, counter-intuitively, as we further decrease the P/G TSV pitch to 50 μm IR-drop noise increases. FFT3 shows even worse IR-drop noise than the baseline design. This is mainly because of the huge TSV size. In our experimental setting, 100 minimum-size inverters can fit into a single TSV area. Previous works did not consider TSV size, thus circuit layouts did not change even though more P/G TSVs are inserted. However, as more P/G TSVs are inserted, more spaces are needed for these P/G TSVs which is shown in Fig. 5.9. As a result, wirelength and power consumption increases as well. These results show that finer P/G TSV pitch in 3D P/G network does not always result in IR-drop noise improvement.

5.5.3 *Non-regular P/G TSV Placement Algorithm*

The results shown in the Sect. 5.5.2 indicate that if too many P/G TSVs are used, IR-drop noise does not improve due to increased area and wirelength. In this section, we compare IR-drop analysis results between regularly placed P/G TSVs and our algorithm on two die-stacked 3D designs. We set the 3D design whose P/G TSVs are regularly placed with 100 μm pitch as a baseline, since 100 μm pitch case achieves high IR-drop improvement with less than 3 % area overhead. We also set the IR-drop constraint for our algorithm to be the worst IR-drop noise of the baseline. With the P/G TSV locations obtained from our algorithm, we design two die-stacked 3D ICs and validate IR-drop noise using Cadence VoltageStorm.

Tables 5.6 and 5.7 show that our algorithm achieves similar IR-drop results within 7 % error, with much smaller number of P/G TSVs compared to the baseline 3D designs. We were able to save the number of P/G TSVs by 59.3 % on average. For instance, FFT3 shows 68.4 % reduction in P/G TSV count. It is observed that as circuit size becomes larger, P/G TSV count reduction percentage increases.

It is possible that conventional 2D array style P/G TSV placement scheme uses more P/G TSVs than necessary, especially in non-power-noisy spots. If an accurate power profile is available in a design phase, the number of TSVs for 3D P/G network can be reduced significantly. With the reduced number of P/G TSVs, footprint area and total wirelength are reduced by 3.4 and 3.5 % on average, respectively.

5.6 TSV RC Variation

Process variations on TSVs are inevitable due to factors such as misalignment, TSV diameter/height and oxide thickness variation, wafer surface cleanliness and roughness. However, extreme misalignment, which causes systematic variations and increases contact resistance, is highly unlikely in state-of-the-art wafer bonding processes [13]. Thus, TSV RC parasitic variation can be modeled as random effects. In this section, we model TSV RC variation based on TSV dimension variation using analytical models. We ignore TSV inductance since inductive voltage drop by TSV is comparable only for frequencies above several GHz [7], which is not the case for PDN.

5.6.1 TSV Resistance Variation

The analytical expression of the dc resistance of TSV is given by

$$R_{TSV} = \frac{\rho l_{TSV}}{\pi r_{TSV}^2} \tag{5.6}$$

where ρ is the resistivity of conducting material, and r_{TSV} and l_{TSV} represent the radius and height of TSV, respectively. With Cu TSV conductor, the resistivity is 16.8 nΩ·m at 20 °C. Also, we adopt a contact resistivity of $0.45\,\Omega \cdot \mu\text{m}^2$ from measured data based on Cu direct bonding [2]. Thus, total TSV resistance is sum of TSV dc resistance and contact resistance.

We use a TSV with 5 μm diameter, 30 μm height, and 120 nm oxide thickness as a baseline TSV structure. Then, we vary both diameter and height by $\pm 10\,\%$ of nominal values to model process variation. TSV diameter shows super-linear relationship with TSV resistance while TSV height has linear dependency. With fixed TSV height of 30 μm and $\pm 10\,\%$ of TSV diameter variation, TSV resistance changes from -13.6 to $+19.3\,\%$ of the nominal TSV resistance.

5.6.2 TSV Capacitance Variation

The nature of the TSV *C-V* characteristics is similar to the planar MOS capacitor such that accumulation capacitance is the oxide capacitance given as [7]

$$C_{TSVacc} = C_{ox} = \frac{2\pi\varepsilon_{ox} l_{TSV}}{ln(\frac{t_{ox}+r_{TSV}}{r_{TSV}})} \quad (5.7)$$

As the TSV bias increases, the depletion capacitance acts in series with the oxide capacitance, which is given by

$$C_{TSVdep} = \frac{2\pi\varepsilon_{si} l_{TSV}}{ln(\frac{t_{ox}+r_{TSV}+d_{dep}}{t_{ox}+r_{TSV}})} \quad (5.8)$$

where t_{ox} is the TSV oxide thickness and d_{dep} is the depletion width in silicon substrate. We assume that substrate doping is $2 \times 10^{15}/\text{cm}^3$.

The effective TSV capacitance is the series combination of oxide and depletion capacitances given by

$$C_{TSV} = \frac{C_{ox} C_{TSVdep}}{C_{ox} + C_{TSVdep}} \quad (5.9)$$

The accumulation and depletion capacitance formulae show that C_{TSV} is proportional to the TSV height and inversely proportional to the TSV oxide thickness. Also, d_{dep} is inversely proportional to substrate doping concentration. We assume that substrate doping is $2 \times 10^{15}/\text{cm}^3$.

Even though TSV height has direct impact on capacitance value, its percent deviation from the nominal value will be smaller compared with TSV diameter and oxide thickness. This is mainly because TSV height is several times larger than TSV diameter. Thus, we focus on the impact of TSV diameter and oxide thickness variation on TSV capacitance variation. We vary both TSV diameter and oxide thickness by $\pm 10\,\%$ of nominal values. Our related experiments show that impact of oxide thickness variation on TSV capacitance is negligible. With fixed oxide thickness and $\pm 10\,\%$ of TSV diameter variation, TSV capacitance changes from -8.6 to $+8.6\,\%$ of an original TSV capacitance.

From this TSV RC variation modeling, we see that TSV resistance is more susceptible to TSV variation than TSV capacitance. This indicates possible variation of static noise, hence degrades the robustness of PDN in 3D ICs. We show TSV RC variation impact on power supply noise in full-chip scale in Sect. 5.7.

5.6.3 Validation Methodology

In this chapter, we build a sign-off level 3D power supply noise analysis flow using existing 2D commercial tools and our own scripts. First, we extract RC parasitics of P/G net (SPEF) of each die separately using StarRC. We also model the coupling capacitance between top metal wires and the substrate of an adjacent die by performing capacitance extraction on 3D structures using Q3D extractor and scale capacitance values for these top metal P/G wires in SPEF file accordingly,

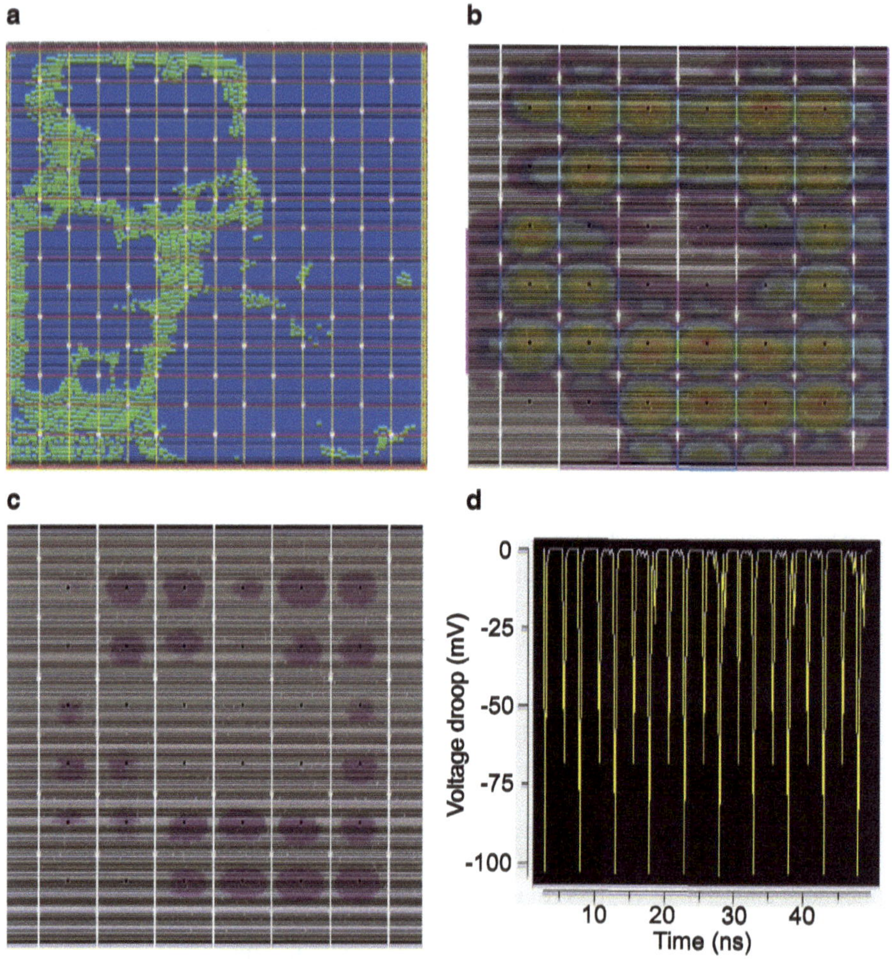

Fig. 5.10 Layout and power supply noise maps in the bottom die of ind1. (**a**) Layout. *White rectangles* are P/G TSVs and *green rectangles* are signal TSVs. (**b**) Dynamic noise map. *Red color* represents noisy spots. (**c**) Static noise map. (**d**) Transient voltage waveform of noisy region

which is not supported by 2D parasitic extraction tools. Then, we use our in-house tool to merge P/G nets from multiple dies to build a single PDN and insert P/G TSV parasitics, which follow given normal distribution, between adjacent dies.

We also modify design and interconnect technology files for an individual die to create a unified 3D design. Next, we generate pseudo-random input vectors which obey a given switching activity. With 3D design files and input vectors, we obtain current wave form of all signal nets in the 3D design using NanoSim. Once these files are ready, we use PrimeRail to perform both static and dynamic noise analysis of 3D PDN. Due to the limitation on the number of layers these tools can process, we restrict our simulations to two-die stacked 3D ICs. Figure 5.10 shows the layout and power supply noise maps from our analysis flow.

5.7 Variation Analysis Results

We use three industrial circuits for our analysis. All circuits are designed with 45 nm technology to two-die stacked 3D ICs, which are listed in Table 5.8. We use top two metal layers to construct regularly distributed power grid for each 2D tier and insert P/G TSV at each grid node shown in Fig. 5.10a. We deploy decaps uniformly across the die with 0.5 fF/um^2 density and worst dynamic noise is reduced by 53 mV for the circuit ind1 in a deterministic simulation, for example. We assume that P/G TSVs are connected directly to C4 bumps. The power to power pitch of TSV, C4 bump, and grid is 100 μm in all our designs. To model C4 bump and package impedance, we use 5 mΩ resistance, 500 pH inductance, and a parallel 30 fF capacitance for each C4 bump. We generate current waveforms using 500 MHz clock frequency, with a switching activity of 0.2 for input signals. We use a TSV with 5 μm diameter, 50 μm height, and 120 nm thick oxide, unless specified. Finally, we perform more than 400 Monte Carlo power supply noise simulations for every case.

5.7.1 *Impact of TSV RC Variation Range*

In this section, we study the impact of TSV RC variation range on power supply noise in 3D PDN. We generate TSV RC parasitics which follow the normal distribution with standard deviation of 10, 20, or 30 % of a nominal value. Table 5.9 shows that for both static and dynamic noise case, there are negligible variations

Table 5.8 Benchmark circuits

Circuit	# gates	area (μm^2)	# signal TSVs	# P/G TSVs	% area by P/G TSVs	% area by decap
ind1	355 K	810^2	1,632	85	0.91	5.32
ind2	1.16 M	1,850^2	14,957	613	1.26	5.32
ind3	2.52 M	2,822^2	22,413	1,405	1.24	5.32

Table 5.9 Impact of TSV RC variation range, where the nominal TSV capacitance and resistance values are 35.5 fF and 45.6 mΩ, respectively. Noise values are reported in mV. The σ/m ratio is shown in %

	TSV resistance variation											
	10 %				20 %				30 %			
	Static noise		Dynamic noise		Static noise		Dynamic noise		Static noise		Dynamic noise	
TSV cap variation(%)	Mean	σ/m	Mean	σ/m	Mean	σ/m	Mean	σ/m	Mean	σ/m	Mean	σ/m
10	10.6	0	133.1	0.00014	10.6	0	133.1	0.00022	10.6	0	133.1	0.00030
20	10.6	0	133.1	0	10.6	0	133.1	0.00030	10.6	0	133.1	0.00024
30	10.6	0	133.1	0.00015	10.6	0	133.1	0.00022	10.6	0	133.1	0.00029

Table 5.10 Impact of number of variation sources

		TSV R $\sim N(45.6, (0.3 \times 45.6)^2)$ (unit: mΩ) TSV C $\sim N(35.5, (0.3 \times 35.5)^2)$ (unit: fF)			
	# P/G	Static noise (mV)		Dynamic noise (mV)	
Circuit	TSVs	Mean	% σ/m	Mean	% σ/m
ind1	85	10.562	0	133.055	0.00029
ind2	613	8.794	0	104.435	0
ind3	1,405	9.234	0	121.042	0.00031

Table 5.11 Impact of number of P/G C4 bumps

	TSV R $\sim N(45.6, (0.3 \times 45.6)^2)$ (unit: mΩ) TSV C $\sim N(35.5, (0.3 \times 35.5)^2)$ (unit: fF)			
	Static noise (mV)		Dynamic noise (mV)	
# bumps	Mean	% σ/m	Mean	% σ/m
21	25.948	0.058	242.897	0.070
42	12.630	0.029	133.864	0.001
85	10.56	0	133.05	0.0002

from mean values for given RC variation range. This is because TSV RC parasitic is much smaller compared with parasitics from 2D P/G grid, decaps, and C4 bumps, hence TSV RC parasitic variation does not affect the quality of 3D PDN.

5.7.2 Impact of Number of Variation Sources

Since TSV RC variation range shows negligible effect on power supply noise, we explore the impact of the number of variation sources, i.e. P/G TSV count, to see if increasing number of variation sources worsen the robustness of PDN. We keep P/G TSV pitch, C4 bump pitch, and P/G grid density same for all circuits to study the impact of P/G TSV count only. Table 5.10 shows that the number of variation sources poses negligible effects on power supply noise. This is again parasitics of P/G TSVs are much smaller than those of 2D P/G grid.

5.7.3 Impact of Number of C4 Bumps

It is known that P/G C4 bump pitch is a critical factor for 3D PDN quality. In this section, we vary the number of C4 bumps with fixed P/G TSV count and location to explore the its impact on power supply noise variation. We use ind1 circuit for this experiment. Table 5.11 shows that both mean and standard deviation of power supply noise increase with decreasing number of bumps. This is because P/G TSVs that are directly connected to bumps carry more current than others, hence more susceptible to TSV RC variation. However, power supply noise variation is still negligible with different number of C4 bumps.

Table 5.12 Impact of TSV size. Aspect ratio is 10 for all TSVs

TSV diameter (μm)	Area (μm²)	# P/G TSVs	TSV RC ~ $N(m,(0.3m)^2)$ Static noise (mV) Mean	% σ/m	Dynamic noise (mV) Mean	% σ/m
2.5	$1{,}610^2$	421	11.762	0.00177	166.166	0
5	$1{,}850^2$	613	8.794	0	104.435	0
10	$2{,}430^2$	1,005	10.193	0.00094	203.645	0.00122

Table 5.13 Impact of TSV parasitics. Aspect ratio is 10 for all TSVs

TSV diameter (μm)	TSV RC ~ $N(m,(0.3m)^2)$ (unit: mΩ, fF) Static noise (mV) Mean	% σ/m	Dynamic noise (mV) Mean	% σ/m
2.5	8.794	0.00162	104.435	0
5	8.794	0	104.435	0
10	8.794	0.00052	104.435	0.00013

5.7.4 Impact of TSV Size

So far, we used TSV diameter of 5 μm for all experiments. In this section, we study the impact of TSV size on power supply noise variation. We use TSV diameter of 2.5, 5, and 10 μm with an aspect ratio of 10. We also redesign the circuit ind2 with different TSV size, since TSV size can affect layout quality significantly. Table 5.12 shows that increasing TSV size expands footprint area significantly and accommodates more P/G TSVs if we keep P/G TSV pitch same. Even though mean values of both static and dynamic noise change due to TSV size, variation of power supply noise is still negligible.

To further study the impact of P/G TSV parasitics of different TSV size, now we only vary P/G TSV parasitics with the same layout. In this case, only P/G TSV RC values change while parasitics of 2D P/G grid remain same. Table 5.13 shows that P/G TSV parasitic change does not affect both mean and standard deviation. This is because nominal values of these TSV RC parasitics are already small, i.e. 21 ~ 86 mΩ and 10 ~ 132 fF, hence the effect of different P/G TSV parasitics on the 3D PDN is not noticeable.

It is natural that fabrication technology will enable TSV to scale down to increase TSV density. However, smaller TSV size might cause more problems in PDN due to increased TSV resistance and its variation. To evaluate the impact of TSV RC parasitics of future nano-scale TSVs, we perform power supply noise analysis with the circuit ind1. We keep TSV height 15 μm for all cases. Table 5.14 shows that variation of power supply noise is still small compared with mean values, even though the magnitude of standard deviation increases. More serious problem is high noise level itself in nano-scale TSVs, not the variation in power supply noise.

Table 5.14 Impact of parasitics of nano-scale TSVs

TSV diameter (μm)	TSV RC variation range	Static noise (mV) Mean	Static noise (mV) % σ/m	Dynamic noise (mV) Mean	Dynamic noise (mV) % σ/m
	$\sigma = 0.1\,m$	10.562	0	133.053	0
1	$\sigma = 0.2\,m$	10.562	0	133.053	0
	$\sigma = 0.3\,m$	10.562	0	133.053	0
	$\sigma = 0.1\,m$	14.706	0.08822	154.170	0.07671
0.5	$\sigma = 0.2\,m$	14.705	0.11981	154.163	0.10521
	$\sigma = 0.3\,m$	14.701	0.17602	154.124	0.15435
	$\sigma = 0.1\,m$	157.872	0.24234	847.569	0.22812
0.1	$\sigma = 0.2\,m$	157.877	0.32161	847.942	0.31513
	$\sigma = 0.3\,m$	157.877	0.43053	847.568	0.42943

5.8 Conclusions

In this chapter, we studied the impact of P/G TSVs on IR-drop noise. Due to the large size of TSVs, both signal and P/G net routing in 3D IC becomes challenging. Experimental results showed that increasing the number of P/G TSV beyond a certain level could worsen IR-drop noise because of a large TSV size. A non-regular P/G TSV placement algorithm is proposed to minimize the number of P/G TSVs used, while satisfying IR-drop noise constraint. Experimental results showed that our non-regular P/G TSV placement algorithm reduces the number of P/G TSVs by 59.3 % as well as footprint area and wirelength compared to the conventional regular P/G TSV placement scheme.

We also explored the effect of TSV RC variation on the robustness of two-die stacked 3D PDN under TSV RC variation. We learned that TSV RC variations cause negligible influence on both static and dynamic noise in 3D PDN due to much smaller RC parasitic values of TSVs compared with that of entire PDN.

References

1. G.V. der Plas et al., Design issues and considerations for low-cost 3D TSV IC technology, in *ISSCC on Digest Technical Papers*, San Francisco, 2010
2. P. Enquist, G. Fountain, C. Petteway, A. Hollingsworth, H. Grady, Low cost of ownership scalable copper direct bond interconnect 3D IC technology for three dimensional integrated circuit applications, in *IEEE International 3D System Integration Conference*, San Francisco, 2009
3. M.B. Healy, S.K. Lim, Power delivery system architecture for many-tier 3D systems, in *IEEE Electronic Components and Technology Conference*, Las Vegas, 2010
4. G. Huang, M. Bakir, A. Naeemi, H. Chen, J.D. Meindl, Power delivery for 3D chips stacks: physical modeling and design implication, in *Proceedings of the IEEE Electrical Performance of Electronic Packaging*, Atlanta, 2007, pp. 205–208

5. M. Jung, S.K. Lim, A study of IR-drop noise issues in 3D ICs with through-silicon-vias, in *IEEE International 3D System Integration Conference*, Münich, 2010
6. M. Jung, S. Panth, S.K. Lim, A study of TSV variation impact on power supply noise, in *IEEE International Interconnect Technology Conference*, San Jose, 2011
7. G. Katti, M. Stucchi, K.D. Meyer, W. Dehaene, Electrical modeling and characterization of through silicon via for three-dimensional ICs. IEEE Trans. Electron Device **57**, 256–262 (2010)
8. N.H. Khan, S.M. Alam, S. Hassoun, System-level comparison of power delivery design for 2D and 3D ICs, in *IEEE Electronic Components and Technology Conference*, San Diego, 2009
9. D.H. Kim, K. Athikulwongse, S.K. Lim, A study of through-silicon-via impact on the 3D stacked IC layout, in *Proceedings of the IEEE International Conference on Computer-Aided Design*, San Jose, 2009
10. Y.-J. Lee, M. Healy, S.K. Lim, Co-design of reliable signal and power interconnects in 3D stacked ICs, in *Proceedings of the IEEE International Interconnect Technology Conference*, Sapporo, 2009
11. D. Stark, M. Horowitz, Techniques for calculating currents and voltages in VLSI power supply networks. IEEE Trans. Comput. Aided Des. Integr. Circuits Syst. **9**, 126–132 (1990)
12. X.D.S. Tan, C.J.R. Shi, Fast power/ground network optimization based on equivalent circuit modeling, in *Proceedings of the ACM Design Automation Conference*, Las Vegas, 2001
13. A.W. Topol, J.D. C.L. Tulipe, L. Shi, D.J. Frank, K. Bernstein, S.E. Steen, A. Kumar, G.U. Singco, A.M. Young, K.W. Guarini, M. Ieong, Three-dimensional integrated circuits. IBM J. Res. Dev. **50**, 491–506 (2006)
14. H. Yu, J. Ho, L. He, Simultaneous power and thermal integrity driven via stapling in 3D ICs, in *Proceedings of the IEEE International Conference on Computer-Aided Design*, San Jose, 2006

Chapter 6
3D Clock Routing for Pre-bond Testability

Abstract Pre-bond testing of 3D stacked ICs involves testing each individual die before bonding. The overall yield of 3D ICs improves with pre-bond testability because manufacturers can avoid stacking defective dies with good ones. However, pre-bond testability presents unique challenges to 3D clock tree design. First, each die needs a complete 2D clock tree to enable pre-bond test. Second, the entire 3D stack needs a complete 3D clock tree for post-bond test and operation. In the case of a two-die stack, a straightforward solution is to have two complete 2D clock trees connected with a single through-silicon-via (TSV). In this chapter, we show that this solution suffers from long wirelength and high clock power consumption. Our algorithm improves on this solution, minimizes the overall wirelength and clock power consumption, and provides both pre-bond testability and post-bond operability with minimum skew and constrained slew. Compared with the single-TSV solution, SPICE simulation results show that our multi-TSV approach significantly reduces the clock power by up to 15.9 % for two-die and 29.7 % for four-die stacks. In addition, the wirelength is reduced by up to 24.4 and 42.0 %.

The materials presented in this chapter are based on [27].

6.1 Introduction

Three-dimensional system integration has emerged as a key enabling technology to continue the scaling trajectory predicted by Moore's Law for future IC generations. With 3D integration technology, both the average and maximum distance between components can be substantially reduced by placing them on different dies, which translates into significant savings in delay, power, and area. Moreover, it enables the integration of heterogeneous devices, making the entire system more compact

S.K. Lim, *Design for High Performance, Low Power, and Reliable 3D Integrated Circuits*, DOI 10.1007/978-1-4419-9542-1_6,

and efficient. Nevertheless, the success of 3D stacked ICs is predicated on the final post-bond yield, i.e., minimizing the number of good dies bonded to defective dies. Therefore, each die must be tested prior to the bonding process.

Recently, the authors of [16, 26] demonstrated that there exists a TSV vs. wirelength (and thus power) tradeoff in 3D clock trees: the more TSVs used in the 3D clock tree, the shorter the total wirelength. This clearly motivates using more TSVs in a 3D clock tree. However, it is also shown in [16, 26] that 3D clock trees containing multiple TSVs have an interesting property: only one die in the stack contains a fully connected 2D clock tree; the other dies contain many small, isolated subtrees. These trees take advantage of TSVs to shorten the total wirelength, but such a design makes pre-bond testing next to impossible because each clock subtree requires its own probe pad. The state-of-the-art testing equipment, e.g., from [23], has more than ± 100 ps overall timing accuracy (OTA). This makes it very challenging to use multiple clock probe pads to provide a low-skew clock signal. In addition, the cost of dedicating so many probes to a single signal is significant. Our work addresses these issues, providing low-cost methods for designing low-power pre-bond testable clock trees for 3D stacked ICs. The following specific topics are covered in this chapter:

- We study the first work on pre-bond testable clock routing. Our pre-bond testable clock tree can be used for both pre-bond test and post-bond operation. We introduce two new circuit elements, a *TSV-buffer* and a *redundant tree*, to enable efficient pre-bond testing while minimizing the overall wirelength and clock power. Compared with the simple pre-bond testability solution of using a single TSV to connect two complete 2D trees, our solution significantly reduces the wirelength and power consumption in both two-die and four-die 3D stacks.
- In order to improve the reliability of our pre-bond testable 3D clock tree, we develop a slew-aware merging and buffering method to keep the slew rate at each clock sink within a given constraint. This method has the added benefit of reducing the wirelength and power consumption of the pre-bond testable 3D clock tree.
- We show that by allocating the clock source in a middle die in the 3D stack, our pre-bond testable clock tree will use significantly fewer TSVs while still achieving power and wirelength savings comparable to other cases.
- We analyze the impact of the parasitic TSV capacitance on pre-bond testable clock trees in terms of wirelength, buffer count, and clock power. We demonstrate that a large capacitance tends to increase the wirelength and the number of buffers required, in turn increasing the clock power.

6.2 Existing Works

The history of clock tree synthesis for 3D ICs is short. Pavlidis et al. [18] presented measurement data from a fabricated 3D clock distribution network. Arunachalam and Burleson [3] used a separate layer for the clock distribution network to reduce power. Minz et al. [16] presented the first work on 3D clock routing with the goal of minimizing wirelength. They also tackled the impact of thermal variations on clock skew. Zhao and Lim [26] presented a comprehensive study of 3D clock tree synthesis and proposed several design techniques for generating reliable and low-power 3D clock tree designs. Kim and Kim [11] proposed a clock embedding method for 3D clock tree synthesis. They focused on minimizing the TSV count and wirelength. None of these works address pre-bond testability, unfortunately.

To tackle the 3D IC testing problem, several testing methods have been investigated. Lee and Chakrabarty presented a comprehensive study of the challenges of testing 3D ICs in [12]. Marinissen and Zorian [15] provided an overview of manufacturing processes in TSV-based 3D stacked ICs and discussed the test challenges. To improve testability with stitching wirelength minimization, 3D scan chain design approaches are proposed by Wu et al. [24]. In the work [25], Wu et al. developed a test-access mechanism (TAM) optimization technique for minimizing the test time of 3D core-based SoCs while constraining the total number of TSVs and the TAM widths. Noia et al. addressed the test-wrapper optimization of TSV-based 3D ICs in [17]. The scan-test time is minimized for a core under the constraint of the total number of TSVs available for testing. All these works focus on the post-bond test in 3D ICs.

Lewis and Lee presented an architectural solution in [13] to the pre-bond testability problem for 3D die-stacked microprocessors. They discussed how to perform testing for functional modules that are partitioned across multiple dies. They also investigated new design and test methods in [14] to address similar issues for 3D circuits. Jiang et al. [9] presented a heuristic method for optimizing test time and routing cost for both post-bond test and pre-bond wafer-level test. In [10], Jiang et al. proposed a layout-driven test-architecture design technique under a constrained pre-bond test pin count.

6.3 Preliminaries

6.3.1 3D Abstract Tree Generation

The authors of [16] proposed the three-dimensional method of means and medians (3D-MMM) algorithm to generate the abstract tree for a set of 3D clock sinks in a top-down manner. An upper bound for the TSV count (hereafter called *TSV bound*) is a user-defined constraint on the maximum number of TSVs the algorithm can use. The basic idea of 3D-MMM is to recursively divide the given sink set into two subsets until each sink belongs to its own set. Figure 6.1a demonstrates the partitioning process based on both $x - y$ coordinates and the TSV bound. At each recursive partitioning step, we divide the given Sink Set S into two Subsets S_A and

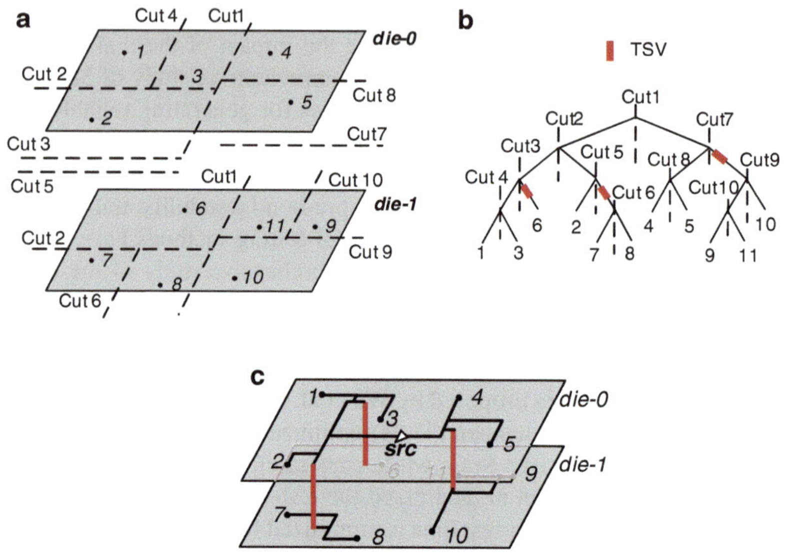

Fig. 6.1 3D clock routing for a two-die stack with a maximum TSV count of three. (**a**) *Top-down* partitioning using 3D-MMM algorithm [16], (**b**) generated abstract binary tree, and (**c**) final 3D clock topology

S_B. The following two cases are considered based on the TSV bound for the current sink set S:

- If the TSV bound is one, the current sink set needs to be partitioned such that the sinks in the same die belong to the same subset. The connection between S_A and S_B needs one TSV.
- If the TSV bound is greater than one, the z-dimension coordinates are ignored, and the set is partitioned geometrically by a straight cut line. Since each subset contains sinks from all the dies, multiple TSVs will be needed to connect them.

At the end of partitioning S, the TSV bound for each Subset S_A and S_B is determined as follows: (1) estimate the number of TSVs required by each subset, and (2) assign a portion of the TSV bound of S to each subset according to the ratio of the estimated TSV counts. The cut direction is set so as to balance the TSV bound across the subsets. After completing 3D-MMM partitioning, we obtain an abstract binary tree, as shown in Fig. 6.1b. This binary tree represents the hierarchical connection among clock sinks, internal nodes, TSVs, and the clock source.

During the embedding and buffering step, the internal nodes and TSVs in the 3D abstract tree are placed in a bottom-up fashion, and the buffers are inserted to maintain the zero-skew property (delay is estimated with the Elmore delay model [5]). The classic deferred-merge and embedding (DME) algorithm [4] is extended to generate a topology embedding for the given abstract tree. A cost function that considers the capacitance of the buffers, TSVs, and wires is used for buffer insertion [16].

6.3.2 3D-MMM Algorithm and Pre-bond Testing

The 3D clock tree generated by the 3D-MMM algorithm poses the following major challenges for pre-bond testing. First, the connected tree on the clock source die is not zero skew pre-bond, as illustrated in Fig. 6.1c. The entire 3D clock tree shown in this figure was constructed for zero-skew operation post-bond. However, without the clock sinks in die-1, the tree in die-0 is missing many of its branches (and thus their parasitic capacitors) which will cause serious timing violations during pre-bond testing that cannot be fixed by slowing down the clock frequency.

Second, each die except the clock source die lacks a complete tree that connects all the sinks in that die, as illustrated in Fig. 6.1c. In this figure, there are three subtrees that connect the sinks in die-1. In order to provide a skew-free clock signal during pre-bond test of die-1, we need three probe pads to provide three synchronized clock signals. This cost only increases as more TSVs are used to form more subtrees in order to minimize wirelength and power consumption.

Thus, the goals of our work are: (1) to construct a 3D clock tree that provides a zero-skew clock signal for both pre-bond test and post-bond test and operation, (2) to limit each die to a single clock source (and thus a single probe pad) during pre-bond test, (3) to minimize the wirelength and silicon costs of the tree, (4) to minimize the overall power consumption, and (5) to bound the clock slew within a given constraint.

6.4 Problem Formulation and Terminology

The pre-bond testable 3D clock routing problem is defined as follows: given a set of clock sinks distributed across N dies (where $N > 1$) and a TSV bound, construct a 3D clock tree such that (a) during post-bond operation, the tree connects all the sinks with a minimum-skew clock signal, and (b) during pre-bond test, a single 2D clock tree exists in each die that provides a minimum-skew clock signal to the sinks in that die. The objective is to minimize the wirelength and clock power given the TSV bound and clock slew bound constraints. The clock sinks may represent flip-flops, clock input pins for IP blocks, or memory blocks. Our pre-bond testable clock routing algorithm can operate under any TSV bound greater than zero, and it constructs a high quality 3D clock tree in terms of clock skew,[1] wirelength, power consumption, and clock slew for both pre- and post-bond testing and operations.

For an N-die stack clock network, we number the die as die-0, die-1, $\cdots$, die-(N − 1) in a top-down order. Given an N-die clock tree design with the clock source located in die-0, the term *"post-3d"* refers to the fully connected 3D clock tree used in post-bond operation; *"pre-die-k"* is the fully connected 2D clock tree in die-k for pre-bond test; *"sub-die-k"* refers to the set of unconnected subtrees in die-k; *"red-die-k"* refers to the redundant tree used to fully connect a sub-die-k set in a given die. We have the following relations among these trees: (1) *pre-die-k = sub-die-k + red-die-k, when k* > 0, (2) *post-3d = pre-die-0 + sub-die-1 + sub-die-2 +* $\cdots$ *+ sub-die-(N − 1)*.

6.5 Pre-bond Testable Clock Routing

6.5.1 Overview

Without loss of generality, we first develop a pre-bond testable clock routing algorithm for a two-die stack. We extend it to the stacks containing more-than-two dies in Sect. 6.5.5. The input to our algorithm includes the location and capacitance of the sinks in each die (die-0 and die-1), a TSV bound (>0), and a slew constraint. Die-0 is assumed to contain the clock source. Our algorithm consists of two main steps:

- **3D tree construction**: we generate a 3D clock tree (*post-3d*) connecting all the sinks in both dies so that (1) the overall 3D tree is zero skew under the Elmore delay model; (2) the total wirelength is minimized; and (3) die-0 contains a fully connected 2D tree (*pre-die-0*) with zero skew. In this case, the 3D tree is used during post-bond test and operation, while the 2D tree in die-0 is used for pre-bond test of die-0. We utilize so called "TSV-buffers" to ensure that the 2D tree in die-0 maintains zero skew in *both* pre-bond and post-bond configurations.
- **Redundant tree routing**: if multiple TSVs are used, the 3D tree construction step generates a 3D tree, where die-1 contains several separate subtrees (*sub-die-1*). In this case, we route a so-called "redundant tree" in die-1 (*red-die-1*) to connect the roots of the subtrees in die-1 and form a single fully connected 2D tree (*pre-die-1*) with (1) an estimated zero skew, and (2) a minimum total wirelength. This 2D tree is used for the pre-bond test of die-1. Transmission gates (*TG*s) are inserted to disconnect the redundant tree for post-bond operation.

[1] In the pre-bond testable clock routing, our algorithm generates zero-skew clock trees based on the Elmore delay model [5]. To obtain accurate clock-related metrics, we then extract the netlist, and report the SPICE simulation results, including delay, skew, slew, and power consumption.

6.5.2 TSV-Buffer Insertion

Testing die-0 pre-bond requires a fully connected clock tree in die-0 so that the clock signal is delivered to all die-0 sinks using a single test probe. As mentioned earlier, if multiple TSVs are used, the 3D tree construction step gives a 3D tree, where die-0 contains a single fully-connected tree and die-1 contains a forest of small subtrees. During pre-bond test, the two dies are separate and tested individually. In this case, the 2D tree in die-0 can be used without any additional modification. However, the skew of this tree may no longer be zero because the downstream capacitances of the subtrees in die-1 are not present. This additional skew will either slow down or corrupt the testing process.

To avoid this high-skew situation, we employ our TSV-buffer, simply a buffer inserted right before a TSV. In our test-aware DME (*TaDME*) algorithm, we add a TSV-buffer for each TSV and route the tree accordingly under the zero-skew constraint. In this case, the TSV-buffers are inserted in die-0, where the clock source is located. Since the buffers shield die-0 from the downstream capacitance, die-0 remains zero-skew when tested pre-bond. The outcome of *TaDME* is a zero-skew 3D tree that contains a zero-skew 2D tree in die-0 for pre-bond test.

In what follows, we describe how our *TaDME* algorithm modifies the traditional DME algorithm to construct a zero-skew 3D clock tree in the presence of TSV-buffers. A key step in *TaDME* is bottom-up recursive tree merging. Given a pair of zero-skew subtrees that must be merged, our goal is to determine the merging segment (the set of potential locations for the merging points) and to connect it to the root nodes of the subtrees so that the new merged tree is also zero-skew. Figure 6.2 shows the traditional merging process as used in the original DME algorithm, where the merging segment of internal node E is determined based on the parasitics of the TSVs, wires, downstream capacitances, and internal delays of the two subtrees. In this case, if the right branch (TSV, edge (E,A), and CT_2) of the overall tree is missing, the delay from E to B will change due to the change in the downstream capacitance at node E. However, if we use a TSV-buffer as shown in Fig. 6.3, the delay from E' to B will not change even if we remove the right branch. This is because the TSV-buffer hides the downstream capacitance at node E'.

The following notations are used in Figs. 6.2 and 6.3: r and c denote the unit-length wire resistance and capacitance, respectively. R_d is the output resistance of a buffer, C_L is the input capacitance of a buffer, and t_d is the intrinsic delay of a buffer. R_{TSV} and C_{TSV} are the resistance and capacitance of a TSV. Die-0 contains a subtree CT_1 with the root B and a loading capacitance C_{LB}. The internal delay from B to the sinks of CT_1 is t_B. Similar symbols are used for CT_2. A clock wire of length l is modeled as a π-type circuit with a resistor (rl) and two capacitors $(cl/2)$. We also model the TSVs with π-type circuits with resistance R_{TSV} and two capacitances $C_{TSV}/2$. Note that the downstream capacitance at the internal node E' in Fig. 6.3 is $cl_{E'B} + C_{LB} + C_L$ *both* before and after the dies are bonded. Thus, TSV-buffers allow us to build a 3D tree for die-0 that is zero-skew both pre-bond and post-bond.

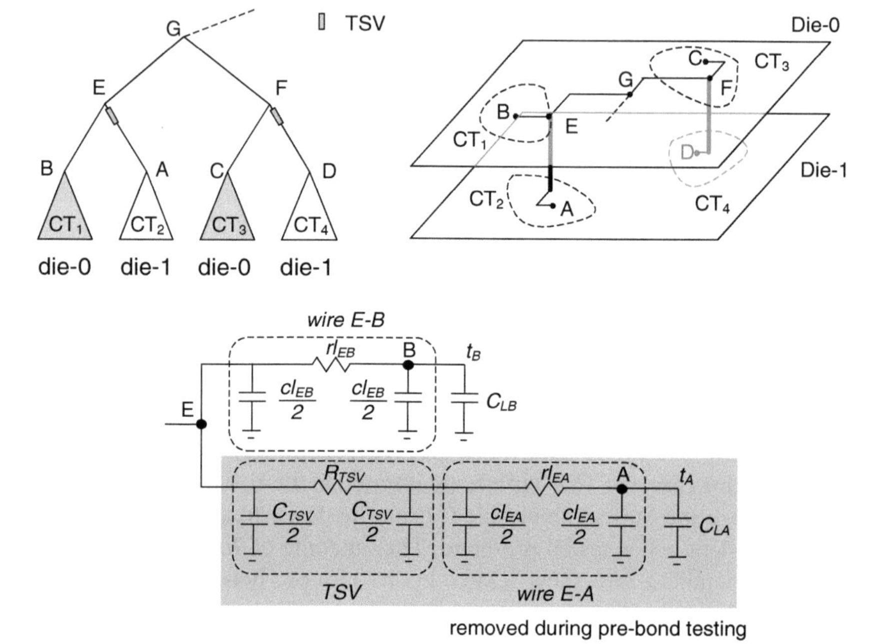

Fig. 6.2 3D clock tree built with TSVs, where the separation of die-0 and die-1 skews the tree in die-0

In the bottom-up merging process, we require that the delay from E' to sinks in CT_1 (through $B = d_{E',CT_1}$) be equal to delay to the sinks of CT_2 (through $A = d_{E',CT_2}$). That is,

$$d_{E',CT_1} = d_{E',CT_2} \tag{6.1}$$

Referring to the merging structure in Fig. 6.3, d_{E',CT_1} and d_{E',CT_2} can be expressed as following:

$$d_{E',CT_1} = rl_{E'B}(cl_{E'B}/2 + C_{LB}) + t_B \tag{6.2}$$

$$\begin{aligned} d_{E',CT_2} &= t_d + R_d(C_{TSV} + cl_{E'A} + C_{LA}) \\ &\quad + R_{TSV}(C_{TSV}/2 + cl_{E'A} + C_{LA}) \\ &\quad + rl_{E'A}(cl_{E'A}/2 + C_{LA}) + t_A \end{aligned} \tag{6.3}$$

where t_A is the internal delay from A to sinks of CT_2, and C_{LA} is the downstream capacitance of Node A. If there is no detour and given the distances between E' and A ($l_{E'A}$) and between E' and B ($l_{E'B}$), it follows that:

$$l_{E'B} + l_{E'A} = L \tag{6.4}$$

where L is the minimum merging distance between A and B. $l_{E'A}$ and $l_{E'B}$ can be determined by solving Eqs. (6.1)–(6.4).

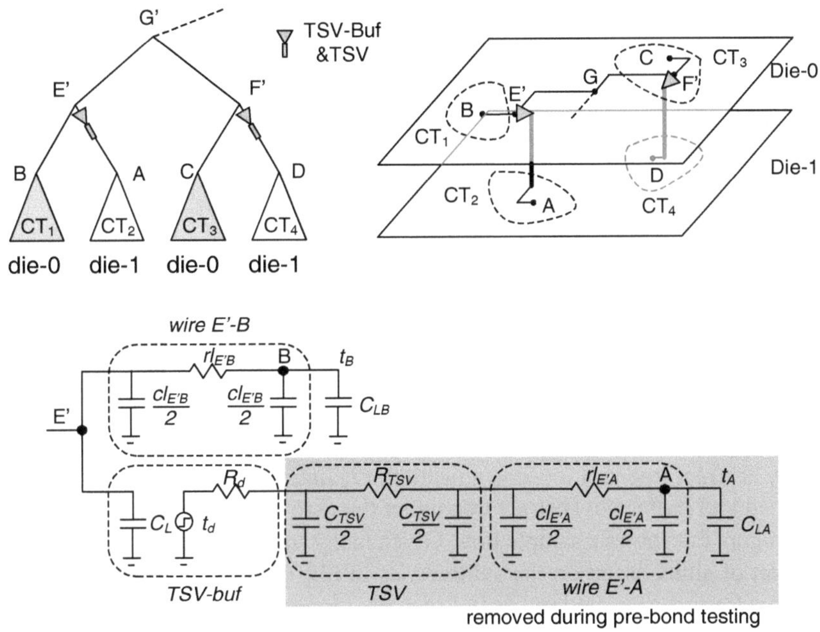

Fig. 6.3 3D clock tree built with TSV-buffers, where the separation of die does not skew the die-0 tree

If $l_{E'A}$ or $l_{E'B}$ is negative, a wire detour is required. For example, when $l_{E'A}$ is negative, $l_{E'B}$ must be longer than L to obtain a zero-skew merging. In this case, $l_{E'A}$ is set to zero, and $l_{E'B}$ is calculated by solving Eqs. (6.1)–(6.3). If the calculated $l_{E'B}$ is too long, we insert a clock buffer along Edge $E'B$. Equation (6.2) is updated correspondingly. The decision to avoid a detour with a buffer is made by a cost function that considers the capacitance of clock wires, buffers, and TSVs; we use a wire detour if the cost is less than that of buffer insertion and satisfies the slew constraint.

6.5.3 *Redundant Tree Insertion*

Pre-bond test of die-1 requires a fully connected clock tree so that the clock signal is delivered to all the sinks in die-1 from just a single test probe. As mentioned earlier, when multiple TSVs are used for wirelength reduction, the 3D tree construction generates a forest of subtrees in die-1. Therefore, our goal is to combine these subtrees into a single fully connected clock tree with zero clock skew and minimum overall wirelength. We accomplish this by adding a redundant tree that connects the roots of the subtrees while maintaining zero skew. We use this fully connected tree during the pre-bond test of die-1. Note that the redundant tree is not used during post-bond test and operation. We use TGs to disconnect the redundant tree.

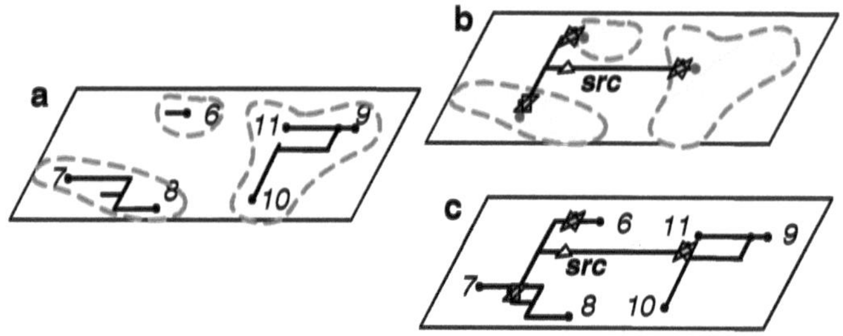

Fig. 6.4 Redundant tree insertion in die-1. (**a**) Extract sinks from subtrees, (**b**) generate a redundant tree and insert transmission gates, and (**c**) final pre-bond testable clock tree in die-1. The extra control signal that connects the transmission gates is not shown here for simplicity

The redundant tree routing is done using a conventional algorithm: (1) construct a binary abstract tree in a top-down fashion; (2) insert a TG at each sink node; and (3) embed and buffer the abstract tree under the zero-skew and minimal wirelength goals. Figure 6.4 shows a sample flow. Given many subtrees in die-1, we first extract a new set of sinks based on the subtrees, as in Fig. 6.4a. Then, we construct a 2D clock tree for this extracted set, as in Fig. 6.4b. Figure 6.4c shows the final pre-bond testable clock tree in die-1 (*pre-die-1*), which consists of three subtrees (*sub-die-1*) and one redundant tree (*red-die-1*). Lastly, we connect the enable input of the TGs using an extra control wire. In order to minimize the routing overhead, we need to minimize the total wirelength of this control signal. We use the Rectilinear Minimum Spanning Tree algorithm (RMST-pack) [20] for this purpose. The cost of this overhead is reported in Sect. 6.7.3.

6.5.4 *Putting It Together*

Upon the completion of our algorithm, we obtain fully connected zero-skew 2D clock trees for both die-0 and die-1 as well as a fully connected zero-skew 3D tree for the entire stack. In die-1, we turn on the TGs to connect the redundant tree to the subtrees for pre-bond test. Once the pre-bond testing is complete, we turn off the TGs to disconnect the redundant tree. By doing this, the original zero-skew 3D tree is used for post-bond test and normal operation. We will show in our experimental results section that our 3D trees with multiple TSVs, TSV-buffers, and TGs plus the control signal consume significantly less power than a simple single-TSV solution.

Figure 6.5a shows an illustration of the entire design flow. In post-bond operation, the TGs are turned off and the *pre-die-0* and *sub-die-1* trees are connected with TSVs to form the *post-3d* tree, as shown in Fig. 6.5b. In pre-bond test, the *pre-die-0* tree can be reused with zero skew to test die-0, as shown in Fig. 6.5c. To test die-1, we turn on the TGs, and the *red-die-1* and *sub-die-1* trees form the zero-skew *pre-die-1* tree, as shown in Fig. 6.5c.

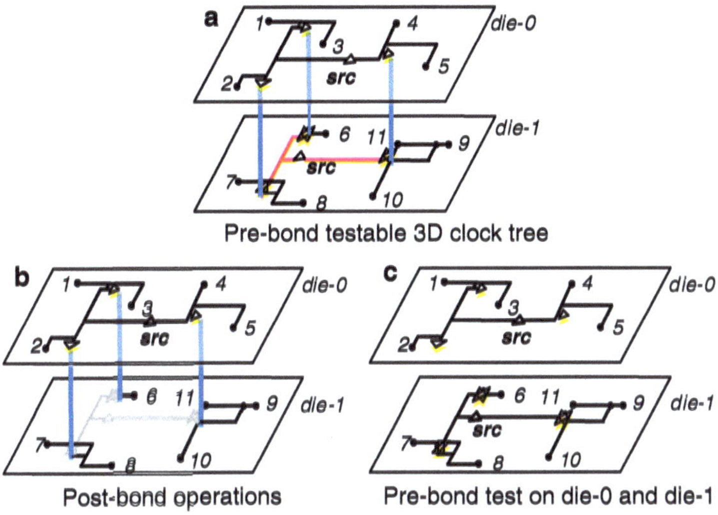

Fig. 6.5 Example of the post-bond operations and pre-bond test using our 3D clock tree. (**a**) A pre-bond testable 3D clock tree, (**b**) a *post-3d* in post-bond operation with TGs turned off, and (**c**) *pre-die-0* and *pre-die-1* in pre-bond test with TGs turned on

6.5.5 Multiple-Die Extension

For a stack with more-than-two dies, we face the same challenges of creating clock trees for pre-bond test. We take a four-die stacked clock tree in Fig. 6.6 as an example. The clock source is located in die-0. If we apply the 3D-MMM algorithm [16], the resulting *post-3d* tree contains the following topology: (1) die-0 has a complete clock tree connecting all the sinks in die-0; (2) the non-source dies (die-1, die-2, and die-3) have each a *sub-die-k* ($k = 1, 2, 3$), which is connected to the clock source through 10 TSVs.

Our pre-bond testable clock routing algorithm for a two-die stack can be easily extended to larger die stacks with an arbitrary clock source location. Our basic 3D tree construction algorithm presented in Sect. 6.3.1 generates a 3D tree, where die-s (defined as containing the clock source; die-0 in Fig. 6.6) has a single, fully-connected tree, while all the other dies have a forest. Our TSV-buffer insertion algorithm is extended as follows. During the bottom-up merging process,

- If a TSV connects die-s and a non-source die-k where ($k \neq s$), we insert a TSV-buffer in die-s.
- If a TSV connects non-adjacent dies and passes through die-s (e.g., connecting die-($s - 1$) and die-($s + 1$)), we insert a single TSV-buffer in die-s.
- If a TSV does not connect to or travel through die-s, no TSV-buffer is required.

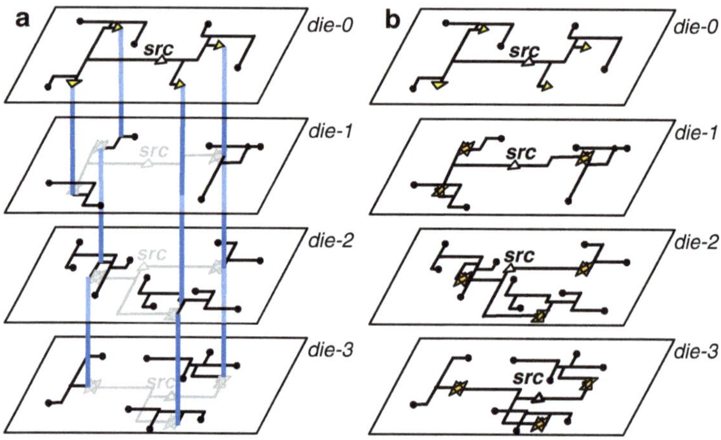

Fig. 6.6 An example of a pre-bond testable clock routing in a four-die stack. (**a**) 3D clock tree, in post-bond. (**b**) 2D clock trees, in pre-bond test

Once the TSV-buffer insertion and embedding and buffering are completed, we add redundant trees to the non-source dies. In addition, we insert TGs at the root of each subtree and add a global control signal to connect all the TG enable inputs in each die. This allows us to use the redundant trees for pre-bond test (TGs on) and disable them during post-bond test and operation (TGs off). The outcome of the whole process is: (1) a single zero-skew 3D clock tree for post-bond test and normal operation; (2) a zero-skew 2D clock tree in each die for pre-bond test; and (3) a global control signal that connects the enable inputs of the TGs in each die. Figure 6.6 shows an illustration of the pre-bond testable and post-bond operational 3D clock tree for a four-die stack.

6.6 Buffering for Wirelength and Slew Control

This section presents our buffering strategies for balancing wirelength and controlling slew.

6.6.1 Wirelength Balancing with Clock Buffers

Our pre-bond testable 3D clock routing algorithm inserts two kinds of buffers: clock buffers and TSV-buffers. Clock buffers, as discussed in Sect. 6.3.1, are mainly used to control delay and skew. These clock buffers are usually inserted close to the clock source and drive large loads to reduce the delay along the clock paths.

The TSV-buffers, as discussed in Sect. 6.5.2, are inserted at every TSV location in the clock source die to ensure that the clock tree in that die is also zero-skew during pre-bond test.

Our observations indicate that TSV-buffers may unbalance the wirelength during the bottom-up merging process. Consider the example of two Subtrees CT_1 and CT_2 in die-0 and die-1, respectively; we must use a TSV-buffer in die-0 to merge these subtrees. As shown in Fig. 6.3, TSV-buffer insertion can increase the delay from E' to CT_2. If the internal delay of CT_2 is already much greater than that of CT_1, adding the TSV-buffer only makes the difference worse. If the difference is too large, wire snaking is required to balance the delays and to achieve a zero-skew merged tree. Thus, the addition of a TSV-buffer has led to a significant clock wirelength overhead in die-0.

To mitigate this overhead, we add extra clock buffers to die-0 to balance the internal delays and eliminate snaking. Specifically, when a TSV-buffer causes significant unbalance in the delay, we insert an extra clock buffer on the other branch as a counter-balance. In Fig. 6.3, we add an extra clock buffer along $E' - B$. We observe that this delay balancing scheme reduces the overall wirelength in die-0. We also observe that few clock buffers are required in this way because such unbalance do not occur frequently.

6.6.2 Slew Rate Control with Clock Buffers

Clock slew rate control is an important reliability issue for high-speed clocking. If the slew rate is too low – that is, if it takes too long for the clock signal to rise or fall – setup and hold times may be violated, a problem which cannot be fixed with a lower clock frequency. Existing work on slew-aware clock tree synthesis relies on buffer insertion [1, 2, 7, 22]. Buffers are added along the clock paths so that the output load of each buffer is limited. This bounding condition, denoted as *cmax* in the literature, is shown to be effective in controlling the slew rate; a smaller *cmax* value improves the slew rate but requires more buffers are inserted. Most existing works insert buffers in a given clock tree as a post-processing step to improve the slew rate under various constraints: buffer area, clock power, etc. This post-synthesis slew-aware buffer insertion must be done carefully to avoid introducing new clock skew. This may constrain the location of the buffers.

Our strategy is to tackle the slew rate issue *during* the construction of the pre-bond testable clock trees by adding buffers to meet the *cmax* constraint. Specifically, we insert clock buffers, together with TSV-buffers, during the bottom-up merging process so that *cmax* is satisfied for both types of buffers. We add clock buffers along the paths from the merging node to the subtree root nodes if the downstream capacitance at the merging node exceeds *cmax*. Depending on the load, we may insert multiple clock buffers to meet the *cmax* requirement.

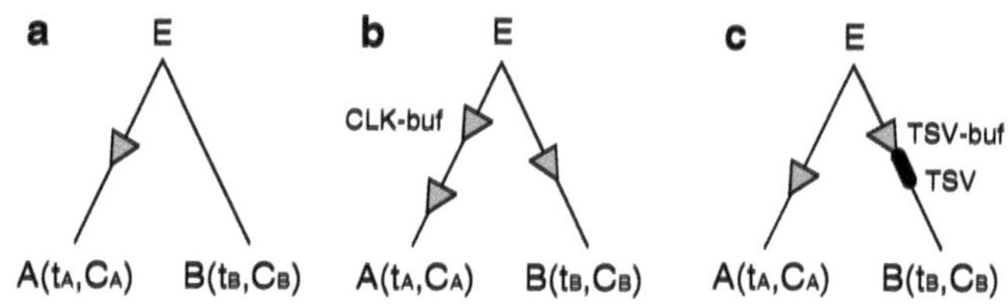

Fig. 6.7 Examples of clock buffer and TSV-buffer insertion. (**a**) A clock buffer is inserted to balance the delay of the two branches, where $t_A < t_B$, (**b**) multiple clock buffers are inserted if the wires are long and/or the download capacitance is large, and (**c**) a clock buffer is inserted along with a TSV-buffer to balance the delay

Figure 6.7 shows several possible scenarios for clock buffer and TSV-buffer insertion. In summary, our clock tree synthesis algorithm uses three criteria for buffer insertion during the bottom-up merging process:

- For pre-bond testability: we add a TSV-buffer for every TSV connecting to the clock source die.
- For wirelength reduction: we add a clock buffer to correct unbalance in the delays of two merging subtrees as discussed in the previous section.
- For slew rate control: we add clock buffers if the downstream capacitance of any buffer exceeds the given limit *cmax*.

6.7 Experimental Results

We implemented our algorithm using C++/STL on Linux. We use five benchmarks from the IBM suite [6] and four from the ISPD clock network synthesis contest suite [8]. Since these designs are for 2D ICs, we obtain 3D designs by randomly partitioning the clock sinks across the multiple dies and scaling the footprint area by $\sqrt{2}$ and $\sqrt{4}$ for two-die and four-die stacks, respectively.

We use technology parameters from the 45 nm Predictive Technology Model (PTM) [19]; the unit-length wire resistance is $0.1\,\Omega/\mu m$, and the unit-length wire capacitance is 0.2 fF/μm. The sink capacitance values range from 5 to 80 fF. The buffer parameters are $R_d = 122\,\Omega$, $C_L = 24$ fF, and $t_d = 17$ ps. We use 10μm × 10μm via-last TSVs with 20 μm height and 0.1 μm liner oxide thickness. By simulating the TSV structure with Synopsys Raphael [21], we determine the TSV parasitics to be $R_{TSV} = 0.035\,\Omega$ and $C_{TSV} = 15.48$ fF. The clock frequency is set to

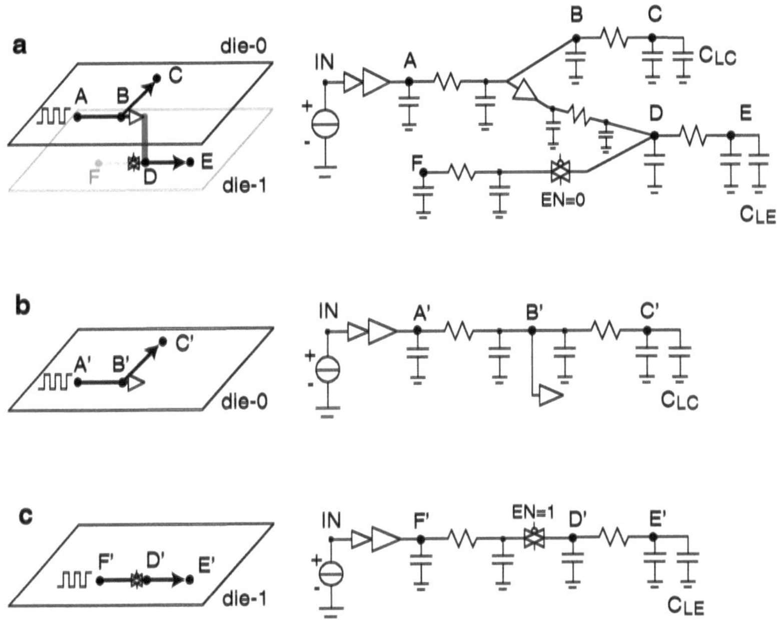

Fig. 6.8 Circuit models for (**a**) the post-bond 3D clock tree, (**b**) the pre-bond testable 2D clock tree in die-0, and (**c**) the pre-bond testable 2D clock tree in die-1

1 GHz and the supply voltage (Vdd) to 1.2 V.[2] The maximum load capacitance for each buffer *cmax* is 300 fF for slew rate control.

In SPICE simulation, wire segments and TSVs are represented as π models, and clock buffers and TSV-buffers are represented as inverter pairs. The simulated clock skew and slew tolerances are 3 and 10 % of the clock period, respectively. We report wirelength in μm, clock power in mW, skew and slew in ps, and capacitance in fF.

6.7.1 TSV-Buffer and TG Model Validation

In pre-bond testable clock routing, we utilize TSV-buffers and TGs to facilitate pre-bond test and post-bond test and operation. Figure 6.8 shows the equivalent circuits used for SPICE validation of the TSV-buffers and TGs. We simulate a post-bond 3D clock tree in a two-die stack and two pre-bond testable 2D clock trees in die-0 and die-1. Node *A* is the clock source for post-bond operation. Sink *C* in die-0

[2]Note that our clock trees with single and multiple TSVs are simulated under the same Vdd, and the power savings mainly come from the capacitance reduction. Therefore, the efficiency of our algorithm in low power and pre-bond testability apply on different Vdd (e.g., from 1.2 to 1.0 V).

and Sink E in die-1 have loading capacitances of C_{LC} and C_{LE}, respectively. Node B and D are connected by a TSV-buffer and a TSV. Edge (D,E) is a subtree in die-1 and is connected to F, the clock source for pre-bond test of die-1, via a TG. C_{LC} and C_{LE} are set to 5 fF. Wires (A,B), (B,C), (D,E) and (F,D) all have 500 μm length.

First, we observe from SPICE simulation that the delay from A to C in Fig. 6.8a is 42.21 ps, which is the same as that from A' to C' in Fig. 6.8b. This verifies that die-0 is zero skew before die-1 is attached and so the TSV-buffer has done its job. Second, the TG has 14.2 fF capacitance between Node D and the ground when it is off. This TG completely blocks the clock signal from A to F. When the TG is on for pre-bond testing on die-1, however, it has 108 Ω between its input and output nodes, 16.4 fF between its input and the ground, and 18.4 fF between its output and the ground. The intrinsic delay of a TG is 1.04 ps. Under this model, the calculated delay from F' to E' is 54.13 ps, which closely matches the simulated delay of 54.14 ps.

6.7.2 *Sample Trees and Waveforms*

Figure 6.9 shows a series of pre-bond testable clock trees for the circuit r_1 from the IBM suite given a TSV bound of 10. The TSVs are shown as black dots, the clock sources as triangles. Figure 6.9a is the zero-skew 3D clock tree for post-bond test and normal operation. This 3D clock tree contains 10 TSVs. The solid and dotted lines represent the clock trees in die-0 and die-1, respectively. Note that die-1 contains many subtrees (dotted lines) that are not connected to each other except through die-0. Figure 6.9b shows the zero-skew pre-bond testable 2D clock tree for die-0, which is identical to the solid line clock tree in Fig. 6.9a. Figure 6.9c shows the zero-skew pre-bond testable 2D clock tree for die-1, which contains all the subtrees (dotted lines) in die-1 and the redundant tree (solid line) which connects them.

Figure 6.10 shows two groups of clock waveforms for benchmark r_5, where each group contains 25 waveforms (one each for the 25 sinks in each tree). The first group (shown on top) is from the post-bond 3D clock tree, the second group (shown on bottom) is from the pre-bond testable 2D clock tree for die-0. We first observe that the 25 waveforms are almost identical, which is desirable. In addition, the two groups have similar waveforms, which demonstrates that the TSV-buffer does maintain the balance of the tree in both pre-bond and post-bond test configurations. Second, the SPICE simulation shows that the clock skew among all sinks in both cases is 29.1 ps, observed by the width of waveforms at 50 % Vdd. Third, the maximum slew rate is 88.4 ps, measured as the rise time from 10 to 90 % of Vdd (or fall time from 90 to 10 % Vdd) at the slowest node. Both the skew and slew values are within the tolerances (3 and 10 % of the clock period, respectively).

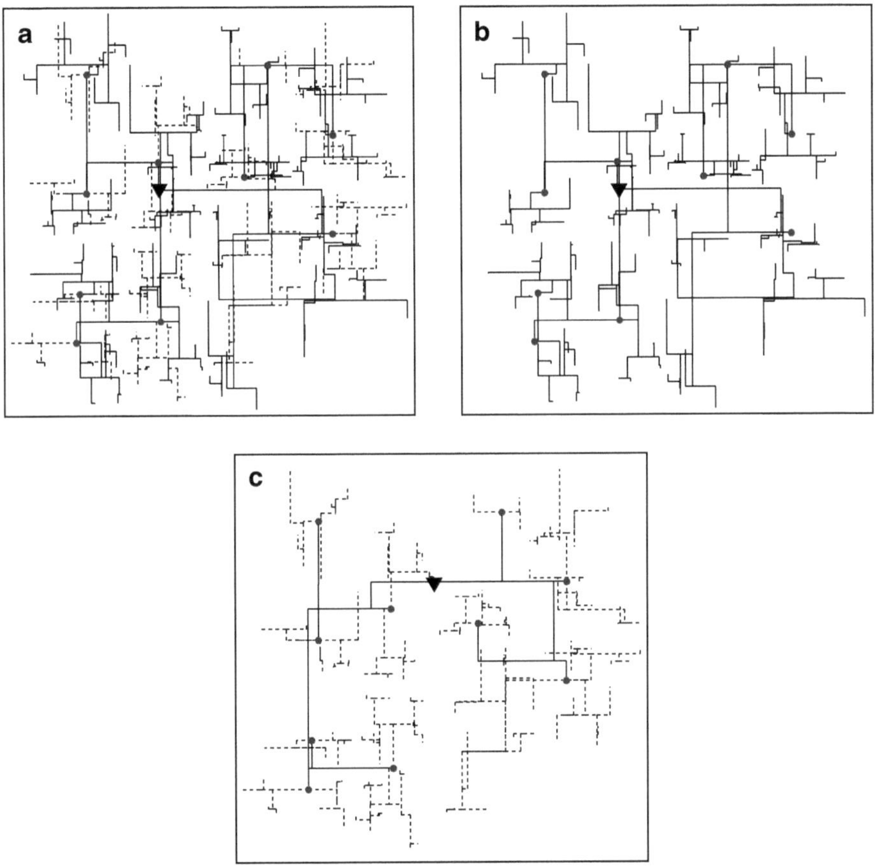

Fig. 6.9 Pre-bond testable clock trees for circuit *r*1 in a two-die stack for a TSV bound of 10. The TSVs and the clock sources are represented by *black dots* and *triangles*, respectively. (**a**) The post-bond 3D clock tree, where the *solid* and *dotted lines* denote the trees in die-0 and die-1, respectively. (**b**) The pre-bond testable 2D clock tree for die-0. (**c**) The pre-bond testable 2D clock tree for die-1, where the redundant tree and the subtrees are drawn in *solid* and in *dotted lines*, respectively

6.7.3 Wirelength, Skew, and Power Results

Tables 6.1 and 6.2 show the wirelength (μm), power consumption (mW), and skew (ps) results for the post-bond 3D clock tree (*post-3d*), the pre-bond testable 2D clock tree for die-0 (*pre-die-0*) and die-1 (*pre-die-1*). For die-1, we report the total wirelength (*WL*) and the wirelength of the subtrees (*WL-sub*), redundant tree (*WL-red*), and TG control signal (*WL-TG*). In this case, the wirelength of the pre-bond testable clock tree for die-1 is equal to the sum of *WL-sub* and *WL-red*. In addition, the wirelength of the post-bond 3D clock tree is the sum of the wirelength of *pre-die-0* and *WL-sub* from *pre-die-1*.

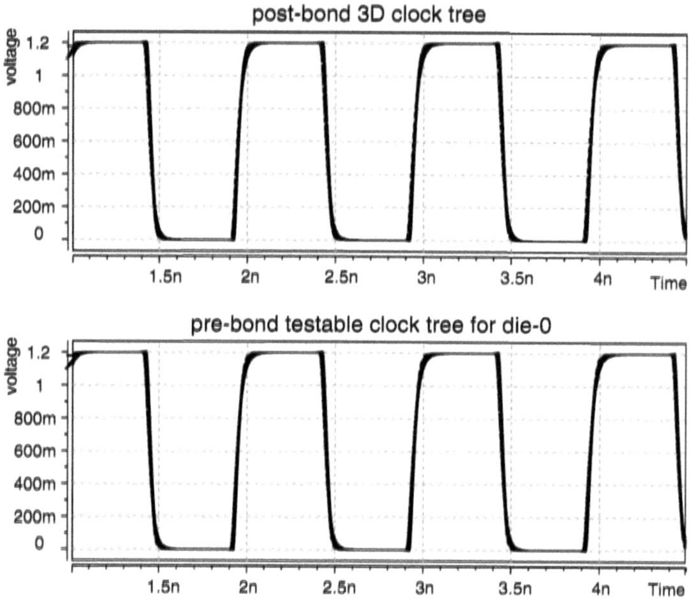

Fig. 6.10 Clock waveforms from the post-bond 3D clock tree and the pre-bond testable 2D clock tree for die-0. We superimpose the waveforms of the 25 clock sinks in r_5. Clock frequency is 1 GHz, skew is 29.1 ps, and maximum slew rate is 88.4 ps

Table 6.1 Wirelength, clock power, and skew results for post-bond testable 3D clock tree

			Post-bond 3D		
Ckt	#sinks	#TSVs	WL	Power	Skew
r_1	267	57	227,141	128.4	13.7
r_2	598	95	488,987	274.1	14.2
r_3	862	183	616,077	361.6	15.5
r_4	1,903	265	1,311,290	763.2	15.5
r_5	3,101	269	1,998,950	1,115.0	29.1
*f*11	121	44	129,391	73.3	9.4
*f*12	117	36	127,763	71.2	6.8
*f*21	117	42	136,676	75.6	5.0
*f*22	91	30	80,977	46.8	15.3
Ratio			1.00	1.00	1.00

Based on the wirelength-related columns, we observe that (1) the total wirelength of *pre-die-0* and *pre-die-1* are comparable (0.72 vs. 0.69 in ratio); (2) in several cases, the wirelength of the redundant tree is about 2× of the total wirelength of the subtrees in die-1 (0.41 vs. 0.28); and (3) in several cases, the wirelength of the TG control signal is about half of the redundant tree in die-1 (0.29 vs. 0.41).

The total clock routing resource cost is equal to the sum of *post-3d* and *WL-red* from *pre-die-1*. Normalized to the wirelength of *post-3d*, the overall wirelength of the pre-bond testable clock tree and its redundant trees is 1.41. We can derive the

Table 6.2 Wirelength, clock power, and skew results for pre-bond testable 2D clock trees. RATIO is computed with respect to the results shown in Table 6.1

			Pre-bond testable die-0			Pre-bond testable die-1					
Ckt	#sinks	#TSVs	WL	Power	Skew	WL	WL-sub	WL-red	WL-TG	Power	Skew
r_1	267	57	166,691	103.0	13.5	150,219	60,450	89,769	62,732	68.2	13.0
r_2	598	95	328,914	196.0	14.1	302,023	160,073	141,950	109,031	148.6	11.8
r_3	862	183	444,156	280.5	15.5	429,950	171,921	258,029	161,561	201.9	16.2
r_4	1,903	265	889,460	536.4	14.9	846,980	421,830	425,151	259,442	422.1	15.1
r_5	3,101	269	1,255,760	715.9	29.1	1,236,417	743,190	493,227	310,855	615.9	20.9
$f11$	121	44	99,393	64.1	9.2	99,169	29,998	69,171	51,214	44.3	6.3
$f12$	117	36	96,093	60.4	6.2	93,625	31,669	61,956	42,134	42.0	5.7
$f21$	117	42	107,834	67.0	4.7	101,968	28,841	73,127	52,241	45.0	7.3
$f22$	91	30	61,504	40.4	15.2	59,870	19,473	40,397	29,449	26.4	14.9
Ratio			0.72	0.79	0.97	0.69	0.28	0.41	0.29	0.57	0.94

following: die-0 and die-1 utilize 51 and 49 % of the total clock routing resource, respectively. In the post-bond operations, the *post-3d* consumes 71 % of the clock routing resource. This means that 29 % of the clock resource is used for the pre-bond test only. Note that the redundant tree and the TG control signal are used only during the pre-bond testing for die-1. This non-negligible overhead is compensated by the significant power savings to be discussed in Sect. 6.7.4.

Lastly, the clock skew values do not exceed 30 ps, satisfying our 3 % of the clock period constraint on the simulated skew. Die-0 consumes more clock power than die-1, primarily due to the TSV-buffers inserted in die-0.

6.7.4 Comparison with the Single-TSV Approach

Our baseline 3D clock tree contains a single, fully-connected zero-skew clock tree in each die; these trees are connected with a single TSV in the two-die stacks and a single column of TSVs in taller stacks. Table 6.3 compares the wirelength (μm), clock power (mW) and skew (ps) results from the SPICE simulation. In the multi-TSV designs, we choose the TSV count that gives us the minimum power by an exhaustive search, wherein we sweep the TSV bound from 2 to infinity, construct a 3D clock tree for each bound, and simulate the power consumption. The clock synthesis time for each tree is less than 1 s in all cases.

We make the following observations. First, our multi-TSV approach significantly outperforms the single-TSV approach in terms of wirelength: 14.8–24.4 % reductions for the two-die stacks and 39.2–42.0 % reductions for the four-die stacks. Similarly, power savings for the clock trees are 10.1–15.9 % for the two-die cases and 18.2–29.7 % for the four-die cases. These results convincingly demonstrate the benefits of our multi-TSV approach. Second, the total number of buffers (#Bufs) used in the clock trees consists of the clock buffers and the TSV-buffers. Table 6.4 shows the detailed buffer usages in the two-die cases, including the total number of buffers (#Bufs), the TSV-buffer count (#TBs), and the clock buffer count (#CBs). We observe that a similar number of buffers is used in both the single- and the multi-TSV trees. In the single-TSV design, buffers are inserted to control the wirelength and slew in each die. In the multi-TSV policy, we need more TSV-buffers to ensure pre-bond testability but use fewer clock buffers. This is because the total wirelength is shorter in the multi-TSV designs and the TSV-buffers have positive impact on slew control.

6.7.5 Impact of TSV Bound on Power

Figure 6.11 shows the impact of the TSV bound on wirelength, buffer count, and clock power consumption. These metrics are normalized to the baseline results from the single-TSV approach. The x-axis corresponds to the TSV bound used

Table 6.3 Comparison between single-TSV and multi-TSV designs. % improvement belongs to the multi-TSV case

		Single TSV				Multi-TSV					% imprv	
ckt	#sinks	#bufs	WL	Power	Skew	#TSVs	#bufs	WL	Power	Skew	WL	Power
Two die results												
r_1	267	327	279,796	145.0	12.7	57	324	227,141	128.4	13.7	18.8	11.4
r_2	598	693	600,880	310.6	12.5	95	684	488,987	274.1	14.2	18.6	11.8
r_3	862	928	765,397	404.3	16.1	183	925	616,077	361.6	15.5	19.5	10.6
r_4	1,903	1,982	1,576,510	848.7	15.3	265	1,963	1,311,290	763.2	15.5	16.8	10.1
r_5	3,101	2,528	2,344,960	1,242.0	22.2	269	2,449	1,998,950	1,115.0	29.1	14.8	10.2
$f11$	121	212	168,500	85.4	7.6	44	201	129,391	73.3	9.4	23.2	14.1
$f12$	117	215	164,966	84.2	5.8	36	193	127,763	71.2	6.8	22.6	15.5
$f21$	117	226	180,867	89.9	9.4	42	211	136,676	75.6	5.0	24.4	15.9
$f22$	91	106	106,401	53.2	15.1	30	111	80,977	46.8	15.3	23.9	12.1
Four die results												
r_1	267	318	272,355	141.8	10.5	248	325	160,394	111.4	13.3	41.1	21.4
r_2	598	700	582,115	304.5	14.4	434	647	353,646	233.9	15.7	39.2	23.2
r_3	862	945	735,299	398.0	14.9	718	922	442,903	317.1	13.7	39.8	20.3
r_4	1,903	1,956	1,532,220	831.1	14.8	1,651	2,011	908,375	675.6	16.5	40.7	18.7
r_5	3,101	2,939	2,312,930	1,272.0	22.2	2,469	3,134	1,368,370	1,041.0	20.3	40.8	18.2
$f11$	121	216	159,752	83.1	8.4	129	176	93,440	60.0	5.8	41.5	27.8
$f12$	117	208	155,542	80.9	8.9	114	160	90,281	56.8	10.2	42.0	29.7
$f21$	117	212	163,816	83.0	17.8	102	160	99,179	58.4	7.8	39.5	29.6
$f22$	91	99	98,123	48.7	18.0	81	88	57,342	36.1	14.7	41.6	25.9

Table 6.4 Buffer usage between the single- and multi-TSV cases. We report the total number of buffers (#bufs), TSV-buffers (#TBs), and clock buffers (#CBs). The number of dies is 2

	Single TSV			Multi-TSV			
ckt	#bufs	#TBs	#CBs	#TSVs	#bufs	#TBs	#CBs
*r*1	327	1	326	57	324	57	267
*r*2	693	1	692	95	684	95	589
*r*3	928	1	927	183	925	183	742
*r*4	1,982	1	1,981	265	1,963	265	1,698
*r*5	2,528	1	2,527	2,69	2,449	269	2,180
*f*11	212	1	211	44	201	44	157
*f*12	215	1	214	36	193	36	157
*f*21	226	1	225	42	211	42	169
*f*22	106	1	105	30	111	30	81

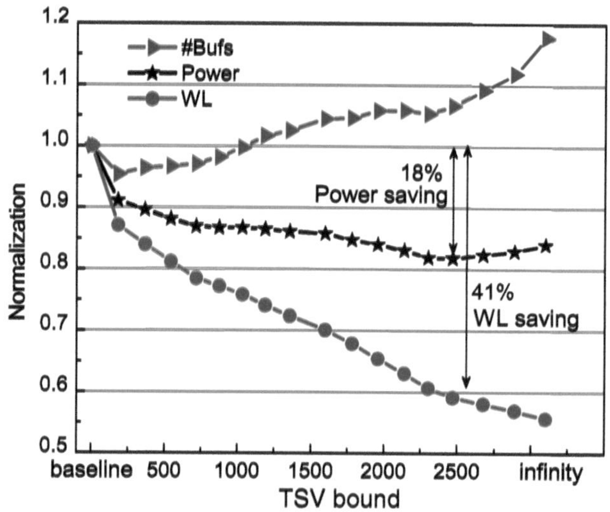

Fig. 6.11 Impact of the TSV bound constraint on wirelength, buffer count, and clock power consumption based on the four-die stack of r_5. The baseline is the single-TSV approach

to build our multi-TSV pre-bond testable 3D clock tree. Note that the actual TSV usage may be less than the TSV bound because the clock tree synthesis algorithm may determine that the optimal number of TSVs is less than the allowed number. For example, when the TSV bound is set to infinity, only 3,097 TSVs are actually used in the four-die stack of benchmark r_5.

We first observe that the wirelength consistently reduces as more and more TSVs are used in our 3D pre-bond testable clock trees. The wirelength savings reach 45 % if the TSV bound is set to infinity. This confirms that, in general, TSVs help to reduce the overall wirelength of 3D clock trees. Second, the total number of buffers (both clock buffers and TSV-buffers) increases as more TSVs are used.

Table 6.5 Impact of TSV-buffer insertion. % increase refers to the wirelength and power consumption increases from TSV-buffer insertion

		No	Yes	% increase	
ckt	#TSVs	Skew	Skew	WL	Power
*r*1	248	47.3	9.6	−0.77	7.95
*r*2	434	34.8	13.6	3.89	6.27
*r*3	718	38.5	12.3	3.11	9.19
*r*4	1,651	45.2	14.9	2.72	11.28
*r*5	2,469	48.0	15.9	4.56	11.17
*f*11	129	30.4	3.9	−1.69	8.28
*f*12	114	33.2	6.8	0.06	7.47
*f*21	102	24.7	4.4	−1.66	3.34
*f*22	81	33.3	13.6	−2.68	2.30

This is mainly due to the insertion of required TSV-buffers for pre-bond testability. Considering both trends, the power consumption decreases consistently but slowly for a time but eventually begins to rise as the cost of the TSV-buffers finally begins to outweigh the wirelength savings. The maximum power saving for r_5 is around 18 %. The corresponding 3D clock tree uses approximately 2,500 TSVs across all four dies. With more than 2,500 TSVs, the power consumption finally rises due to the excessive number of TSV-buffers. This trend gives us an optimum TSV bound for a given power budget: for the four-die stack r_5, the TSV bound should be set to 300 for a power consumption savings of 10 %.

6.7.6 Impact of TSV-Buffer Insertion

As discussed earlier, TSV-buffers help trees maintain low clock skew during pre-bond test of the clock source die. Table 6.5 shows the impact of TSV-buffer insertion, where we compare two clock trees in the clock source die, one with TSV-buffers and one without. We observe that the skew in the source die increases 3× to 10× if TSV-buffers are not used. However, as discussed in Sect. 6.6.1, TSV-buffers cause minor increases in wirelength and overall power consumption in the range of 2.3–11.28 %.

6.7.7 Impact of Clock Source Location

Next we consider the placement of the clock source. Table 6.6 compares two cases for a four-die stack: locating the clock source in the top die (die-0) versus in a middle die (die-1). We observe that by locating the clock source in the middle die, we use fewer TSVs while achieving comparable power consumption. The middle-die cases use 7.8–20.2 % fewer TSVs than the top-die cases. Meanwhile, power and wirelength differences are within ±2 % in most cases.

Table 6.6 Impact of the clock source location in four-die 3D stacks

	src in the top die (die-0)					src in the middle die (die-1)					% imprv	
ckt	#TSVs	#bufs	WL	Power	Skew	#TSVs	#bufs	WL	Power	Skew	#TSVs	Power
*r*1	248	325	160,394	111.4	13.3	208	300	163,249	110.0	12.1	16.1	1.3
*r*2	434	647	353,646	233.9	15.7	394	631	352,561	232.6	19.7	9.2	0.6
*r*3	718	922	442,903	317.1	13.7	620	891	435,177	312.3	13.9	13.6	1.5
*r*4	1,651	2,011	908,375	675.6	16.5	1,449	1,976	893,178	667.9	20.8	12.2	1.1
*r*5	2,469	3,134	1,368,370	1,041.0	20.3	2,208	3,022	1,349,334	1,027.0	22.4	10.6	1.3
*f*11	129	176	93,440	60.0	5.8	103	164	94,034	59.3	9.3	20.2	1.2
*f*12	114	160	90,281	56.8	10.2	97	163	88,850	57.0	7.3	14.9	−0.4
*f*21	102	160	99,179	58.4	7.8	94	162	95,920	58.1	7.5	7.8	0.6
*f*22	81	88	57,342	36.1	14.7	71	85	59,417	37.1	19.7	12.3	−2.7

Table 6.7 Stacked-TSV distribution in four-die 3D stacks. The clock source is located in the top die (die-0) or the middle die (die-1). Note that we do not need 3-stack TSVs for the middle-die case

	Clock src in the top die				Clock src in the middle die		
	Stacked TSV distribution				Stacked TSV distribution		
ckt	#TSVs	1-stack	2-stack	3-stack	#TSVs	1-stack	2-stack
$r1$	248	1×106	2×41	3×20	208	1×144	2×32
$r2$	434	1×239	2×69	3×19	394	1×282	2×56
$r3$	718	1×303	2×137	3×47	620	1×406	2×107
$r4$	1,651	1×665	2×307	3×124	1,449	1×901	2×274
$r5$	2,469	$1 \times 1{,}125$	2×444	3×152	2,208	$1 \times 1{,}464$	2×372
$f11$	129	1×40	2×28	3×11	103	1×67	2×18
$f12$	114	1×41	2×29	3×5	97	1×65	2×16
$f21$	102	1×39	2×24	3×5	94	1×58	2×18
$f22$	81	1×27	2×15	3×8	71	1×43	2×14
Ratio (%)	100	40.5	39.4	20.1	100	65.5	34.5

Table 6.7 presents a detailed list of TSV usage (stacked vs. non-stacked). When connecting two clock sinks in non-adjacent dies (e.g., die-1 and die-3), we can use either two 1-stack TSVs (TSVs are not stacked) or one 2-stack TSVs (TSVs are stacked). For both cases, the number of TSVs (#TSVs) is counted as two. If a clock network uses k N-stacked TSVs, the resulting #TSVs is calculated as $k \times N$. We observe that if the clock source is located in the middle, we use more 1-stack TSVs. In addition, we do not need to use 3-stack TSVs.

6.7.8 Impact of Buffer Load Constraint on Power and Slew

Table 6.8 shows the impact of *cmax* on wirelength (μm), power (mW), skew (ps), buffer count, and maximum slew (ps) as *cmax* increases from 150 to 300 fF. We use the four-die stack of benchmark r_1 and compare the single-TSV clock trees with the multi-TSV clock trees. We also report the reductions in wirelength and clock power.

We first observe that in the pre-bond testable clock tree design, bounding the maximum load capacitance for each buffer remains an efficient way to control the maximum slew. As *cmax* increases, the maximum slew in both the single-TSV and the multi-TSV cases increases. In other words, a tighter (smaller) *cmax* bound means better (smaller) slew. All of the slew values are below the 10 % constraint (100 ps). Second, the power and wirelength benefits of the multi-TSV design remain consistent regardless of the value of *cmax*. The multi-TSV approach achieves more than 40 % wirelength reduction and more than 21 % power reduction across the full range of *cmax*. Third, for all *cmax*, a multi-TSV tree uses fewer buffers but still achieves a slightly better maximum slew. Lastly, clock skew is less than 30 ps for both the single-TSV and the multi-TSV cases for all values of *cmax*. There is no obvious skew trend for the single-TSV case, but the skew tends to reduce with

Table 6.8 Impact of the *cmax* (upper bound for the buffer output load). We use the four-die stack of r_1

	Single TSV					Multi-TSV						% imprv	
cmax	#bufs	WL	Power	Skew	Slew	#TSVs	#bufs	WL	Power	Skew	Slew	WL	Power
150	676	272,732	180.8	22.6	37.1	262	545	157,908	134.3	5.6	37.4	42.10	25.72
175	578	271,403	168.9	22.0	43.9	259	486	159,395	128.0	6.3	44.0	41.27	24.22
200	488	272,995	159.7	8.8	51.5	251	428	158,020	121.0	6.7	50.5	42.12	24.23
225	431	269,901	152.1	11.3	58.7	251	386	158,926	117.3	7.3	54.0	41.12	22.88
250	387	268,709	146.9	9.7	67.4	250	359	158,860	114.1	8.3	59.7	40.88	22.33
275	357	275,939	146.6	12.4	76.4	248	334	161,954	112.7	11.4	71.0	41.31	23.12
300	318	272,355	141.8	10.5	86.6	248	325	160,394	111.4	13.3	80.8	41.11	21.44

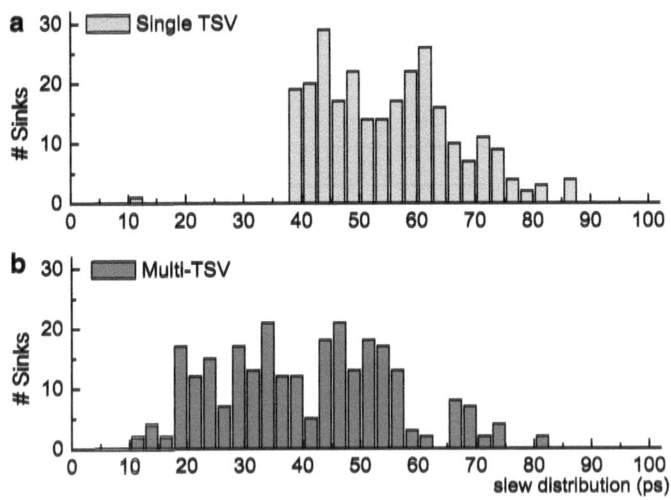

Fig. 6.12 Slew distribution for the four-die stack of r_1. The slew constraint is set to 10 % of the clock period. *cmax* is 300 fF. (**a**) A single-TSV clock tree, (**b**) a multi-TSV clock tree with 248 TSVs. We observe that the slew values are smaller for (**b**)

tighter *cmax* values for the multi-TSV cases. The main reason is that the wirelengths are shorter in the multi-TSV cases, which results in the clock buffers added for slew control having a positive impact on delay and skew as well.

Figure 6.12 shows the detailed slew distribution of the single-TSV and the multi-TSV clock trees based on the four-die stack of r_1. *cmax* is set to 300 fF. In the single-TSV case, slew varies from 12.3 to 86.6 ps with an average slew of 54.8 ps. The slew distribution of the multi-TSV case is 11.1–80.8 ps with an average slew value of 40.6 ps. Compared with the single-TSV case, the multi-TSV tree reduces the maximum slew and average slew by 5.8 and 14.2 ps, respectively, and shows a narrower slew distribution.

Figure 6.13 shows the impact of *cmax* on the clock power consumption and the slew distribution (minimum, average, and maximum). We use the four-die stack implementation of r_1 for this experiment. We observe that multi-TSV designs have a positive impact on the maximum and average slew, showing nice reductions in these metrics.

6.7.9 Impact of TSV Capacitance

As the TSV liner oxide thickness decreases, the TSV capacitance can increase to as much as 100 femto-farads. Tables 6.9 and 6.10 show comparison of wirelength (μm), buffer count, clock power consumption (mW), and clock skew (ps) as the TSV capacitance increases from 0 to 100 fF. We focus on the four-die stack

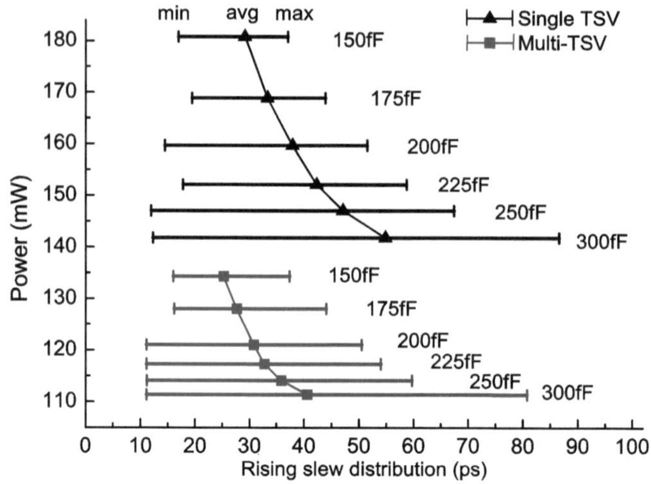

Fig. 6.13 Comparisons of slew variations and clock power between the single-TSV and multi-TSV clock trees based on the four-die stack of r_1. *cmax* varies from 150 to 300 fF

Table 6.9 Impact of the TSV capacitance on single vs. multiple TSV (= 183 TSVs). The results are normalized to the single-TSV case

	Single TSV				Multi-TSV (#TSVs = 183)				% imprv	
TSVCap	#bufs	WL	Power	Skew	#bufs	WL	Power	Skew	WL	Power
0	2,939	2,312,770	1,273.3	22.3	2,788	2,012,360	1,154.9	20.5	13.0	9.3
15	2,939	2,312,930	1,272.0	22.2	2,803	2,014,790	1,159.1	20.3	12.9	8.9
25	2,939	2,313,010	1,272.4	21.8	2,814	2,021,910	1,167.4	20.7	12.6	8.3
50	2,939	2,313,230	1,273.2	21.8	2,834	2,033,640	1,180.8	19.9	12.1	7.3
100	2,941	2,313,700	1,274.7	19.4	2,890	2,071,800	1,215.0	16.0	10.5	4.7

Table 6.10 Impact of the TSV capacitance on single vs. multiple TSV (= 2,469 TSVs). The results are normalized to the single-TSV case

	Single TSV				Multi-TSV (#TSVs = 2,469)				% imprv	
TSVCap	#bufs	WL	Power	Skew	#bufs	WL	Power	Skew	WL	Power
0	2,939	2,312,770	1,273.3	22.3	2,970	1,337,980	972.4	23.3	42.1	23.6
15	2,939	2,312,930	1,272.0	22.2	3,134	1,368,370	1,041.0	20.3	40.8	18.2
25	2,939	2,313,010	1,272.4	21.8	3,237	1,404,560	1,087.3	18.6	39.3	14.5
50	2,939	2,313,230	1,273.2	21.8	3,603	1,489,930	1,220.9	21.0	35.6	4.1
100	2,941	2,313,700	1,274.7	19.4	4,249	1,719,590	1,499.7	25.7	25.7	−17.7

implementation of r_5. We observe that the clock tree with 2,469 TSVs has the lowest power when the TSV capacitance is small (0, 15, or 25 fF). The clock tree with 183 TSVs obtains the lowest power if the TSV capacitance is high (50 or 100 fF). Therefore, we compare three TSV-usage cases: single TSV, multi-TSV with 183 TSVs, and multi-TSV with 2,469 TSVs. We also report the wirelength and power consumption.

We first observe that, for a fixed number of TSVs, a larger TSV capacitance leads to longer wirelength, more clock buffers, and higher clock power consumption. For example, for the 183 TSVs case, as the TSV capacitance increases from 0 to 100 fF, wirelength, buffer count and clock power increase by 3.0, 3.7, and 5.3 %, respectively. There are two reasons for these trends. First, a larger TSV capacitance increases the difference between the internal delays of subtrees on different dies. As a result, longer wires and additional clock buffers are required to re-balance these subtrees. The larger the TSV capacitance, the longer the wirelength, and the greater the clock buffer count required to equalize the delays. Second, in order to meet the slew constraint, the load capacitance of each clock buffer is constrained below *cmax*. This means that as the capacitance of the clock network increases, more clock buffers must be inserted to control the slew. Therefore, more clock buffers are also required for slew control.

Our second observation is that the TSV capacitance diminishes the advantages of our multiple-TSV approach in terms of wirelength and power reduction. As the TSV capacitance increases from 0 to 100 fF, the wirelength reduction decreases from 13.0 to 10.5 % for the 183 TSVs designs and from 42.1 % down to 25.7 % for the 2,469 TSVs designs. Similarly, power savings decrease from 9.3 to 4.7 % for 183 TSVs and from 23.6 to −17.7 % in 2,469 TSVs cases. Finally, we observe that the TSV count and the TSV parasitics have little effect on the effectiveness of our algorithm. The clock skew is well controlled under 30 ps for all cases.

6.7.10 Impact of TSV Bound and Capacitance

Figure 6.14 shows the impact of the TSV capacitance (TSVCap) and TSV bound on clock power, wirelength, and buffer count (#Bufs) trends. We use the four-die stack implementation of r_5. These metrics are normalized to the results from a design with a single column of TSVs. The TSV capacitance increases from 0 to 100 fF. Given both a TSVCap and a TSV bound, we construct a pre-bond testable 3D clock tree, run SPICE simulation on the tree, and report the clock power, wirelength, and buffer count.

We observe that using multiple TSVs affects the clock power in different ways, which depends on the TSV capacitance. First, when the TSV capacitance is small (from 0 to 25 fF), we observe that using many TSVs helps to reduce the wirelength, buffer count, and clock power. We obtain the lowest power using 2,469 TSVs. In the ideal case when using 0 fF TSVs, we can achieve up to a 23.6 % power reduction compared with the single-TSV case, and wirelength is reduced by more than 42 %. For the 15 or 25 fF TSVs, power is reduced by 18.2 and 14.5 %, respectively.

Second, when the TSV capacitance is large (such as 50 or 100 fF), clock power first decreases and then increases when using more TSVs. In Fig. 6.14, when TSVCap is 100 fF, the lowest clock power (a 4.7 % power reduction) comes from the clock tree with 183 TSVs. When thousands of TSVs are used, power increases significantly.

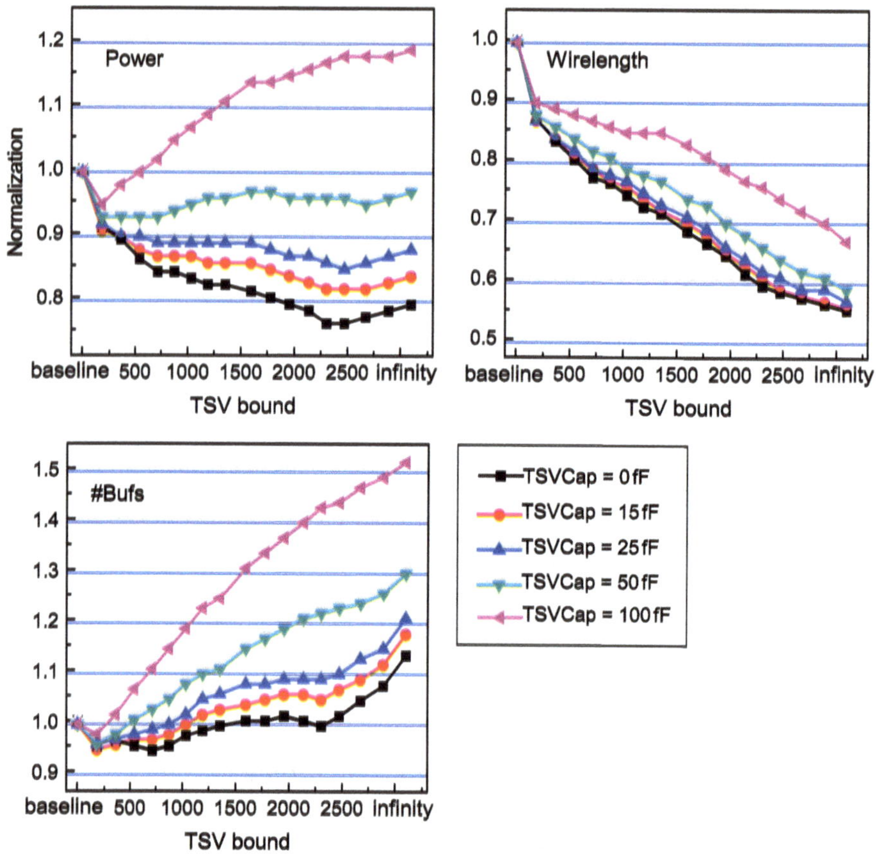

Fig. 6.14 Impact of the TSV capacitance and the TSV usage on the clock power consumption, wirelength, and buffer count trends based on the four-die stack of r_5. The baselines are the single-TSV clock tree for each value of the TSV capacitance

Third, as the TSV capacitance increases, it becomes more challenging to achieve a low-power clock network. Based on 0 fF TSVs, the multi-TSV policy is able to obtain a low-power design with 23.6 % power saving; for 100 fF TSVs, the multi-TSV strategy can only achieve 4.7 % power reduction.

Those observations result mainly from the following factors. First, TSV usage and the TSV capacitance have opposite effects on wirelength: using more TSVs tends to reduce the size of each subtree in the non-clock source dies, reducing the wirelength. However, TSVs with large capacitance tend to unbalance the subtrees, increasing wire snaking. Depending on which factor dominates – the wirelength increase from the large TSV capacitance or the wirelength reduction from multiple TSVs – the trend of the total wirelength changes dramatically. The same discussion applies to the buffer count.

Table 6.11 Comparisons with [11]

	MMM-3D + ZCTE-3D [11]			Ours		
ckt	#TSVs	WL	Delay	#TSVs	WL	Delay
r1	83	1,441,849	1.64	74	1,567,927	1.7
r2	197	2,831,346	4.34	176	3,133,533	4.44
r3	276	3,725,294	6.37	245	4,036,177	6.89
r4	653	7,424,886	19.28	566	8,162,013	19.95
r5	1,052	10,940,984	35.2	943	11,806,895	36.21

Lastly, clock power is consumed by the capacitance of the wires, buffers, and TSVs. The multi-TSV strategy helps to reduce the power consumed by the wires but at the cost of increasing the power consumed in the TSVs. When using large capacitance TSVs, the TSV power consumption increases faster than wire power consumption decreases, so the total clock power increases. Therefore, as the TSV capacitance grows, the lowest-power design is achieved with just a few TSVs. In general, a large TSV capacitance makes it hard to achieve a low-power pre-bond testable 3D clock tree.

6.7.11 Comparison with Existing Work

In Table 6.11, we show the comparison of our work with [11]. Note that [11] does not support pre-bond testability, insert buffers, nor provide any SPICE simulation results. However, we attempted a comparison with [11] by disabling our support for pre-bond testing and buffer insertion. We use the same benchmark settings and report the skew/delay values in the Elmore delay model. We observe that our method uses 10.4–13.3 % fewer TSVs than [11] while using 7.9–10.7 % more wirelength. Note that in our work we can control the TSV count versus wirelength tradeoff by tweaking the TSV bound. In addition, these results come from unbuffered clock trees. Our pre-bond testable algorithm supports buffer insertion, which helps to properly control wire snaking and therefore better minimizes the wirelength.

6.8 Conclusions

In this chapter, we studied how to construct a clock tree for a 3D stacked IC so that both enables test of each die before bonding and provides a minimum-power clock network after bonding. Our solution utilizes many TSVs to reduce wirelength and clock power but necessitates the use of new circuit elements – TSV-buffers and transmission gates – in the clock tree to support the low-skew and low-power characteristics. We studied the impact of buffer insertion on slew rate in 3D stacked ICs clocking. SPICE results show that our method of inserting multiple TSVs into the clock tree significantly reduces the wirelength and power consumption of the 3D clock tree as compared against a single-TSV baseline. We also studied the impact of the TSV parasitic capacitance on power consumption and wirelength. It shows that a larger TSV capacitance makes it harder to optimize 3D pre-bond testable clock trees.

Some designs allow (or even necessitate) multiple clock probe pads for each die for pre-bond test. When this happens, the test equipment must provide multiple clock probes with good OTA. If multiple clock domains are used, we will need a separate clock probe pad for each domain in each die for pre-bond testing. In this case, the wirelength of the redundant trees is likely to reduce because they need to connect fewer subtrees. This will likely lead to more power savings.

The authors in [15] discussed the importance of testing on the partial stacks, where testing is done not only to individual die before bonding or to the entire 3D stack but also to the partially bonded dies. For stacks containing more than two dies, applying test after each bonding step helps to enhance the yield, but significantly increases the cost of test. In addition, the clock network in these partial stacks will suffer from high clock skew during testing. Therefore, clock delivery becomes challenging for this test method.

References

1. C. Albrecht, A.B. Kahng, B. Liu, I.I. Mandoiu, A.Z. Zelikovsky, On the skew-bounded minimum-buffer routing tree problem. IEEE Trans. Comput. Aided Des. Integr. Circuits Syst. **22**(7), 937–945 (2003)
2. C.J. Alpert, A.B. Kahng, B. Liu, I.I. Mandoiu, A.Z. Zelikovsky, Minimum buffered routing with bounded capacitive load for slew rate and reliability control. IEEE Trans. Comput. Aided Des. Integr. Circuits Syst. **22**(3), 241–253 (2003)
3. V. Arunachalam, W. Burleson, Low-power clock distribution in a multilayer core 3D microprocessor, in *Proceedings of Great Lakes Symposium on VLSI* (ACM, New York, 2008), pp. 429–434
4. K.D. Boese, A.B. Kahng, Zero-skew clock routing trees with minimum wirelength, in *Proceedings of Fifth Annual IEEE International ASIC Conference and Exhibit, 1992* (IEEE, Piscataway/New York, 1992), pp. 17–21
5. W.C. Elmore, The transient analysis of damped linear networks with particular regard to wideband amplifiers. J. Appl. Phys. **19**(1), 55–63 (1948)
6. GSRC Benchmark, http://vlsicad.ucsd.edu/GSRC/bookshelf/Slots/BST
7. S. Hu, C.J. Alpert, J. Hu, S.K. Karandikar, Z. Li, W. Shi, C.N. Sze, Fast algorithms for slew-constrained minimum cost buffering. IEEE Trans. Comput. Aided Des. Integr. Circuits Syst. **26**(11), 2009–2022 (2007)
8. ISPD Contest (2009), http://www.sigda.org/ispd/contests/ispd09cts.html
9. L. Jiang, L. Huang, Q. Xu, Test architecture design and optimization for three-dimensional SoCs, in *Proceedings of Design, Automation and Test in Europe*, (IEEE, Los Alamitos/California, 2009), European Design and Automation Association, Belgium, pp. 220–225
10. L. Jiang, Q. Xu, K. Chakrabarty, T.M. Mak, Layout-driven test-architecture design and optimization for 3D SoCs under pre-bond test-pin-count constraint, in *Proceedings of IEEE International Conference on Computer-Aided Design* (ACM, New York, 2009), pp. 191–196

11. T.-Y. Kim, T. Kim, Clock tree embedding for 3D ICs, in *Proceedings of Asia and South Pacific Design Automation Conference* (IEEE, Piscataway, 2010), pp. 486–491
12. H.-H.S. Lee, K. Chakrabarty, Test challenges for 3D integrated circuits. IEEE Des. Test Comput. **26**(5), 26–35 (2009)
13. D.L. Lewis, H.-H.S. Lee, A scan-Island based design enabling pre-bond testbility in die-stacked microprocessors, in *IEEE International Test Conference* (IEEE, New York; International Test Conference, Washington, DC, 2007), pp. 1–8
14. D.L. Lewis, H.-H.S. Lee, Testing circuit-partitioned 3D IC designs, in *Proceedings of International Symposium on VLSI* (IEEE, Piscataway, 2009), pp. 139–144
15. E.J. Marinissen, Y. Zorian, Testing 3D chips containing through-silicon vias, in *IEEE International Test Conference* (International Test Conference, Washington, DC, 2009), pp. 1–11
16. J. Minz, X. Zhao, S.K. Lim, Buffered clock tree synthesis for 3D ICs under thermal variations, in *Proceedings of Asia and South Pacific Design Automation Conference* (IEEE, Piscataway, 2008), pp. 504–509
17. B. Noia, K. Chakrabarty, Y. Xie, Test-wrapper optimization for embedded cores in TSV-based three-dimensional SOCs, in *Proceedings of IEEE International Conference on Computer Design* (IEEE, Piscataway, 2009), pp. 70–77
18. V.F. Pavlidis, I. Savidis, E.G. Friedman, Clock distribution networks for 3-D integrated circuits, in *Proceedings of IEEE Custom Integrated Circuits Conference* (IEEE, Piscataway, 2008), pp. 651–654
19. Predictive Technology Model, http://ptm.asu.edu/
20. RMST-Pack, http://vlsicad.ucsd.edu/GSRC/bookshelf/Slots/RSMT/RMST/
21. Synopsys, Raphael, http://www.synopsys.com
22. G.E. Tellez, M. Sarrafzadeh, Minimal buffer insertion in clock trees with skew and slew rate constraints. IEEE Trans. Comput. Aided Des. Integr. Circuits Syst. **16**(4), 333–342 (1997)
23. Verigy V93000 SOC Series Pin Scale Digital Cards, http://www1.verigy.com
24. X. Wu, P. Falkenstern, Y. Xie, Scan chain design for three-dimensional integrated circuits (3D ICs), in *Proceedings of IEEE International Conference on Computer Design* (IEEE, Los Alamitos/California/Piscataway, 2007), pp. 208–214, 2007
25. X. Wu, Y. Chen, K. Chakrabarty, Y. Xie, Test-access mechanism optimization for core-based three-dimensional SOCs, in *Proceedings of IEEE International Conference on Computer Design* (IEEE, Los Alamitos/California/Piscataway, 2008), pp. 212–218
26. X. Zhao, S.K. Lim, Power and slew-aware clock network design for through-silicon-via (TSV) based 3D ICs, in *Proceedings of Asia and South Pacific Design Automation Conference* (IEEE, Piscataway, 2010), pp. 175–180
27. X. Zhao, D.L. Lewis, H.-H.S. Lee, S.K. Lim, Pre-bond testable Low-Power Clock Tree Design for Pre-Bond Testing of 3-D Stacked ICs, in *Proceedings of IEEE International Conference on Computer-Aided Design of Integrated Circuits and Systems* **30**(5), 732–745, (ACM, New York, 2011)

Part II
Electrical Reliability in 3D IC Designs

This part of the book covers design-for-electrical-reliability for 3D ICs. In Chap. 7, we study the TSV-to-TSV coupling issues and investigate various ways to alleviate the associated problems. In Chap. 8, we investigate the current crowding problem at the wire-to-TSV junction in the power delivery network and its impact on IR-drop. In Chap. 9, we study the electromigration failure mechanisms in TSVs caused by the current density, mechanical stress, and thermal gradient issues in 3D ICs.

Chapter 7
TSV-to-TSV Coupling Analysis and Optimization

Abstract In this chapter we study TSV-to-TSV coupling in 3D ICs [4]. A full-chip signal integrity (SI) analysis flow is developed based on the proposed coupling model. Analysis results show that TSVs cause significant coupling noise and timing problems despite the fact that TSV count is much smaller compared with the gate count. Two approaches are proposed to alleviate TSV-to-TSV coupling, namely TSV shielding and buffer insertion. Analysis results show that both approaches are effective in reducing the TSV-caused-coupling and improving timing.

The materials presented in this chapter are based on [4].

7.1 Introduction

Through-Silicon-Via (TSV) and 3D stacking technology are currently being actively evaluated as a potential solution to alleviate the interconnect delay problems in giga-scale circuits and systems [8]. Some studies have been done to show that 3D ICs have advantages in total wire length [1] and timing performance [5] compared with 2D ICs.

However, signal integrity (SI) is another key challenge caused by the advance of nano-scale interconnect technologies because of the rising number of analog effects. Due to the big TSV size it is highly possible that TSVs will introduce new coupling sources, which are bad to the circuit's SI performance. A big coupling noise between interconnections has two major impacts on the circuit performance. First, it increases the path delay due to the Miller effect. When the aggressor and the victim signals switch in the opposite direction, the effective coupling capacitance between them doubles and thus degrades timing. Second, the coupling noise can result in a wrong logic function. For dynamic logic, the coupling noise causes charge-sharing,

S.K. Lim, *Design for High Performance, Low Power, and Reliable 3D Integrated Circuits*, DOI 10.1007/978-1-4419-9542-1_7,

which may flip the signal unintentionally. For static logic, the coupling noise can change the state of the sequential element by flipping the cross-coupled inverter loop.

To gain comprehensive understanding of SI issues in 3D ICs, we still need to answer the following two questions: (1) How much SI issues do the TSVs cause to the 3D IC design from a full-chip perspective? (2) If the impact of TSVs to the full-ship SI is non-negligible, how should we alleviate the TSV-caused coupling problem from a designer's perspective? We answer these two questions in this chapter. The following specific topics are covered in this chapter:

- We study the on-chip TSV-to-TSV coupling and present a compact circuit model for full-chip SI analysis. In addition, we, for the first time, observed that changing the distance between TSVs is inefficient in reducing TSV-to-TSV coupling level for low frequency signals (under a few GHz).
- We study full-chip 3D SI analysis using an accurate TSV-to-TSV coupling circuit model. Analysis results show that TSV-to-TSV coupling has a big impact on the full-chip coupling noise and timing performance.
- We study two approaches for full-chip optimization to alleviate the TSV-caused coupling problem, namely, buffer insertion and TSV shielding.

7.2 Existing Studies

Several Studies have been done to illustrate the impact of TSVs on SI in 3D ICs [7,9]. However, these studies only look at simple individual coupling cases in device level. Recently, there have been several studies presented to investigate TSV-to-TSV coupling from the device level. Weerasekera et al. [9] studied a specific case, where 9 TSVs are placed as a 3×3 array. Savidis and Friedman [7] gave an analytical model for the coupling capacitance between TSVs. However, these models ignore the TSV liner, which has big contributions to TSV coupling.

7.3 TSV-Induced Coupling Model

7.3.1 Coupling Sources Due to TSVs

TSVs introduce several new coupling sources to 3D ICs. The first coupling source is the big TSV landing pad to the wires and devices. Considering the TSV landing pad is big (typically 25 μm^2), which occupies several standard cell rows, the metal wire running above or next to it will suffer from significant coupling capacitance. Fortunately, this coupling source can be analyzed by existing SI tools easily, because it is essentially a traditional wire coupling problem.

Another coupling source is TSV-to-device coupling. This coupling happens between the TSV and the S/D region of the MOSFET through the substrate. The coupling path is mainly the silicon-bulk surface, which can be well controlled by substrate contact. Therefore, by adding sufficient substrate contacts, the surface can be strongly tied to ground, thus alleviating TSV-to-device coupling.

The third coupling source is TSV-to-TSV coupling. Different from TSV-to-device coupling, TSV-to-TSV coupling happens not only on the silicon-bulk surface, but also deep inside the substrate, because TSV is a via that goes through the entire substrate. Considering the height of the TSV (typically 50–100 μm), simply adding substrate contact cannot guarantee to eliminate this coupling. Therefore, TSV-to-TSV coupling usually cannot be ignored. Moreover, TSV-to-TSV coupling scheme is totally different from the traditional wire coupling. In wire coupling case, two wires and the dielectric between them form a capacitor, through which the two wires are coupled. In contrast, TSV-to-TSV coupling is more complicated. Two TSVs are coupled through two liner layers and the silicon substrate, which cannot be treated as a single capacitor. Because of this difference, it is difficult for existing SI analysis tools to directly handle TSV-to-TSV coupling. Due to these reasons, we focus on the TSV-to-TSV coupling issues.

7.3.2 TSV-to-TSV Coupling Modeling

In this chapter we study a TSV-to-TSV coupling model for full-chip coupling analysis. Unlike vias in PCBs and packages, TSVs inside ICs are surrounded by a thin liner. In addition, silicon substrate is very lossy, and has not only resistive components, but also capacitive components. Therefore, the TSV-to-TSV structure must contain all components in the coupling path, including TSV copper, liner layer, silicon substrate and I/O drivers. Figure 7.1 shows an equivalent circuit model for the TSV-to-TSV coupling structure. A similar TSV modeling work [2] also considered these components. However, this model was devised to analyze signal transmission, and only considered one signal TSV with two ground TSVs.

A TSV can be modeled by a resistor (R_{TSV}) and an inductor (L_{TSV}) in series, and the liner which surrounds the TSV could be modeled as a capacitor (C_{TSV}). Silicon substrate could be modeled by a resistor (R_{si}) and a capacitor (C_{si}) that is in parallel. We use the following equations to calculate the value of these components:

$$C_{\mathrm{TSV}} = \frac{1}{4}\frac{2\pi\varepsilon_0\varepsilon_{\mathrm{r}}}{\ln\frac{r_{\mathrm{TSV}}+t_{\mathrm{ox}}}{r_{\mathrm{TSV}}}} \cdot l_{\mathrm{TSV}} \tag{7.1}$$

$$C_{\mathrm{si}} = \frac{\pi\varepsilon_0\varepsilon_{\mathrm{r}}}{\ln\{\frac{d}{2r_{TSV}} + \sqrt{(\frac{d}{2r_{TSV}})^2 - 1}\}} \tag{7.2}$$

$$R_{\mathrm{si}} = \frac{\varepsilon}{C_{\mathrm{si}}\sigma} \tag{7.3}$$

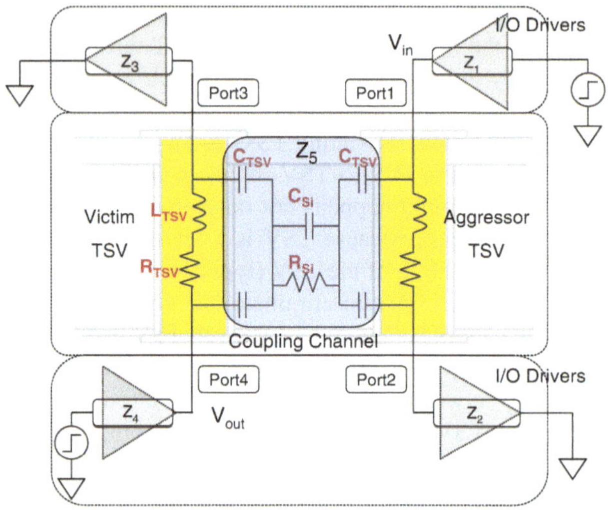

Fig. 7.1 TSV-to-TSV coupling model

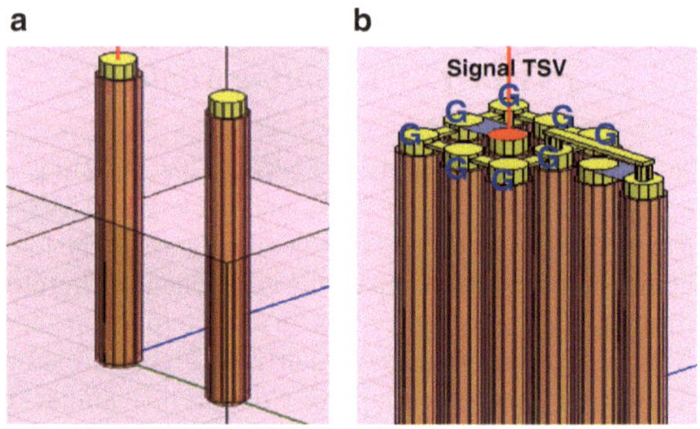

Fig. 7.2 Coupling structure for HFSS simulation. (**a**) two signal TSVs and (**b**) one unshielded and one shielded signal TSV (= surrounded by eight ground TSVs) shown in *red*

where r_{TSV} is the TSV radius, l_{TSV} is the TSV height, t_{ox} is the thickness of the liner, and d is the pitch between two TSVs.

This lumped circuit model is validated by a commercial 3D electromagnetic simulator (Ansoft HFSS) using S-parameter simulation. A simulation structure built for HFSS is shown in Fig. 7.2a. Figure 7.3a shows the S-parameter result comparison between the HFSS structure and the lumped model. We see the model is very accurate in the simulated frequency range. We use this lumped model to perform transient simulation and measure the coupling noise on the victim TSV. The simulation is performed using 45 nm technology with 1.2 V power supply. Simulation result shows that the peak-to-peak coupling noise can reach up to 200 mV, which is non-negligible, as shown in Fig. 7.3b.

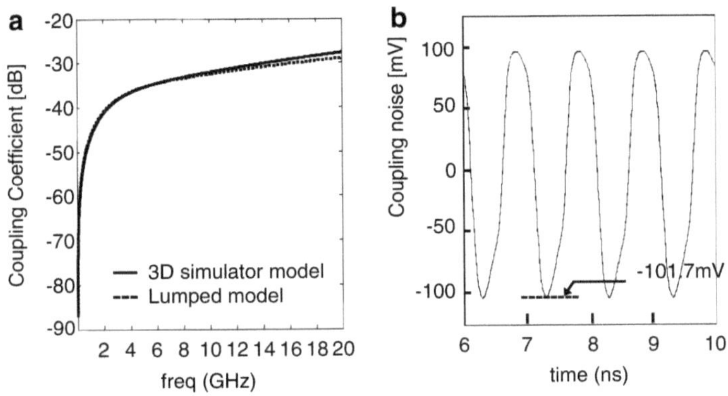

Fig. 7.3 (**a**) S-parameter simulation for coupling coefficient. (**b**) Transient response for the victim TSV in a coupled TSV pair shown in Fig. 7.2a

7.4 Full-Chip Signal Integrity Analysis

By studying the simple TSV coupling pair, we showed that TSV-to-TSV coupling is non-negligible in Sect. 7.2. However, compared with the standard cell count, the TSV count is much smaller in a realistic design. Therefore, whether TSV coupling will cause troubles in a real digital design will still be a question. In this section, we try to answer this question by performing full-chip SI analysis while considering TSV-related coupling. The TSV-related coupling we are dealing in this chapter is mainly TSV landing-pad related coupling and TSV-to-TSV coupling. The former can be handled by existing tools (CeltIC, Primetime, etc.). We use the coupling model developed in Sect. 7.2 to help analyze TSV-to-TSV coupling.

7.4.1 Full Chip 3D SI Analysis Flow

Currently, existing SI analysis tools cannot well handle 3D circuits. There are two major reasons. First, 3D SI analysis tool must consider all nets and all TSVs in all the tiers simultaneously. This is because the total noise experienced by a 3D net may come from coupling within the same tier as well as neighboring tiers. Second, current SI analysis tools can only handle simple wire-to-wire capacitive coupling. As discussed in Sect. 7.2, TSV-to-TSV coupling consists of complicated coupling network, which cannot be handled by existing SI analysis tools.

To solve these two problems, we designed our 3D SI flow, which utilizes our own scripts in combination with the existing circuit simulation (= HSPICE) tools and timing analysis (= PrimeTime) tools. First, we use RC extraction tool to obtain the SPEF files containing the interconnect RC information for each die. Then a top-level

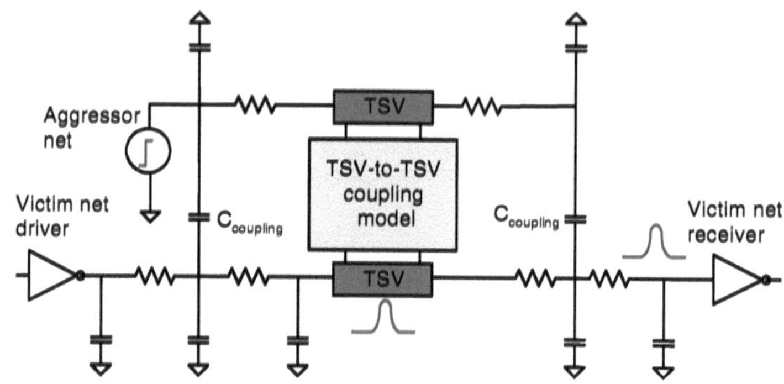

Fig. 7.4 An example of a SPICE netlist for coupling noise analysis

Verilog file and a top-level SPEF file are generated containing the information of all the dies using our in-house tool. We also make a script to find out which TSVs interfere with each other based on their locations and record the TSV-to-TSV coupling information.

Once these files are ready, we use PrimeTime to read in Verilog files and SPEF files in incremental mode, and generate a new stitched SPEF file containing the RC information of all the dies and the TSVs. Then we use our script to analyze the stitched SPEF file and generate the SPICE netlist for each individual net for coupling noise simulation. Note that each individual net contains the wire coupling information obtained from RC extraction. During SPICE netlist generation, the script also automatically plugs in the TSV-to-TSV coupling circuit model based on the TSV-coupling model in Sect. 7.2. Then the aggressor signal and the victim driver model are also applied to the SPICE netlist. Using the generated SPICE netlist as shown in Fig. 7.4, we perform SPICE simulation on each victim nets one by one and record the peak noise at each port.

7.4.2 *Design and Analysis Results*

We use FIR32, which is a 32-bit FIR filter as a test circuit. The circuit has 35 K gates and 548 TSVs. The design is a Two-die 3D IC based on 45 nm technology. Our TSV is 4 μm in diameter and 60 μm in height. The TSV landing pad is 5 ×5 μm, which occupies three standard-cell rows. Each TSV also has a 0.5 μm keep-out zone, where no standard cells can be placed inside. We use our Cadence Encounter-based tool flow to generate 3D layouts [3]. The 3D timing optimization is performed using the timing-scaling method in [3]. In the following experiments, we use both original design and timing-optimized design for comparison.

After the designs are ready, we perform coupling noise analysis using the proposed flow. The analysis compares two cases with and without considering

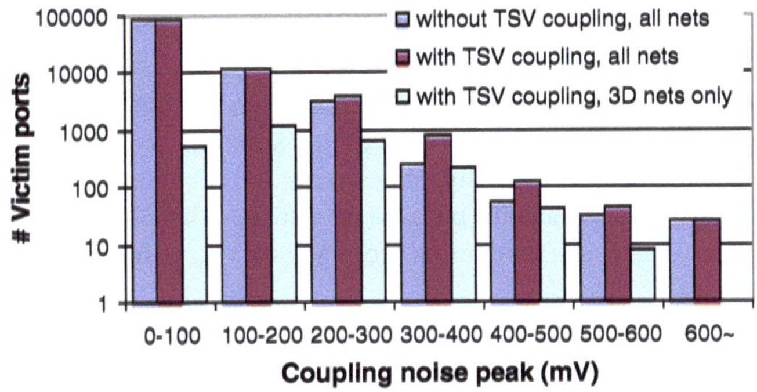

Fig. 7.5 Glitch analysis results comparison

Table 7.1 Impact of TSV-to-TSV coupling on crosstalk and timing

	Original design		Timing optimized	
	No coupling	w/ coupling	No coupling	w/ coupling
Footprint (μm^2)	402×402	402×402	402×402	402×402
Wirelength (μm)	7.506×10^5	7.506×10^5	7.516×10^5	7.516×10^5
Total coupling noise (V)	4,518.75	4,955.15	4,230.74	4,548.17
Longest path delay (ns)	13.09	22.79	5.54	9.24
Total negative slack (ns)	600.498	1,175.14	335.076	836.88

TSV-to-TSV coupling. Based on the analysis results, we have two major observations. First, TSV-to-TSV coupling increases the total coupling-noise. The total noise for the original design increases from 4,518 to 4,955 V after considering TSV-to-TSV coupling. The total coupling noise on 3D nets is 471 V, which is responsible for most of the total noise increase. Second, the contribution of TSV-to-TSV coupling is more on the high noise region. Figure 7.5 shows the coupling-noise distribution comparison. We also show the noise distribution only on 3D nets. We observe that after considering TSV-to-TSV coupling, the design has more victim ports with noise above 300 mV. The average coupling noise on a 3D net is 170 mV, which is three times more than that on the a 2D net (43 mV). In summary, although the TSV count is much smaller than the gate count, it can still cause non-negligible coupling noise problem, especially in the high-noise region.

Besides coupling noise, TSV coupling also has significant contribution to timing degradation. We perform timing analysis on both original design and timing optimized design. The results show that due to TSV coupling, the longest path delay (LPD) almost doubles compared with the design without TSV-to-TSV coupling. A similar trend exists for the total negative slack (TNS). Table 7.1 summarizes the overall analysis results on the impact of TSV-to-TSV coupling.

7.5 TSV-to-TSV Coupling Reduction

After realizing that TSV-to-TSV coupling has significant contributions to the SI and timing performance degradation, we need to find solutions to reduce TSV coupling. We start from analyzing the coupling model in Fig. 7.1. For simplification, we ignore the TSV resistance and inductance because they are very small. Using this simplified model, we derive the transfer function from V_{in} to V_{out} using Kirchhoff's law, as shown in Eq. (7.4):

$$V_{out} = V_{\text{in}} \cdot \frac{Z_2 Z_3 Z_4}{Z_1 \cdot Z_A + Z_2 Z_3 Z_4 + Z_5 \cdot Z_B} \tag{7.4}$$

where

$$Z_A = Z_2 Z_3 + Z_2 Z_4 + Z_3 Z_4 + Z_3 Z_5 \tag{7.5}$$

$$Z_B = Z_1 Z_4 + Z_2 Z_3 + Z_2 Z_4 \tag{7.6}$$

$$Z_5 = Z_{C_{si}} // Z_{R_{\text{si}}} + 2Z_{C_{\text{TSV}}} \tag{7.7}$$

Equation (7.4) shows that the coupling level between two TSVs depends on the coupling-channel impedance (Z_5), the termination condition (Z_2, Z_3, Z_4) and the driver condition (Z_1). In another words, to reduce the coupling level, we can either increase the channel impedance, decrease the port impedance at the victim net, or increase the impedance at the aggressor driving port. From a designer's perspective, possible methods include: (1) increase TSV distances (to increase Z_5), (2) shield the victim TSVs (to cutoff the coupling path and increase Z_5), (3) insert buffers at the victim net (to decrease Z_4), (4) decrease the driver size at the aggressor net. (to increase Z_1), and (5) increase the load at both victim and aggressor net (to decrease Z_3 and Z_2). Since option 4 and 5 have negative implications to timing performance, our focus is on the first three options.

7.5.1 *Why TSV Spacing Is Inefficient*

To solve the traditional wire coupling problem, the most intuitive way is to increase the distance between the coupled wires. However, for TSV-to-TSV coupling, increasing the TSV distance is not an effective method, which is completely different from wire coupling. This is because in low frequency region (under a few GHz), the coupling channel impedance Z_5 is mainly determined by C_{TSV}. Increasing the TSV distance has big impact on R_{si} and C_{si}, but has little impact on C_{TSV}. Therefore, the total coupling channel impedance Z_5 is not sensitive to the TSV distance. To verify this, we perform transient simulations to examine the coupling noise variation with different TSV distances. The signal frequency is 300 MHz with 1.2 V power supply. Figure 7.6a, b show the victim transient response with

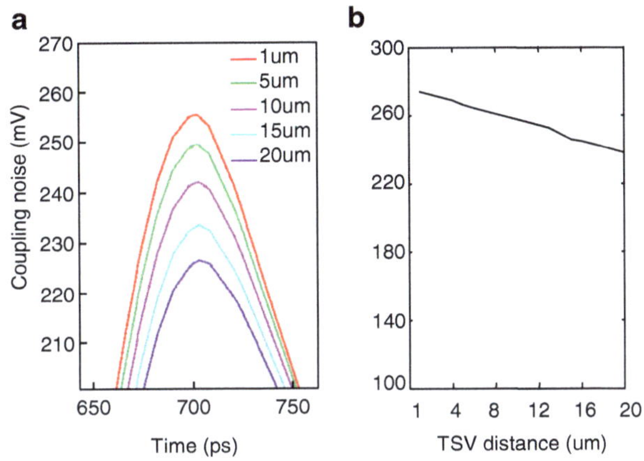

Fig. 7.6 Glitch peak with different TSV distances. (**a**) Shows the transient response of the victim TSV and (**b**) shows the relationship between coupling noise peak and TSV distance

different TSV distances. When TSV distance varies from 1 to 20 μm, the glitch noise only decreases from 255 to 224 mV. Therefore, trying to reduce TSV coupling by increasing the TSV distance proves to be inefficient. Thus, we exclude TSV repositioning from the potential solutions.

7.5.2 TSV Shielding to Alleviate Coupling

Similar to the coaxial cable, we use ground TSVs to shield a sensitive signal TSV as shown in Fig. 7.2b. By doing this, the coupling path through the substrate is cutoff so that the coupling from other signal TSVs is minimized. To verify how effective the shielded structure is in terms of coupling noise reduction, we create an HFSS structure, which consists of a shielded TSV and an aggressor TSV as shown in Fig. 7.2b, and perform S-parameter and transient simulations. The shielding structure we use contains eight ground TSVs. We apply an aggressor signal to the aggressor TSV nearby. S-parameter simulation result in Fig. 7.7a shows that the coupling level between the two signal TSVs is below −60 dB. Transient simulation result in Fig. 7.7b shows that the coupling noise is below 10 mV, which agrees with the S-parameter simulation. Therefore, we conclude that with the proposed shielding structure, the coupling between the shielded TSV and neighboring signal TSV is negligible.

Based on this observation, we study a design optimization flow utilizing shielded TSVs. This flow is performed after cell placement. The basic idea is to gradually replace TSVs which suffer from severe coupling with shielded TSVs. To perform this optimization flow, we need to define a new shielded TSV cell in the standard cell library. Since the shielded TSV cell is much larger than a regular TSV, we need to pay for a bigger footprint area.

In our flow, all the TSV pins are converted into TSV cells first. Using the coupling model in Fig. 7.1, TSVs are then sorted by the coupling-path impedance. As we

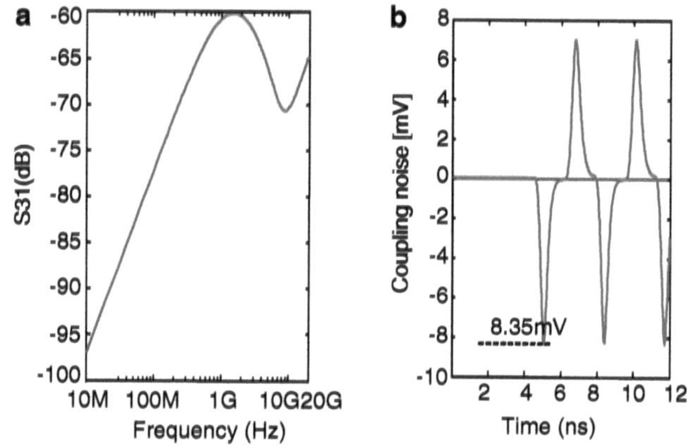

Fig. 7.7 (**a**) S-parameter results of coupling coefficient (**b**) transient simulation for the coupling noise on the shielded TSV

discussed in the beginning of this section, the smaller the total impedance is, the bigger the coupling level is. Then, we generate a TSV list, which contains TSVs to be replaced with shielded TSV cells. To generate the list, we start from the TSV with the highest coupling level and gradually choose the TSVs based on the coupling level order until we reach the coupling level threshold. After one TSV is chosen, we mark all its neighbors so that they will not be chosen. The reason we skip the neighbors is that we do not want the shielded TSVs to gather together because it is likely to cause over compensation. After we obtain the TSV list, we recalculate the chip area based on the number of TSVs shielded and redo floorplanning. Then we replace the TSVs in the list with shielded TSVs and perform ECO placement to remove the overlaps. We perform this flow iteratively until total TSV coupling level is below the desired value.

Figure 7.8b shows the layout after TSV shielding. There are 118 TSV cells replaced with shielded TSV cells. As a result, the total chip area increases from 402 × 402 to 421 ×421 μm. Based on this layout, we perform routing and perform full-chip noise analysis and timing analysis. Table 7.2 summarizes the analysis results. We see that TSV shielding reduces total coupling noise from 4,955.15 to 3,376.98 V for the original design, and from 4,548.17 to 3,032.16 V for the timing-optimized design. Note that this noise reduction is not only from the 3D nets, but also from the 2D nets because of the less congested routing resulted from the increased area. If we only look at the coupling noise on 3D nets, the total coupling noise decreases from 473.07 to 273.46 V for the original design. Table 7.3 shows the noise distribution comparison for 3D nets between the original design and the TSV-shielded design. We see that compared with the original design, the noise distribution moves to the low-noise region. We observe that the same trend exists in the timing-optimized design.

Besides coupling noise reduction, the timing performance also improves. As shown in Table 7.2, the longest path delay reduces from 22.79 to 12.86 ns

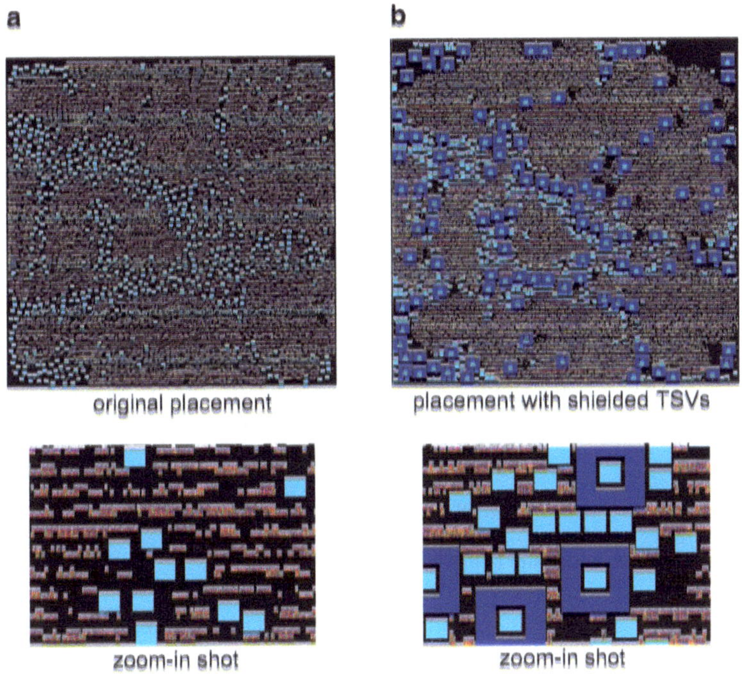

Fig. 7.8 Various die shots using Virtuoso. *Blue squares* denote M1 TSV landing pads. (**a**) Original design. (**b**) Design with shielded TSVs

Table 7.2 TSV shielding results

	Original design		Timing optimized	
	No shielding	w/ shielding	No shielding	w/ shielding
Area (μm^2)	402 × 402	421 × 421	402 × 402	421 × 421
Shielded-TSV count	0	118	0	118
Total noise (V)	4,955.15	3,376.98	4,548.17	3,032.16
Total noise (3D nets) (V)	471.091	273.46	329.967	226.525
Longest path delay (ns)	22.79	12.86	9.24	6.34
Total negative slack (ns)	1,175.14	706.581	806.88	371.175

Table 7.3 Coupling noise peak distribution. We report # of 3D victim net ports before and after TSV shielding

Noise (mV)	0–100	100–200	200–300	300–400	400–500	500–600
Before	539	1,254	659	210	41	8
After	1,539	727	314	119	12	0

for the original design and from 9.24 to 6.24 ns for the timing-optimized design. We observe the same trend on the total negative slack. Therefore, we conclude that TSV shielding is an effective way in alleviating TSV-caused crosstalk and timing problems. However, the cost we need to pay is the increased area.

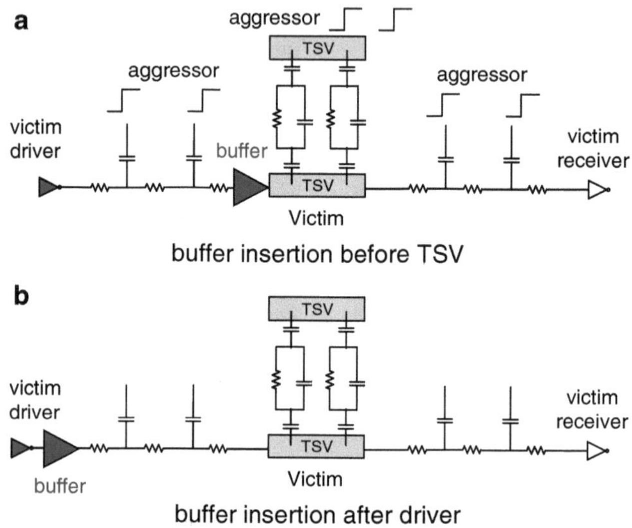

Fig. 7.9 Coupling reduction with buffering

Table 7.4 Impact of buffer insertion on the 3D net. We report the glitch noise and delay

	Original net	Buffering near TSV	Buffering near driver
Noise @ receiver (V)	0.4058	0.1207	0.1624
Noise @ TSV (V)	0.4059	0.1207	0.1624
Delay (ns)	0.278	0.09701	0.09899

7.5.3 *Buffer Insertion to Alleviate Coupling*

Another effective way to alleviate TSV-to-TSV coupling problem is buffer insertion. As discussed earlier, TSV-to-TSV coupling-caused glitch level is strongly sensitive to the port impedance. Buffer insertion before TSVs helps reduce the victim driving port impedance. To demonstrate the effectiveness of buffer insertion, we choose a 3D net, which is extracted from the SPEF file with the TSV coupling model in Fig. 7.1 plugged in. This 3D net is originally driven by a 2X driver. We insert a 4X buffer before the TSV and perform SPICE simulation, as shown in Fig. 7.9a. Table 7.4 lists the coupling noise simulation results. We see that the coupling noise reduces by 70 %, and the path delay also reduces by 65 %.

Despite the fact that buffer insertion is effective in reducing TSV-to-TSV coupling noise, we still face the following question. Since timing optimization engine will insert a lot of buffers for timing purpose, is it enough to use timing optimization engines only to solve the TSV-to-TSV coupling problems? Here, we give a negative answer because of the following two reasons. First, the timing engine cannot see the TSV-to-TSV coupling, and will not consider the TSV-to-TSV coupling-caused delay in timing optimization. Second, even if the timing

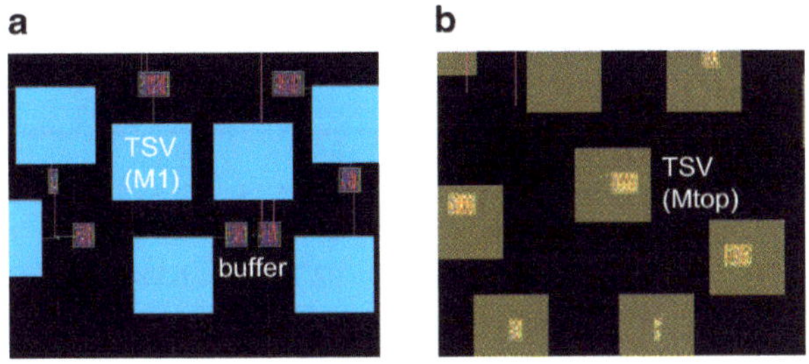

Fig. 7.10 Buffers inserted in the layout of (**a**) *top die* (**b**) *bottom die*. *Yellow squares* are TSV landing pads on Mtop, which can overlap with buffers in the device layer of the *bottom die*

engine is able to consider TSV-to-TSV coupling for timing optimization, it is still not enough to solve coupling noise problem. This is because coupling-noise aware buffer insertion requires buffers to be inserted close to the TSV, while timing-aware engine does not necessarily insert buffers close to the TSV. This is because delay is not sensitive to the buffer locations in the 3D net, which is very different from the 2D net. In short, for a 3D net, coupling noise is very sensitive to buffer-to-TSV distance, while timing is not. To illustrate this, we use the circuit in Fig. 7.9 to study the impact of buffer-to-TSV distance. In this experiment, we compare two cases where buffer is close to TSV and buffer is close to the original driver as shown in Fig. 7.9a, b. We perform both glitch noise simulation and delay simulation on these two cases.

Table 7.4 shows the simulation results. We see that after we move the buffer from the driver end to the TSV end, the glitch at the receiver end reduces by 26 %, while the delay decreases by 1.9 %. This phenomenon is because of the resistive shielding effect [6]. A 3D net is a non-uniform net because of the TSV. If we model a TSV as a big capacitance, the resistive shielding effect from the wire will be applied to this capacitance. Therefore, the Elmore delay model is not effective. Our further experiment shows that a TSV has about 200 um freedom to move between buffers without significantly changing delay.

Since timing-aware buffer insertion is not enough in reducing the TSV-to-TSV coupling noise, we study an SI-aware buffering approach to co-optimize timing and SI. First, we perform coupling analysis for all the TSVs. Based on their coupling levels, we insert buffers with different sizes right before the TSVs. Then we perform timing optimization considering the TSV and its buffer as a single cell. One merit of this approach is that each TSV is shielded by the buffer so that we can use 2D optimization tool to optimize the design with proper timing constraints.

Figure 7.10 shows the buffers inserted in both dies associated with the TSV landing pads. Table 7.5 shows the crosstalk and timing analysis results for four designs: original design, original design with SI-aware buffer insertion, timing-optimized design and SI-timing co-optimized design. The results in Table 7.5 show that buffer insertion is very effective in reducing the coupling-noise for 3D nets. Using the buffer-before-timing approach, we obtained the best critical path delay number. Of course, we need to the pay for the cost of higher power consumption due to the inserted buffers.

Table 7.5 Buffer insertion results

	Original design	SI-aware buffering	Timing-aware buffering	SI+timing buffering
Total buffer count	357	722	556	808
Total noise (V)	4,955.15	4,513.11	4,548.17	4,301.6
Total noise (3D nets) (V)	471.091	82.8308	329.967	73.0874
Longest path delay (ns)	22.79	6.98	9.24	5.64
Total negative slack (ns)	1,175.14	469.625	806.88	431.712

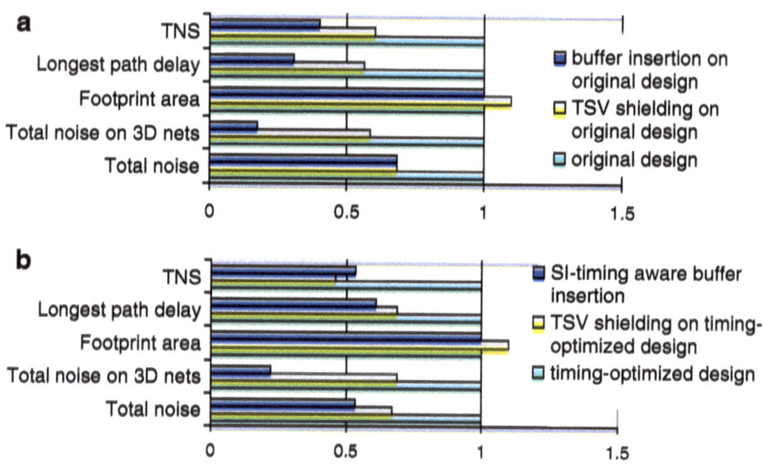

Fig. 7.11 Design and TSV-coupling optimization summary of (**a**) original design (**b**) timing-optimized design

7.5.4 *Overall Comparison*

Figure 7.11 presents an overall comparison between various optimization methods. We see that both buffer insertion and TSV-shielding are effective in alleviating TSV-to-TSV coupling caused problems. However, for 3D-net noise reduction, buffer insertion is more effective. This is because we can afford to insert buffers before every TSVs for noise reduction, but we can only afford to choose some TSVs for shielding due to the increased area cost. If we shield every TSV in this design, the total area increases significantly, which is not affordable. On the other hand, TSV shielding has the advantage of lowering the total coupling noise. The 2D net noise also reduces due to the increased chip area. In terms of timing performance, buffer insertion works better than TSV-shielding. This is not only because of the shielded-TSV number constraints, but also because TSV-shielding results in longer wirelength due to the larger chip area. Finally, TSV shielding achieves lower power consumption than buffer insertion. This is simply because adding more buffers will increase the power consumption significantly. Considering the larger chip area, TSV shielding also has the advantage of lower power density.

7.6 Conclusions

In this chapter we studied the impact of TSV-to-TSV coupling issues in 3D ICs. Based on HFSS and SPICE simulations, we demonstrated that TSV-to-TSV coupling is more sensitive to terminal impedance than TSV distance. A compact TSV-to-TSV coupling model was developed for full-chip 3D signal integrity analysis. Using this model, a SPICE-based full-chip coupling analysis flow was developed. Our 3D SI results showed that TSV-to-TSV coupling has a big contribution to the total glitch noise and timing degradation. To alleviate TSV-to-TSV coupling, we studied two design solutions. Experimental results showed that both TSV shielding and buffer insertion are useful to improve SI as well as timing performance.

References

1. D.H. Kim, K. Athikulwongse, S.K. Lim, A study of through-silicon-via impact on the 3D stacked IC layout, in *Proceedings of the IEEE International Conference on Computer-Aided Design*, San Jose, 2009
2. J. Kim, E. Song, J. Cho, J.S. Pak, J. Lee, H. Lee, K. Park, J. Kim, Through silicon via (TSV) equalizer, in *Proceedings of the IEEE Electrical Performance of Electronic Packaging*, Portland, 2009
3. Y.-J. Lee, S.K. Lim, Timing analysis and optimization for 3D stacked multi-core microprocessors, in *IEEE International 3D System Integration Conference*, Münich, 2010
4. C. Liu, T. Song, J. Cho, J. Kim, J. Kim, S.K. Lim, Full-chip TSV-to-TSV coupling analysis and optimization in 3D IC, in *Proceedings of the ACM Design Automation Conference*, San Diego, 2011
5. M. Pathak, Y.-J. Lee, T. Moon, S.K. Lim, Through-silicon-via management during 3D physical design: when to add and how many? in *Proceedings of the IEEE International Conference on Computer-Aided Design*, San Jose, 2010
6. J. Qian, Modeling the effective capacitance for the RC interconnect of CMOS gates. IEEE Trans. Comput. Aided Des. Integr. Circuit Syst. **13**, 1526–1535 (1994)
7. I. Savidis, E.G. Friedman, Closed-form expressions of 3-D via resistance, inductance, and capacitance. IEEE Trans. Electron Device **56**, 1873–1881 (2009)
8. K.L. Shepard, V. Narayanan, Noise in deep submicron digital design, in *Proceedings of the IEEE International Conference on Computer-Aided Design*, San Jose, 1996
9. R. Weerasekera, M. Grange, D. Pamunuwa, H. Tenhunen, L.-R. Zheng, Compact modelling of through-silicon vias (TSVs) in three-dimensional (3-D) integrated circuits, in *IEEE International 3D System Integration Conference*, San Francisco, 2009

Chapter 8
TSV Current Crowding and Power Integrity

Abstract Due to the large geometry of through-silicon-vias (TSVs) and their connections to the power grid, significant current crowding can occur in 3D ICs. Prior works model TSVs and power wire segments as single resistors, which cannot capture the detailed current distribution and may miss trouble spots associated with current crowding. This chapter studies DC current crowding and its impact on 3D power integrity. First, we explore the current density distribution within a TSV and its power wire connections. Second, we build and validate effective TSV models for current density distributions. Finally, these models are integrated with global power wires for detailed chip-scale power grid analysis.

The materials presented in this chapter are based on [16].

8.1 Introduction

Power delivery network (PDN) design has become a challenging task in ICs as technology scales. Since the supply voltage scales slower than transistors and interconnects, the current density has been rapidly increasing. The increased current density, along with the high temperature, accelerates transistor and wire degradation and shortens the lifetime of both devices and wires. Today, the current density can reach to several hundred thousands of amperes per square centimeter. At this current density magnitude, electro-migration (EM) becomes significant. PDN design needs to be accurately checked for excessive current density to insure EM limits are not exceeded and voltage drops (IR) are within specifications before releasing to manufacturing.

3D IC power delivery network (PDN) provides power supply to all devices in the entire 3D stack. The inter-die power delivery interconnects, formed by power/ground (P/G) through-silicon-vias (TSVs) or micro-bumps, are unique

S.K. Lim, *Design for High Performance, Low Power, and Reliable 3D Integrated Circuits*, DOI 10.1007/978-1-4419-9542-1_8,

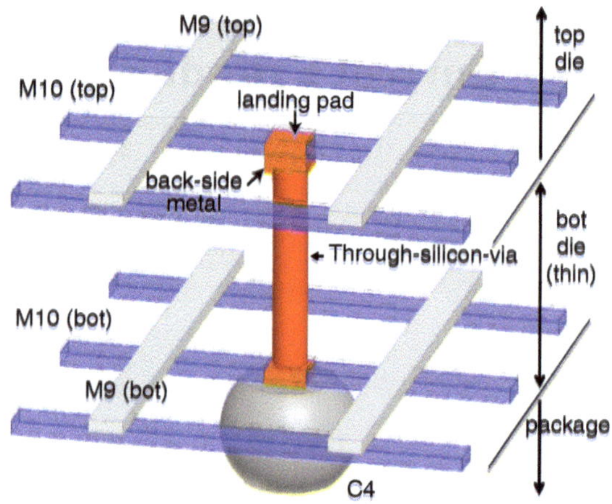

Fig. 8.1 3D connection in a global power delivery network

components in 3D power grids. These vertical connections carry large amounts of current and may suffer from Electro-migration (EM) degradation due to an excessive current density as well as have large IR drops. Therefore, detailed and accurate analysis on the 3D PDN is important to predict the performance and improve the power integrity as necessary.

The purpose of this chapter is to study the DC current crowding in TSVs and its impact on power integrity of generic 3D PDNs. A small cross-section of the global 3D PDN is illustrated in Fig. 8.1. Two dies (top and bottom) are bonded face-to-back and are connected using vias-last TSVs. Voltage is supplied from the package through C4s. For the bottom die, current is delivered directly to Metals 10 and 9; however, for the top die, current is delivered to Metals 10 and 9 through TSVs. Intermediate and local sections of the PDN (using Metals 1–8) are connected using local vias to the global PDN. Inter-die connection can be achieved by either directly bonding the landing pads of the bottom-die backside metal with the top-die Metal 10, or through micro-bumps. The TSV has 5 um diameter and 30 um height, which is similar to the structure described in [4, 12]. The TSV landing pads are 6×6 um [6] and the global power wires are 2 um thick. This generic structure is used throughout the chapter for both isolated TSV modeling and large-scale 3D PDN modeling.

The following specific topics are covered in this chapter:

- We study the DC current density distribution in 3D IC PDNs with specific focus on the current crowding inside the TSVs and at the connections between TSVs and power wires.
- We study the current crowding impact on the IR drop of full-scale 3D PDN.
- The proposed TSV models are integrated with the chip power wires and simulated using a power simulator (PSIM) for chip-scale TSV-based power grid analysis. The simulation results show that PSIM is able to efficiently analyze

detailed current density distribution within TSVs in the context of chip-scale PDNs. PSIM can identify regions of excessive current density or IR drop to help the designer optimize 3D PDNs.

8.2 Existing Works

Some recent chapters discussed TSV EM modeling and analysis [11, 14] and TSV-based 3D PDN analysis [2, 3, 5, 7, 9, 13, 15]. However, none of these works investigates detailed current density distribution or current crowding inside P/G TSVs, where some of the edges may suffer from a large current gradient and are subject to a potential EM reliability issue.[1] Moreover, prior works model TSVs and powers wire segments as single resistors, which are insufficient to accurately analyze the detailed current density distribution inside P/G TSVs and 3D PDNs.

8.3 Current Crowding in 3D IC

In this section, the current crowding inside a TSV and its local connections to the global PDN are investigated. We discuss the impact of current crowding on the effective TSV resistance and the corresponding voltage drop. ANSYS Q3D is used to simulate the DC current density distribution and the voltage drop.

8.3.1 Current Density Distribution Inside a TSV

The test case used to investigate the current density distribution inside a TSV is shown in Fig. 8.2. This corner case was chosen specifically to study a highly asymmetric current distribution and consists of the following components: (1) a TSV with 5 um diameter and 30 um height, (2) landing pads (6×6 um), (3) two power wires on the top, each 2 um wide, (4) one 6 um wide large power wire at the bottom.[2] The power wire has 2 um thickness with copper resistivity of $18\Omega\cdot$ nm. Two current sources are inserted at the top-left corner, each consuming 50 mA current; the current sink is defined at the bottom-right corner. In a 3D PDN, power wires connect to landing pads in both directions. This test case constrains the current flow direction and helps us to investigate the current density distribution in the TSV and its connection points to the PDN. ANSYS Q3D is used to simulate the DC current density distribution and the voltage drop.

---

[1] EM may become significant with the current density near or higher than 10 mA/um^2 [1].

[2] In PDN designs, depending on the dimension of power wire and landing pads, either multiple power wires or a single power wire can connect to a landing pad.

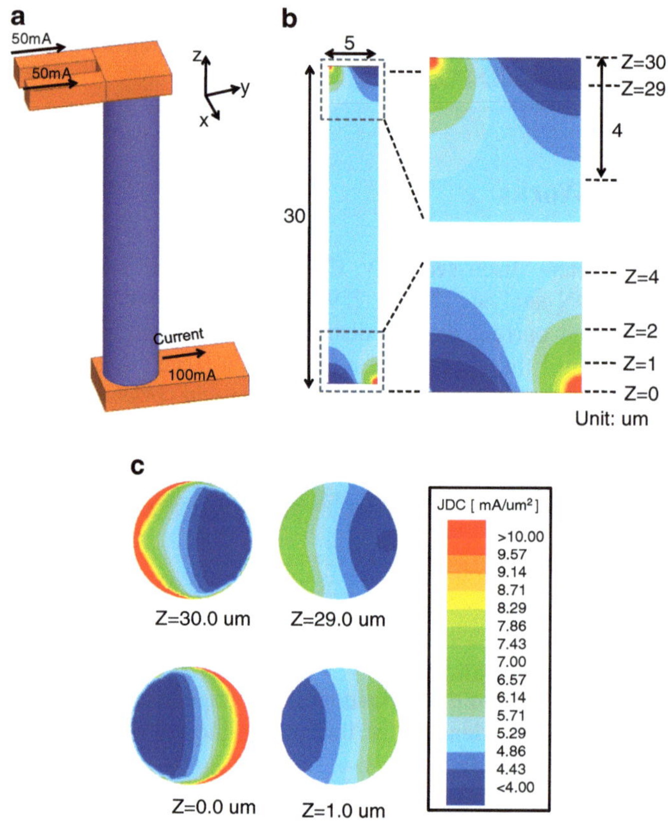

Fig. 8.2 Current crowding in a test case of a TSV and power wires. The current density distribution is shown in a ZY plane (side view) (**b**), and in *top-down* XY planes for Z = 30.0, 29.0, 1.0, and 0.0 um (**c**)

The magnitude of current density is plotted for several cross-sections in Fig. 8.2b, c. In Fig. 8.2b, we observe that a large portion of current from the power wires dives into the top-left TSV edge and flows out at the bottom-right edge. Compared with the average current density inside the TSV which is 5.1 mA/um^2, the edge current density is approximately 10 mA/um^2, nearly twice as large. For the current density distribution along the ZY plane, significant current crowding is observed approximately 4 um into the TSV from both the top and bottom interfaces in Z direction. In the center region of the TSV (4 um < Z < 26 um), the current is uniformly distributed inside the TSV. Figure 8.2c is a plot showing the magnitude of current density distribution in the XY planes, when Z = 30.0, 29.0, 1.0, and 0.0 um. Most of the current is concentrated at the connection between power wires and the TSV.

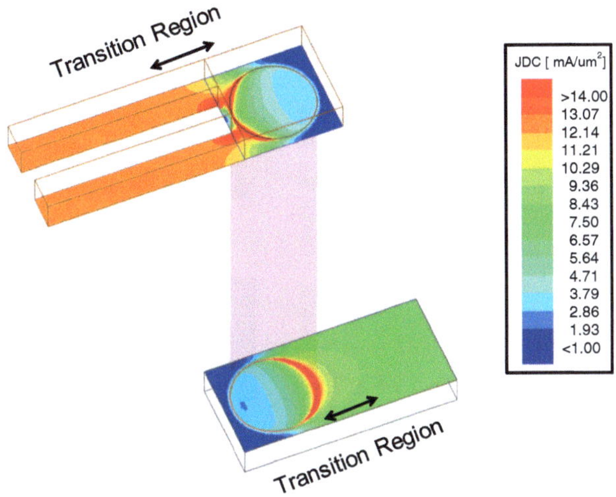

Fig. 8.3 The transition region between power wires and TSVs

8.3.2 Power-Wire-to-TSV Interface

The current density gradient not only occurs at the edge of the TSV, but also at the connection between power wires and the TSV landing pad as shown in Fig. 8.3. Before connecting to the landing pad, the current density inside power wires is relatively uniform. In the transition region, the current concentrates toward the nearest interface between the TSV and the landing pad, especially for the case where multiple power wires connect to the landing pad as indicated by the red in Fig. 8.3.

8.3.3 TSV-Diameter-to-Wire-Thickness Ratio

The amount of current crowding depends on the relative ratio of the TSV diameter to the power wire thickness. The thinner the power wires compared with the TSV diameter, the more crowding the current at the corner. Figure 8.4 shows current density distributions, where the TSV diameter is fixed at 5 um and the wire thickness is changed from 1, 2, to 3 um. The test case structure in Sect. 8.3.1 is used. For thicker wires current, a significant amount of current is shunted over the power cap instead of concentrating at the edge. This is due to the lower resistive path in the thicker wire compared with the high resistive path for a thin wire. For the cases with the same ratio between TSV diameters and wire thicknesses, similar current density distributions will be expected, for example, the 5 um diameter and 1 um wire thickness will result in similar current crowding as 10 um diameter and 2 um wire thickness.

The maximum current density inside the TSV due to crowding is shown in Table 8.1, where the TSV diameter varies from 16 to 2 um, and the wire thickness is

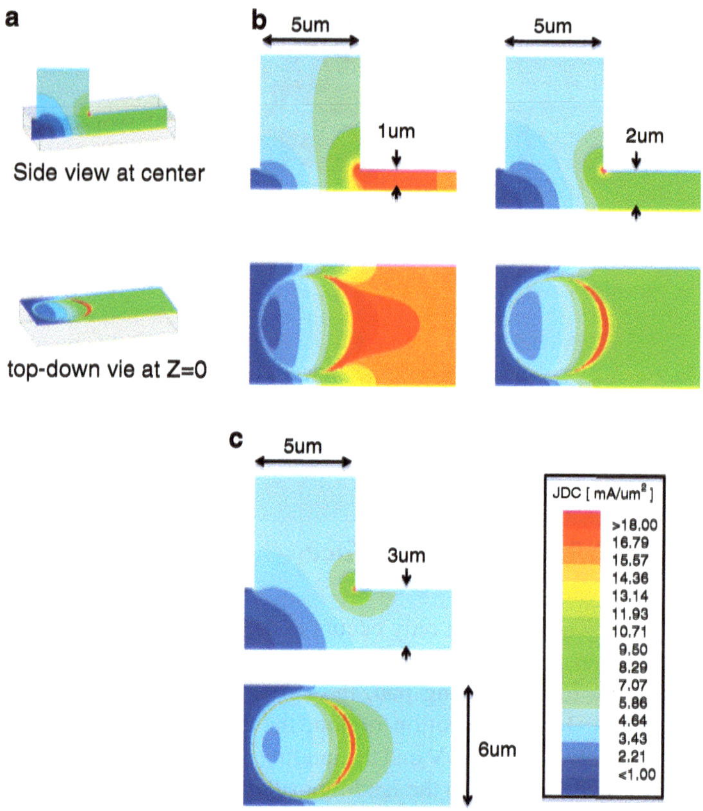

Fig. 8.4 The ratio of the TSV diameter to the wire thickness affects the current crowding at the connection corner. The TSV diameter is set to 5 um, and the power thickness increases from 1 um (**a**), 2 um (**b**), to 3 um (**c**)

Table 8.1 Impact of the TSV diameter on the current crowding. The TSV delivers 100 mA current, and the power wire thickness is 2 um

	Case 1	Case 2	Case 3	Case 4	Case 5
TSV diameter (um)	16	8	5	4	2
TSV height (um)	48	48	30	24	12
Power Cap length (um)	18	10	6	5	3
Javg (mA/um^2)	0.5	2.0	5.1	8.0	31.8
Jmax (mA/um^2)	5.5	10.4	19.2	25.8	62.0
Jmax/Javg	11.1	5.2	3.8	3.2	2.0
TSV diameter: wire thickness	8:1	4:1	2.5:1	2:1	1:1

held constant at 2 um. Comparing Jmax/Javg for different TSV diameters, it clearly appears that current crowing increases with increasing TSV diameter. For the 16 um diameter TSV, the maximum current density is more than ten times larger than

Table 8.2 Impact of current crowding on voltage drop through a TSV. The thickness of power wire varies from 1 to 3 um

Wire thickness (um):	1.0	2.0	3.0
Voltage drop w/ current crowding (mV)	3.33	3.11	3.02
Voltage drop w/o current crowding (mV)	2.75	2.75	2.75
Increase by current crowding (%)	21.1	13.1	9.8

the average value; for the results of the 2 um-diameter TSV, the maximum current density reaches twice of the average value. Therefore, we conclude that a higher maximum current density can occur at the edge of the TSV with a larger diameter.

8.3.4 Impact of Current Crowding on IR Drop

Current crowding inside the TSV changes the effective resistance of the TSV as well as the voltage drop across the TSV. The spreading resistance [8] is caused by the nonparallel current between two spatially separated contacts. Considering the TSV current crowding, its effective resistance is larger than the value obtained using $R_0 = \rho \cdot l/A$, where ρ is the resistivity, l is the length, and A is the cross sectional area of the TSV.

We can use Q3D to find the voltage drop across the TSV taking into account the spreading resistance. For these simulations we hold the TSV dimensions constant and increase the power wire thickness from 1 to 3 um. The resulting voltage drop through the TSV is shown in Table 8.2. For a 100 mA current, the voltage drop through R_0 is $IR_0 = 2.75$ mV, which is not affected by the wire thickness. However, since current crowding is sensitive to the power wire thickness, as the power wire thickness increases from 1 to 3 um, the voltage drop decreases from 3.33 to 3.02 mV, corresponding to 21.1–9.8% greater voltage drop than the calculated value.[3]

8.4 TSV Current Crowding Model

Commercial FEM tools provide accurate simulation on current density distribution. However, using a FEM tool for chip-scale PDN analysis requires unaffordable runtime and computational resources. Simulations on a single TSV can take up to 1 h. Therefore, a better modeling approach for the TSV with reasonable accuracy and fast runtime is required for chip-scale power grid analysis.

[3] The impact of the TSV-diameter-to-wire-thickness ratio on current crowding is discussed in Sect. 8.3.3.

In traditional PDN modeling, power wire segments and TSVs are modeled as single resistors. This model can only represent uniform current density within a wire segment or a TSV, and is insufficient to accurately capture non-uniform current distributions caused by current crowding. Likewise, it is also insufficient to accurately calculate voltage drop that is related to the spreading resistance and depends on current distributions. Here we describe a TSV model that allows non-uniform current densities within a TSV and its transition regions. The model can be easily integrated into netlists for chip-scale PDN analysis and is simple enough that runtime remains reasonable. Our TSV model has been integrated into a power grid simulator (PSIM) [10]. This simulator is used to solve the power resistance network and analyze current and voltage.

8.4.1 3D Resistance Network for TSV Modeling

An illustration on our TSV model and the corresponding resistance network is shown in Fig. 8.5. A TSV is composed of rectangular mesh boxes having the basic structure shown in Fig. 8.5a, where each mesh box consists of six resistors: east, south, west, north, up, and down. The resistance value depends on the geometry of each box and how the TSV overlays the box. These rectangular mesh boxes connect with the neighboring boxes using the center as a connecting point. The 3D mesh structure of a TSV is generated as follows: (1) *Z-mesh*: The TSV is divided into multiple short cylinders with the same diameter but various thicknesses; (2) *XY-mesh*: Each short cylinder is then meshed into a 2D resistance network on a virtual XY plane, which is located at the center of each cylinder.

Along the Z-dimension, several virtual XY planes are created by partitioning this dimension. The Z locations of these XY-planes, referred to as the Z-mesh, are determined by the current gradient in the ZY-plane shown in Fig. 8.2b. More cylinders are created in the regions of larger current crowding and fewer in the regions of uniform current density. Specifically, the Z-mesh size is finer near the top/bottom of the landing pads and coarser in the middle of the TSV. As a result, more virtual XY planes will be created at the top and bottom of the TSV. A side view of the 3D resistance network is shown in Fig. 8.5c, where two virtual XY-planes are generated for one Z-mesh. The resulting model is a non-uniform 3D resistance network consisting of two types of resistors: (1) The resistors along Z-axis (Rz1, Rz2, and Rz3 in Fig. 8.5c, d) connecting the neighboring virtual XY planes; (2) the resistors in virtual XY planes (Rx/y1 and Rx/y2 in Fig. 8.5c, d).

If a mesh tile is completely covered by the real TSV shape, Rz and Rx/y are directly obtained referring to the XY-mesh and Z-mesh size. However, around the TSV boundary, mesh tiles partially overlap with the real shape as shown in Fig. 8.5b.

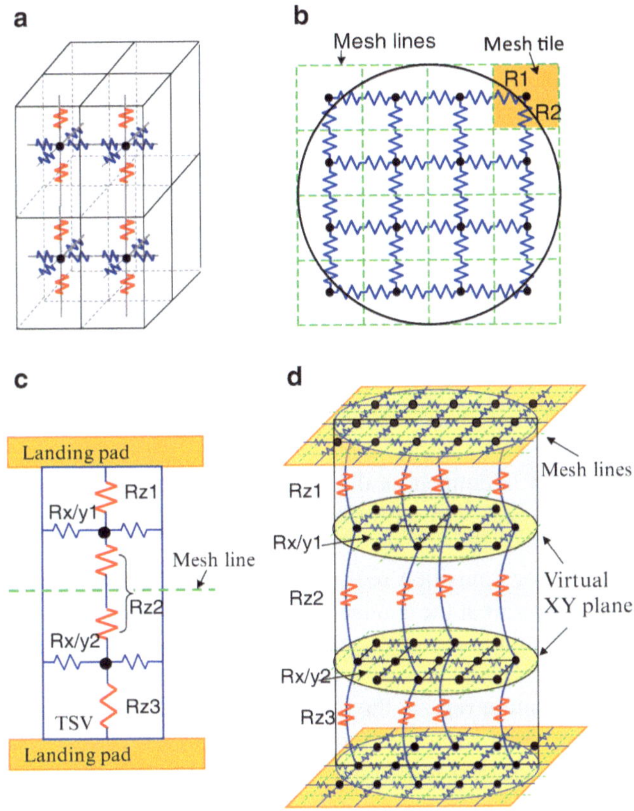

Fig. 8.5 TSV modeling approach of a three-dimensional resistance network. (**a**) Basic rectangular box after 3D meshing; (**b**) XY-mesh and partial overlap mesh tiles; (**c**) side view; (**d**) 3D view of the network

For this case, the overlap area is calculated as the cross sectional area for Rz calculation, and the effective length along X- and Y-axis is then obtained for R1 and R2 calculation.

Figure 8.5d is a schematic in nature. Most virtual planes and resistors are not shown for readability. The non-uniform Z-mesh actually used in the model is a tradeoff between complexity and accuracy. The 30 um-height TSV is partitioned at $Z = 0.1, 0.4, 0.9, 2.0, 5.0, 16.0, 27.0, 28.9, 29.4, 29.7, 29.9, 30.0$ um, where the TSV bottom is at $Z = 0$. We implemented three different XY-mesh sizes for comparisons: 0.25, 0.5, and 1.0 um.

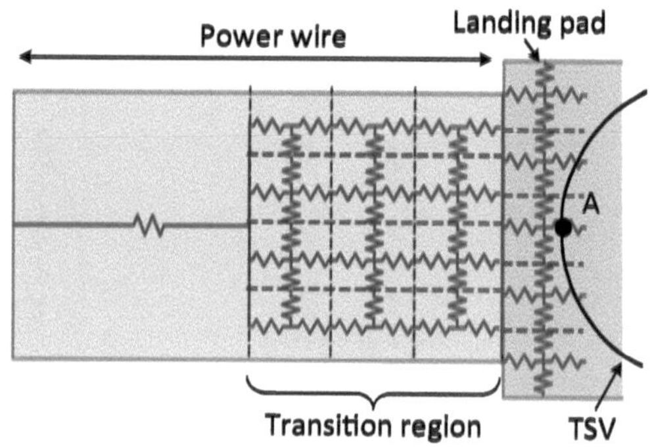

Fig. 8.6 Meshing on the transition region. Using a single resistor in the transition region leads to all the current injected into the Point A

8.4.2 *Modeling of the Transition Region*

The transition region is defined as the connection area between the power wires and the TSV landing pad. Non-uniform current gradient can occur in the transition region.[4]

The meshing on the transition region is shown in Fig. 8.6, where non-uniform current gradient can occur at the connections between power wires and TSV landing pads. Without meshing the transition region, the entire current would flow into the single Point A, which results in a large but incorrect current at the edge of the TSV. By meshing the transition region, the current spreads evenly along the power wire and then flows into the landing pad and the TSV along the edge, which results in better accuracy. A transition region approximately 6 um long is found to be long enough.

8.4.3 *Modeling Accuracy*

Detailed comparisons between ANSYS Q3D and PSIM using the new TSV model are shown in Fig. 8.7. The XY-mesh size is 0.25 um. For the PSIM results, the current in each mesh tile is extracted and divided by the effective area. For the Q3D results, we first simulate the current gradient using its internal mesh generator and solver. Next, the current values are mapped into our mesh tile structure. The current density distribution from Q3D and PSIM are plotted in the top half of Fig. 8.7 for the virtual plane located at $Z = 0.1$ um. The error histogram of the current density between Q3D and PSIM for all 357 tiles is shown at the bottom of Fig. 8.7, where the error for each tile is defined as the absolute difference between Q3D and PSIM.

This comparison at $Z = 0.1$ um is for the closest virtual XY plane to the landing pad in this model, where the largest current crowding is observed. PSIM has very good accuracy compared with the Q3D results. The relative error for each tile is less

[4] Detailed analysis results are shown in Sect. 8.3.2.

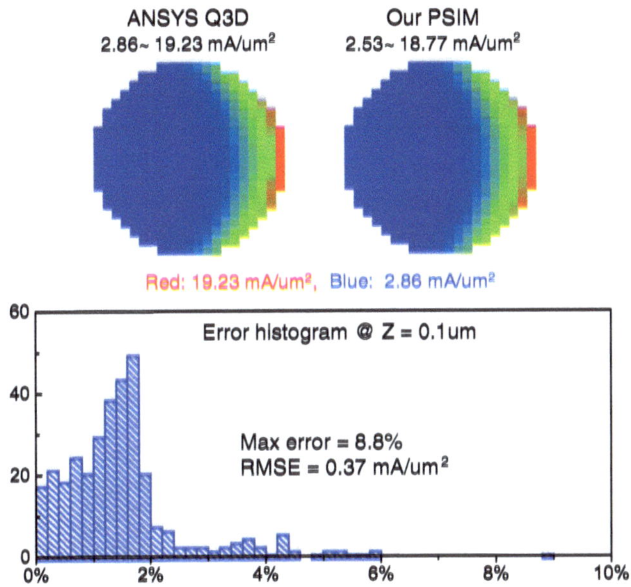

Fig. 8.7 Current density distributions and the error histogram of ANSYS Q3D and our TSV modeling approach in PSIM at Z = 0.1 um virtual XY plane. The error in each tile is the absolute difference between Q3D and PSIM

than 10 %, and most of them are within 5 %. The root-mean-square error (RMSE) is 0.36 mA/um^2, defined as the square root of the arithmetic mean of the square of the error for each tile:

$$\sqrt{\Big(\sum_{i=1 \text{ to } n} (J_i^{Q3D} - J_i^{PSIM})^2\Big)/n}$$

where i is the ith tile and n is the total number of tiles. The voltage drop from PSIM is 3.07 mV, which is 0.33 % different from the Q3D value of 3.08 mV.

Differences between Q3D and PSIM are mainly due to constraining the XY-mesh and Z-mesh sizes and mesh structure. While we use low density orthogonal meshing boxes for simplicity, Q3D supports more sophisticated meshing structures such as triangular and tetrahedral shapes. However, PSIM takes less than 1 s simulation time, whereas Q3D takes up to 1 h. This demonstrates that our modeling approach has potential for chip-scale power analysis with reasonable accuracy and acceptable runtime.

8.4.4 Impact of XY-Mesh Size

The impact of varying the XY-mesh size on the accuracy of current density and voltage drop is shown in Table 8.3. Here the Z-mesh size is held constant and the

Table 8.3 Impact of the XY-mesh size on the current density (mA/um^2) and the voltage drop (mV)

mesh (um)	#tiles	Max. Current density				Voltage drop		
		RMSE	Q3D	PSIM	err (%)	Q3D	PSIM	err (%)
0.25	4,641	0.25	19.2	18.8	−2.1	3.1	3.10	0.3
0.5	1,313	0.34	18.0	20.8	15.6	3.1	3.09	0.7
1.0	325	0.55	12.2	15.6	27.9	3.1	2.99	3.9
None	1	–	19.2	5.1	73.4	3.1	2.75	11.3

XY-mesh size is increased from 0.25, 0.5, to 1.0 um. Using larger meshing tiles, the RMSE of the current density increases from 0.25 to 0.55 mA/um^2, which is equal to 4.9–10.7 % of the average current density. Note that, to report the maximum current density in Q3D for a given mesh size, the Q3D simulation result is mapped into each mesh tile. Thus, the maximum current density of Q3D reduces with different mesh sizes in Table 8.3 as well. The error of the maximum current density increases from 2.1 to 27.9 %, and the voltage drop error increases from 0.3 to 3.9 %. The cost of using finer size is that the total number of mesh tiles increases from 325 to 4,641. Simulation results using a single resistor are also shown in the table, which results in average current density of 5.1 mA/um^2 (73.4 % smaller than the maximum current density from Q3D) and lower voltage drop of 2.75 mV.

8.4.5 Chip-Scale PDN Circuit Model

In this section, we integrate our TSV models with the chip power grid and use PSIM to analyze the entire network. A circuit model of a partial 3D PDN for two dies is illustrated in Fig. 8.8. Both power wire segments and local vias are represented as lumped resistors. The 3D power connection, including TSVs and transition regions, is modeled using the proposed approach. We assume ideal voltage of 1 V supplied from C4s. The current sinks are located at the intersections of the power grids in each die. Power wires are 2 um thick and 5 um wide. The TSV has 30 um height, 5 um diameter, and 6×6 um^2 landing pads. The goal is to analyze global PDNs that have high current density and contain TSV connections.

We first focus on analyzing the 3D power integrity in a two-die stacked 3D PDN design from Sects. 8.5–8.5.3. We discuss the impact of the mesh size and offset between TSVs and C4s on the maximum current density and the IR drop. The impact of the power wire density, the TSV diameter, and the TSV and C4 count on 3D power integrity are also discussed. We then analyze the power integrity using PSIM on five large-scale 3D PDNs in Sect. 8.5.7.

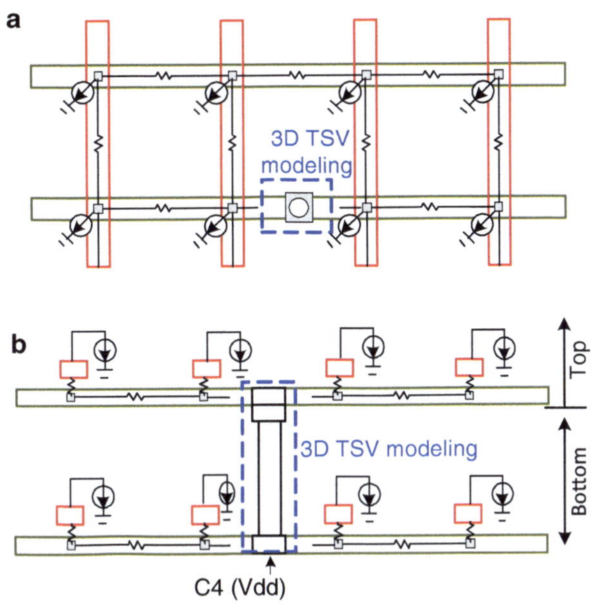

Fig. 8.8 A circuit model for a two-die TSV-based PDN using the proposed 3D TSV modeling approach in top-down view (**a**) and side view (**b**)

8.5 Experimental Results

8.5.1 Chip-Scale Noise Analysis

The global PDN and the power map in the bottom die are shown in Fig. 8.9c. The footprint area is $1.4 \times 1.4\,\text{mm}^2$. Each die has 16×16 power wires (in purple and yellow) and a thick power ring around the boundary with 15 um wide. 4×4 TSVs and C4s are aligned in the bottom die, which are enlarged as the white blocks for readability. Each black box at the intersection of power wires represents the current sink. The XY-mesh size of the TSV model is 0.25 um, and the Z-mesh size is kept the same as described in Sect. 8.4.1.

Power maps in both top die (S2) and bottom die (S1) have a cool spot in the bottom-left corner and a hot spot in the top-right corner. In the center of each die, another two narrow cool spots are placed in the left and right. These power maps result in different current density patterns connecting to the TSVs: (1) symmetric current density, e.g., TSV1 in Fig. 8.9c, where all the power wires have high current density; (2) asymmetric current density, e.g., TSV2 in Fig. 8.9c, where left power wires have much lower current density than the right power wires.

The voltage drop maps in S2 and S1 are shown in Fig. 8.9a, b. The top-right corner has the maximum IR drop: 23.0 mV in top and 19.0 mV in bottom. The IR drops in S2 are larger than S1 due to the TSV parasitic resistance. Since TSVs and C4s are aligned, the region closer to TSVs has a smaller IR drop than the region far from TSVs.

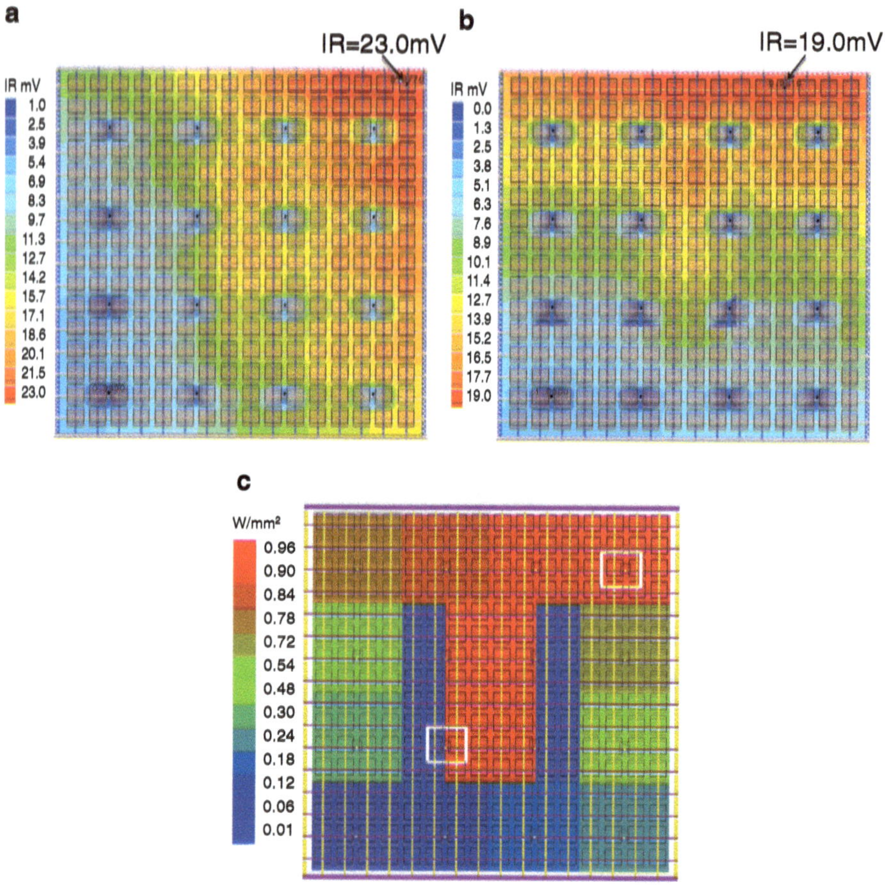

Fig. 8.9 (**a**) Voltage drop map of the *top* die (max = 23.0 mV) and (**b**) voltage drop map of the *bottom* die (max = 19.0 mV), (**c**) power map of the *bottom* die, where TSVs and C4s are aligned and enlarged for readability

Detailed current density distribution in two TSVs (TSV1 and TSV2 in Fig. 8.9c) are shown in Fig. 8.10, where TSV1 located in the center of the hot region has a fairly symmetric current density along power wires, and TSV2 located at the boundary between a power hot region and a cool region has asymmetric current densities in the power wires. Figure 8.10 shows plots of the current density distribution in the XY direction (Jxy) and Z direction (Jz) on different layers. Jxy in metal layers S2-M10, S1-BM, and S1-M10 are plotted in Fig. 8.10a, c, f, respectively. Figure 8.10b, d, e plot Jz through the interface between S1-BM and S2-M10, the top surface of the TSV, and the bottom surface of the TSV, respectively.

First, PSIM is effective to capture the detailed current density distribution inside the 3D power connection. Symmetric current crowding is observed in both edges of TSV1; most of the current crowding occurs at the right edge of TSV2. Second, large current crowding inside TSVs is observed. For TSV1, the maximum current

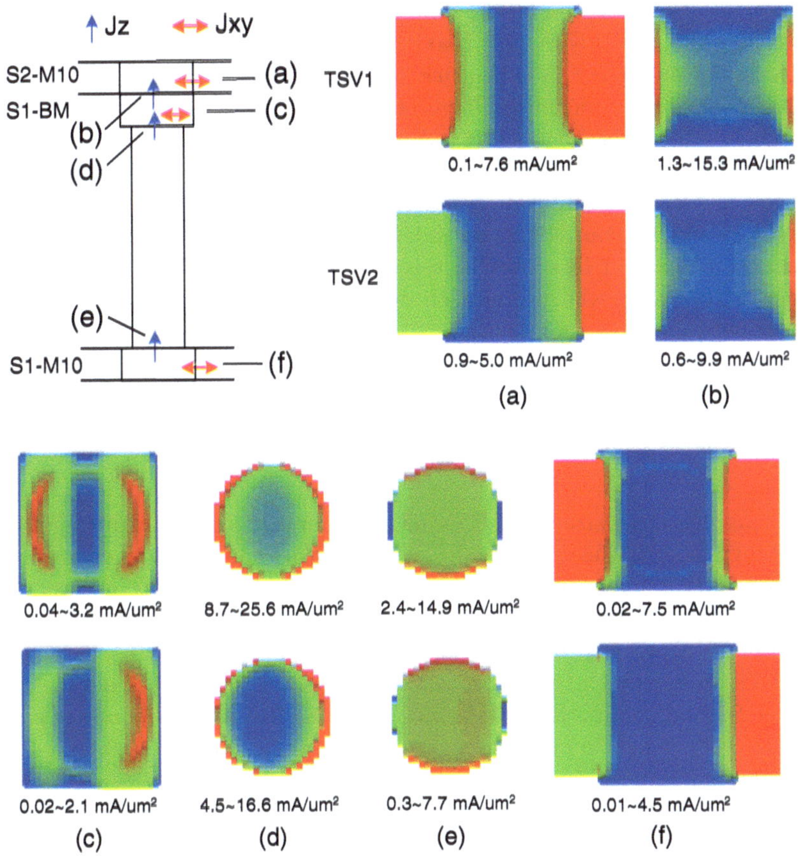

Fig. 8.10 Current density distribution in XY (Jxy) and Z (Jz) direction of TSV1 and TSV2 in Fig. 8.9. TSV1 has symmetric current density along power wires; TSV2 has asymmetric current

density (J_{max}) in Fig. 8.10a is 7.6 mA/um^2, where most current concentrates at the connection between the power wire and the landing pad; the J_{max} through the TSV in Z-direction (Fig. 8.10d) can reach to 25.6 mA/um^2, which is approximately 2.4 times larger than the wire J_{max}.

Third, larger current crowding occurs at the TSV top surface than the bottom surface due to the aligned TSVs and C4s. Jz through the TSV bottom surface (Fig. 8.10e) is 14.9 mA/um^2 compared with the top surface of 25.6 mA/um^2. Fourth, in the bottom TSV surface (Fig. 8.10e), Jz current crowds at the top and bottom edges instead of concentrating at the left and right edges. This is because large amount of Jxy currents flow out from the left and right edges to feed the S1 current sinks. As a result, the current, delivered to the S2 power grid, concentrates at the top and bottom edges. Moreover, current crowding leads to 5.7 mV IR drop through TSV1, which is 3.7 % larger than the IR drop without considering the crowding value of 5.5 mV.

The next subsections contain the following results: (1) Maximum current density (J_{max}) along the power wires. (2) Maximum and average current density (J_{avg}) of the TSVs. For multiple TSVs, J_{max} and J_{avg} through each TSV are first measured. We then compare the percentage increase (J_{inc}) of J_{max} over J_{avg} for each TSV, as $J_{inc} = (J_{max} - J_{avg})/J_{avg}$. These current densities of the TSV with the maximum J_{max} are reported, as well as the min/max/avg J_{inc} among all the TSVs. (3) Min/max/avg IR drops in top and bottom dies. (4) IR drop through the TSVs. The IR drop with (IR_c) and without (IR_n) considering current crowding are reported. The percentage increase of IR_c over IR_n is also included. The baseline PDN design contains a 16×16 power gird in each die, 4×4 TSVs and C4s, and TSVs with 5 um diameter and 0.25 um mesh size.

8.5.2 Impact of TSV Mesh Size

We increase the mesh size of the TSV model from 0.25, 0.5, to 1.0 um, and fix the other design factors. The results of the current density and the IR drop are shown in Table 8.4. First, using a larger mesh size in the TSV model results in lower J_{max} in each mesh tile compared with using a finer mesh size. As the mesh size increases from 0.25 to 1.0 um, J_{max} reduces from 25.6 to 14.3 mA/um^2 and J_{inc} reduces from 151–192 % to 41–48 %. This is because the coarse mesh averages out the current gradient. Second, the mesh size does not affect the IR drop of the power grid and wire J_{max} very much. In contrast, significant current crowding is observed in the TSVs for smaller mesh sizes. For the mesh size of 0.25 um, the J_{max} in the TSV is 25.6 mA/um^2, which is 110 % larger than the TSV J_{avg} of 10.2 mA/um^2.

8.5.3 Impact of TSV and C4 Offset

Previous simulations assume aligned TSVs and C4s. To study the offset impact on power integrity, we leave 175 um distance between the TSV and the C4. The offset design has 12 C4 and 16 TSVs. The simulation results are shown in Table 8.5. Detailed illustration is shown in Fig. 8.11.

Current crowding has larger impact on TSV IR drop in the offset design than the aligned design. The six columns from the right compare the IR drop through the TSV with (IR_c) and without (IR_n) considering current crowding. Current crowding in the offset design results in 5.9–10.6 % larger IR drop than the IR_n; whereas, in the aligned design, the TSV IR drop caused by current crowding is 3.4–5.2 % larger than the IR_n. This is mainly because large current crowding occurs in both the top and bottom surfaces of TSVs in the offset design; whereas, in the aligned design, only the top interface between the TSV and the backside metal has large current crowding, where the voltage at the bottom interface between TSV and S1-M10 is constantly supplied by C4s.

Table 8.4 Impact of the TSV mesh size on current density (mA/um^2) and IR drop (mV). TSV diameter is 5 um

Power grid	#TSV &#C4	Mesh (um)	Wire J_{max}	TSV w/ max(J_{max})			J_{inc} (%) of TSVs			IR_Bottom			IR_Top		
				J_{max}	J_{avg}	J_{inc}(%)	Min	Avg	Max	Min	Avg	Max	Min	Avg	Max
16 × 16	4 × 4	0.25	10.5	25.6	10.2	151	151	161	192	2.1	9.5	19.1	3.8	12.7	23.0
16 × 16	4 × 4	0.50	10.4	20.2	10.1	100	100	105	124	2.4	10.0	19.8	4.1	13.3	23.7
16 × 16	4 × 4	1.00	10.5	14.3	10.2	41	41	42	48	2.1	9.4	18.9	3.9	12.9	23.1

Table 8.5 Impact of TSV and C4 offset on current density (mA/um^2) and IR drop (mV) through TSVs

		Aligned	Offset
#TSVs		16	16
#C4s		16	12
Wire J_{max}		10.5	25.2
TSV with max(J_{max})	J_{max}	25.6	22.0
	J_{avg}	10.2	8.3
	J_{inc}(%)	151	165
J_{inc}(%) of TSVs	Min	151	149
	Avg	161	164
	Max	192	190
IR_Bottom	Min	2.1	10.2
	Avg	9.5	26.5
	Max	19.1	48.3
IR_Top	Min	3.8	29.7
	Avg	12.7	46.7
	Max	23.0	65.5
TSV w/ max(IR)	IR_c	5.7	4.8
	IR_n	5.5	4.5
	Inc.(%)	3.7	6.4
Inc(%) of TSV IR	Min	3.4	5.9
	Avg	4.0	8.0
	Max	5.2	10.6

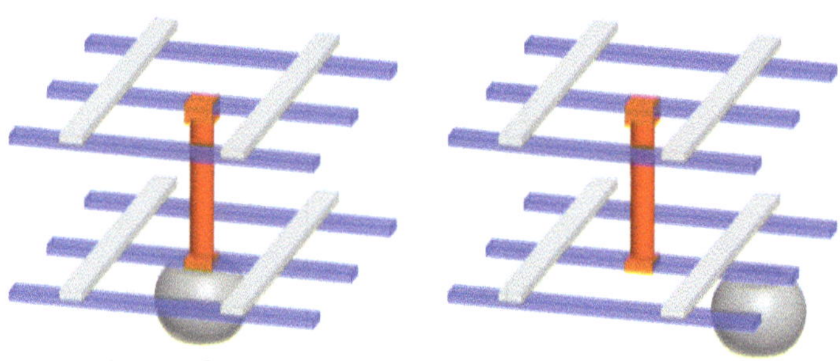

Fig. 8.11 Zoom-in for partial PDNs with aligned vs offset TSV and C4

8.5.4 Impact of Power Wire Density

For the baseline PDN design and power maps described in Sect. 8.5, we increase the power grid from 8×8, 12×12, 16×16, to 20×20 and fix other design factors. The power wire density increases from 2.9 to 7.1 % over the footprint area. The simulation results of current density and IR drop are shown in Table 8.6. Using more power wires helps to reduce the IR drop in both dies, but reduces the J_{max} of TSVs and wires in small scale. The maximum IR drop in bottom and top die reduces from 36.0 to 15.2 mV and 37.8 to 20.6 mV, respectively; the maximum J_{max}

Table 8.6 Impact of the power wire density on current density (mA/um^2) and IR drop (mV). TSV mesh size is 0.25 um, and TSV diameter is 5 um

Power grid	P-wire den (%)	#TSV &#C4	Wire J_{max}	TSV w/ max(J_{max})			J_{inc} (%) of TSVs			IR_Bottom			IR_Top		
				J_{max}	J_{avg}	J_{inc}(%)	Min	Avg	Max	Min	Avg	Max	Min	Avg	Max
8 × 8	2.9	4 × 4	11.8	28.8	11.2	157	147	169	223	3.1	15.7	36.0	4.8	18.5	37.8
8 × 8	4.3	4 × 4	11.6	27.9	10.6	162	151	167	200	2.6	11.1	22.8	4.1	15.1	29.3
16 × 16	5.7	4 × 4	10.5	25.6	10.2	151	151	161	192	2.1	9.5	19.1	3.8	12.7	23.0
8 × 8	7.1	4 × 4	10.5	25.2	9.9	154	150	160	183	1.9	8.0	15.2	3.6	11.7	20.6

of the TSVs only reduces from 28.8 to 25.2 mA/um^2. This is mainly due to the fixed placement of TSVs and C4s, where the current through each TSV in Z-direction is related to the TSV count.

8.5.5 Impact of TSV and C4 Count

For the baseline PDN design and power maps described in Sect. 8.5, we increase the TSV and C4 count from 2×2, 3×3, 4×4, to 5×5 and fix other design factors. The simulation results of current density and IR drop are shown in Table 8.7. Using more TSVs and C4s significantly reduces both the J_{max} and IR drop. With TSV and C4 count increases from 2×2 to 5×5, the J_{max} of wires reduces from 30.3 to 8.4 mA/um^2; the J_{max} of the TSV reduces from 74.7 to 19.8 mA/um^2; the worst IR in bottom and top die reduces from 59.5 to 15.1 mV, and 77.1 to 18.3 mV, respectively. This is mainly because using more TSVs leads to less current per TSV and thus lower J_{max}, and using more C4s helps to improve the IR. Furthermore, the J_{max} of the TSVs caused by current crowding is still around 150–190 % larger over the J_{avg}, and approximately 140 % larger over the J_{max} of the wires.

8.5.6 Impact of TSV Diameter

For the baseline PDN design and power maps described in Sect. 8.5, we increase the TSV diameter from 4, 5, 8, to 16 um, with mesh size of 0.25, 0.25, 0.5, and 0.5 um, respectively. Other design factors are fixed. The simulation results of current density and IR drop are shown in Table 8.8. We observe that larger TSVs significantly reduce J_{avg} from 15.7 to 10 mA/um^2 and J_{max} of the TSVs from 33.5 to 10.6 mA/um^2. However, J_{max} of TSVs reduces slower than J_{avg} of TSVs. As a result, for the 16 um-diameter TSVs, J_{max} of TSVs is even 930–1,180 % larger than J_{avg} of the TSVs. In addition, the TSV diameter only affects IR drops in the top die. IR drops in the bottom die are insensitive to the TSV diameter, where the voltage is directly supplied by C4s from the package. The top die has lower IR drops when using larger TSVs, which is due to the reduced TSV effective resistance and the IR through TSVs.

8.5.7 Power Integrity in Large-Scale 3D PDN

Five large-scale two-die stacked PDNs are designed for 3D power analysis using PSIM. The power wire utilization, defined as the power wire area per die over the footprint area, is set to 5 %. Assume the bottom die needs $N_P \times N_P$ power wires,

Table 8.7 Impact of the TSV count on current density (mA/um^2) and IR drop (mV). TSV mesh size is 0.25 um, and TSV diameter is 5 um

Power grid	#TSV &#C4	Wire J_{max}	TSV w/ max(J_{max})			J_{inc} (%) of TSVs			IR_Bottom			IR_Top		
			J_{max}	J_{avg}	J_{inc}(%)	Min	Avg	Max	Min	Avg	Max	Min	Avg	Max
16 × 16	2 × 2	30.3	74.7	29.5	153	153	154	156	18.7	44.8	59.5	30.0	58.5	77.1
16 × 16	3 × 3	16.9	41.0	16.0	156	155	157	162	5.7	18.4	34.1	9.6	24.0	40.8
16 × 16	4 × 4	10.5	25.6	10.2	151	151	161	192	2.1	9.5	19.1	3.8	12.7	23.0
16 × 16	5 × 5	8.4	19.8	7.7	158	147	161	193	1.1	6.1	15.1	2.1	8.1	18.3

Table 8.8 Impact of the TSV diameter (um) on current density (mA/um^2) and IR drop (mV). TSV mesh size is 0.25 um

Power grid	#TSV &#C4	TSV (um)	Wire J_{max}	TSV w/ max(J_{max})			J_{inc} (%) of TSVs			IR_Bottom			IR_Top		
				J_{max}	J_{avg}	J_{inc}(%)	Min	Avg	Max	Min	Avg	Max	Min	Avg	Max
16 × 16	4 × 4	4	10.5	33.5	15.7	113	109	113	117	2.2	9.5	19.2	4.3	13.6	24.1
16 × 16	4 × 4	5	10.5	25.6	10.2	151	151	161	192	2.1	9.5	19.1	3.8	12.7	23.0
16 × 16	4 × 4	8	10.4	19.0	4.0	372	372	394	463	2.3	9.9	19.7	3.3	11.8	21.7
16 × 16	4 × 4	16	10.7	10.6	1.0	928	928	986	1,177	2.2	9.5	19.2	2.3	9.8	18.8

Table 8.9 3D power integrity analysis for large-scale PDNs. We show the footprint (mm^2), power density (W/mm^2), current density (mA/um^2), and IR drop (mV)

Design		PDN1	PDN2	PDN3	PDN4	PDN5
Footprint		5 × 5	6 × 6	9 × 9	11 × 11	15 × 15
power grid		50 × 50	60 × 60	90 × 90	110 × 110	150 × 150
#TSVs		144	225	484	729	1,369
#C4s		144	225	484	729	1,369
Pwr dens.	Top	0.57	0.4	0.8	0.71	0.47
	Bot	0.57	0.75	0.8	0.91	0.49
Jmax_wire	Top	7.0	3.5	13.6	8.7	16.2
	Bot	7.2	6.6	12.1	11.4	17.4
TSV with max(J_{max})	J_{max}	9.6	5.0	18.5	11.1	23.3
	J_{avg}	6.8	3.6	13.1	7.5	16.3
	J_{inc}(%)	41	40	41	47	43
IR_Bottom	Min	5.1	6.0	4.4	8.1	1.8
	Avg	8.7	9.9	11.3	12.8	6.8
	Max	15.9	13.3	24.2	25.2	34.9
IR_Top	Min	7.9	5.0	6.8	9.7	2.7
	Avg	11.7	7.2	15.6	13.5	8.8
	Max	19.6	9.2	37.8	24.2	49.6
TSV with max(IR)	IR_c	4.1	2.1	7.9	5.1	9.9
	IR_n	3.7	1.9	7.1	4.5	8.8
	Inc.(%)	11.4	11.4	11.5	11.4	12.2

the aligned TSV and C4s count are determined as ($25\%N_P$) × ($25\%N_P$). The power TSVs have 5 um diameter with 1.0 um mesh size. Power wires have 5 um width and 2 um thickness. The local and global power density refers to the 3D core-to-memory PDN designs [2,9], Intel microprocessors, and the power density estimation in ITRS 2005.

The results of power analysis on large-scale PDNs are shown in Table 8.9. First, excessive current density through the TSVs is observed. The J_{max} of the TSV is 40–47 % larger than the J_{avg} through the TSV. Note that the 1 um mesh size is used for this simulation. As previously discussed, using coarser mesh size usually underestimates the J_{max} on 3D power connection compared with using finer mesh or using FEM tools. It is reasonable to expect even larger current density through the TSVs. Second, the wire J_{max} in S1 and S2 are affected by the power density in each die, where larger power density (PDN2 and PDN4) in the bottom die results in comparable J_{max} between wires and TSVs; for other designs with lower power density in the bottom die, TSV J_{max} is 13 % larger than wire J_{max}. Third, current crowding also increases the IR drop through the TSVs, which is 11.4–12.2 % larger than the IR drop without considering current crowding. Moreover, IR drops in S1 and S2 power grid are also affected by the power density in each die. When each die has comparable power density, the maximum IR in the top die is usually larger than the bottom die; allocating the high power density close to C4s (in the bottom die) helps to reduce the IR drops in the top die.

In summary, (1) PSIM with the proposed simple TSV model can analyze the chip-scale 3D PDN for detailed current density distribution and voltage drop; (2) by identifying the current crowding corner inside each TSV, PSIM helps the designer to assign reasonable current limits and voltage drop limits for 3D PDN design and optimization; and (3) PSIM can select a different mesh size depending on the resolution of the power analysis: for a large-scale PDN, 1.0 um mesh size can be firstly used to quickly identify the hotspots associated with the maximum current density and IR drop; then in a bounded hotspot region, a finer mesh size is used to identify the detailed current density distribution and optimize the power grid, correspondingly.

8.6 Conclusions

In this chapter, we studied current crowding inside TSV-based 3D power connection. First, we studied the current density distribution inside the 3D TSV-based power grids. A large current gradient, current crowding, near the interface between power wires and TSVs has been observed. In addition, the current crowding also increases the effective resistance of the TSV and thus voltage drop in the PDN. Second, a 3D TSV model was implemented and simulated using PSIM. The model has good accuracy and far less complexity compared with FEM tools. Third, PSIM was used for chip-scale 3D power grid analysis, which was able to generate the current density distribution and the resulting IR drop.

References

1. J. Abella, X. Vera, Electromigration for microarchitects. ACM Comput. Surv. **42**(2), 9:1–9:18 (2010)
2. M.B. Healy, S.K. Lim, Distributed TSV topology for 3-D power-supply networks. IEEE Trans. VLSI Syst. **20**(11), 2066–2079 (2012)
3. G. Huang, M. Bakir, A. Naeemi, H. Chen, J.D. Meindl, Power delivery for 3D chip stacks: physical modeling and design implication, in *Proceedings of the IEEE Electrical Performance of Electronic Packaging*, Atlanta, 2007, pp. 205–208
4. C. Huyghebaert, J. Van Olmen, Y. Civale, A. Phommahaxay, A. Jourdain, S. Sood, S. Farrens, P. Soussan, Cu to Cu interconnect using 3D-TSV and wafer to wafer thermocompression bonding, in *Interconnect Technology Conference (IITC), 2010 International*, Burlingame, 2010, pp. 1–3
5. M. Jung, S.K. Lim, A Study of IR-drop noise issues in 3D ICs with through-silicon-vias, in *IEEE International 3D System Integration Conference*, Münich, 2010, pp. 1–7
6. M. Jung, J. Mitra, D. Pan, S.K. Lim, TSV stress-aware full-chip mechanical reliability analysis and optimization for 3D IC, in *Proceedings of the ACM Design Automation Conference*, San Diego, 2011, pp. 188–193
7. M. Jung, S. Panth, S.K. Lim, A study of TSV variation impact on power supply noise, in *IEEE International Interconnect Technology Conference*, San Jose, 2011, pp. 1–3

8. S. Karmalkar, P. Mohan, H. Nair, R. Yeluri, Compact models of spreading resistances for electrical/thermal design of devices and ICs. IEEE Trans. Electron Device **54**(7), 1734–1743 (2007)
9. N. Khan, S. Alam, S. Hassoun, Power delivery design for 3-D ICs using different through-silicon via (TSV) technologies. IEEE Trans VLSI Syst. **19**(4), 647–658 (2011)
10. S.R. Nassif, J.N. Kozhayz, Fast power grid simulation, in *Proceedings of the ACM Design Automation Conference*, Anaheim, Los Angeles, 2000, pp. 156–161
11. J. Pak, M. Pathak, S.K. Lim, D.Z. Pan, Modeling of electromigration in through-silicon-via based 3D IC, in *IEEE Electronic Components and Technology Conference*, Lake Buena Vista, Las Vegas, 2011, pp. 1420–1427
12. A. Redolfi, D. Velenis, S. Thangaraju, P. Nolmans, P. Jaenen, M. Kostermans, U. Baier, E. Van Besien, H. Dekkers, T. Witters, N. Jourdan, A. Van Ammel, K. Vandersmissen, S. Rodet, H. Philipsen, A. Radisic, N. Heylen, Y. Travaly, B. Swinnen, E. Beyne, Implementation of an industry compliant, 550 um, via-middle TSV technology on 300 mm wafers, in *IEEE Electronic Components and Technology Conference*, Lake Buena Vista, Las Vegas, 2011, pp. 1384–1388
13. A. Shayan, X. Hu, H. Peng, C.-K. Cheng, W. Yu, M. Popovich, T. Toms, X. Chen, Reliability aware through silicon via planning for 3D stacked Ics, in *Proceedings of the Design, Automation and Test in Europe*, Nice, 2009, pp. 288–291
14. Y.C. Tan, C.M. Tan, X.W. Zhang, T.C. Chai, D.Q. Yu, Electromigration performance of Through Silicon Via (TSV), A modeling approach. Microelectron. Reliab. **50**(9–11), 1336–1340 (2010)
15. J. Xie, M. Swaminathan, Electrical-thermal co-simulation of 3D integrated systems with micro-fluidic cooling and joule heating effects. IEEE Trans. Compon. Packag. Manuf. Technol. **1**(2), 234–246 (2011)
16. X. Zhao, M. Scheuermann, S.K. Lim, Analysis of DC current crowding in through-silicon-vias and its impact on power integrity in 3D ICs, in *Proceedings of the ACM Design Automation Conference*, San Francisco, 2012 pp. 157–162

Chapter 9
Modeling of Atomic Concentration at the Wire-to-TSV Interface

Abstract Electro-migration (EM) in through-silicon-vias (TSVs) could become a serious design issue for TSVs having regions of high current density. In these regions, grain boundaries play a large role in EM dominating atomic transport. The focus of this chapter is to study EM in a comprehensive mutli-physics context including atomic transport due to grain boundaries. Atomic concentration and its time evolution is modeled. Our simulations show how voids and hillocks evolve in a TSV due to current crowding. We also investigate the impact of current, temperature, and various grain sizes on the EM reliability. These results and discussions provide guidance for designers to better understand and avoid EM reliability failures in 3D ICs.

9.1 Introduction

Electro-migration decreases the reliability of integrated circuits (ICs). It may eventually cause shorts or opens in circuits and interconnects which can reduce IC lifetimes, or worst, cause field fails. EM is driven by multiple physical mechanisms, including electric current, temperature gradient, stress gradient, and atomic concentration gradient. The evolution of atomic concentration or the mean time to failure (MTTF) are two important parameters to investigate the EM reliability. This analysis requires a transient analysis of the atomic concentration. Atomic diffusion is significantly different within a metal grain and along grain boundaries, each having different activation energies. Atomic transport is dominated by grain boundary diffusion and must be included in any realistic EM simulation.

Through-silicon-via (TSV)-based 3D integration has gained a lot of interest due to its potential to overcome conventional CMOS scaling limitations and its potential to enable heterogeneous integration. Reliability of TSV-based 3D ICs is an important issue for main stream acceptance. In particular, the reliability related to EM in 3D TSVs and TSV connections is a critical issue to explore. TSVs, especially the power/ground (P/G) TSVs in 3D power delivery networks (PDNs),

S.K. Lim, *Design for High Performance, Low Power, and Reliable 3D Integrated Circuits*, DOI 10.1007/978-1-4419-9542-1_9,

carry large amounts of current. Specifically, P/G TSVs which typically have a high average current density can have much higher local current densities due to current crowding [14]. These regions of high local current density are much more susceptible to EM degradation. Moreover, the large power density with high temperature or large thermal gradient inside 3D ICs due to multi-tier stacking or joule heating can accelerate atomic migration. Therefore, analyzing the evolution of atomic concentration and the EM lifetime for the 3D connection is important.

The purpose of this chapter is to model and analyze the atomic concentration in TSVs and to investigate the impact of current crowding, grain structure, and temperature on EM lifetime using a multi-physics simulation. A test case to study the EM reliability of wire-to-TSV interface is shown in Fig. 9.1. A TSV having no grain structure is illustrated in Fig. 9.1a. Here the entire TSV is considered as a perfect crystal. However, in reality, most metal materials have polycrystalline structures with grains having a characteristic average size. These grains are separated by grain boundaries having a characteristic thickness. The TSVs shown in Fig. 9.1b, c have simplified grains structures with sizes of 2.0 and 1.0 um, respectively. The grain structure significantly affects the atom diffusion and the EM lifetime.

This chapter focuses on modeling and investigating the atomic concentration for EM in 3D ICs with a specific interest at the interface between the TSV and the landing wire. The following specific topics are covered in this chapter:

- We model transient analysis of the atomic concentration and its evolution;
- We model grain and grain boundary atomic transport;
- We observe that current crowding at the wire-to-TSV interface accelerates atomic migration and reduces the lifetime of TSVs. In addition, the impact of current, temperature, and grain structure on the EM lifetime of TSVs are studied.

9.2 Existing Works

EM modeling and analysis for interconnects have been extensively studied for many decades [1, 12]. However, the history of modeling on TSV reliability is very short. A recent chapter analyzed and modeled the DC current crowding inside the TSVs and at the connections between TSVs and power wires [14]. This crowding increases the TSV effective resistance and voltage drop in the power delivery network of 3D ICs. Some chapters modeled the thermal-mechanical stress at the interface between TSVs and the substrate [5, 9], and its impact on device performance [13]. The impact of TSV stress on back end of line (BEOL) interconnects and its EM lifetime was modeled and discussed [7, 8]. A modeling approach on TSV EM reliability was also proposed in [10]. However, none of these works present transient analysis on atomic concentration for EM lifetime. Moreover, the grain structure and the atom diffusion along grain boundaries are not considered, which may underestimate the EM degradation.

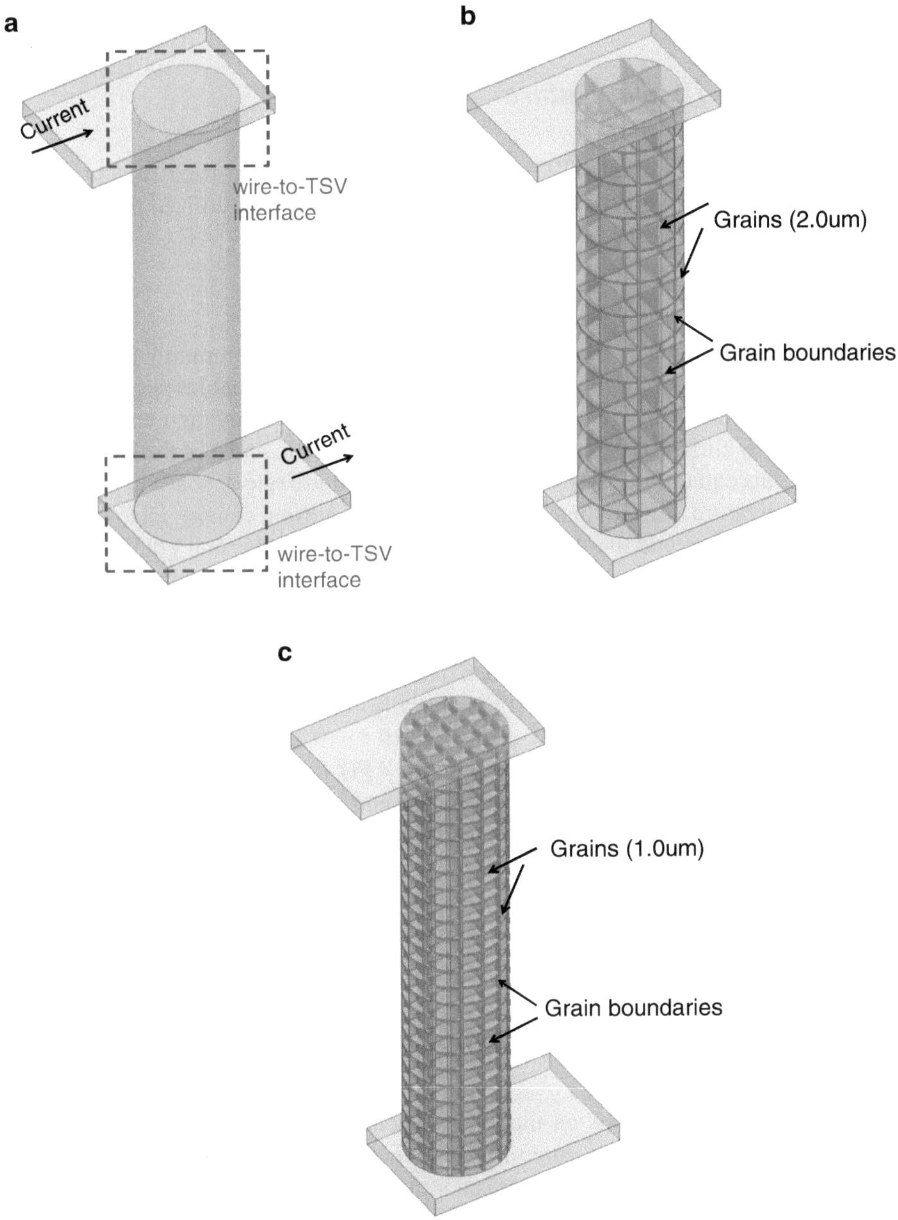

Fig. 9.1 A test case to study the EM reliability of wire-to-TSV interface, (**a**) with no grain structure, (**b**) 2.0 um grain size, and (**c**) 1.0 um grain size

9.3 Preliminaries

9.3.1 Mean Time to Failure

The mean-time-to-failure (MTTF) is an important parameter used to characterize the time to potential failures during operation. Previous work utilized the following criteria to identify the MTTF subject to EM.

$$\mathrm{MTTF} = Aj^{-n}e^{E_{\mathrm{A}}/kT} \tag{9.1}$$

Equation (9.1), known as Black's equation [1] is the most commonly used method to predict the life span of integrated circuits due to EM. It enables accelerated EM testing, where the coefficient A, scaling factor n, and the activation energy E_{A} are determined by fitting the model to the experimental data, and k is the Boltzmann's constant. This equation clearly shows that the EM failure rate depends exponentially on the temperature T and depends on a power of the current density j. However, this model does not include the thermal migration caused by thermal gradients, and is not based on a specific physical model. Thus, it is hard to identify the potential failure locations.

Another method to locate the EM sensitive regions is to calculate the atomic flux divergence (AFD) [6, 10] at each location using the finite element model (FEM) approach. The maximum AFD is usually considered as a likely failure site. The atomic fluxes are calculated using Eqs. (9.4)–(9.7), where the initial atomic concentration is N_0. However, the maximum AFD is a stationary analysis result, which cannot predict the atomic concentration and its evolution over time.

Our modeling approach is based on the FEM approach. The atomic concentration is solved by using the partial differential equations, Eqs. (9.2)–(9.7). Detailed discussions will be presented in Sect. 9.4. In this chapter, we report the MTTF when the atomic concentration has 10% deviation of the initial value.

9.3.2 Grains and Grain Boundaries

The grain structure depends on the pretreatment of copper and on the conditions of deposition. This grain structure of the conductive material has a strong influence on the lifetime. Theoretically, the crystal structure of a solid material, e.g., copper, should have a perfect periodic arrangement of "face centered cubic" structure. However, in reality, most metals never consist of only one crystal, but contain a collection of small crystals, so called polycrystalline structure as shown in Fig. 9.2. Each small crystal, called a grain, has periodic arrangement of atoms. The average diameter of grains is called grain size. Inside each grain, momentum exchange between the electrons and atoms is small because of the uniform lattice structure of

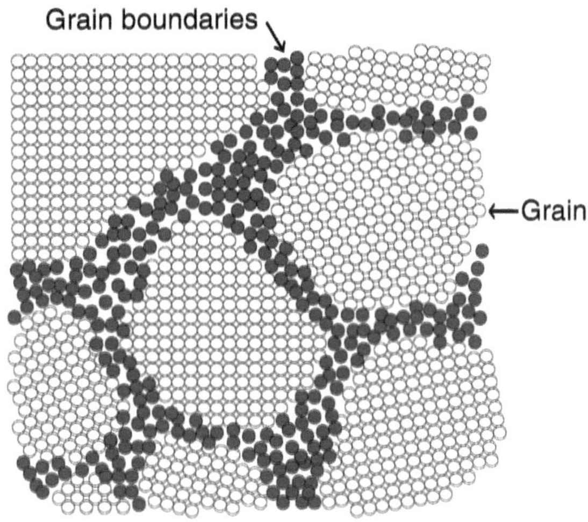

Fig. 9.2 Illustrations of grains and grain boundaries in polycrystalline

metal ions. However, the periodic pattern is broken at the interface between two grains, called a grain boundary. The atoms in this transition region cannot match up perfectly with both crystal lattice, so the momentum transfer in the grain boundary is much larger.

Since the atoms are bounded weakly in the grain boundaries, once a strong force is applied, such as concentration gradient, thermal gradient, current, or stress gradient, the atoms become mobile. The diffusion caused by EM includes lattice diffusion, surface diffusion, and grain boundary diffusion. Since the diffusion barrier layer between the TSV and the silicon dioxide typically helps minimize the migration of TSV metal into the silicon, our model mainly focuses on the lattice diffusion and grain boundary diffusion of the TSVs. Correspondingly, the lattice has high activation energy E_A, whereas the grain boundaries have low E_A.

9.4 Modeling Approach and Settings

A set of partial differential equations (PDE) is used to obtain the atomic concentration, $N(x, y, z, t)$, at each location (x, y, z) at time t, where the atomic concentration evolution is described in a continuity equation (9.2), and the atomic flux $\mathbf{J}$ is determined by the combined mechanisms of concentration gradient ($\mathbf{J}_N$ in Eq. 9.4), current density ($\mathbf{J}_c$ in Eq. 9.5), thermal gradient ($\mathbf{J}_T$ in Eq. 9.6), and stress gradient ($\mathbf{J}_s$ in Eq. 9.7).

These PDEs are formulated and solved in COMSOL multi-physics simulation tool [3]. In addition, the current density, temperature, and hydrostatic stress can be also obtained by electrical-thermal and thermal-mechanical coupling simulations in COMSOL.

Table 9.1 Notations and meanings in EM PDEs

Term	Meaning
N	Atomic concentration in atoms/m^3
$\mathbf{j}$	Current density in mA/um^2
E_A	Activation energy in eV
k	Boltzmann constant in J/K
T	Absolute temperature in K
e	Electric charge in C
Z^*	Effective valence charge
ρ	Electrical resistivity in Ω·m
D_0	Self-diffusion coefficient in m^2/s
Q^*	Heat of transport
Ω	Atomic volume in m^3
σ_H	Hydrostatic stress in Pa

9.4.1 Electro-Migration Equations

The PDEs used to obtain the atomic concentration are shown as follows.

$$\frac{\partial N}{\partial t} + \nabla \cdot \mathbf{J} = 0 \tag{9.2}$$

$$\mathbf{J} = \mathbf{J_c} + \mathbf{J_T} + \mathbf{J_s} + \mathbf{J_N} \tag{9.3}$$

$$\mathbf{J_N} = -D\,\nabla N \tag{9.4}$$

$$\mathbf{J_c} = \frac{N}{kT} eZ^* \rho \mathbf{j}\, D \tag{9.5}$$

$$\mathbf{J_T} = -\frac{NQ^*}{kT^2}\, D\,\nabla T \tag{9.6}$$

$$\mathbf{J_s} = \frac{N\Omega}{kT}\, D\,\nabla \sigma_H \tag{9.7}$$

$$D = D_0 \exp\left(\frac{-E_A}{kT}\right) \tag{9.8}$$

where N is the atomic concentration per unit volume, which is the variable in the PDEs. N_0 is the initial concentration. ∇N is the concentration gradient. The atomic diffusion D is described as $D_0 \exp(\frac{-E_A}{kT})$, where D_0 is the self-diffusion coefficient, k is the Boltzmann constant, T is the absolute temperature. The diffusion $\mathbf{J}$ is the total atomic flux at a location, which includes the flux caused by concentration gradient $\mathbf{J_N}$, current density $\mathbf{J_N}$, thermal gradient $\mathbf{J_T}$, and stress gradient $\mathbf{J_S}$. $\mathbf{j}$ is the current density. ∇T is temperature gradient. $\nabla \sigma_H$ is the hydrostatic stress gradient. The meanings of other notations are summarized in Table 9.1. Detailed discussions of these PDEs will be presented in the following subsections of this chapter.

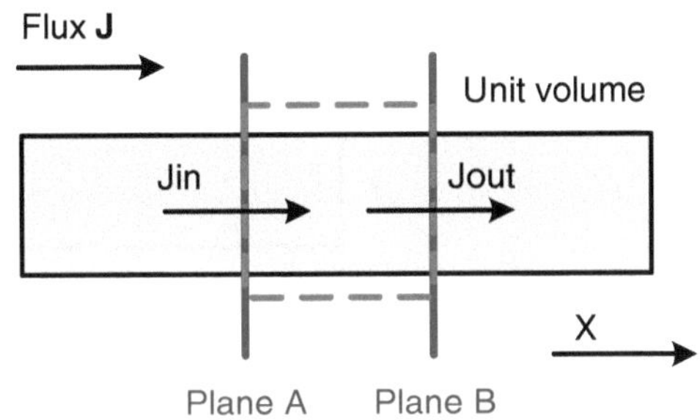

Fig. 9.3 Illustration of the atomic flux and divergence

9.4.2 Atomic Flux and Atomic Flux Divergence

Atomic flux, **J**, describes the total number of atoms that flow across a unit area per unit time. A large atomic flux means the atoms move fast across the unit area. Atomic flux divergence, $\nabla \cdot \mathbf{J}$, describes the changes of atomic number through unit volume per unit time, which is the spatial difference between the inward and outward flux at the boundary planes of the unit volume.

Equation (9.2) is the continuity equation that describes the atomic concentration evolution over time and also insures that atoms are conserved. It governs the atomic flux divergence over spatial dimensions and determines how the atomic concentration evolves over time. As shown in Fig. 9.3, along the direction x, when the inward atomic flux $\mathbf{J_{in}}$ is larger than the outward flux $\mathbf{J_{out}}$, the atoms in the unit volume tends to increase. From Eq. (9.2), the atomic flux decreases over the x direction, which corresponds to a negative atomic flux divergence. As a result, the $\frac{\partial N}{\partial t}$ will be positive for mass conservative. This means the atomic concentration N in this unit volume tends to increase over time.

9.4.3 Effect of Activation Energy and Atomic Concentration

If we set $\mathbf{J_c}$, $\mathbf{J_T}$, $\mathbf{J_s}$ to zero, atomic diffusion can still occur due to the atomic flux $\mathbf{J_N}$ from the atomic concentration gradient and the difference of activation energy in grains and grain boundaries. Equation (9.4) is analogous to the Fick's first law. The flux is proportional to the negative concentration gradient $-\nabla N$. The atoms are likely to flow from high-concentration regions to low-concentration regions.

The activation energy E_A is different in grains $E_A(\mathrm{g})$ and grain boundaries $E_A(\mathrm{gb})$. Smaller $E_A(\mathrm{gb})$ in grain boundaries results in high diffusion, whereas larger $E_A(\mathrm{g})$ in grains results in low diffusion. Because D exponentially depends on $-E_A$ (shown in Eq. 9.8), a large divergence of diffusion in grains and grain boundaries can be observed, which leads to large atomic flux divergence. Therefore, atomic accumulation or depletion may be observed around the grain boundaries.

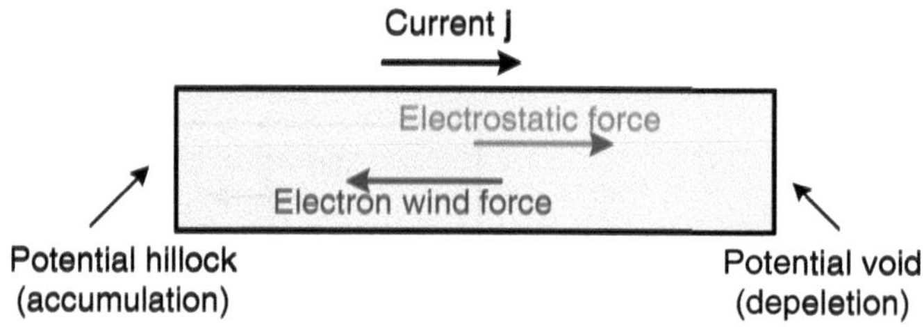

Fig. 9.4 The electrostatic force and electron wind force on the atoms, and the weak positions of void and hillock formation

9.4.4 Effect of Current

Atomic flux caused by electric current density is governed by Eq. (9.5), where e is the electron charge, ρ is the resistivity of the conductor, $\mathbf{j}$ is the local current density.

In the present of non-zero local current density, thermally activated metal ions are acted on by two opposing forces as shown in Fig. 9.4, which can be described as the effective valence Z^*. The Z^* consists of two parts, $Z^* = Z^*_{el} + Z^*_{wd}$, where $Z^*_{el}eE$ is the direct electrostatic force on the positive ions as a result of the electric field E. This force has the same direction as electric filed, but opposite to the electron flow. The $Z^*_{wd}eE$ is called the electron wind force, which is caused by the momentum exchange between conducting electrons colliding with the activated metal ions, and is in the opposite direction as $Z^*_{el}eE$. The diffusion of the atom is found to be enhanced in the direction of electron wind, so the momentum exchange effect is much greater than the electrostatic filed effect for electro-migration in metals [11].

9.4.5 Effect of Thermal and Stress

Atomic flux caused by thermal gradient is shown in Eq. (9.6), where Q^* is the heat of transport. This means that the atomic flux is proportional to the negative gradient of temperature ($-\nabla T$), and atoms likely to move from high-temperature regions to low-temperature regions.

In addition to the effect of thermal gradient, the temperature profile also exponentially impacts the atom diffusion D in each atomic flux $\mathbf{J_c}$, $\mathbf{J_N}$, and $\mathbf{J_s}$. That is, the high temperature will accelerate the atomic diffusion, thus shortening the lifetime.

Atomic flux caused by hydrostatic stress gradient is shown in Eq. (9.7), where Ω is the atomic volume, $\sigma_H = (\sigma_x + \sigma_y + \sigma_z)/3$, σ_x, σ_y, σ_z are the corresponding normal stresses in the Cartesian coordinates system (x, y, z). The $\nabla\sigma_H$ is the stress gradient that results from material accumulation and depletion due to electro-migration. The stress involves both EM-induced back-flow mechanical stress [2] and the residual stress that is generated in thermal processing when there is a difference in thermal expansion coefficients (CTE) in the TSV structure [7]. We found $\mathbf{J_s}$ which related to CTE mismatch is small compared with the other atomic flux ($\mathbf{J_T}$, $\mathbf{J_c}$, $\mathbf{J_N}$) components and will therefore ignore $\mathbf{J_s}$ in the simulations in the rest of this chapter.

9.4.6 Model Settings

The structure to investigate the atomic concentration evolution and EM lifetime is shown in Fig. 9.1, which consists of the following components: (1) a copper TSV with 5 um diameter and 25 um height, (2) landing pads are 6×10 um and 1 um thick, (3) a TSV is composed of regular cubic grains and grain boundaries. A current sources is inserted at the top-left corner; the current sink is defined at the bottom-right corner. This test case constrains the current flow direction and helps us investigate the current density impact on atomic concentration and EM reliability with the presence of grain boundaries.

COMSOL multi-physics is used to simulate the DC current density distribution, temperature distribution, and stress distribution, and to solve the partial differential equations to obtain the atomic concentration over time.

Our special interest is to investigate the atomic concentration and EM reliability at the wire-to-TSV interface, where large current crowding happens. The MTTF is defined as the 10 % deviation of atomic concentration over the initial concentration inside the TSV, where the initial concentration N_0 is 1.53e28 Atoms/um^3. Since we focus at the TSV-to-wire interface, the wire is assumed a perfect diffusion model, where no grain is modeled inside the wire. In reality, if the depletion or accumulation of atomic concentration is expected at a specific TSV-wire interface, then both the TSV and the wire are expected to have voids or hillocks. Meanwhile, the grain structure of the wire will also affect the atomic concentration.

The TSV structure contains regular cubic grains and grain boundaries. Since no measurement data has been reported on the TSV grain structure, which can be different from Cu interconnects, we vary the grain size from 2.0 to 1.0 um to study its impact on MTTF. The default grain size is 0.9 um with a grain boundary thickness of 0.1 um. The activation energy of the lattice (grain) is 2.1 eV; and the grain boundary has default activation energy as 0.8 eV, which will vary from 0.7 to 0.9 eV [4] to investigate its impact on MTTF.

Unless specified, other default values in the models are as follows: the input current density inside the TSV is 3.1 mA/um^2; the temperature is 350 K; the Z^* is -4; k is 1.38e–23 J/K, e is 1.6e–19C, $\rho = \rho_0(1 + \alpha(T - T_0))$, where ρ_0 is 1.68e–8Ω·m, α is 0.0039, T_0 is 293 K. D_0 is 1e–7 m^2/s, Q^* is 1.387e–20, Ω is 1.182e–29 m^3.

9.5 Experimental Results

9.5.1 Impact of Current Crowding

A recent work [14] analyzed the current density distribution for the 3D connection of wires and P/G TSVs. They discovered that for some geometries significant current crowding can occur giving rise to high local current densities at the wire-to-TSV interface. To analyze the impact of current crowding on atomic concentration, in this chapter, we assume 60 mA current flows from the top-left landing pad, through a TSV and flows out of the bottom-right landing pad. The atomic concentration is affected by both atomic flux from current density ($\mathbf{J_c}$ in Eq. 9.5) and from atomic concentration gradient ($\mathbf{J_N}$ in Eq. 9.4). Therefore, we include these two terms of flux in the continuity equation Eq. (9.2) and set $\mathbf{J_s}$ and $\mathbf{J_T}$ to zero.

The atomic concentration on the top and bottom wire-to-TSV interfaces at time 1e5s, 1e7s, and 1e8s are shown in Fig. 9.5, where the color legend displays the percentage difference of the atomic concentration compared with the initial concentration (N_0 =1.53e28 Atoms/m^3). First, we observe that the atoms begin to accumulate/deplete along the grain boundary at time = 1e5s. This accumulation or depletion penetrate to the neighboring grains over a short distance. This is because that grain boundaries provide fast paths with low activation energy for atom diffusion. Second, we observe that most accumulation (red) occurs at the top-left interface; and most depletion (blue) happens at the bottom-right interface. For example, at time = 1e5s, the maximum atomic concentration is 2.2 % larger than the initial value, whereas the minimum atomic concentration is 1.5 % smaller than the initial concentration. In addition, these accumulation and depletion densities grow over time, and will very likely cause hillocks and voids, respectively. From time 1e5s to 1e8s, the maximum deviation of the atomic concentration compared with initial value increases from 2.2 to 4.2 %. That is mainly due to the current crowding at each location and fast diffusion along grain boundaries. The local high current density increases the atomic flux and enlarges the atomic flux divergence as indicated in Eq. (9.5).

Meanwhile, the current crowding at the wire-to-TSV interface is significantly affected by the thickness of landing wires [14]. A thinner landing wire causes larger current crowding than a thicker landing wire at the corners of wire-to-TSV interface. As a result, the maximum current density increases at these interfaces.

To investigate the current crowding impact on atomic concentration, we vary the landing wire thickness from 0.5, 1.0, 1.5, 2.0, to 3.0 um. Meanwhile, the TSV diameter is kept at 5.0 um, and the total current is 60 mA. The impact of wire thickness on current density distribution and atomic concentration at time 1e7s for 0.5 and 3.0 um thick wires are also shown in Figs. 9.6 and 9.7. Figure 9.6a, c are the 3D structure, current density distribution in side view and top/bottom wire-to-TSV interfaces, and atomic concentration in side view and top-bottom wire-to-TSV

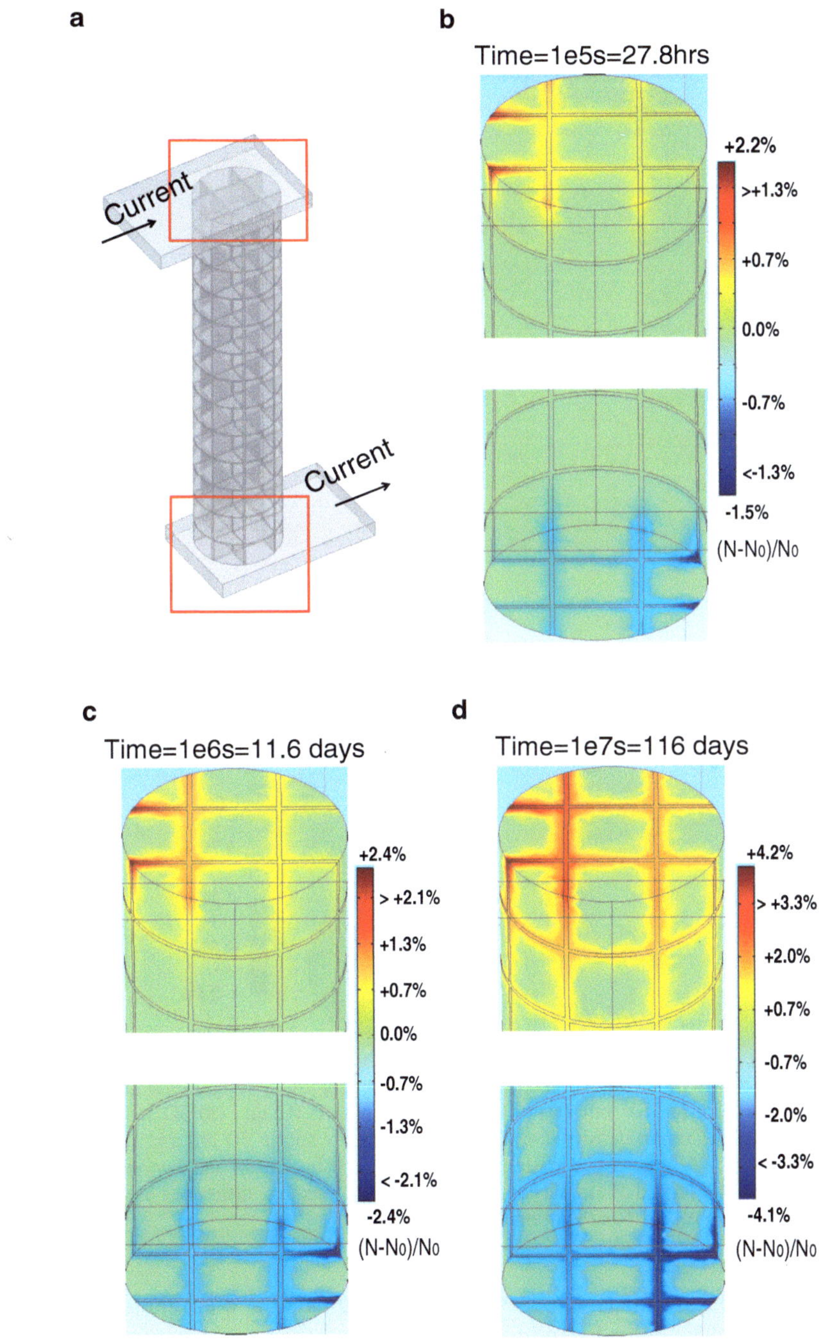

Fig. 9.5 Atomic concentration at (**a**) *top* and *bottom* wire-to-TSV interface, (**b**) time = 1e5s, (**c**) time = 1e6s, and (**d**) time = 1e7s. The color legend displays the percentage difference of atomic concentration normalized to the initial concentration ($N_0 = 1.53\text{e}28$ Atoms/m^3)

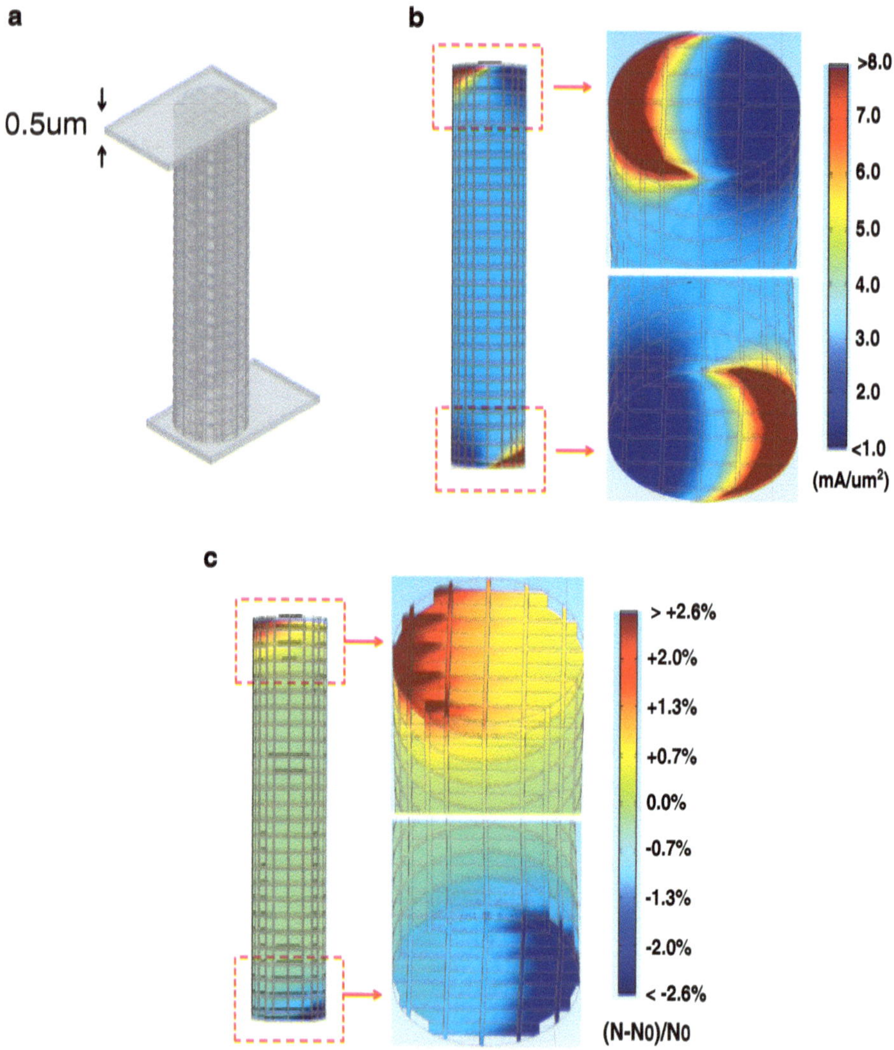

Fig. 9.6 Impact of wire thickness on current crowding and atomic concentration at time 1e7s for *top* and *bottom* wire-to-TSV interfaces. The wire thickness is 0.5 um. (**a**) 3D views for 0.5 um wire thickness. (**b**) Current density distributions in side view and in 3D *top* and *bottom* wire-to-TSV interfaces. (**c**) Atomic concentrations in side view and in 3D *top* and *bottom* wire-to-TSV interfaces. The color legend of atomic concentration is the percentage difference normalized to the initial concentration $N_0 = 1.53\text{e}28$ Atoms/m^3

interfaces for 0.5 um thick wire, respectively. Figures 9.7a, c are those for 3.0 um wire thickness. The color legend of atomic concentration is the percentage difference of the concentration over the initial value.

In Figs. 9.6b and 9.7b, we observe significant current crowding at both top and bottom corners of wire-to-TSV interfaces. Thinner wires result in more current

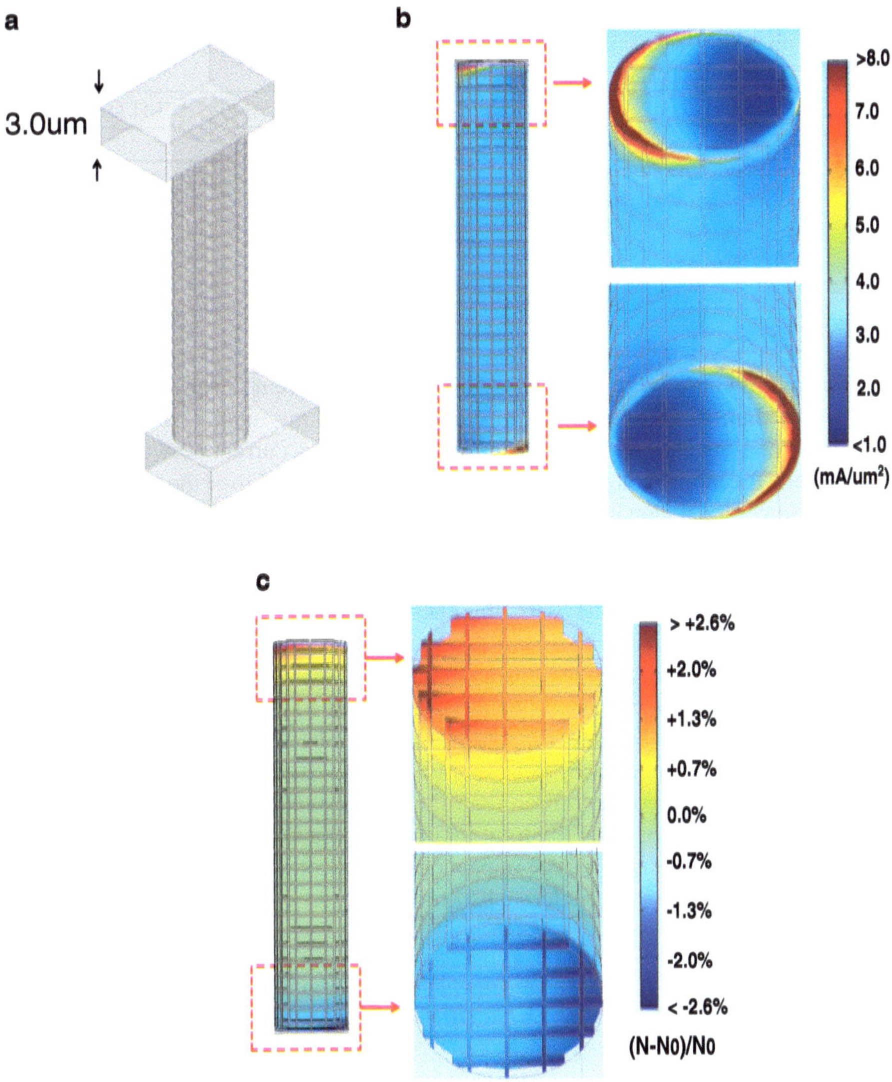

Fig. 9.7 Impact of wire *thickness* on current crowding and atomic concentration at time 1e7s for *top* and *bottom* wire-to-TSV interfaces. The wire *thickness* is 3.0 um. (**a**) 3D views for 0.5 um wire *thickness*. (**b**) Current density distributions in side view and in 3D *top* and *bottom* wire-to-TSV interfaces. (**c**) Atomic concentrations in side view and in 3D *top* and *bottom* wire-to-TSV interfaces. The color legend of atomic concentration is the percentage difference normalized to the initial concentration $N_0 = 1.53e28$ Atoms/m^3

crowding. These results are consistent with the current crowding discussed in other chapters [14]. The atomic concentration distributions, shown in Figs. 9.6c and 9.7c, demonstrate that more atoms accumulate at the top-left and deplete at the

Table 9.2 Impact of wire thickness on current density inside the TSV (mA/um^2), atomic concentration (Atoms/m^3) at time = 1e7(s), and MTTF (s). Initial concentration is 1.53×10^{28} Atoms/m^3

Wire	Current density		Atomic conc.($\times 10^{28}$)		MTTF
thickness (um)	Jmax	Javg	Max	Min	($\times 10^8$)
0.5	37.1	3.1	1.63	1.44	0.3
1.0	32.0	3.1	1.60	1.46	1.6
1.5	22.6	3.1	1.59	1.47	2.1
2.0	13.5	3.1	1.58	1.48	2.5
3.0	11.0	3.1	1.57	1.49	3.0

bottom-right, where current crowding gives higher current densities. This implies that using thin wires may result in earlier EM failures than using thick wires. In addition, in the case of 3.0 um-thick wires, since less current crowding occurs at the corners, the atom accumulation and depletion are spread over the entire interface but with lower local density.

Detailed results of maximum current density (Jmax) and average current density (Javg) inside the TSV, the atomic concentration at time t = 1e7s, and MTTF are shown in Table 9.2. As the wire thickness decreases from 3.0 to 0.5 um, the maximum current density inside the TSV increases from 11.0 to 37.1 mA/um^2, however, the average current density remains at 3.1 mA/um^2. Meanwhile, the maximum atomic concentration increases from 1.57 to 1.63e28 Atoms/m^3, which corresponds to 2.6–6.5 % larger concentration than the initial value; the minimum atomic concentration decreases from 1.49 to 1.44e28 Atoms/m^3, which corresponds to 2.6–5.9 % smaller concentration than the initial one; and the MTTF decreases from 3.0 to 0.3e8s. Note that the total input current is kept constant for each case. Current crowding can have a large impact on atomic concentration generating voids and hillocks and therefore accelerate EM failure.

9.5.2 *Impact of Current Direction and Density*

The current direction determines the location of voids and hillocks. From Fig. 9.5, we observe that the bottom-right wire-to-TSV interface has smaller concentration than the initial value (atom depletion) and the top-left wire-to-TSV interface has larger concentration than the initial value (atom accumulation). This means atoms move from the bottom-right corner to the top-left corner, which is opposite to the direction of positive current. This makes sense from a physical point of view and is due to the momentum exchange from the electrons to the atoms, which is the dominant force in EM, the atoms are pushed in the same direction as electrons (i.e., the opposite direction of current). Over time, atoms accumulate forming voids where current is injected. Likewise, atoms deplete forming voids where current is removed.

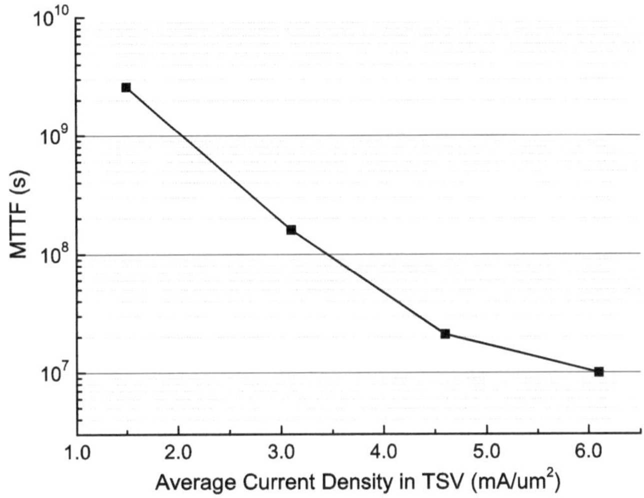

Fig. 9.8 MTTF vs. average current density. The average current density increases from 1.5 to 6 mA/um^2, T = 350 K

As the average current density inside the TSV is increased from 1.5 to 6 mA/um^2 with the temperature set to 350 K, the resulting MTTF is shown in Fig. 9.8. The EM lifetime of a TSV with high current dramatically reduces from 2.6e9s to 1.0e7s. A TSV, carrying a high current density accelerates the depletion and accumulation of atoms and decreases the EM lifetime. For P/G TSVs, which can carry the current density larger than 5 mA/um^2, the EM reliability may become critical.

9.5.3 Impact of Temperature

Temperature also plays an important role in atomic concentration and EM reliability. From Eqs. (9.3) to (9.4), the diffusivity D is exponentially related to the temperature. Moreover, Eq. (9.6) also shows that atomic flux is affected by the thermal gradient. Note that, in 3D operation, the temperature can vary from tens of degree C to a hundred of degree C.

Joule heating from high current density inside a TSV causes high temperature. However, due to high thermal conductivity of copper, the thermal gradient is very small inside the TSV. The thermal gradient caused by joule heating of the TSV with 60 mA input current is shown in Fig. 9.9. The structure consists of three silicon layers (each is 25 um thick), two inter-layer dielectric (ILD) layers (each is 4 um thick), a TSV liner (SiO_2 with 0.2 um thick), and a copper TSV with two landing wires. The heat sink is assigned at the top surface with the heat transfer coefficient of 25e3W/(m^2·K). Figure 9.9b shows a small thermal gradient in the ILD layers,

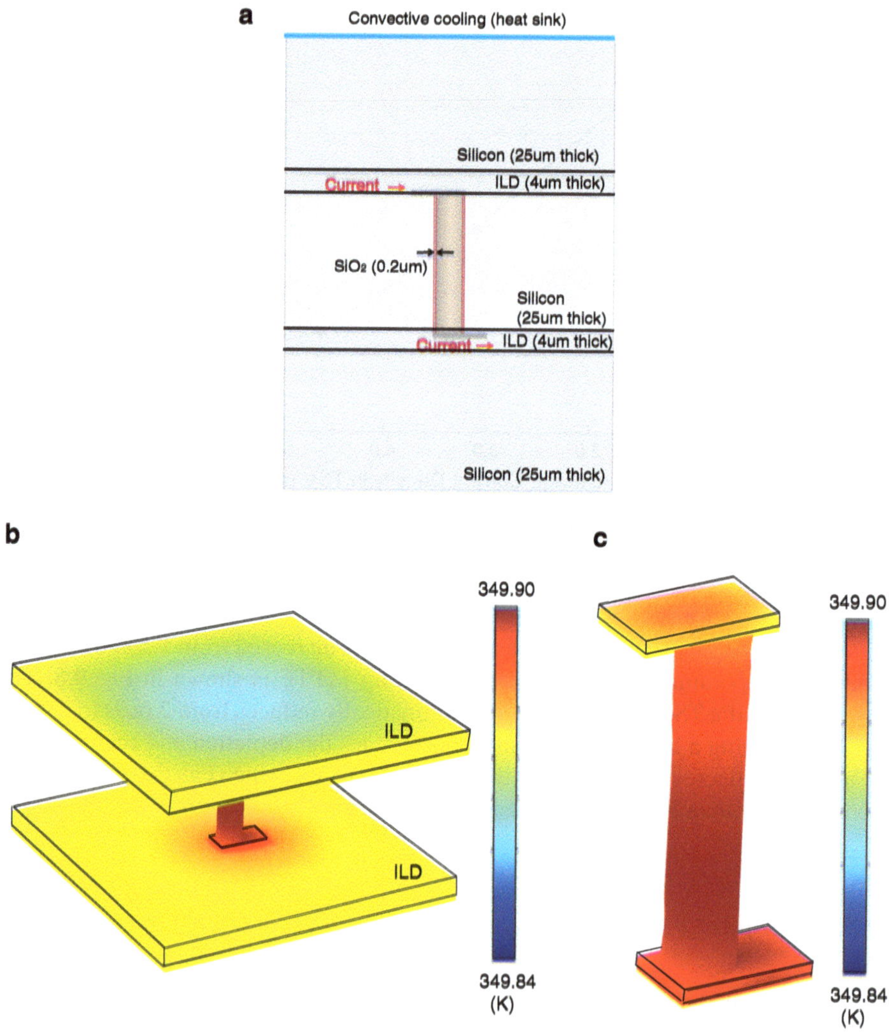

Fig. 9.9 Simulation of joule heating for a TSV with 60 mA input current. The structure (**a**) consists of three silicon layers, two ILD layers, a TSV liner (SiO_2), and a TSV with two landing wires. Heat sink is assigned at the *top* surface. (**b**) Is the thermal gradient in ILD layers, landing wires, and the TSV. (**c**) Is the thermal gradient inside the TSV which is negligible with a small range of 349.90–349.86 K

landing wires, and the TSV, where the temperature varies from 349.84 to 349.90 K. Figure 9.9c shows the thermal gradient inside the TSV and landing wires, which covers a small range of 349.86–49.90 K. Therefore, we include the flux caused by current $\mathbf{J_c}$ and concentration gradient $\mathbf{J_N}$ in continuity Eq. (9.2), and set other two terms of temperature gradient $\mathbf{J_T}$ and stress gradient $\mathbf{J_s}$ to zero.

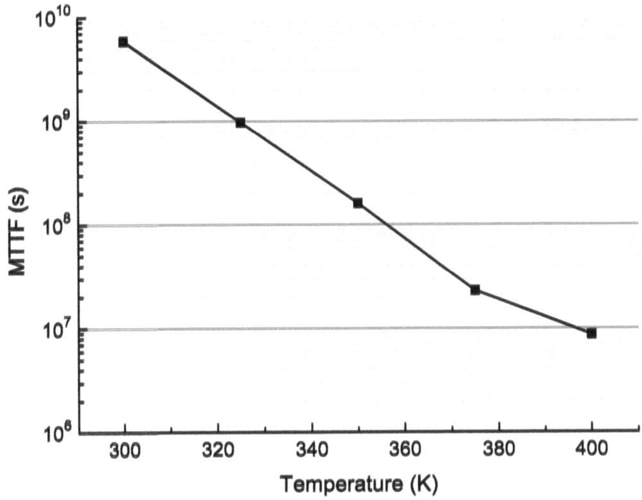

Fig. 9.10 MTTF vs. temperature. The temperature is varied from 300 to 400 K, and the current density is 3.1 mA/um^2

To analyze the impact of temperature on migration, the current value is kept constant, and the temperature is increased from 300 to 400 K. This temperature range is affected by both the power density from neighboring devices and the joule heating of the TSV. The impact of temperature on EM lifetime is shown in Fig. 9.10. As the temperature increases from 300 to 400 K, the MTTF is dramatically reduced from 5.9e9s to 8.7e6s.

9.5.4 *Impact of Grain Size*

The grain structure and size is mainly determined by the manufacturing process, and can vary over a wide range. To study this, we vary the grain size of the TSV from 1.9 to 0.9 um, while the grain boundary thickness is kept at 0.1 um. The total current is 60 mA, and the temperature is 350 K.

The resulting MTTF is shown in Fig. 9.11. With the grain size increases, the MTTF is increased from 1.6 to 3.1e8s. A TSV with larger grains helps to increase the lifetime. This is because the total grain boundaries with fast diffusion path decreases. Of course the average grain sizes and average grain boundary thicknesses can vary more than we have shown in this simple simulation. However, all these details can all be added to the model as needed.

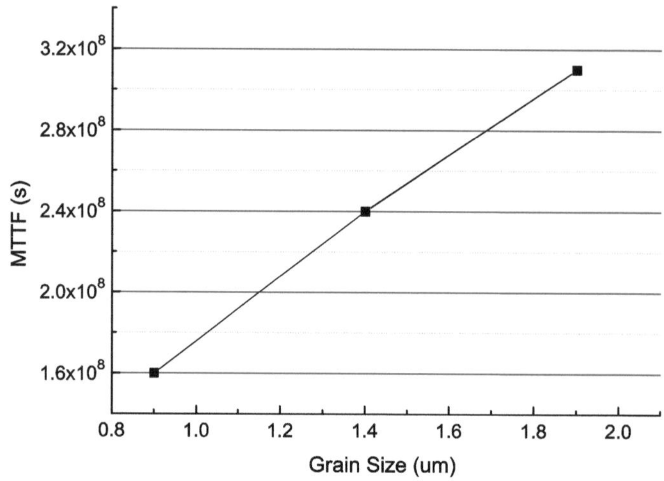

Fig. 9.11 MTTF vs. grain size

9.5.5 *Impact of Activation Energy*

For these simulations, the activation energy of the grains and grain boundaries may also vary a lot. Especially, the small activation energy of grain boundaries determines the EM lifetime. Therefore, we increase the activation energy of grain boundaries from 0.7 to 0.9 eV to investigate its impact. The resulting MTTF is shown in Fig. 9.12. We observe that with the activation energy reduces from 0.9 to 0.7 eV, the MTTF dramatically reduces from 3.55e9s to 5.2e6s. This demonstrates the exponential impact of E_{A} on the atomic flux.

9.6 Conclusions

In this chapter, we studied electro-migration (EM) in 3D IC by modeling atomic concentration in TSVs. From a set of extensive investigations, we made the following observations: (1) Atoms depleted or accumulated at the corner of wire-to-TSV interfaces, where the high current density are crowded; (2) Potential hillocks and voids inside the TSV have been simulated at the corner of wire-to-TSV interfaces; (3) High temperature, large current density, small grain size, or low activation energy of grain boundaries can accelerate the electro-migration, thus shortening the lifetime of the TSV. Our work can help designers find the EM risk locations of the 3D connection and analyze the atomic concentration evolution over time. The modeling approach can be extended to include irregular grain structures in the TSVs, grain boundaries in both the wires and TSVs, surface diffusion, grain boundary thickness, grain boundary non-uniformity, and other physical details.

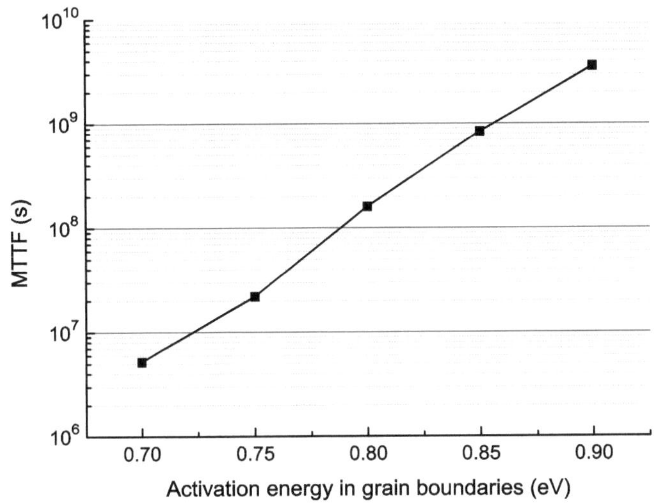

Fig. 9.12 MTTF vs. activation energy in grain boundaries. Grain size and grain boundary size is 0.9 and 0.1 um

References

1. J.R. Black, Electromigration – a brief survey and some recent restuls. IEEE Trans. Electron Device **ED-16**(4), 338–347 (1969)
2. I.A. Blech, Diffusional back flows during electromigration. Acta Mater. **46**(11), 3717–3723 (1998)
3. COMSOL, http://www.comsol.com/
4. C.-K. Hu, L.M. Gignac, E. Liniger, E. Huang, S. Greco, P. McLaughlin, C.-C. Yang, J.J. Demarest, Electromigration challenges for nanoscale Cu wiring. AIP Conf. Proc. **1143**(1), 3–11 (2009)
5. M. Jung, J. Mitra, D. Pan, S.K. Lim, TSV stress-aware full-chip mechanical reliability analysis and optimization for 3D IC, in *Proceedings of the ACM Design Automation Conference*, San Diego, 2011, pp. 188–193
6. W. Li, C.M. Tan, Enhanced finite element modelling of Cu electromigration using ANSYS and matlab. Microelectron. Reliab. **47**(9–11), 1497–1501 (2007)
7. J. Pak, M. Pathak, S.K. Lim, D.Z. Pan, Modeling of electromigration in through-silicon-via based 3D IC, in *IEEE Electronic Components and Technology Conference*, Las Vegas, Lake Buena Vista, 2011, pp. 1420–1427
8. M. Pathak, J. Pak, D.Z. Pan, S.K. Lim, Electromigration modeling and ull-chip reliability analysis for BEOL interconnect in TSV-based 3D ICs, in *Proceedings of the IEEE International Conference on Computer-Aided Design*, San Jose, 2011, pp. 555–562
9. S.-K. Ryu, K.-H. Lu, X. Zhang, J.-H. Im, P. Ho, R. Huang, Impact of near-surface thermal stresses on interfacial reliability of through-silicon vias for 3-D interconnects. IEEE Trans. Device Mater. Reliab. **11**(1), 35–43 (2011)
10. Y.C. Tan, C.M. Tan, X.W. Zhang, T.C. Chai, D.Q. Yu, Electromigration performance of through silicon via (TSV), A modeling approach. Microelectron. Reliab. **50**(9–11), 1336–1340 (2010)
11. K.N. Tu, Electromigration in stressed thin films. Phys. Rev. B **45**, 1409–1413 (1992)
12. K.N. Tu, Recent advances on electromigration in very-large-scale-integration of interconnects. J. Appl. Phys. **94**(9), 5451–5473 (2003)

13. J.-S. Yang, K. Athikulwongse, Y.-J. Lee, S.K. Lim, D. Pan, TSV stress aware timing analysis with applications to 3D-IC layout optimization, in *Proceedings of the ACM Design Automation Conference*, Anaheim, 2010, pp. 803–806
14. X. Zhao, M. Scheuermann, S.K. Lim, Analysis of DC current crowding in through-silicon-vias and its impact on power integrity in 3D ICs, in *Proceedings of the ACM Design Automation Conference*, San Francisco, 2012

Part III
Thermal Reliability in 3D IC Designs

This part of the book covers design-for-thermal-reliability for 3D ICs. In Chap. 10, we study thermal-aware architectural floorplanning for 3D IC and its impact on other metrics such as area, wirelength, and performance. In Chap. 11, we study gate-level placement techniques to alleviate thermal problems in 3D IC designs. In Chap. 12, we investigate the issues in co-design and co-analysis of thermal, power delivery, and performance targeting a 3D IC that employees micro-fluidic channels for cooling.

Chapter 10
Multi-objective Architectural Floorplanning for 3D IC

Abstract In this chapter, we study the multi-objective micro-architectural floorplanning algorithm for high performance processors implemented in IC. Our floorplanner takes a micro-architectural netlist and determines the dimension as well as the placement of the functional modules into single or multiple device layers while simultaneously achieving high performance and thermal reliability. The traditional design objectives such as area and wirelength are also considered. Our 3D floorplanning algorithm considers the following 3D-specific issues: vertical overlap optimization and bonding-aware layer partitioning. Our hybrid floorplanning approach combines Linear Programming and Simulated Annealing, which is shown to be very effective in obtaining high-quality solutions in short runtime under the multi-objective goals. We provide comprehensive experimental results on making tradeoffs among performance, thermal, area, and wirelength for 3D ICs.

The materials presented in this chapter are based on [25].

10.1 Introduction

Future processors implemented in nano-meter technologies will spend more time in communicating data operands or exchanging control information than actually performing useful computation. Meanwhile, the impact of power and thermal densities on these nanometer devices and interconnects continue to increase, thereby raising the cost for cooling solutions, eroding performance gains, and threatening overall circuit reliability. Micro-architectural floorplanning has drawn significant interests from both the computer architecture and EDA communities recently [6, 11, 21, 35, 39]. The main motivation is to tackle the ever-worsening wire delay problem of high-performance processors [1, 26] with a collaborative effort between micro-architecture and physical CAD.

S.K. Lim, *Design for High Performance, Low Power, and Reliable 3D Integrated Circuits*, DOI 10.1007/978-1-4419-9542-1_10,

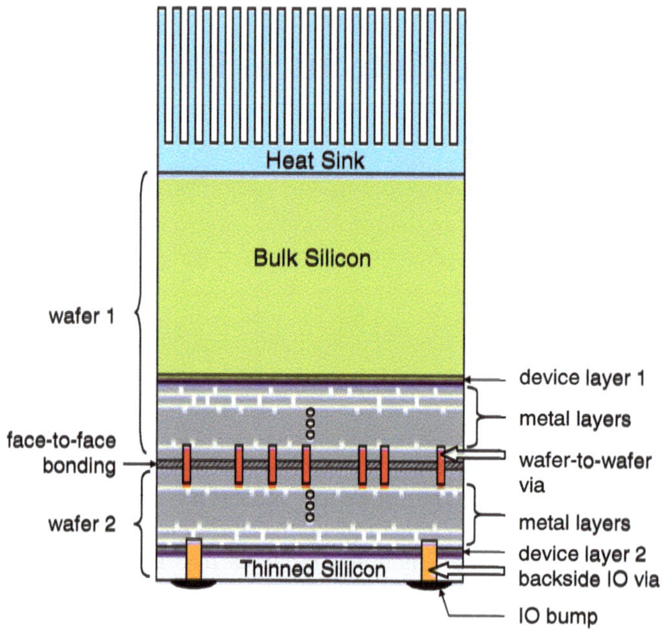

Fig. 10.1 Two-die 3D IC with face-to-face bonding

The 3D integrated circuit (3D IC) is an emerging technology that vertically stacks multiple die with a die-to-die interconnect as illustrated in Fig. 10.1. The die-to-die via pitch is very small and provides the possibility of arranging digital functional unit blocks across multiple die at a very fine level of granularity. This results in a decrease in the overall wire length, which translates into less wire delay and less power. Thus, 3D ICs can address the wire delay problem effectively by replacing the long and slow global interconnects with short and fast vertical routes. Advances in 3D integration and packaging are undoubtedly gaining momentum and have become of critical interest to the semiconductor community. These 3D integrated circuit and package manufacturing technologies are rapidly being adopted by several leading companies for commercial applications.

The location of individual micro-architectural modules plays a significant role on many important metrics. First, floorplanning has a huge impact on the performance of a given micro-architecture [measured by instructions per cycle (IPC)] as the global interconnects between modules are likely to be pipelined in order to meet high target clock frequencies. This may increase or decrease the access latency on all inter-module interconnects. Second, the thermal and leakage profile is highly correlated to the floorplan. This is because the temperature of each micro-architectural module is not solely dependent on the heat generation rate of each individual module but also the heat coupling between its neighboring modules. Moreover, the leakage power of each transistor is exponentially proportional to the temperature. Third, floorplanning affects the dynamic power consumption of the buses and clock distribution network. The total number of flip-flops (FFs) inserted on global interconnects changes the dynamic power consumed by the clock distribution network. However, the performance and thermal objectives are conflicting with each other since shorter distance among the hot modules improves the performance while exacerbating the thermal issue. To address the different design constraints of different domains, we need a goal-directed, automated floorplanner that allows users to weight their own design requirements and make effective design tradeoffs. The following specific topics are covered in this chapter:

- We study multi-objective floorplanning for nano-scale processors at the *micro-architectural level*. Our 3D floorplanners simultaneously consider performance, thermal reliability, footprint area, and interconnect length objectives, providing various tradeoff points.
- We study micro-architectural thermal modeling that considers the thermal and leakage inter-dependence for effective thermal runaway avoidance. Our micro-architectural power analysis, integrated with our thermal analyzer, models the dynamic and leakage power consumed by functional modules, global interconnects, and the clock distribution network for higher modeling accuracy.
- We study the following important 3D-specific problems: *vertical overlap optimization*, and *bonding-style aware layer partitioning*. We study how the vertical overlap among modules in 3D floorplanning affects performance, thermal, and area objectives. In addition, we discuss how layer partitioning is done under different inter-die via requirements existing in face-to-face, face-to-back, and back-to-back bonding in 3D stacked ICs.
- We discuss our floorplanning optimizer that consists of two steps: initial solution construction via Linear Programming and stochastic refinement via Simulated Annealing. This hybrid approach proves to be very effective in obtaining high-quality solutions in short runtime.

10.2 Existing Works

Recent studies have focused on traditional 2D micro-architectural floorplanning for performance optimization but not on thermal concerns [6, 11, 21, 35, 39]. For example, [39] uses a statistical design of experiments to approximate the effect on IPC of various wire lengths and then uses this approximation during simulated annealing to improve performance. Several micro-architecture studies on thermal [4, 29, 44] and leakage power [16, 17, 24, 31, 33] provide runtime management of the functional modules but do not perform floorplanning. In [24], the most recently published, they present a system level leakage power model and discuss dynamic management to reduce the thermal problem, as well as discussing thermal runaway and showing that a dynamic management scheme must include consideration of leakage power to be effective.

Most existing floorplanning and placement studies on thermal [3, 7, 10, 12, 30, 40, 47] target circuit designs, not micro-architectural designs. For example, [12] presents a 3D temperature driven floorplanner based on TCG and a novel bucket structure to represent module overlap. They use various thermal analyzers to trade off runtime with accuracy and overall performance. In addition, recently developed physical design tools for 3D ICs [3, 8, 9, 12–15, 22, 23, 32, 36, 37, 41, 46, 50] target gate-level netlists, are inefficient, and not suitable for evaluating different micro-architecture options during the early design stage. Thus, our work is the first to simultaneously consider performance, thermal, and leakage for the automated floorplanning of an entire processor micro-architecture with full simulation of the results of floorplanning.

10.3 Simulation Infrastructure

10.3.1 Micro-architectural Model

The micro-architecture used in our experiment is illustrated in Fig. 10.2. Each block represents a micro-architectural module used by our floorplanner. In order to model performance more faithfully for modern processors, we isolate and model each wire as a separate *resource* which consumes power and has a delay in proportion to its length. Note that architectural simulators that ignore inter-module communication latencies will no longer be useful for evaluating high frequency processors designed with nano-scale technologies due to wire delays, floorplan constraints, and thermal concerns. Essentially, the inter-module latency is a function of the distance and the number of flip-flops between modules and must be taken into account in both

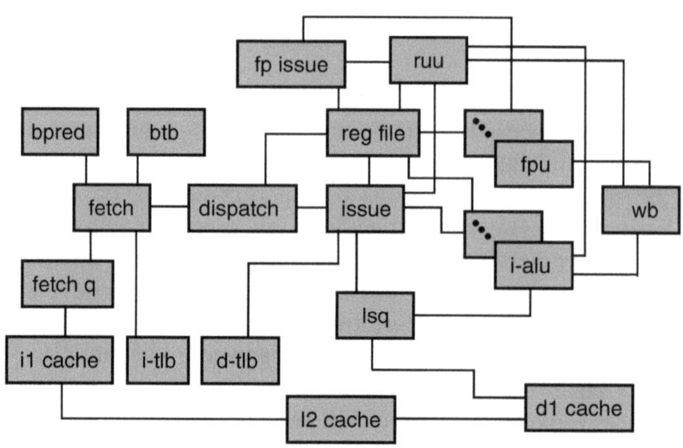

Fig. 10.2 Processor micro-architecture model

performance evaluation and floorplanning. For this reason, we use the distances generated by the floorplanner to determine the latency-related parameters such as pipeline depth and communication/forwarding latencies for our performance simulation.

The micro-architectural configuration used in our study[1] is summarized as follows: the machine width is 8. We use a 1,024-entry gshare branch predictor, a 512-entry register update unit (RUU) [45] that combines the functionality of a reservation station and a reorder buffer, 16 KB instruction and data L1 caches, a 256 KB unified L2 cache and no L3 cache, 128-entry instruction and data TLBs, 8 ALUs, 4 FPUs, and a 64-entry load store queue.

10.3.2 Dynamic Power Modeling

While collecting the inter-module traffic, we also generate the power consumption profile for each micro-architectural module cumulatively for every hundred thousand cycles. The rationale for such sampling is that the temperature is very unlikely to elevate abruptly within a processor's operation period of a few hundred thousand cycles. Note that these detailed traffic activity and dynamic power profiles are only collected once at the very beginning of the entire design flow. The thermal analyzer then uses these power statistics to provide the thermal profile. The floorplanner generates a new floorplan for the given thermal profile and module netlist.

We assume that the intra-module dynamic power consumption remains the same for different floorplans as the module activity factors primarily depend on the program behavior rather than the relative positions. Since the new floorplan may lead to different interconnect lengths between modules, our tool recomputes all of the inter-module interconnect power based on the new lengths and adds it to the dynamic per-module power collected earlier.

The number of flip-flops inserted on the wires for an extremely high clock frequency can create a larger load on the clock distribution network. This combined with the increasing percentage of the power budget that the clock distribution network consumes necessitates modeling the clock power at a finer granularity. Toward this, we use the accurate clock power model from [18]. This model considers clock distribution network power for memory structure precharge arrays, distribution wiring and drivers, pipeline flip-flops, and the phase locked loop.

---

[1]Our algorithm is general enough to take in many different configurations. For the sake of expediency, one configuration was chosen for experimentation.

10.3.3 Leakage Power Modeling

The leakage power is modeled in a separate process within our design flow. The model based on [48] considers different bias conditions, though it only estimates subthreshold leakage power. For array-like structures, such as caches and TLBs, the number of bits (or SRAM cells) stored is multiplied by the amount of leakage current per bit and by the supply voltage to calculate the total leakage power for the structure. To calibrate our model, we also calculate the subthreshold leakage currents using the method in eCACTI [20]. Our model closely matches the leakage power estimated from eCACTI. For logic structures, we assume CMOS gates where half the transistors are leaking at any given time. The number of transistors in these structures is estimated using the area values from GENESYS [19].

The following equation shows the relation between the subthreshold leakage current I_{sub} and a given temperature θ:

$$I_{sub} = k \cdot W \cdot e^{-V_{th}/nV_{\theta}}(1 - e^{-V_{dd}/V_{\theta}})$$

where k and n are experimentally derived, W is the gate width, V_{th} is the threshold voltage, V_{dd} is the supply voltage. V_{θ} is the thermal voltage that increases linearly as temperature elevates. Due to the temperature dependence on the subthreshold leakage current, we first use our model to estimate the leakage power based on an initial temperature. The results are then fed to our thermal analyzer so that it will estimate the temperature and the leakage power more accurately. This is done within the thermal analyzer by modeling their interdependence. First a baseline temperature is calculated with a static leakage estimation, then a leakage power based on those temperatures is calculated, then a new temperature based on the previous iterations leakage power, and so on until convergence or thermal runaway is detected. We follow the criteria [34] for detecting the scenarios of thermal runaway: (1) the maximum module temperature T_{max} is increasing, (2) the increment of power is larger than the increment of package's heat removal ability. The heat removal ability of the package is defined as $(T_{max} - T_a)/R_t$, where T_a and R_t are ambient temperature and thermal resistance of the package.

10.3.4 Thermal Modeling

The linearized differential equation ($k \cdot \nabla^2 T + P = 0$) for steady state heat flow was the basis of our thermal model, as described in [47]. In the equation, k is the thermal conductivity, T is the temperature, and P is the power density of heat sources. The chip is divided into a 3D grid as shown in Fig. 10.3 to apply a finite difference approximation to the differential equation. We rewrite the thermal equation into the following matrix form: $\mathbf{R} \cdot \overrightarrow{P} = \overrightarrow{T}$, where $\mathbf{R}$ is the thermal resistance matrix ($\mathbf{R}_{i,j}$ is the thermal resistance between node i and node j), $\overrightarrow{P}$ is the power profile vector

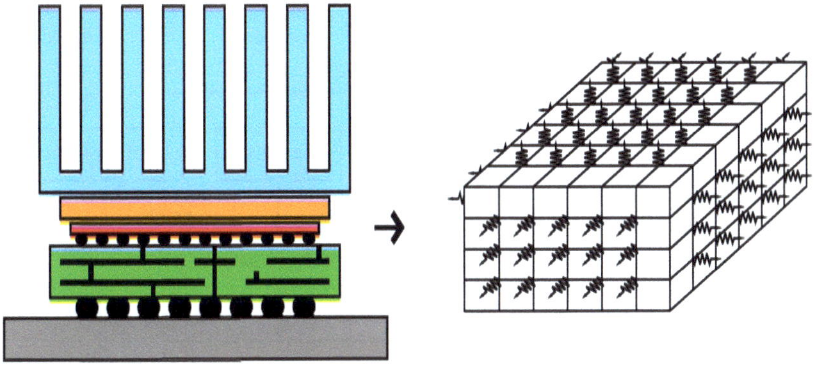

Fig. 10.3 3D grid of a chip for thermal modeling

($\vec{P}_i$ is the power dissipation of node i), and $\vec{T}$ is the temperature profile vector ($\vec{T}_i$ is the temperature of node i). Thus, the temperature of all the active nodes can now be calculated from the power profile using a single matrix-vector multiplication. The clock power is distributed evenly across the modules according to their areas. The bus power for each net is added to the total power of the source block. Then, the leakage power and temperature of each module are calculated iteratively using our model until they either converge or thermal runaway is detected.[2]

In order to facilitate fast but reasonably accurate temperature calculation, we use a non-uniform 3D thermal resistor mesh, where grid lines are defined at the center of each micro-architectural module. These grid lines are defined for the X and Y directions and extend through the Z direction to form planes. The intersection of grid lines in the X and Y directions define the thermal nodes of the resistor mesh. Each thermal node models a rectangular prism of silicon that may dissipate power if it covers some portion of a block. The total power of each block is distributed according to and among the $X - Y$ area of the nodes that block covers.

10.3.5 Integrated Design Flow

Our design flow incorporates the dynamic power, leakage power, performance, and thermal analysis discussed earlier into our floorplanner. An overview of this design flow is illustrated in Fig. 10.4. First, we use technology parameters and an architectural description to estimate the area and delay of the micro-architectural modules using the following analytical tools: CACTI [43] and GENESYS [19]. Then a cycle-accurate simulation using SimpleScalar [2] combined with Wattch [5]

[2]The average number of iterations needed was found to be approximately seven for the largest number of layers. A smaller number of layers requires fewer iterations.

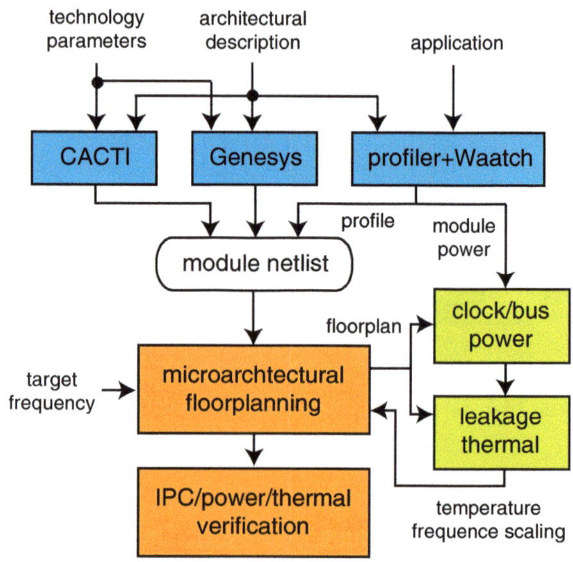

Fig. 10.4 Overview of our micro-architectural floorplanning framework

is done in order to collect and extract the amount of traffic between modules and estimate dynamic power consumption for each benchmark. From these tools we extract a profile weighted module netlist and power consumption information and feed all of this data into our multi-objective floorplanner. We also integrated the clock power estimation from [18] and the leakage estimation from [48] as described above with our thermal analyzer.

Our floorplanner consists of two steps: initial solution construction via Linear Programming (LP) and stochastic refinement via Simulated Annealing (SA). We recursively bipartition the floorplan area until each module is confined in its own partition. Each bipartitioning solution is optimized by an LP-based approach, where performance and thermal objectives are simultaneously considered under the leakage power constraint. We then call our thermal/leakage analyzer upon each bipartitioning to update the thermal and leakage profile. The interdependence between leakage power and temperature creates the possibility of thermal runaway [24], in which the temperature and leakage are caught in a positive feedback loop and both continue to exacerbate. If the floorplanner decides that thermal runaway is unavoidable given the current clock frequency then it scales the frequency down until it succeeds in avoiding runaway. Once the recursive bipartitioning is finished, we further optimize the current solution during our SA-based refinement. We perform low-temperature annealing to fine-tune the LP-based solution, where thermal/leakage analyzer is again used to guide our optimization. When the final solution is obtained, we use SimpleScalar, Wattch, and our thermal/leakage analyzer to evaluate the final solution for IPC, power, and thermal metrics.

10.4 2D Micro-architectural Floorplanning

Given a set of micro-architectural modules and a netlist that specifies the connectivity among these modules, our multi-objective 2D micro-architectural floorplanner tries to determine the width and height of each module and to place it into a single chip such that (1) there is no overlap among modules, (2) a user-specified clock frequency constraint is satisfied, and (3) thermal runaway does not occur under the constraint. Our objective is to provide a floorplan that effectively maximizes the performance of a processor while simultaneously minimizing the footprint area of the floorplan and maximum module temperature for better thermal reliability. We discuss our LP-based floorplan construction and simulated annealing-based refinement in this section.

10.4.1 LP-Based 2D Floorplanning

Algorithm 8 describes our slicing floorplanning algorithm. The basic idea behind our algorithm is to perform recursive bipartitioning until each partition contains a single module as shown in Fig. 10.5. In our approach the slicing operation determines the overall relative location among the modules, while an LP fine-tunes the location and determines the dimension of the modules. After we choose a partition to be divided, we perform thermal/leakage analysis to obtain module temperature. The first iteration of the recursive bi-partitioning contains no temperature objective because there is no way to obtain block temperatures without a floorplan. All subsequent iterations use temperatures calculated from the previous iteration's block positions. We then use our LP-based floorplanning to simultaneously optimize the performance and thermal distribution under the target frequency, leakage, center of gravity constraints (to remove overlap among the modules), and boundary constraints. An *iteration* in our algorithm combines a single bipartitioning and a subsequent LP-based floorplanning of all modules. Thus, we perform $k-1$ iterations if there are k modules in the netlist. Note that each iteration can be repeated multiple times to obtain different cutlines. This is because there exist multiple solutions that satisfy the boundary and center of gravity constraints during each bipartitioning. Thus, we perform each bipartitioning several times and pick the best solution in terms of performance and thermal profile.

The following variables are used for our LP-based floorplanning formulation:

- N: set of all modules in the netlist.
- E: set of all nets in the netlist.
- x_i, y_i: location of module i.
- w_i, h_i: half width and half height of module i
- a_i, g_i: area and delay of module i
- $w_m(i)$, $w_x(i)$: minimum/maximum width of module i
- $\lambda_{i,j}$: normalized profile weight on wire (i,j)

Algorithm: Slicing 2D floorplanning
Input: block-level netlist
Output: floorplanning with module location and shape determined
while *there exists a partition with multiple modules* **do**
    Choose a partition j to be divided;
    Call thermal/leakage analysis;
    for *number_of_repetitions* **do**
        Insert a cutline and compute center of gravity;
        Solve LP with inserted cutline;
    end
    Pick the best cutline from the set of repetitions;
    Update centers of gravity and bounding boxes;
end
return $x_i, y_i, w_i, h_i, z_{ij}$ for all modules;

Algorithm 8: Description of our slicing 2D floorplanning algorithm. We perform a top-down recursive bipartitioning and solve LP-based floorplanning at each iteration.

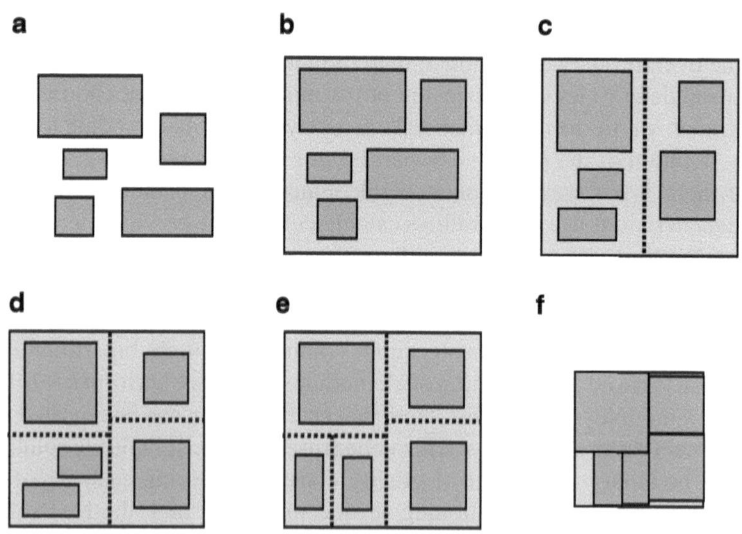

Fig. 10.5 Illustration of our 2D micro-architectural floorplanning. (**b–e**) LP-based slicing floorplan, (**f**) non-slicing floorplan refinement

- $z_{i,j}$: number of flip-flops on wire (i,j) after insertion
- $X_{i,j} = |x_i - x_j|$ and $Y_{i,j} = |y_i - y_j|$
- $T_{i,j}$: normalized product of the temperature of modules i and j
- A: aspect ratio of the chip
- X_x : maximum x_i, Y_x: maximum y_i

- C: target cycle time
- d_r: unit length delay of repeated interconnects

Our LP floorplanner determines the values for the following decision variables: x_i, y_i, w_i, h_i, and z_{ij}. The following are the variables used for bipartitioning:

- $B(u)$: set of all modules at iteration u
- $M_j(u)$: set of all modules in partition j at iteration u
- $S_{j,k}(u)$: set of modules assigned to subpartition k ($k \in \{1,2\}$ for bipartitioning) in partition j at iteration u
- $(\bar{x}_{jk}, \bar{y}_{jk})$: center of subpartition k contained in partition j
- r_j,v_j,t_j,b_j: the right, left, top, and bottom boundaries of partition j

Our LP formulation is used to perform floorplanning at iteration u of the main algorithm shown in Algorithm 8. Our LP-based slicing floorplanning is formulated as follows:

Minimize:

$$\sum_{(i,j)\in E} (\alpha \cdot \lambda_{ij} \cdot z_{ij} + \beta \cdot (1 - T_{ij})(X_{ij} + Y_{ij}) + \gamma \cdot X_x) \tag{10.1}$$

Subject to:

$$z_{ij} \geq \frac{g_i + d_r(X_{ij} + Y_{ij})}{C},\ (i,j) \in E \tag{10.2}$$

$$X_{ij} \geq x_i - x_j \text{ and } X_{ij} \geq x_j - x_i,\ (i,j) \in E \tag{10.3}$$

$$Y_{ij} \geq y_i - y_j \text{ and } Y_{ij} \geq y_j - y_i,\ (i,j) \in E \tag{10.4}$$

$$z_{ij} \geq 0,\ (i,j) \in E \tag{10.5}$$

$$w_m(i) \leq w_i \leq w_x(i),\ i \in N \tag{10.6}$$

$$x_i,\ y_i \geq 0,\ i \in N \tag{10.7}$$

$$X_x \geq x_i \text{ and } A \cdot X_x \geq y_i,\ i \in N \tag{10.8}$$

Boundary Constraints:

$$x_i + w_i \leq r_j,\ i \in M_j(u), j \in B(u) \tag{10.9}$$

$$x_i - w_i \geq v_j,\ i \in M_j(u), j \in B(u) \tag{10.10}$$

$$y_i + m_i w_i + k_i \leq t_j,\ i \in M_j(u), j \in B(u) \tag{10.11}$$

$$y_i - m_i w_i - k_i \geq b_j,\ i \in M_j(u), j \in B(u) \tag{10.12}$$

Center of Gravity Constraints: for $k \in \{1,2\}, j \in B(u)$

$$\sum_{i \in S_{jk}(u)} a_i x_i = \sum_{i \in S_{jk}(u)} a_i \times \bar{x}_{jk} \tag{10.13}$$

$$\sum_{i \in S_{jk}(u)} a_i y_i = \sum_{i \in S_{jk}(u)} a_i \times \bar{y}_{jk} \tag{10.14}$$

Our objective function shown in Eq. (10.1) contains three terms: profile-weighted wirelength ($= \lambda_{ij} \cdot z_{ij}$), thermal-weighted wirelength ($= (1 - T_{ij})(X_{ij} + Y_{ij})$), and footprint area ($= X_x$), where λ_{ij} is the profiled activity factor of the wire between modules i and j.[3] The minimization of the first term improves IPC while the minimization of the second term stretches the distance of two modules, thereby reducing thermal coupling.

$(1 - T_{ij})(X_{ij} + Y_{ij})$ was chosen as the temperature dependant portion of the cost function because it satisfies several properties: It is linear with respect to distance between module i and module j, it considers the temperatures of both module i and module j, and it grows smaller when considering hot blocks and larger when considering cool blocks. Because the cost function is being minimized in the LP, it is necessary to only consider minimizing the distance between cool blocks and not maximizing the distance between hot blocks, as would be preferable. Since minimizing $X_x \cdot Y_x$ (=floorplan area) is non-linear, we only minimize X_x since the constraint (10.8) enforces $A \cdot X_x$ to be greater than all y values. Note that α, β, and γ are user-defined parameters for weighing the performance, thermal, and area objectives. In case $\alpha = 0$, our floorplanner optimizes thermal + area only. In case $\beta = 0$, our floorplanner optimizes performance + area objective only. Lastly, the conventional area/wirelength-driven floorplanner uses the following new objective function:

$$\gamma \cdot X_x + \delta \cdot \sum_{(i,j) \in E} (X_{ij} + Y_{ij}) \tag{10.15}$$

We provide an extensive comparison among these four different floorplanning objectives in Sect. 10.6.3[4]:

- Performance + thermal + area
- Performance + area
- Thermal + area
- Area + wirelength

[3] Since we add performance and thermal-related weights to the pure wirelength, we do not explicitly consider non-weighted pure wirelength objective. However, we report the wirelength metric in all of our experiments to show the impact of this multi-objective on wirelength.

[4] Note that the area objective is used in all of these variations. The area objective has a positive impact on performance and wirelength objectives and a negative impact on thermal objective.

Constraint (10.2) is obtained from the definition of latency. If there is no FF on a wire (i, j), the delay of this wire is calculated as $d(i, j) = d_r(X_{ij} + Y_{ij})$. Then, $g_i + d(i, j)$ represents the latency of module i accessing module j, where $d(i, j)$ denotes the delay between i and j. Since C denotes the clock period constraint, $(g_i + d(i, j))/C$ denotes the minimum number of FFs required on (i, j) in order to satisfy C. Absolute values on x and y distance are given in (10.3)–(10.4). Constraint (10.5) requires that the number of FFs on each edge is non-negative. The block boundary constraints (10.9)–(10.12) require that all modules in the block be enclosed by these block boundaries. The center of gravity constraints (10.13)–(10.14) require that the module area-weighted mean (= center of gravity) among all modules in each sub-block corresponds to the center of the sub-block.

10.4.2 Stochastic Refinement

The standard LP relaxation of the floorplanning problem introduces several non-optimalities. The recursive bipartitioning process also yields only slicing floorplans. In order to address these issues we implemented a simulated annealing based refinement engine for our floorplanner. This allows us to search around the local space and find a local minimum without being constrained by linearity. We use three intra-layer moves during the simulated annealing refinement: swapping in positive sequence, swapping in both positive and negative sequences, and rotation. We derive a sequence pair from the LP floorplanning result and perform low temperature annealing with them. We use the *gridding* scheme described in [38] to derive the corresponding sequence pair representation from the slicing floorplan. Specifically, we draw the positive and negative loci for each module and order these loci to obtain the sequence pair. Next we compute the initial annealing temperature by setting the probability of accepting bad moves to a low value. This reduces the runtime required for the annealing process significantly and focuses on results that are near the LP based result, which is assumed to be fairly close to optimal. We use the following cost function during our annealing:

$$cost = \alpha \cdot perf\_wire + \beta \cdot max\_temp + \gamma \cdot area$$

where $perf\_wire$ is the profile-weighted wirelength and max_temp is the maximum module temperature. We use the same weighting constants α and β used in Eq. (10.1) between the performance and thermal objectives. It is important, however, to note that our temperature is *not* the weighted distance between two hot blocks but the *actual* temperature we obtain from our thermal analyzer. Thus, our thermal analysis is the runtime bottleneck during our refinement since we need to perform the analysis for potentially many candidate solutions during the annealing process. The consideration of performance is done in both the Simulated Annealing and Linear Programming approaches by the inclusion of profile weighted wirelength in the cost function.

Assuming that the thermal conductivity of functional modules are similar (they are mostly silicon), swapping the location of modules would not change the thermal resistance matrix **R**. This means that matrix **R** only needs to be computed once in the beginning. To calculate the temperature profile of a new floorplan, the power vector $\overrightarrow{P}$ needs to be updated and then multiplied by **R**. Alternatively, a change in power profile $\Delta\overrightarrow{P}$ can be defined. Multiplying **R** and $\Delta\overrightarrow{P}$ will give change in temperature vector $\Delta\overrightarrow{T}$. Adding $\Delta\overrightarrow{T}$ to the old temperature vector will give the new temperature profile. Swapping two blocks usually has a small effect on the power profile, so $\Delta\overrightarrow{P}$ is usually sparse. This reduces the number of multiplications required by the second method at the expense of doing extra additions and subtractions. This approach may not give us the most accurate temperature numbers but does provide high fidelity to distinguish good solutions from bad ones. Our related experiments shown in Sect. 10.6.6 support this claim. Lastly, the leakage and clock power updates are done faster since it basically involves evaluating a set of equations based on the new module locations and temperature values.

10.5 Extension to 3D Floorplanning

The extension to 3D floorplanning requires a new approach in floorplanning as well as updates on the architectural simulation for performance, power, and thermal evaluation. Our 3D floorplanning algorithm considers the issues that are specific to 3D: vertical overlap optimization and bonding-aware layer partitioning. We solve this problem using our LP-based 3D slicing floorplanning plus stochastic non-slicing floorplan refinement.

10.5.1 3D Extension of Architectural Simulation

In order to support the performance, power, and thermal simulation for 3D micro-architecture floorplanning, we extend the simulation engines discussed in Sect. 10.3 as follows:

- Performance: the IPC computation for 3D is not too much different from 2D case except that the access latency on each interconnect is calculated based on a 3D floorplan that involves delay in the z-dimension.
- Dynamic power: we again assume that the module power is independent of floorplanning. However, bus and clock power are heavily dependant on floorplanning and given the reduction of interconnect lengths in a 3D floorplan. The existing bus power calculator is extended to consider inter-layer interconnects. We assume that an H-tree is used for each layer, and these H-trees are connected by through-silicon-vias. The number of FFs and buffers included in the 3D clock tree is calculated based on the area of each layer.

- Temperature/leakage: the thermal analysis for 3D becomes more complex because of the multiple die structure. Thus, we add more layers in our 3D mesh to model the multiple sets of device, metal, and bonding layers. The leakage power computation is straightforward in our model once the temperature for each module is known.

Finally, the architecture-to-floorplan design flow shown in Fig. 10.4 remains the same except that all the related boxes now are 3D-aware.

10.5.2 Vertical Overlap Optimization

A unique challenge in 3D floorplanning is the issue of *vertical module overlap*. The primary benefit that a 3D IC provides is the ability to place the tightly connected modules *on top of* each other instead of *adjacent to* each other as in 2D case. This reduces the length and thus the delay/power of related interconnects significantly. Since the parasitics associated with the inter-die vias is similar to those of short interconnects, the additional freedom in z-dimension promises higher quality floorplans in terms of footprint area, performance, and power consumption. In addition, the shorter interconnects naturally mitigate the interconnect congestion problems. More specifically, the vertical overlap affects the quality of 3D micro-architectural floorplanning in the following ways:

- Performance: the performance of a 3D micro-architectural floorplan tends to improve when the vertical overlap is maximized among the blocks with higher access frequencies. This is mainly caused by the shorter interconnect and thus the lower access latency among the frequently communicating modules.
- Thermal: the thermal profile of a 3D micro-architectural floorplan tends to deteriorate due to compressed space. More hotspots are created when the vertical overlap is maximized among the hot modules. This harmful thermal coupling causes the leakage power to increase, raising the likelihood of thermal runaway.
- Power: the dynamic module power and clock power are rarely affected by the vertical overlap. However, the overall bus power consumption tends to decrease with more vertical overlap among the modules with higher switching activities. This is because the dynamic power saving is greater when highly active modules drive shorter interconnects. Note that this contradicts to the thermal objective since highly active modules tend to become hotter.

In summary, our 3D floorplanning tries to maximize the vertical overlap among the frequently communicating and highly switching modules while minimizing the vertical overlap among the hot modules.[5] Since these objectives are competing each other, trading one objective off the others is inevitable.

[5]Note that it is possible to impose the vertical overlap constraints among the related groups of modules. The investigation of this direction is out of the scope of this chapter, which may require the extension of floorplanning encoding scheme such as Sequence Pair [38].

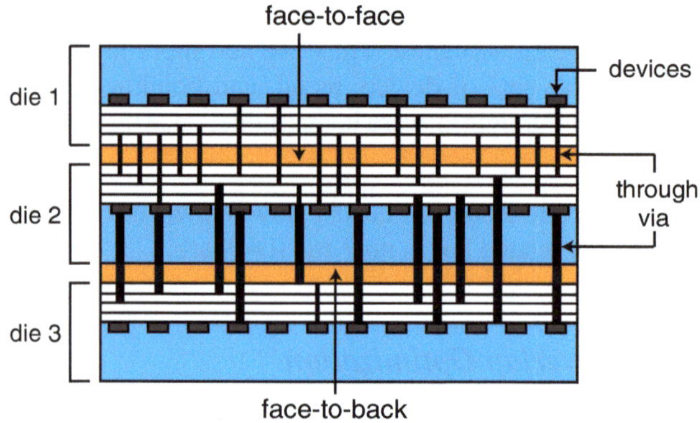

Fig. 10.6 Through vias in 3D ICs with face-to-face and face-to-back bonding. Back-to-back style forms when the two substrate sides are attached (not shown in this figure)

10.5.3 Bonding-Aware Layer Partitioning

A 3D IC requires special kinds of vias for inter-die connection called *through-silicon-vias* (TSVs). There are three kinds of through-silicon-vias depending on the style of bonding mechanism used to bond two dies together: *face-to-face* (F2F), *face-to-back* (F2B), and *back-to-back* (B2B) through-silicon-vias, as illustrated in Fig. 10.6. The "face" refers to the metal layer side of a die, whereas the substrate side is called "back". F2F through-silicon-vias ($\approx 0.5 \times 0.5\,\mu$) have a smaller pitch than F2B ($\approx 5 \times 5\,\mu$) and B2B through-silicon-vias ($\approx 15 \times 15\,\mu$) [27]. In addition, too many F2B/B2B through-silicon-vias fabricated on a single thinned wafer may adversely affect its reliability [49] since these vias actually penetrate the substrate. Thus, it is desirable to *reduce* the number of inter-die connections in F2B/B2B bonding. In the case of F2F bonding, however, it is desirable to *increase* the number of inter-die connections since the via density is much higher (almost the same as intra-die via density) and thus enables a significantly higher bandwidth for inter-layer communication. Note that F2B/B2B bonding is inevitable if the number of die exceeds two. Moreover, in the case that all three bonding styles are used in a single 3D IC, the 3D floorplanning has to be done carefully to exploit both bonding styles.

In our two-step approach for 3D floorplanning, we first partition the modules into layers (= die) and then floorplan these layers. The goal during our layer partitioning is to exploit the bonding style and vertical overlap opportunities, whereas our floorplanning optimizes the vertical overlap for performance, footprint area, and thermal objectives. During our layer partitioning, we assign a layer to each module such that the connection at the F2F boundary is maximized while the F2B/B2B connection is minimized. Next, we split the pair of modules connected via high profile-weighted edge into two layers with F2F bonding, so that we can vertically overlap them during the subsequent floorplanning step for achieving better performance. In addition, we split highly active modules in the same way, i.e., two

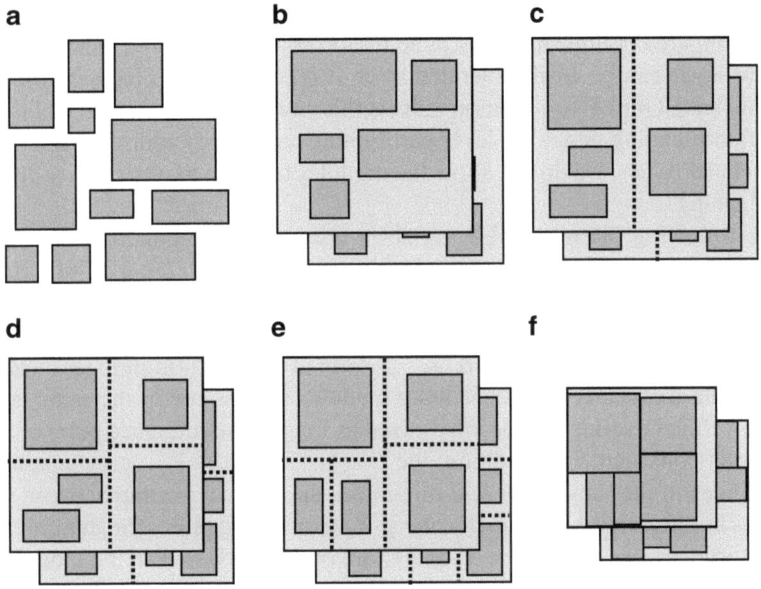

Fig. 10.7 Illustration of our 3D micro-architectural floorplanning. (**b**) layer partitioning, (**c**–**e**) LP-based 3D slicing floorplan, (**f**) non-slicing floorplan refinement

layers with F2F bonding, such that the shorter interconnect connected to these modules help reduce the dynamic power. Since the temperature of the modules requires floorplanning, our layer partitioning is not temperate-aware. Finally, we separate the modules with large area such as the RUU into different layers to help minimize the footprint area and reduce the amount of white space. In our greedy construction algorithm, we sort the modules according to their size, power density, and switching activity. We then assign the best possible layer for each module based on the performance, power, and area objectives mentioned earlier.

10.5.4 LP-Based 3D Floorplanning

In our LP-based 3D floorplanning, we extend the slicing floorplanning discussed in Sect. 10.4.1 to handle multiple layers simultaneously. Specifically, we insert each slicing cutline to cut all layers simultaneously as illustrated in Fig. 10.7. The goal of our slicing 3D floorplanning remains the same as the 2D case: to determine the dimension and relative position among the modules so that the multi-objective function is minimized. In addition, these locations will be refined via our 3D non-slicing floorplanning during our post refinement. The major difference between the 2D and 3D slicing floorplan is the interaction with different layers, which is the key element for an effective 3D floorplan. More specifically, the vertical overlap discussed in Sect. 10.5.2 has a high impact on performance and thermal objectives. In addition, area optimization has to be footprint-aware: the area increase from the smallest layer can be easily tolerated since it is less likely to increase the overall footprint area. Our LP formulation reflects this new optimization goal that is unique to 3D floorplanning. Since the layer partitioning has already addressed the bonding-style-related issues, we do not allow the modules to move to other layers during the floorplanning.

The following 3D-related LP variables are used in conjunction with the 2D-related variables shown in Sect. 10.4.1: l_i: layer of module i, $L_{ij} = |l_i - l_j|$, d_v: delay of inter-layer vias. It is crucial to note that the LP objective function used for 2D floorplanning, i.e., Eq. (10.1), can be used *as is* so long as we consider *all* layers simultaneously. Specifically, the $\alpha \cdot \lambda_{ij} \cdot z_{ij}$ term in Eq. (10.1) minimizes the distance between the frequently communicating modules if these are in the same layer—if not, the vertical overlap will be maximized as long as the reference point of module location is consistent.[6] In addition, the $\beta \cdot (1 - T_{ij})(X_{ij} + Y_{ij})$ term separates two hot modules in the same layer *and* minimizes the vertical overlap between two hot modules in different layers. Finally, the $\gamma \cdot X_x$ term still captures the minimization of 3D footprint area as long as the X_x and Y_x are computed based on the modules in all layers. The only difference between the LP formulations of 2D and 3D floorplanning is the latency constraint, for which we update the Eq. (10.2) with the following:

$$z_{ij} \geq \frac{g_i + d_r(X_{ij} + Y_{ij}) + d_v L_{ij}}{C}, \ (i,j) \in E \tag{10.16}$$

This latency constraint considers the delay of inter-layer via delay as well as interconnect delay during the computation of FFs needed to satisfy the clock period constraint C. We assume that d_r (= unit length delay of repeated interconnects) is larger than d_v (= delay of inter-layer vias).

10.5.5 3D Stochastic Refinement

The goal of our 3D stochastic refinement is to improve the 3D slicing floorplanning solution we obtain from our LP-based construction algorithm. Our basic approach is the same as the 2D case discussed in Sect. 10.4.2: non-slicing floorplanning with low-temperature simulated annealing to simultaneously refine the performance, thermal, and area objectives. The major difference between the 2D and the 3D case is that we use one sequence pair per layer to represent the entire 3D solution. In addition, our perturbation scheme does not allow inter-layer module movement to maintain the bonding-aware layer separation. Finally, the temperature calculation takes even longer since our thermal model needs to be expanded to consider

[6] We use the lower left corner of each module in our case.

multiple die. Thus, the annealing schedule is adjusted in such a way not to increase the runtime too much, which involves tuning such parameters as the initial/final annealing temperature, total number of moves each annealing temperature, cooling ratio, and annealing termination criteria.

10.6 Experimental Results

10.6.1 Experimental Setting

Our experiments were performed on ten programs from the SPEC2000 benchmark suite. We chose 4 from the floating point and 6 from the integer benchmark suites. For IPC evaluation, we ran each benchmark on the average case floorplan using a modified SimpleScalar 3.0 [2] by fast-forwarding four billion instructions and simulating the next four billion instructions. The reported temperature is simulated after all floorplanning steps and is adjusted relative to a 45 °C ambient temperature. We report the maximum temperature among all blocks in the floorplan. Our 3D floorplan is based on a four-layer stacked IC. We assume face-to-face bonding between layer 0 (topmost) and 1 and layer 2 and 3. A back-to-back bonding is used between layer 1 and 2. The heat sink is attached to layer 3. Wirelength is reported in mm. The "area" in our results refers to the footprint area (= maximum width × maximum height) of the four-layer floorplan and is reported in mm^2. The runtime of our framework was collected on Pentium Xeon 2.4 GHz dual-processor systems. The runtime of profiling four billion instructions after fast-forwarding four billion instructions was about 4 h per benchmark as was the power collection simulation for the same sets of instructions. The floorplanning steps took approximately 25 min and the simulations for the reported values of temperature and IPC took approximately 2 min and 1 h per benchmark, respectively.

10.6.2 Comparison to Existing 3D Floorplanner

Table 10.1 shows the comparison of our floorplanner to CBA-T [12]. Here we tested our floorplanner with the MCNC and GSRC benchmark circuits that were used in [12]. Since the power density values are randomly generated in [12], a fair temperature comparison is not possible. Since MCNC/GSRC benchmarks are not micro-architecture designs, we cannot compute the power density using our tool. We note, however, that our floorplanner obtains comparable results in terms of area, wirelength, and temperature. In addition, tuning the weighting constants among the objectives may result in different results.

Table 10.1 Comparison with CBA-T [12]. Our floorplanner is LP + SA with A + W + T objective. The baseline is CBA-T

	CBA-T [12]			LP + SA (ratio)		
Bench	Area	Wire	Temp	Area	Wire	Temp
ami33	4.14e + 05	24,442	160	0.94	1.11	0.96
ami49	1.84e + 07	477,646	151	0.79	1.21	0.94
n100	6.56e + 04	92,450	158	1.27	0.95	0.93

Table 10.2 Multi-objective 2D floorplanning results with performance (P), maximum block temperature (T), area (A), wirelength (W) objectives. The LP + SA-based floorplanner is used. Temperature is in °C

	A		A + W		A + P		A + T		A + P + T	
Bench	IPC	Temp	IPC	Temp	IPC	Temp	IPC	Temp	IPC	Temp
gzip	2.04	80.4	2.01	78.3	2.83	100.4	2.03	75.2	2.69	86.2
Swim	0.48	66.9	0.52	64.3	0.85	78.4	0.54	63.0	0.66	70.5
vpr	0.77	90.4	0.95	87.6	1.19	113.8	0.82	82.3	1.15	95.9
Art	0.34	64.4	0.38	67.9	0.62	83.3	0.39	65.4	0.51	74.4
mcf	0.03	64.1	0.07	63.0	0.09	76.9	0.07	62.1	0.10	69.4
Equake	0.34	65.5	0.40	62.7	0.47	76.3	0.41	61.8	0.43	69.0
Lucas	0.58	101.3	0.63	95.6	0.75	123.2	0.64	88.3	0.80	103.5
Gap	1.19	70.9	1.17	70.1	1.24	87.8	1.18	68.1	1.32	77.3
bzip2	1.43	82.1	1.42	80.4	1.90	103.6	1.47	77.1	1.65	88.4
twolf	0.59	97.4	0.60	92.3	0.94	120.8	0.61	85.8	0.61	101.1
Avg.	0.78	78.4	0.81	76.2	1.09	96.46	0.82	72.9	0.99	83.6
Area (mm^2)	50.5		52.46		57.23		58.66		60.37	
Wire (mm)	380.23		345.20		412.15		358.86		449.67	
Time (s)	168		174		188		1,116		1,064	
Pipeline stage	22		22		19		27		23	
Whitespace %	7		10		20		23		21	

10.6.3 Floorplanning Results

Table 10.2 presents various tradeoffs existing in multi-objective 2D floorplanning. We use our LP + SA method. One can see that the maximum module temperature increased markedly for A + P compared to the baseline A + W. The IPC result of A + P is the best among the four algorithms with an average IPC improvement over A + W by 35%. A + T decreases the temperature by about 24% over A + P while the IPC decreases by 25%. The hybrid A + P + T decreases the temperature by 14% over A + P while maintaining a high IPC value of 22% above the baseline A + W. In general, as the IPC increases, the block-level dynamic power also increases due to the higher activity, which results in a high temperature. This is a reason why A + W obtained lower temperature than A + P and A + P + T. This can be also seen from the fact that A + P obtains the highest IPC as well as temperature. Thus, the temperature drop in A + T compared to A + P is the result of smart floorplanning and lower IPC.

Table 10.3 Multi-objective 3D floorplanning results with performance (P), maximum block temperature (T), area (A), wirelength (W) objectives. The LP + SA-based floorplanner is used

	A		A + W		A + P		A + T		A + P + T	
Bench	IPC	Temp	IPC	Temp	IPC	Temp	IPC	Temp	IPC	Temp
gzip	2.40	108.8	2.74	104.7	3.98	125.9	2.75	98.9	2.85	104.7
Swim	0.72	91.9	0.71	92.9	0.85	106.9	0.72	84.1	0.92	88.0
vpr	0.98	120.7	1.30	111.5	1.40	137.0	1.25	107.1	1.29	114.4
Art	0.58	95.6	0.52	95.6	0.59	111.4	0.52	87.9	0.61	92.0
mcf	0.21	97.8	0.10	92.0	0.11	105.4	0.10	83.1	0.07	86.6
Equake	0.59	89.7	0.54	91.7	0.58	105.0	0.55	82.6	0.67	86.2
Lucas	0.88	127.2	0.87	116.9	0.92	145.3	0.88	113.0	1.19	123.0
Gap	1.47	96.5	1.59	97.0	1.59	114.2	1.62	89.6	1.61	94.5
bzip2	1.75	115.0	1.94	106.8	2.05	129.0	1.98	101.5	2.33	107.4
twolf	0.84	119.1	0.81	114.6	1.03	142.2	0.84	111.0	1.02	118.9
Avg.	1.04	105.3	1.11	102.4	1.31	122.2	1.12	95.8	1.26	101.6
Area (mm^2)	21.6		22.20		23.63		25.45		26.45	
Wire (mm)	247.25		217.20		323.43		252.08		247.02	
Time (s)	175		180		438		16,913		20,016	
Pipeline stage	22		22		17		24		21	
Whitespace %	7		9		16		25		23	

For the 3D case shown in Table 10.3, 3D A + W achieves a 37% increase in IPC and a 34% increase in temperature over 2D A + W while decreasing the total wirelength by almost 40%. The area result of 3D A + W is the best among all objective functions. A + P increases the IPC by 18% over A + W and increases the temperature by 19%. As expected, A + T decreases the temperature result of A + P significantly and achieves the best temperature results among all four 3D algorithms. The 4× increase in grid size for the temperature simulations in the 3D case causes the runtime of those objectives incorporating temperature calculations to increase dramatically.[7] The hybrid A + P + T retains a temperature close to that of A + W while increasing the IPC by 14%. In summary, A + P + T (1) obtains results that are between those of A + T and A + P and (2) outperforms A + W in terms of performance with comparable temperature results for both 2D and 3D. In case the temperature should be more emphasized, the thermal weight can be increased, which will likely lead to performance degradation.

Also shown in Tables 10.2 and 10.3 are the pipeline depth and whitespace percentages for the various objective functions. First, the pipeline depth ranges from 17 to 23, which agrees with current trends in commercial processor designs, e.g. 90 nm Pentium 4 back-end pipeline has 31 stages, the Intel's NGMA has 14 stages, etc. Despite the increase in the pipeline depth from FF insertion, our

[7]Our recent study [42] shows that Random Walk method can improve the runtime of thermal simulation significantly. Our future work includes the integration of this scheme in our micro-architectural floorplanning.

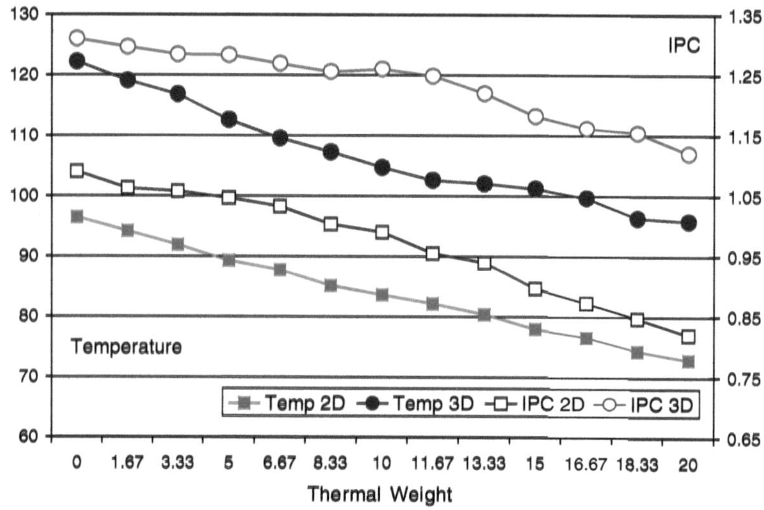

Fig. 10.8 Tradeoff between performance and temperature. Performance and area weights are held constant while thermal weight varies

strategy to add FFs on non-critical wires do not degrade the performance while removing FFs from critical wires improved performance. Second, whitespace ranges from 7 to 23%. In case of area-only objective, the whitespace is 7% for both 2D and 3D cases. This whitespace keeps increasing as we consider other objectives. The whitespace increase caused by wirelength consideration is only 2–3% while performance and thermal objectives cause the whitespace to increase by 9–13% and 16–18%, respectively. Due to the unbalance in block area, it becomes more difficult to optimize whitespace while placing frequently communicating blocks closer (= performance) or separating hot blocks apart (= temperature).

A tradeoff between performance and temperature is shown in Fig. 10.8. Temperature and IPC are reported as averages over the ten benchmarks. The performance and area weights are held constant while the thermal weight is varied. As expected, the graph shows that as the thermal weight is given more consideration by the floorplanner, the performance drops. Ideally there would be some separation between the curves to indicate that high reduction in temperature could occur with little degradation in IPC value. The sweet spot of the curve appears when the thermal weight is around ten. The IPC drops sharply after this and so would be undesirable for the reduction in temperature achieved. One can observe that there is a 15% reduction in IPC and a 22% reduction in temperature between the performance-only objective (0) and the highest weight hybrid objective (20) for the 3D case. As expected and also shown in Table 10.2, the multi-layer floorplans increase both the temperature and IPC over the single layer floorplans. Also of note is that the highest thermal weight multi-layer floorplan has a temperature close to that of the lowest thermal weight single layer floorplan while achieving a higher IPC. This demonstrates the benefits rendered by moving to multi-layer ICs.

Table 10.4 Comparison among pure-SA, pure-LP, and LP + SA approaches. The objective used is a linear combination of performance, temperature, and area all with equal weight. Area is in mm^2, wirelength in mm, and temperature in °C

	2D floorplan						3D floorplan					
	Pure SA		Pure LP		LP + SA		Pure SA		Pure LP		LP + SA	
Bench	IPC	Temp	IPC	Temp	IPC	Temp	IPC	Temp	IPC	Temp	IPC	Temp
gzip	2.38	102.2	1.94	80.19	2.69	86.2	2.74	109.5	2.31	97.5	2.85	104.7
Swim	0.61	83.5	0.66	69.3	0.66	70.5	0.71	91.8	0.70	86.7	0.92	88.0
vpr	0.93	113.1	1.24	86.9	1.15	95.9	1.07	119.8	1.24	103.4	1.29	114.4
Art	0.45	87.5	0.48	71.9	0.51	74.4	0.52	95.7	0.51	89.0	0.61	92.0
mcf	0.08	82.0	0.09	68.3	0.10	69.4	0.10	90.4	0.10	85.9	0.07	86.6
Equake	0.47	81.6	0.49	68.1	0.43	69.0	0.54	90.0	0.53	85.7	0.67	86.2
Lucas	0.75	122.6	0.79	93.8	0.80	103.5	0.87	128.7	0.85	108.1	1.19	123.0
Gap	1.38	91.1	1.34	73.7	1.32	77.3	1.59	98.9	1.49	90.9	1.61	94.5
bzip2	1.68	105.2	1.59	81.8	1.65	88.4	1.94	112.2	1.81	99.4	2.33	107.4
twolf	0.70	118.6	0.68	90.1	0.61	101.1	0.81	124.8	0.77	106.2	1.02	118.9
Avg.	0.94	98.7	0.93	78.4	0.99	83.6	1.09	106.2	1.03	95.3	1.26	101.6
Area	60.90		314.72		60.37		21.59		70.64		26.45	
Wire	388.13		524.81		449.67		230.47		207.57		247.02	
Time	1,225		826		1,064		25,157		18,207		20,016	

10.6.4 Optimization Method Comparison

Experimental results were also gathered across the three floorplanning algorithms; linear programming only, simulated annealing, and the combined approach of linear programming followed by simulated annealing refinement. Table 10.4 presents a comparison of the IPC, temperature, area, wirelength, and runtime of these three floorplanning algorithms for the 2D and 3D cases. One can observe for the 2D case from the table that the LP floorplanner does very poorly on the area of the floorplan and is not as good as the combined approach for IPC. The wirelength values are within an acceptable range for all approaches, though it is interesting to note that while the LP-only approach creates large area the wirelength values are actually less. This is because while wirelength was an objective during the recursive bipartitioning phase of the LP, the area is not, because the formulation has no way to constrain overlap. This was a large part of the motivation to use simulated annealing to refine the LP-only solution. In summary, LP + SA improves LP and outperforms SA consistently in terms of both performance and thermal objectives. The runtime of all approaches was roughly equivalent, showing that in a similar amount of time the combined approach produces better solution quality. These trends are consistent for the 3D cases with increased overall temperature averages and runtime. Again the large runtime increase was due mainly to the increase in simulation time for the temperature.

Table 10.5 shows a comparison among three different layer partitioning styles: area-greedy, bonding-prof, bonding-area. In area-greedy, the blocks are sorted in a

Table 10.5 Comparison between the different layer partitioning styles. The hybrid A + P + T objective is used with combined LP + SA approach. Area is in mm^2, wirelength in mm, and temperature in °C

	Area-greedy		Bonding-prof		Bonding-area	
Bench	IPC	Temp	IPC	Temp	IPC	Temp
gzip	2.98	108.9	2.88	108.8	2.85	104.7
Swim	0.77	93.0	0.87	96.8	0.92	88.0
vpr	1.16	117.8	1.54	112.9	1.29	114.4
Art	0.57	95.6	0.65	99.7	0.61	92.0
mcf	0.11	91.8	0.12	92.9	0.07	86.6
Equake	0.59	89.8	0.66	95.3	0.67	86.2
Lucas	0.96	127.2	1.06	117.1	1.19	123.0
gap	1.77	99.8	1.88	100.2	1.61	94.5
bzip2	2.14	110.4	2.29	109.2	2.33	107.4
twolf	0.90	126.9	0.95	118.0	1.03	118.9
Avg.	1.20	106.1	1.29	105.1	1.26	101.6
Area	22.68		52.54		26.45	
Wire	270.73		263.26		247.02	
Time	19,872		20,102		20,016	

decreasing order of their area and assigned to each layer so that the overall area is balanced among the layers. Cutsize is not optimized in this case. In bonding-prof, our goal is to optimize the profile-weighted cutsize among the modules. Lastly, bonding-area is the algorithm introduced in Sect. 10.5.3. We observe that the bonding-area partitioning outperforms a pure area based approach on IPC and temperature. It has slightly lower IPC than the bonding-prof partitioning, but the area is completely unacceptable in bonding-prof. The wirelength and runtime of all approaches was comparable.

10.6.5 Architectural Analysis

Figures 10.9 and 10.10 show snapshots of our floorplanning solution. We use LP + SA with area, performance, and temperature objectives. The whitespace of the floorplan is somewhat less than optimal but this is due to the higher weights placed on the performance and temperature optimization.[8] Our flow provides the users with the ability to modify the objective weights to suit their needs. This figure demonstrates that there is indeed thermal coupling between adjacent modules and that the thermal portion of the objective has attempted to separate the hottest modules while the performance portion of the objective has caused some of the hottest modules to remain grouped. This stays in line with the rapid dropoff in performance with decreased temperature shown in Fig. 10.8.

[8] These floorplans also highlight the challenge in area optimization for multi-objective, multi-layer floorplanning problem. Our future work tries to address this problem more effectively. A possible solution is to utilize the whitespace for decoupling capacitors, thermal vias, buffers, etc.

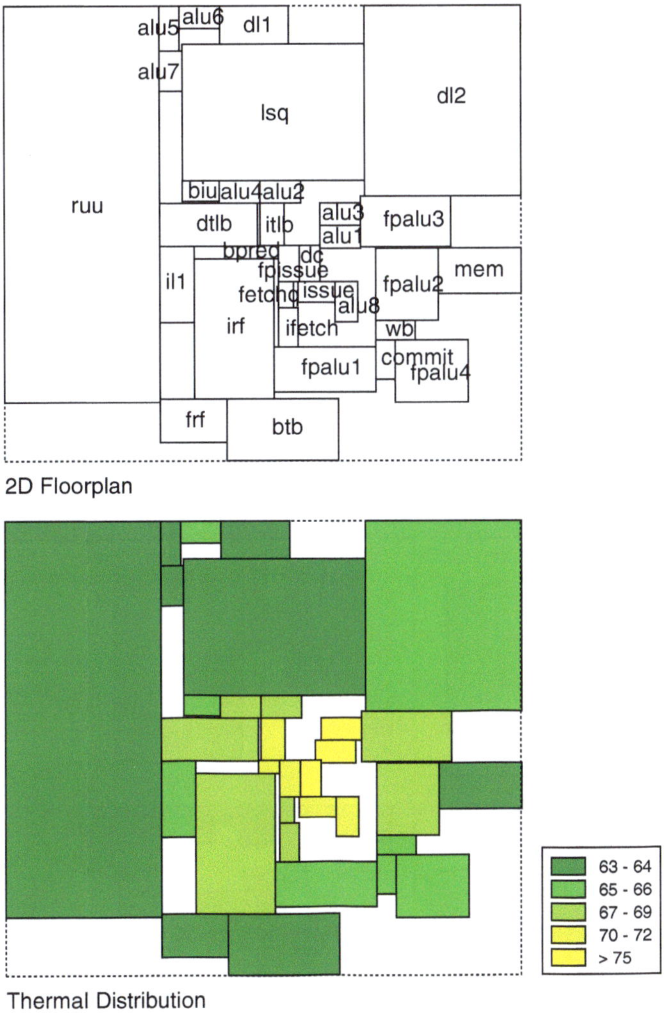

Fig. 10.9 Snapshots of our 2D floorplanning

Table 10.6 shows the top ten micro-architectural modules under various metrics. Physical designers often are only able to view the modules at the floorplan level as little more than rectangles. Here we provide some more detailed information about each of the modules that make up the floorplan. This can provide better opportunities for optimization at the physical design level. The register update unit (RUU) [45] with a large number of read/write ports is larger in area than the next two largest modules combined, which is why it was split up for the multi-layer floorplans. The power density of the ALUs is higher than most of the other modules and hence their temperatures are also generally among the highest in the floorplan. The 3D

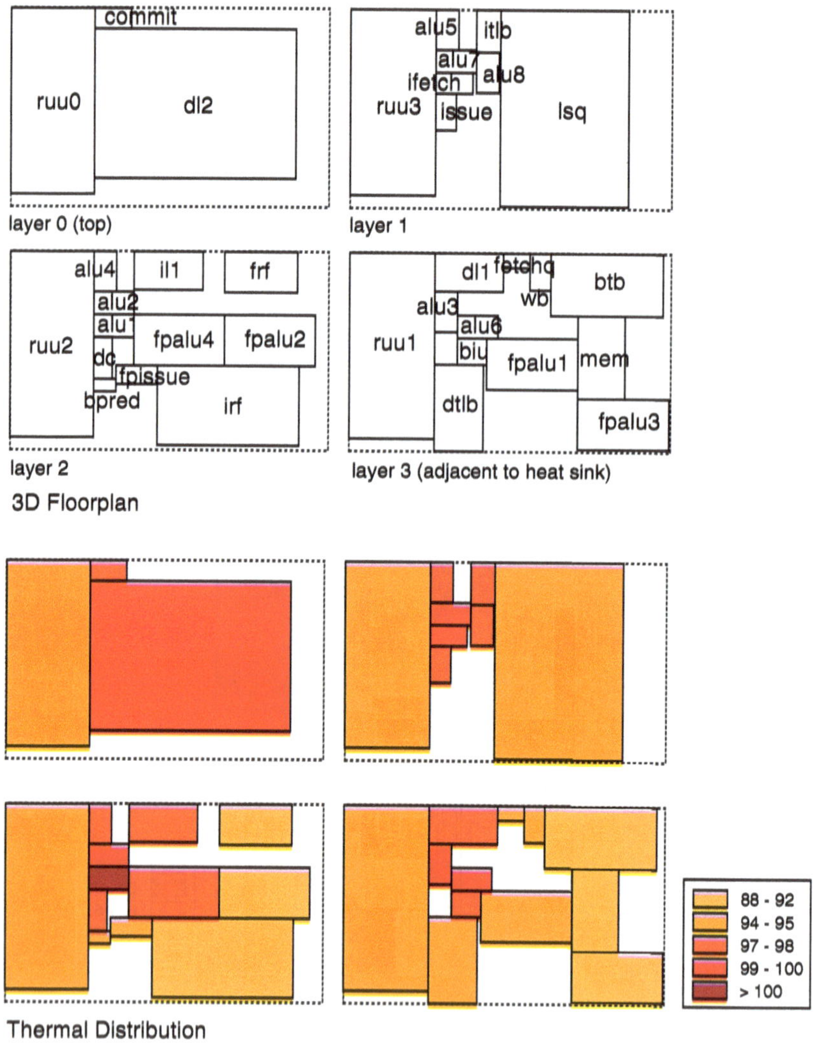

Fig. 10.10 Snapshots of our 3D floorplanning

floorplan is able to mitigate this by placing ALUs in different layers. Though several modules can have similar power consumption, their temperatures may be different, because their nearest neighbors can have a large impact on their final temperature. The leakage power profile among the modules is identical between the 2D and 3D floorplan except for the last two entries. This is because the logic styles of each module are more important in determining the relative leakage power than the variations in temperature. Table 10.7 shows the top ten buses and interconnects under various metrics. It is interesting to note that the longest wire in the multi-layer

Table 10.6 The top ten list of blocks in 3D floorplan under various metrics

Rank	Area (mm^2)		Power (mW/mm^2)		Temp (°C)		Leakage (mW)	
1	RUU	16.38	IALU1	15,408	IALU1	104.7	L2 cache	0.9343
2	L2 cache	7.83	BPRED	1,971	MEM	103.7	ITLB	0.2559
3	LSQ	6.53	COMMIT	1,930	IALU5	103.1	DTLB	0.2559
4	IRF	2.94	FPISSUE	1,930	ITLB	103.1	L1 icache	0.0609
5	BTB	1.81	ITLB	1,049	L2 cache	102.2	L1 dcache	0.0609
6	FPALU 2	1.20	IALU2	1,034	IALU4	102.0	BTB	0.0091
7	FPALU 3	1.20	IALU3	884	FPALU4	101.7	FETCHQ	0.0036
8	FPALU 4	1.20	IALU4	746	IALU8	100.0	FPALU3	0.0015
9	DTLB	1.10	L1 cache	730	IALU2	99.6	FPALU1	0.0015
10	MEM	1.00	IALU5	630	IALU3	97.3	FPALU2	0.0015

Table 10.7 The top ten list of wires 3D floorplan under various metrics

Rank	Access frequency		Long wires (mm)		Short wires (mm)	
1	ITLB-FETCHQ	1.0	IALU6-RUU	4.696	IALU1-FETCHQ	0.23
2	IF-DC	1.0	FPALU3-RUU	4.479	IALU5-IALU1	0.33
3	BTB-IF	1.0	IRF-IALU6	3.962	IALU5-IALU2	0.35
4	IL1-FETCHQ	1.0	WB-COMMIT	3.959	IALU8-IALU3	0.36
5	FETCHQ-IF	1.0	DTLB-RUU	3.688	IRF-FPALU1	0.57
6	DC-ISSUE	1.0	DL1-RUU	3.613	IALU4-IALU1	0.65
7	DL2-DL1	1.0	IRF-IALU5	3.482	IALU8-IALU1	0.67
8	WB-COMMIT	1.0	IRF-IALU2	3.462	IALU2-IALU1	0.67
9	DTLB-RUU	1.0	RUU-FPALU1	3.423	IALU2-IALU4	0.67
10	DL1-RUU	1.0	DL2-IL1	3.395	IALU4-IALU5	0.69

floorplan is almost half as long as the longest wire in the single layer floorplan. The shortest wire list is dominated by inter-ALU connections. This is partly because the ALUs are generally small units and so the center to center distance for them is smaller, but also because there are many data passing lines among the ALUs so they are very tightly connected.

10.6.6 Fidelity Study

Our fidelity study is twofold. First, Table 10.8 shows a comparison of the temperatures provided by our 3D mesh-based model and those provided by Hotspot v3.0 [28] across ten benchmarks. One can observe that our model provides a similar temperature. Second, we study the impact of the frequency of the thermal resistance matrix **R** update (= inversion of thermal conductance matrix) on the final temperature and IPC results. Under the "every move" column, we update **R** at every move during the SA-based refinement. The "no update" column contains the results based on our current implementation, where **R** stays constant throughout the SA refinement. Note that when we update **R** every time, we add a slicing

Table 10.8 Comparison with Hotspot v3.0 [28]

Bench	HotSpot	Ours	Bench	HotSpot	Ours
Equake	86.1	86.2	gzip	109.6	104.7
mcf	86.5	86.6	bzip2	112.7	107.4
Swim	88.3	88.0	vpr	123.2	114.4
Art	93.6	92.0	twolf	130.0	118.9
Gap	97.2	94.5	Lucas	134.6	123.0

Table 10.9 Impact of the frequency of thermal resistance matrix update on IPC and temperature. We use 3D LP + SA floorplanner with A + P + T objectives

	Every move		No update	
Bench	IPC	Temp	IPC	Temp
gzip	2.76	108.2	2.85	104.7
Swim	0.74	89.5	0.92	88.0
vpr	1.08	115.2	1.29	114.4
Art	0.52	95.6	0.61	92.0
mcf	0.09	89.2	0.07	86.6
Equake	0.55	87.6	0.67	86.2
Lucas	0.87	128.7	1.19	123.0
Gap	1.60	98.4	1.61	94.5
bzip2	1.95	109.6	2.33	107.4
twolf	0.79	122.4	1.02	118.9
Avg.	1.10	104.4	1.26	101.6
Area	27.89		26.45	
Wire	245.68		247.02	
Time	433,529		20,016	

cutline in our LP-based floorplan construction. From Table 10.9 we observe that the accurate computation of temperature values (= updating **R** at every move) does not necessarily translate into better results. In fact, we obtained comparable IPC and thermal results within a fraction of runtime with our "no update" method. Thus, we conclude that our thermal analysis and the way we make use of it in SA optimization prove to be highly effective and efficient.

10.7 Conclusions

In this chapter, we studied a multi-objective *micro-architecture-level* floorplanning algorithm for high-performance, high-reliability microprocessors targeting both 2D and 3D ICs. We simultaneously considered performance and thermal objectives such that our automated floorplanner can provide a balanced or goal-directed processor organization that achieves user-specified design objectives. Moreover, we integrated leakage modeling into our thermal analyzer and monitored the temperature/leakage interaction to prevent thermal runaway. We studied how vertical overlap among the modules in 3D floorplanning affects the performance, thermal, and area objectives. In addition, we partitioned the modules into multiple layers while considering the through-via requirements for face-to-face and face-to-back bonding styles. Our hybrid approach that combines Linear Programming and Simulated Annealing proved to be very effective in obtaining a high-quality solution in short runtime.

References

1. V. Agarwal, M.S. Hrishikesh, S.W. Keckler, D. Burger, Clock rate versus IPC: the end of the road for conventional microarchitectures, in *Proceedings of the IEEE International Conference on Computer Architecture*, Vancouver, 2000
2. T.M. Austin, Simplescalar tool suite. SimpleScalar LLC, http:/www.simplescalar.com
3. K. Balakrishnan, V. Nanda, S. Easwar, S.K. Lim, Wire congestion and thermal aware 3D global placement, in *Proceedings of the Asia and South Pacific Design Automation Conference*, Shanghai, 2005
4. D. Brooks, M. Martonosi, Dynamic thermal management for high-performance microprocessors, in *Proceedings of the Seventh International Symposium on High-Performance Computer Architecture*, Monterrey (IEEE Computer Society, Los Alamitos, 2001), p 171
5. D. Brooks, V. Tiwari, M. Martonosi, Wattch: A framework for architectural level power analysis and optimizations, in *Proceedings of the IEEE International Conference on Computer Architecture*, Vancouver, 2000
6. M. Casu, L. Macchiarulo, Floorplanning for throughput, in *Proceedings of the International Symposium on Physical Design*, Phoenix, 2004
7. G. Chen, S. Sapatnekar, Partition-driven standard cell thermal placement, in *Proceedings of the International Symposium on Physical Design*, San Jose, Monterey, CA, 2003
8. L. Cheng, W. Hung, G. Yang, X. Song, Congestion estimation for 3-D circuit architectures. IEEE Trans. Circuit Syst. II Express Brief **51**, 655–659 (2004)
9. L. Cheng, L. Deng, M. Wong, Floorplan design for 3-D VLSI design, in *Proceedings of the Asia and South Pacific Design Automation Conference*, Shanghai, 2005
10. C.N. Chu, D.F. Wong, A matrix synthesis approach to thermal placement. IEEE Trans. Comput. Aided Design Integr. Circuit Syst. **17**, 1166–1174 (1998)
11. J. Cong, A. Jagannathan, G. Reinman, M. Romesis, Microarchitecture evaluation with physical planning, in *Proceedings of the ACM Design Automation Conference*, Anaheim, 2003
12. J. Cong, J. Wei, Y. Zhang, A thermal-driven floorplanning algorithm for 3D ICs, in *Proceedings of the IEEE International Conference on Computer-Aided Design*, San Jose, 2004
13. J. Cong, Y. Zhang, Thermal-driven multilevel routing for 3-D ICs, in *Proceedings of the Asia and South Pacific Design Automation Conference*, Shanghai, 2005
14. S. Das, A. Chandrakasan, R. Reif, Design tools for 3-D integrated circuits, in *Proceedings of the Asia and South Pacific Design Automation Conference*, San Jose, Kitakyushu, Japan, 2003
15. Y. Deng, W. Maly, Physical design of the 2.5D stacked system, in *Proceedings of the IEEE International Conference on Computer Design*, San Jose, 2003
16. S. Dropsho, V. Kursun, D. Albonesi, S. Dwarkadas, E. Friedman, Managing static leakage energy in microprocessor functional units, in *Proceedings of the Annual International Symposium on Microarchitecture*, München, Istanbul, Turkey, 2002
17. D. Duarte, Y. Tsai, N. Vijaykrishnan, M. Irwin, Evaluating run-time techniques for leakage power reduction, in *Proceedings of the Asia and South Pacific Design Automation Conference*, Bangalore, 2002
18. D. Duarte, N. Vijaykrishnan, M. J. Erwin, A clock power model to evaluate the impact of architectural and technology optimizations. IEEE Trans. VLSI Syst. **10**(6), 844–855 (2002)
19. J.C. Eble, V.K. De, D.S. Wills, J.D. Meindl, A generic system simulator (GENESYS) for ASIC technology and architecture beyond 2001, in *International ASIC Conference*, Rochester, NY, 1996

20. eCACTI, http://www.ics.uci.edu/~maheshmn/eCACTI/main.htm.
21. M. Ekpanyapong, J. Minz, T. Watewai, H.-H. Lee, S.K. Lim, Profile-guided microarchitectural floorplanning for deep submicron processor design, in *Proceedings of the ACM Design Automation Conference*, Yokohama, San Diego, 2004
22. B. Goplen, S. Sapatnekar, Efficient thermal placement of standard cells in 3D ICs using a force directed approach, in *Proceedings of the IEEE International Conference on Computer-Aided Design*, San Jose, 2003
23. B. Goplen, S. Sapatnekar, Thermal via placement in 3-D ICs, in *Proceedings of the International Symposium on Physical Design*, San Francisco, 2005
24. L. He, W. Liao, M. Stan, System level leakage reduction considering leakage and thermal interdependency, in *Proceedings of the ACM Design Automation Conference*, Yokohama, San Diego, 2004
25. M. Healy, M. Vittes, M. Ekpanyapong, C. Ballapuram, S.K. Lim, H.-H.S. Lee, G.H. Loh, Multi-objective microarchitectural floorplanning for 2D and 3D ICs. IEEE Trans. Comput. Aided Design Integr. Circuit Syst. **26**(1), 38–52 (2007)
26. R. Ho, K.W. Mai, M.A. Horowitz, The future of wires, *Proceedings of the IEEE*, **89**(4), 490–504 (2001)
27. S.B. Horn, Vertically integrated sensor arrays VISA, in *Defense and Security Symposium*, Orlando, 2004
28. HotSpot, http://lava.cs.virginia.edu/HotSpot
29. M. Huang, J. Renau, S.-M. Yoo, J. Torrellas, A framework for dynamic energy efficiency and temperature management, in *Proceedings of the 33rd Annual ACM/IEEE International Symposium on Microarchitecture*, Monterey, 2000, pp. 202–213
30. W. Hung, Y. Xie, N. Vijaykrishnan, C. Addo-Quaye, T. Theocharides, M. Irwin, Thermal-aware floorplanning using genetic algorithms, in *Proceedings of the International Symposium on Quality Electronic Design*, San Jose, 2005
31. S. Kaxiras, Z. Hu, M. Martonosi, Cache decay: exploiting generational behavior to reduce cache leakage power. in *Proceedings of the 28th Annual International Symposium on Computer Architecture*, pages 240–251, Göteborg, 2001
32. I. Kaya, M. Olbrich, E. Barke, 3-D Placement considering vertical interconnects, in *Proceedings of the IEEE International SOC Conference*, Portland, 2003
33. N. Kim, K. Flautner, D. Blaauw, T. Mudge, Drowsy instruction caches: Leakage power reduction using dynamic voltage scaling and cache sub-bank prediction. In *Proceedings of the Annual International Symposium on Microarchitecture*, Los Alamitos, Istanbul, Turkey, 2002
34. W. Liao, F. Li, L. He, Microarchitecture level power and thermal simulation considering temperature, in *Proceedings of the International Symposium on Low Power Electronics and Design*, Seoul, 2003
35. C. Long, L. Simonson, W. Liao, L. He, Floorplanning optimization with trajectory piecewise-linear model for pipelined interconnects, in *Proceedings of the ACM Design Automation Conference*, San Diego, 2004
36. J. Minz, S. K. Lim, C.K. Koh, 3D Module placement for congestion and power noise reduction, in *Proceedings of the Great Lakes Symposium on VLSI*, Chicago, 2005
37. J. Minz, E. Wong, S.K. Lim, Thermal and power integrity-aware floorplanning for 3D circuits, in *Proceedings of the IEEE International SOC Conference*, Williamsburg, Herndon, VA, 2005
38. H. Murata, K. Fujiyoshi, S. Nakatake, Y. Kajitani, Rectangle packing based module placement. in *Proceedings of the IEEE International Conference on Computer-Aided Design*, San Jose, 1995, pp. 472–479
39. V. Nookala, Y. Chen, D. Lilja, S. Sapatnekar, Microarchitecture-aware floorplanning using a statistical design of experiments approach, in *Proceedings of the ACM Design Automation Conference*, Anaheim, 2005
40. B. Obermeier, F. Johannes, Temperature-aware global placement, in *Proceedings of the Asia and South Pacific Design Automation Conference*, Yokohama, 2004
41. V. Pavlidis, E. Friedman, Interconnect delay minimization through interlayer via placement in 3-D ICs, in *Proceedings of the Great Lakes Symposium on VLSI*, Chicago, 2005

42. S. Salewski, E. Barke, An upper bound for 3D slicing floorplans, in *Proceedings of the Asia and South Pacific Design Automation Conference*, Bangalore, 2002
43. P. Shivakumar, N.P. Jouppi, CACTI 3.0: an integrated cache timing, power, and area model. Technical Report 2001.2, HP Western Research Labs, Palo Alto, 2001
44. K. Skadron, M. Stan, W. Huang, S. Velusamy, K. Sankaranarayanan, D. Tarjan, Temperature-aware microarchitecture, in *Proceedings of the IEEE International Conference on Computer Architecture*, San Diego, 2003, pp. 2–13
45. G. Sohi, S. Vajapeyam, Instruction issue logic for high performance interruptable pipelined processors. *Proceedings of the 14th Annual International Symposium on Computer Architecture*, Pittsburgh, 1987
46. T. Tanprasert, An analytical 3-D placement that reserves routing space, in *Proceedings of the IEEE International Symposium on Circuits and Systems*, Geneva, 2000
47. C. Tsai, S. Kang, Cell-level placement for improving substrate thermal distribution. IEEE Trans. Comput. Aided Design Integr. Circuit Syst. **19**, 253–266 (2000)
48. Y. Tsai, A. Ankadi, N. Vijaykrishnan, M. Irwin, T. Theocharides, ChipPower: an architecture-level leakage simulator, in *Proceedings of the IEEE International SOC Conference*, Santa Clara, 2004
49. M. Umemoto, K. Tanida, Y. Nemoto, M. Hoshino, K. Kojima, Y. Shirai, K. Takahashi, High-performance vertical interconnection for high-density 3D chip stacking package, in *IEEE Electronic Components and Technology Conference*, Las Vegas, 2004
50. R. Zhang, K. Roy, C.-K. Koh, D.B. Janes, Exploring SOI device structures and interconnect architectures for 3-dimensional integration, in *Proceedings of the ACM Design Automation Conference*, Las Vegas, 2001

Chapter 11
Thermal-Aware Gate-Level Placement for 3D IC

Abstract In this chapter, we study two methods used in 3D IC placement that effectively exploit the die-to-die thermal coupling in the stack. First, TSVs are spread on each die to reduce the local power density and vertically aligned across dies simultaneously to increase thermal conductivity to the heatsink. Second, we move high-power logic cells to the location that has higher conductivity to the heatsink while moving TSVs in the upper dies so that high-power cells are vertically overlapping below the TSVs. These methods are employed in a force-directed 3D placement successfully and outperform several state-of-the-art placers published in recent literature.

The materials presented in this chapter are based on [1].

11.1 Introduction

Increasing functionality while miniaturizing footprint of integrated circuits (ICs) is today's trend of electronic industry. Moving to smaller technology node is a traditional approach toward that goal; however, investing in new production lines needs to be economically justified. Three-dimensional (3D) stacking of thinned dies provides feasibility to keep the trend while staying at current technology node. Polymer adhesive is a popular material used to bond thinned dies together [6]. Interleaving layers of thinned dies and polymer adhesive are, therefore, commonly found in 3D ICs.

Stacking thinned dies in 3D ICs results in increasing power density, thus rising temperature, which leads to other reliability problems, such as electro-migration and negative-bias-temperature instability. Because of low thermal conductivity, polymer adhesive exacerbates the problem. Moreover, if the thinned dies are silicon on insulator, an extremely high temperature can be expected. Heat must be removed

S.K. Lim, *Design for High Performance, Low Power, and Reliable 3D Integrated Circuits*, DOI 10.1007/978-1-4419-9542-1_11,

from the dies quickly; otherwise, reliability problems may arise. The following specific topics are covered in this chapter:

- We study two effective heuristics, namely TSV spread and alignment method (TSA) and thermal coupling-aware placement (CA), that exploit the die-to-die thermal coupling in 3D ICs in force-directed temperature-aware placement. We present new forces, and discuss how to manage them to obtain high quality placements.
- We perform extensive experiments to show the tradeoff among wirelength, delay, power, and temperature results obtained from GDSII layouts. Our placers outperform several state-of-the-art placers published in recent literature [2–5, 7].

11.2 Existing Works

A few recent works on temperature-aware placement for 3D ICs have been published. In [3], a force-directed approach was proposed for 3D thermal placement; however, it did not include through-silicon vias (TSVs), which are commonly found in 3D ICs. In [4], a partitioning-based approach was proposed for 3D thermal placement. The work considered the impact of parasitic resistance and capacitance of signal TSVs on power, but failed to include thermal properties of TSVs. Failing again to acknowledge TSV area, it also reported unreasonably large numbers of TSVs even for small circuits. The work in [2] considered TSV thermal properties; however, it assumed that adhesive is an ideal insulator. In reality, heat can still flow through (silicon and) adhesive because of its thinness. Based on the assumption, the work balanced only the number of TSVs in a bin to heat dissipated from cells in the same bin and bins vertically below.

11.3 Motivation

Because of their occupied area and high thermal conductivity of copper, widely used fill material, TSVs have significant impact on temperature. In a 3D IC layout, logic gates cannot overlap with TSVs. Area occupied by TSVs becomes "power whitespace" because no power is consumed and thus no heat is generated. In addition, TSVs conduct majority of heat through polymer adhesive between dies toward the heatsink as shown in Fig. 11.1. In the figure, the hotspot D on the top metal layer of the top die is caused by the TSVs in spot B from the bottom die. Heat flows through TSVs so intensely that its effect still remains on the top die. Thus, the temperature distribution of the top die results from the combination of power profile of the top die and heat flowing from the bottom die through TSVs. Our TSV spread and alignment method presented in this chapter exploits these thermal properties of TSVs by distributing TSVs evenly to reduce power density in local power hotspots and vertically aligning TSVs of adjacent dies to establish direct paths to the heatsink.

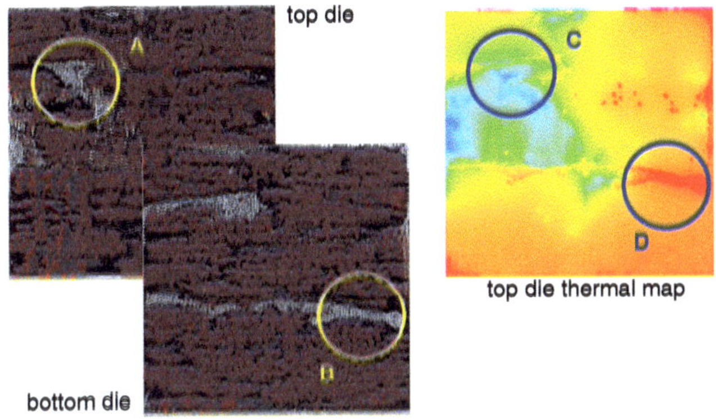

Fig. 11.1 Die-to-die heat coupling from TSVs. TSVs are shown in *white*. The *top die* is closer to heatsink. The cold spot C is caused by the TSVs in spot A on the same die. The hot spot D is caused by the TSVs in spot B from the *bottom die*

Using Ansys FLUENT, we simulate a part of bulk silicon with and without TSVs (and their related structures, e.g., landing pad and liner). We fix the temperature on the top side of the models, apply constant power density on the bottom side, and obtain the temperature distribution. The simulation results indicate that heat flowing through a TSV increases temperature far less than the same amount of heat flowing through bulk silicon and adhesive. We also observe that the temperature slowly increases in bulk silicon with TSVs. On the other hand, in bulk silicon without any TSV, low thermal conductivity of bonding adhesive results in steep temperature rise at first, but temperature does not rise as much inside the silicon. We compute the average thermal conductivity of bulk silicon with and without TSVs, and use them to guide our thermal coupling-aware placer presented in this chapter.

11.4 Evaluation Flow

In this section, we present our framework to evaluate the impact of TSVs on temperature of 3D ICs. The main components of the framework are power analysis and GDSII-level thermal analysis for 3D ICs. The presented evaluation flow allows us to evaluate the effectiveness of our 3D temperature-aware global placement algorithms in reducing temperature. The result of our study is analyzed, and reported in detail in Sect. 11.7.

Our evaluation flow for temperature-aware 3D-IC global placement is shown in Fig. 11.2. After obtaining 3D temperature-aware global placement result, we perform detail placement and detail routing. We report traditional metrics, e.g., area and routed wirelength, of the final GDSII-level layout. We then perform 3D static timing analysis, power analysis (Sect. 11.4.1), and GDSII-level thermal analysis

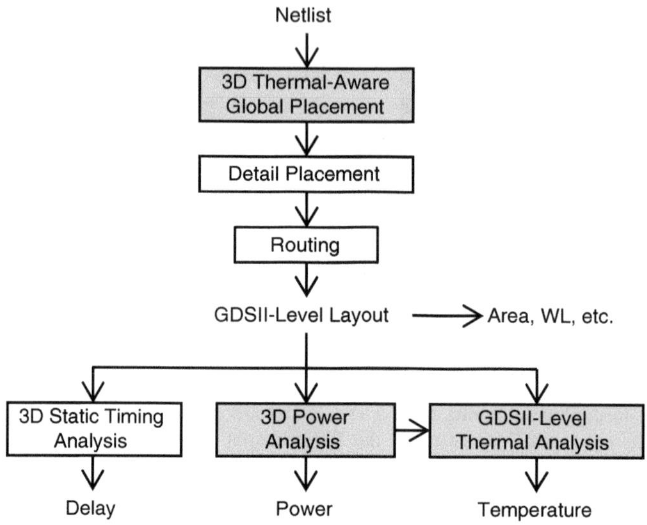

Fig. 11.2 Evaluation flow for temperature-aware 3D-IC global placement

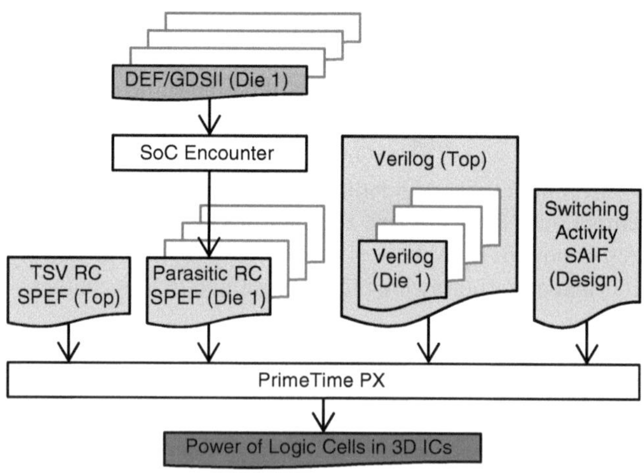

Fig. 11.3 Power analysis flow for 3D ICs

(Sect. 11.4.2) to report delay, power, and temperature, respectively. Note that the result from power analysis needs to be presented to GDSII-level thermal analysis because logic cell power is the heat source in 3D ICs during thermal analysis.

11.4.1 Power Analysis for 3D ICs

The power analysis flow for 3D ICs developed in this chapter starts by obtaining the layout of all dies in a 3D IC in DEF or GDSII format (see Fig. 11.3).

Next, we feed them to Cadence SoC Encounter to extract parasitic resistance and capacitance in SPEF format. We generate a separate SPEF file for parasitic resistance and capacitance of TSVs. The top-level Verilog connects the Verilog of all dies together, and the connection of all dies inside this top-level Verilog represents TSVs. The switching activity of all logic cells in the whole design can be obtained by propagating switching probability, as well as static state probability, from all primary inputs into all nets of the design. Additional accuracy can be gained by performing functional simulation of the whole design. Finally, we use PrimeTime PX to perform static power analysis, and report power dissipation of each logic cell. By stitching all the dies in this method, the parasitic resistance and capacitance of TSVs and wires running across dies also account for the total power of the 3D IC.

11.4.2 GDSII-Level Thermal Analysis

Steady-state temperature of a point $\mathbf{p} = (x, y, z)$ inside a 3D structure can be obtained by solving the heat equation

$$\nabla \cdot (k(\mathbf{p}) \nabla T(\mathbf{p})) + S_{\mathrm{h}}(\mathbf{p}) = 0, \tag{11.1}$$

where k is thermal conductivity in W/m·K, T is temperature in K, and S_{h} is volumetric heat source in W/m^3. This model can be implemented by meshing analyzed structure of a 3D IC into elements as shown in Fig. 11.4. Each element, called a thermal cell, is a volume of specific width and height, and its thickness is the same as each physical layer inside the 3D IC.

To solve Eq. (11.1), boundary conditions must be given on the six surfaces of a 3D chip stack. Generally, a 3D chip stack is very thin and flat, and packaged inside molding materials, which are not good thermal conductor. The majority of heat flows from the stack toward the heatsink. Therefore, we apply adiabatic boundary condition on bottom and four sides of the stack, and apply convective boundary condition on the top side, which is the heatsink.

The thermal analysis flow developed in this chapter is shown in Fig. 11.5. It starts by presenting the layout of all dies in a 3D IC in GDSII format and power dissipation of each logic cell to the layout analyzer that we develop for this work. The position of all TSVs is also presented to our layout analyzer so that all TSV related elements, e.g., landing pad and liner, are included into consideration. Our layout analyzer automatically generates meshed structure of the 3D IC along with thermal conductivity and volumetric heat source of each thermal cell.

A thermal cell can be composed of several different materials, for example, polysilicon, tungsten in vias, copper in TSVs, and dielectric (see Fig. 11.6). With sufficiently fine thermal cell size, equivalent thermal conductivity based on thermal resistive model can be used [11]. In theory, if a thermal cell size is very small, material inside it is homogeneous, and its thermal conductivity is isotropic. However, using small cell size requires high computing resource and long run time. For

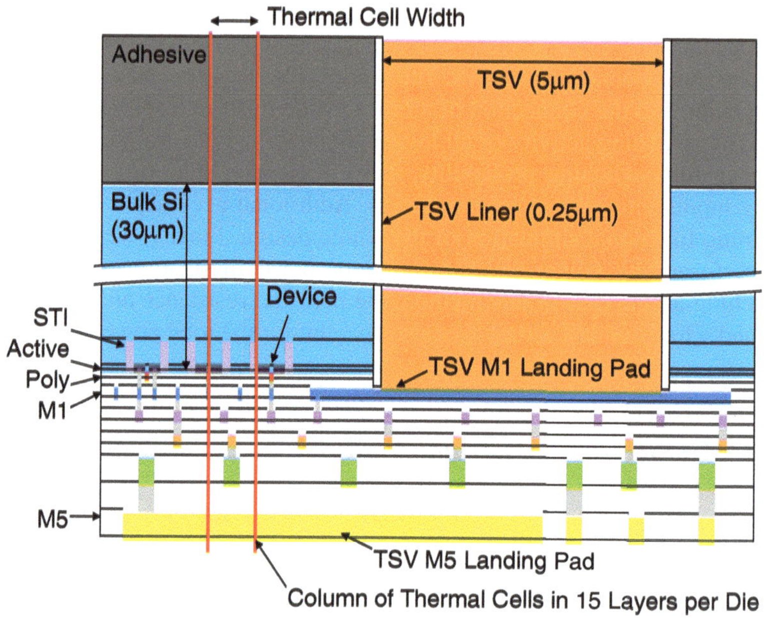

Fig. 11.4 Analyzed structure of a TSV-based 3D IC. Each die is modeled with 15 layers of different materials. The entire four-die structure contains 62 layers

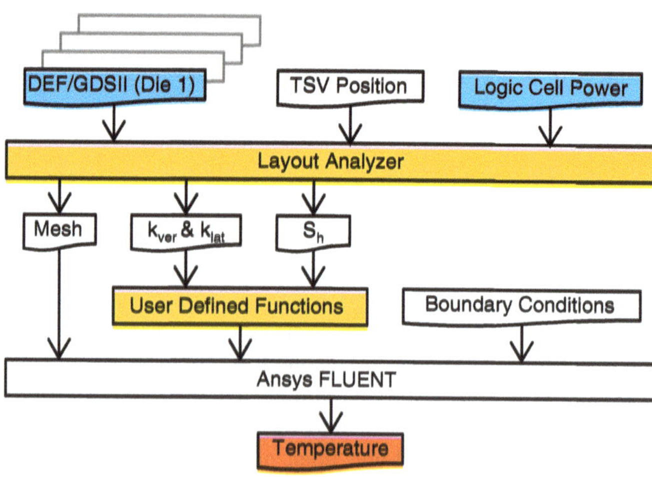

Fig. 11.5 GDSII layout-level thermal analysis flow

practical purpose, large thermal cell size can be used. Because of typical structural geometries found in GDSII layouts, thermal conductivity of each thermal cell is anisotropic. Vertical thermal conductivity k_{ver} and lateral thermal conductivity k_{lat} of a thermal cell consisting of N materials can be computed from

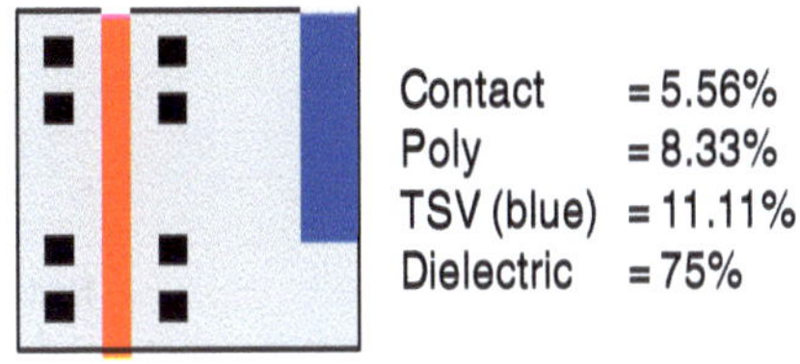

Fig. 11.6 Material composition inside a thermal cell

$$k_{\text{ver}} = r_1 \cdot k_1 + r_2 \cdot k_2 + \cdots + r_N \cdot k_N, \tag{11.2}$$

$$1/k_{\text{lat}} = r_1/k_1 + r_2/k_2 + \cdots + r_N/k_N, \tag{11.3}$$

where r_i is the ratio of material i volume to thermal cell volume, and k_i is the thermal conductivity of material i. Our layout analyzer computes r_i directly from GDSII layout of all dies in the 3D chip stack.

From the power dissipation and position of each logic cell, we can compute total power dissipated inside a thermal cell P_{cell}. Then, volumetric heat source S_{h} can be computed from

$$S_{\text{h}} = \frac{P_{\text{cell}}}{W_{\text{cell}} \cdot H_{\text{cell}} \cdot T_{\text{cell}}}, \tag{11.4}$$

where W_{cell}, H_{cell}, and T_{cell} are width, height, and thickness of the thermal cell, respectively.

We solve Eq. (11.1) by using Ansys FLUENT, a commercial tool. Meshed structure generated from our layout analyzer can be presented to FLUENT directly. However, k_{ver}, k_{lat}, and S_{h} need to be presented to FLUENT through user defined functions because they vary with thermal cell position. Finally, with the boundary conditions described earlier, we can run FLUENT to obtain steady-state temperature of all positions inside a 3D chip stack.

11.5 Global 3D Placement Algorithms

In this section, we describe our two 3D temperature-aware global placement algorithms that are based on force-directed methodology [9]. We extend this placer in two ways to perform thermal optimization in 3D ICs. In the first algorithm, we laterally spread TSVs in each die to form even thermal conductivity while perturbing TSV position to increase vertical overlap among TSVs across the dies in 3D stack. In the second algorithm, the logic cells on each die are positioned by using thermal conductivity-based force while TSVs are positioned by using power density-based force.[1]

[1] We attempted combining these two methods, but the results were not consistent.

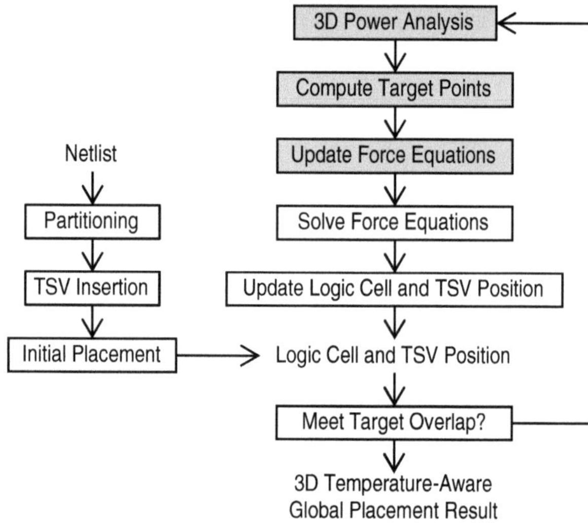

Fig. 11.7 Design flow for our 3D IC global placement

11.5.1 Design Flow

Figure 11.7 shows the overall flow of our placement, where the position of cells and TSVs is determined simultaneously. Given a netlist, we partition cells into dies if the partition is not also given. Then, we insert the minimum number of TSVs required to connect cells on different dies. Once this die partitioning is fixed, we do not move cells across dies during placement. The reason is that changing cell partition results in change in the number of TSVs, and this change causes the complexity of problem to become unmanageable. Next, we minimize wirelength to obtain initial placement, which may contain high overlap among cells and TSVs. In the main loop to resolve the overlap, we use TSV density and TSV position to compute target point for TSVs in the first algorithm. In the second algorithm, we periodically perform 3D power analysis (explained in Sect. 11.4.1) based on current cell and TSV position. Then, we use the cell power, TSV density, and average thermal conductivity of bulk silicon obtained from the simulation results in Sect. 11.3 to compute target points for cells and TSVs to move towards. After updating force equations and solving them, we update the position of cells and TSVs. This loop continues until the overlap is sufficiently reduced.

11.5.2 Force-Directed 3D Placement

In a quadratic placement [9], quadratic wirelength Γ_x and Γ_y along x- and y-axis are separately minimized to obtain the placement result. Treated Γ_x as spring energy, its derivative can be regarded as net force $\mathbf{f}_x^{net}$. By setting $\mathbf{f}_x^{net}$ to zero, the minimum Γ_x and the corresponding placement are found; however, cells may overlap in few

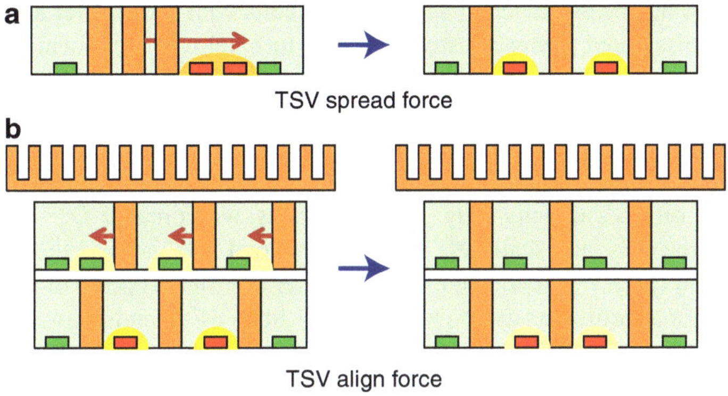

Fig. 11.8 TSV spread and TSV align forces

small areas. Hold force $\mathbf{f}_x^{hold}$ prevents $\mathbf{f}_x^{net}$ from pulling cells back to the initial placement. In addition, density-based force $\mathbf{f}_x^{den}$ reduces the overlap by spreading cells in high density region.

To extend [9] for 3D ICs, cells are not moved across dies during placement in [5] because they are already assigned into dies by the partitioner. In addition, $\mathbf{f}_x^{den}$ is computed die-by-die based on the placement density D_d of each die d, which is defined as

$$D_d(x,y) = D_d^{cell}(x,y) - D_d^{die}(x,y) \tag{11.5}$$

where D_d^{cell} is the cell density on die d, and D_d^{die} is the die capacity scaled to match the total cell area on the die. Then, the placement potential Φ_d is computed by solving Poisson's equation

$$\Delta \Phi_d(x,y) = -D_d(x,y) \tag{11.6}$$

The target point $\hat{x}_i^d$ to connect density-based spring of cell i is computed by

$$\hat{x}_i^d = x_i' - \frac{\partial}{\partial x}\Phi_d(x,y)\bigg|_{(x_i',y_i')} \tag{11.7}$$

where x_i' is the x-position of cell i on die d from the last iteration. Lastly, for each placement iteration, the placement result can be obtained by setting total force $\mathbf{f}_x$ to zero, and solve

$$\mathbf{f}_x = \mathbf{f}_x^{net} + \mathbf{f}_x^{hold} + \mathbf{f}_x^{den} = \mathbf{0} \tag{11.8}$$

11.5.3 TSV Spread and Alignment

In this algorithm, we exploit one of thermal properties of TSVs to help alleviate thermal problems as shown in Fig. 11.8a. TSVs occupy placement area, but do not dissipate power. The existence of TSVs among cells with high power dissipation reduces local dissipated power density, which in-turn helps reduce local temperature. Therefore, spreading TSVs evenly on each die should help reduce intra-die thermal variation in 3D ICs. We study this algorithm because it is simple yet effective. It can be viewed as a method to mimic uniform TSV position. Instead of moving TSVs based on the placement density computed from both TSV and cell area, we move TSVs based on TSV density only. In other words, we compute D_d^{cell} in Eq. (11.5) from TSV area only, and scale D_d^{die} to match the total TSV area on the die.

In addition to TSV spread, we exploit another thermal property of TSVs to help alleviate thermal problems as shown in Fig. 11.8b. TSVs conduct majority of heat through polymer adhesive between dies, causing local hot spots on the adjacent die between the TSVs and heatsink. Therefore, aligning TSVs on each die to TSVs on the adjacent die should help prevent this kind of hot spots, and direct the heat toward the heatsink quickly, resulting in overall temperature decrease. To align TSVs during global placement, we introduce an additional force for TSVs, alignment force denoted $\mathbf{f}_{\text{x}}^{\text{align}}$, into Eq. (11.8). This force can be represented by alignment springs connected to TSVs, and defined as

$$\mathbf{f}_{\text{x}}^{\text{align}} = \mathring{\mathbf{C}}_{\text{x}}^{\text{a}}(\mathbf{x} - \mathring{\mathbf{x}}^{\text{a}}) \tag{11.9}$$

where vector $\mathring{\mathbf{x}}^{\text{a}}$ represents the x-position of target points to connect alignment springs to TSVs, and diagonal matrix $\mathring{\mathbf{C}}_{\text{x}}^{\text{a}}$ collects spring constants $\mathring{w}_{\text{x},i}^{\text{a}}$ of the alignment spring connected to TSV i.

We apply alignment force to TSV i only when its closest TSV j on the adjacent die farther from the heatsink is within a certain range so that we do not excessively increase wirelength. The range is set to the size of TSV because of the high probability of aligning the TSVs in few iterations. We balance $\mathbf{f}_{\text{x}}^{\text{align}}$ against other forces by setting $\mathring{w}_{\text{x},i}^{\text{a}}$ to density-based spring constant $\mathring{w}_{\text{x},i}^{\text{d}}$ of $\mathbf{f}_{\text{x}}^{\text{den}}$ and setting alignment target point $\mathring{x}_i^{\text{a}}$ to x'_j, the x-position from last iteration of TSV j (on the adjacent die farther from heatsink) closest to TSV i. This method naturally balances $\mathbf{f}_{\text{x}}^{\text{align}}$ against $\mathbf{f}_{\text{x}}^{\text{den}}$.

The intuition is that because of the high cell overlap in the early placement iterations, the target point $\mathring{x}_i^{\text{d}}$ is farther away from TSV i than the alignment target point $\mathring{x}_i^{\text{a}}$. Thus, $\mathbf{f}_{\text{x}}^{\text{den}}$ dominates. When cells are evenly distributed in the late iterations of placement, $\mathring{x}_i^{\text{d}}$ is closer to TSV i. Then, $\mathbf{f}_{\text{x}}^{\text{den}}$ becomes weaker, and $\mathbf{f}_{\text{x}}^{\text{align}}$ affects the TSV position more.

11.6 Thermal Coupling-Aware Placement

In this algorithm, we consider the die-to-die thermal coupling during placement. The basic approach is to introduce two new forces, the first that moves cells and the second that moves TSVs, both in an attempt to place high-power cells closer to the

Table 11.1 Notations used for thermal coupling-aware placement

P_d^{cell}	Cell power density of each die d
K_d^{sink}	Effective thermal conductivity from die d to heatsink
K_d^{die}	Thermal conductivity across the opposite sides of die d
p_i	Power of cell i
N_d^{TSV}	Total number of TSVs on die d
N_{die}	Number of dies
B_d^{cond}	Balance factor for the thermal conductivity-based force on die d
s_d^{cond}	Scaling factor to match the effective thermal conductance to heatsink to cell power on die d
B_d^{pow}	Balance factor for the power density-based force on die d
s_d^{pow}	Scaling factor to match the cell power of die d and below to the thermal conductance of die d
s_d^{PD}	Scaling factor to normalize the cell power to the cell area on die d
s_d^{KD}	Scaling factor to normalize the thermal conductance of die d to the cell area on die d
α	Weighting constant for thermal coupling forces

TSV-to-heatsink path. Since the heat dissipated by a cell must flow toward heatsink, we place cells based on their power density and the effective thermal conductivity computed using the same die *and the dies above*. In addition, since TSV conducts heat without raising temperature too much, we place TSVs based on the total power density of the same die *and the dies below*.

Our basic approach is that the area with high power density and low thermal conductivity leads to high temperature. Thus, the temperature at a certain position depends on the difference (or imbalance) between power density and thermal conductivity. The force that moves cells (TSVs) on a die also changes the power density (thermal conductivity) distribution of the die. Our goal is to use these forces to balance the power density and the thermal conductivity at each position on the die. The force in an area with high difference should be stronger than the force in an area with low difference. The strength of a spring force depends on the distance to the connection point, so we set the strength based on this difference. Based on this concept, we first build a map of the difference, and smooth the map in an iterative fashion. Table 11.1 shows the notations used in this section.

11.6.1 For Cell Movement

Our thermal conductivity-based force $\mathbf{f}_x^{\text{cond}}$ is illustrated in Fig. 11.9a. It moves high-power cells toward the position with high thermal conductivity to heatsink, and is defined as

$$\mathbf{f}_x^{\text{cond}} = \mathbf{\mathring{C}}_x^{\text{c}}(\mathbf{x} - \mathbf{\mathring{x}}^{\text{c}}) \tag{11.10}$$

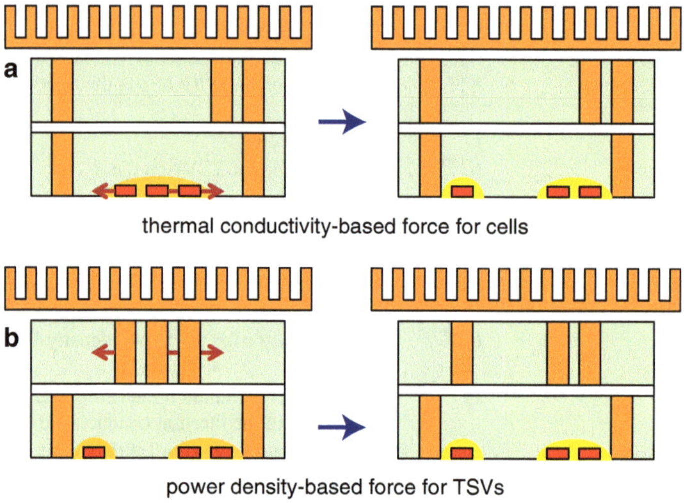

Fig. 11.9 Thermal conductivity-based vs. power density-based forces

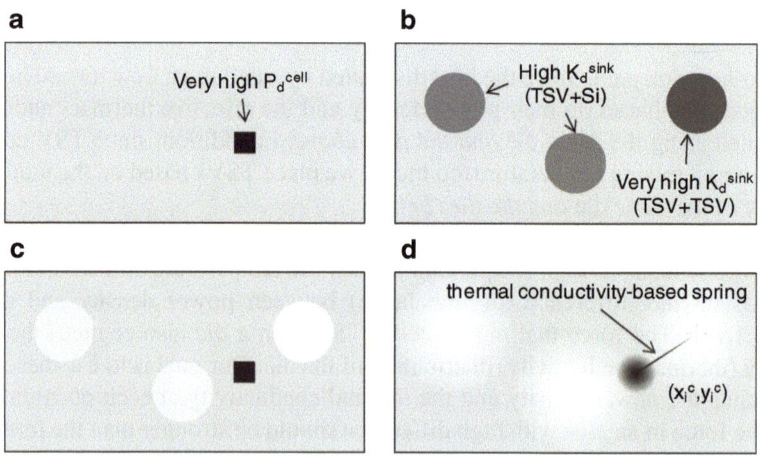

Fig. 11.10 Illustration of B_d^{cond}. **(a)** P_d^{cell}, **(b)** $s_d^{\mathrm{cond}} \cdot K_d^{\mathrm{sink}}$, **(c)** B_d^{cond}, **(d)** potential for B_d^{cond} after solving Poisson's equation

where the vector $\mathring{\mathbf{x}}^{\mathrm{c}}$ represents the x-position of target points to connect thermal conductivity-based springs to cells, and the diagonal matrix $\mathring{\mathbf{C}}_{\mathrm{x}}^{\mathrm{c}}$ contains spring constants $\mathring{w}_{\mathrm{x},i}^{\mathrm{c}}$ of the spring connected to cell i.

We compute $\mathbf{f}_{\mathrm{x}}^{\mathrm{cond}}$ die-by-die by balancing the cell power density P_d^{cell} of each die d against its effective thermal conductivity to heatsink, denoted K_d^{sink}. Under the demand-supply system of the force-directed framework in [9], P_d^{cell} and K_d^{sink} represent the demand and supply to remove the heat from die d in the 3D stack. We define the thermal conductivity-based balance factor B_d^{cond} for die d as (see Fig. 11.10)

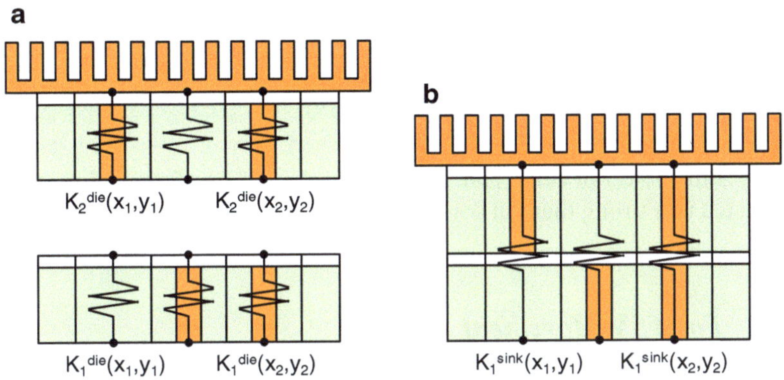

Fig. 11.11 Computation of K_d^{sink}. (**a**) K_1^{die} and K_2^{die}, (**b**) K_1^{sink}

$$B_d^{\mathrm{cond}}(x,y) = P_d^{\mathrm{cell}}(x,y) - s_d^{\mathrm{cond}} \cdot K_d^{\mathrm{sink}}(x,y) \tag{11.11}$$

where s_d^{cond} is a scaling factor to match K_d^{sink} to P_d^{cell} across the die. We use s_d^{cond} to balance the total supply (K_d^{sink}) and the total demand (P_d^{cell}), and compute it by

$$s_d^{\mathrm{cond}} = \frac{\int\int P_d^{\mathrm{cell}}(x,y)\,dx\,dy}{\int\int K_d^{\mathrm{sink}}(x,y)\,dx\,dy} \tag{11.12}$$

Here, K_d^{sink} is computed as

$$K_d^{\mathrm{sink}}(x,y) = \frac{1}{\sum_{j=d}^{N_{\mathrm{die}}} \frac{1}{K_j^{\mathrm{die}}(x,y)}} \tag{11.13}$$

where K_j^{die} is the thermal conductivity of die j, and die N_{die} is the die closest to the heatsink (see Fig. 11.11). Here, $K_{N_{\mathrm{die}}}^{\mathrm{die}}$ includes the thermal conductivity of thick substrate (not shown in the figure) and heatsink, and K_j^{die} is computed based on the TSV density at each position on the die and the average thermal conductivity of bulk silicon with and without TSVs, obtained from the simulation results in Sect. 11.3.

The potential Φ_d^{cond} for B_d^{cond} is computed by solving Poisson's equation

$$\Delta \Phi_d^{\mathrm{cond}}(x,y) = -B_d^{\mathrm{cond}}(x,y) \tag{11.14}$$

The target point $\mathring{x}_i^{c}$ of cell i is computed by

$$\mathring{x}_i^{c} = x_i' - \frac{\partial}{\partial x}\Phi_d^{\mathrm{cond}}(x,y)\bigg|_{(x_i',y_i')} \tag{11.15}$$

where x_i' is the x-position of cell i on die d from the last iteration. We set spring constant $\mathring{w}^{\mathrm{c}}_{\mathrm{x},i}$ for cell i based on cell power and the total cell power by

$$\mathring{w}^{\mathrm{c}}_{\mathrm{x},i} = p_i / \sum_{\forall j} p_j \tag{11.16}$$

where p_i is the power of cell i, and j is a cell on die d. Therefore, a high-power cell is connected to a strong thermal conductivity-based spring.

11.6.2 For TSV Movement

We introduce power density-based force $\mathbf{f}^{\mathrm{pow}}_{\mathrm{x}}$ as illustrated in Fig. 11.9b. It moves TSVs toward the position with high cell power density on the same die and the dies below. We define $\mathbf{f}^{\mathrm{pow}}_{\mathrm{x}}$ as

$$\mathbf{f}^{\mathrm{pow}}_{\mathrm{x}} = \mathring{\mathbf{C}}^{\mathrm{p}}_{\mathrm{x}}(\mathbf{x} - \mathring{\mathbf{x}}^{\mathrm{p}}) \tag{11.17}$$

where the vector $\mathring{\mathbf{x}}^{\mathrm{p}}$ represents the x-position of target points to connect power density-based springs to TSVs, and the diagonal matrix $\mathring{\mathbf{C}}^{\mathrm{p}}_{\mathrm{x}}$ contains spring constants $\mathring{w}^{\mathrm{p}}_{\mathrm{x},i}$ of the spring connected to TSV i.

We compute $\mathbf{f}^{\mathrm{pow}}_{\mathrm{x}}$ die-by-die by balancing the thermal conductivity K^{die}_d of each die d against the total power density $\sum P^{\mathrm{cell}}_j$ that flows through the die toward heatsink. Under the demand-supply system of the force-directed framework in [9], K^{die}_d and $\sum P^{\mathrm{cell}}_j$ represent the demand and supply to conduct heat from the same die and dies below to heatsink. We define the power density-based balance factor B^{pow}_d for die d as

$$B^{\mathrm{pow}}_d(x,y) = K^{\mathrm{die}}_d(x,y) - s^{\mathrm{pow}}_d \cdot \sum_{j=1}^{d} P^{\mathrm{cell}}_j(x,y) \tag{11.18}$$

where s^{pow}_d is a scaling factor to match $\sum P^{\mathrm{cell}}_j$ to K^{die}_d across the die. We use s^{pow}_d to balance the total supply ($\sum P^{\mathrm{cell}}_j$) and the total demand (K^{die}_d), and compute it by

$$s^{\mathrm{pow}}_d = \frac{\int\int K^{\mathrm{die}}_d(x,y)\,dx\,dy}{\int\int \sum_{j=1}^{d} P^{\mathrm{cell}}_j(x,y)\,dx\,dy} \tag{11.19}$$

The potential Φ^{pow}_d for B^{pow}_d is computed by solving Poisson's equation

$$\Delta \Phi^{\mathrm{pow}}_d(x,y) = -B^{\mathrm{pow}}_d(x,y) \tag{11.20}$$

The target point $\mathring{x}^{\mathrm{p}}_i$ of TSV i is computed by

$$\mathring{x}^{\mathrm{p}}_i = x_i' - \frac{\partial}{\partial x}\Phi^{\mathrm{pow}}_d(x,y)\bigg|_{(x_i',y_i')} \tag{11.21}$$

where x_i' is the x-position of TSV i on die d from the last iteration. We set spring constant $\mathring{w}^{\mathrm{P}}_{\mathrm{x},l}$ to $1/N_d^{\mathrm{TSV}}$, where N_d^{TSV} is the total number of TSVs on die d. Therefore, the power density-based spring for each TSVs has the same strength.

11.6.3 Balancing the Forces

We balance the new forces against $\mathbf{f}_{\mathrm{x}}^{\mathrm{den}}$ because $\mathbf{f}_{\mathrm{x}}^{\mathrm{den}}$ is the main force that moves cells and TSVs. First, we scale the new forces so that they have the same magnitude as $\mathbf{f}_{\mathrm{x}}^{\mathrm{den}}$. Then, we apply weighting constants to $\mathbf{f}_{\mathrm{x}}^{\mathrm{den}}$, $\mathbf{f}_{\mathrm{x}}^{\mathrm{cond}}$, and $\mathbf{f}_{\mathrm{x}}^{\mathrm{pow}}$ so that we can control their contribution to the total force.

First, to scale $\mathbf{f}_{\mathrm{x}}^{\mathrm{cond}}$ to $\mathbf{f}_{\mathrm{x}}^{\mathrm{den}}$, we normalize P_d^{cell}, the demand for B_d^{cond} in Eq. (11.11), to D_d^{cell} by a scaling factor s_d^{PD} defined as

$$s_d^{\mathrm{PD}} = \frac{\int\int D_d^{\mathrm{cell}}(x,y)\,dx\,dy}{\int\int P_d^{\mathrm{cell}}(x,y)\,dx\,dy} \tag{11.22}$$

Then, we replace P_d^{cell} in Eqs. (11.11) and (11.12) by $s_d^{\mathrm{PD}} \cdot P_d^{\mathrm{cell}}$. Second, to scale $\mathbf{f}_{\mathrm{x}}^{\mathrm{pow}}$ to $\mathbf{f}_{\mathrm{x}}^{\mathrm{den}}$, we normalize K_d^{die}, the demand for B_d^{pow} in Eq. (11.18), to D_d^{cell} by a scaling factor s_d^{KD} defined as

$$s_d^{\mathrm{KD}} = \frac{\int\int D_d^{\mathrm{cell}}(x,y)\,dx\,dy}{\int\int K_d^{\mathrm{die}}(x,y)\,dx\,dy} \tag{11.23}$$

Then, we replace K_d^{die} in Eqs. (11.18) and (11.19) by $s_d^{\mathrm{KD}} \cdot K_d^{\mathrm{die}}$.

We scale both $\mathbf{f}_{\mathrm{x}}^{\mathrm{cond}}$ and $\mathbf{f}_{\mathrm{x}}^{\mathrm{pow}}$ to $\mathbf{f}_{\mathrm{x}}^{\mathrm{den}}$ based on D_d^{cell}, not on the gradient of Φ_d because of the stability issue. After normalizing P_d^{cell} and K_d^{die} to D_d^{cell} as shown in Eqs. (11.22) and (11.23), the magnitude of B_d^{cond} and B_d^{pow} and gradient of their potential are properly normalized. At an equilibrium, a small magnitude of the gradients results in a small magnitude of $\mathbf{f}_{\mathrm{x}}^{\mathrm{cond}}$ and $\mathbf{f}_{\mathrm{x}}^{\mathrm{pow}}$. If we scale $\mathbf{f}_{\mathrm{x}}^{\mathrm{cond}}$ and $\mathbf{f}_{\mathrm{x}}^{\mathrm{pow}}$ to $\mathbf{f}_{\mathrm{x}}^{\mathrm{den}}$ based on the gradient of Φ_d instead, the magnitude of the gradient of potential of B_d^{cond} and B_d^{pow} would be exaggerated after the normalization, which in turn causes instability.

In summary, $\mathbf{f}_{\mathrm{x}}^{\mathrm{cond}}$ moves cells in such a way that high power density flows through the paths with high thermal conductivity to heatsink. In addition, $\mathbf{f}_{\mathrm{x}}^{\mathrm{pow}}$ moves TSVs in such a way that each TSV establishes a heat path for the high-power cells in the same die and the dies below. Our overall force equation is as follows:

$$\mathbf{f}_{\mathrm{x}} = \mathbf{f}_{\mathrm{x}}^{\mathrm{net}} + \mathbf{f}_{\mathrm{x}}^{\mathrm{hold}} + (1-\alpha)\mathbf{f}_{\mathrm{x}}^{\mathrm{den}} + \alpha(\mathbf{f}_{\mathrm{x}}^{\mathrm{cond}} + \mathbf{f}_{\mathrm{x}}^{\mathrm{pow}}) = \mathbf{0} \tag{11.24}$$

By increasing α, the forces $\mathbf{f}_{\mathrm{x}}^{\mathrm{cond}}$ and $\mathbf{f}_{\mathrm{x}}^{\mathrm{pow}}$ dominate the movement of cells and TSVs for more thermal optimization. The impact of α is studied in Sect. 11.7.4.

11.7 Experimental Results

We use 45-nm technology from FreePDK45 for our experiments. TSV diameter is 5 μm, and the landing pad width is 7 μm. TSV liner thickness is 250 nm [10]. We use copper TSVs with SiO_2 liner [10] and 2.6-μm-thick benzocyclobutene bonding adhesive [6] for our experiments. Each die in the 3D chip stack is thinned to 30 μm except that the topmost die, which is attached to heatsink, retains its thickness at 530 μm. The ambient temperature on top of the heatsink is 300 K. The TSV parasitic resistance and capacitance are 0.1 Ω and 125 fF, respectively. We base all our experiments on four-die chip stacks.

We use IWLS 2005 benchmarks and several industrial circuits from OpenCores. We synthesize the circuits using Synopsys Design Compiler to obtain gate-level netlist, and use the target clock period of each circuit when performing all analyses. The benchmark characteristics are listed in Table 11.2. The numbers of TSVs are based on partitioning results from our own implementation of [4]. We use the same die partitioning results for all algorithms for fair comparison in Sect. 11.7.2. Because [4] does not consider TSV area, it inserts high number of TSVs, resulting in low placement utilization.

We do not optimize the circuits after placement because buffers and sized gates can change power profile, thus affecting temperature. The results reported in this chapter are from commercial tools. We use Cadence Encounter to route the layouts, Synopsys PrimeTime to analyze timing and power, and Ansys FLUENT to analyze temperature. We report all our temperature results in terms of the increase from the ambient temperature measured at the top of the heatsink.

11.7.1 Impact of TSV Density Uniformity

In this experiment, we show how TSV density uniformity impacts thermal profile. Our two baseline 3D placements are wirelength-driven placement with uniform TSV position [5] and wirelength-driven placement with non-uniform TSV position [5]. First, we obtain both baseline placements using our own implementation of [5]. Then, we perform power and thermal analyses on both placement results. The routed wirelength, longest path delay, and power are shown in Table 11.3, and temperatures

Table 11.2 Benchmark circuits

Ckt.	#gates	#TSVs	Util.	Footprt (mm^2)	Profile
Ckt1	119,040	5,725	0.66	0.50 × 0.50	Data encryption
Ckt2	191,420	24,540	0.63	0.90 × 0.90	Graphic accelerator
Ckt3	280,933	17,362	0.49	0.98 × 0.98	Video compression
Ckt4	383,329	17,436	0.53	1.04 × 1.04	Signal processing
Ckt5	644,357	15,024	0.53	1.16 × 1.16	Image encoder

Table 11.3 Routed wirelength, longest path delay, and power of placements with uniform [5] and non-uniform [5] TSV position

	Uniform			Non-uniform		
Ckt.	rWL (m)	D_{max} (ns)	P (W)	rWL (m)	D_{max} (ns)	P (W)
Ckt1	3.897	5.320	0.752	3.014	4.836	0.728
Ckt2	11.718	16.510	2.661	7.744	13.694	2.463
Ckt3	13.532	8.814	2.353	9.326	6.535	2.288
Ckt4	19.355	20.788	2.710	12.457	12.515	2.640
Ckt5	22.708	19.772	3.209	18.711	13.798	3.122
Ratio	1.405	1.350	1.039	1.000	1.000	1.000

Table 11.4 Temperature (°C) of placements with uniform [5] and non-uniform [5] TSV position. ($\Delta T_{ja} = T_{ja,max} - T_{ja,min}$)

	Uniform			Non-uniform		
Ckt.	$T_{ja,max}$	ΔT_{ja}	$T_{ja,ave}$	$T_{ja,max}$	ΔT_{ja}	$T_{ja,ave}$
Ckt1	71.55	17.60	64.50	74.13	18.33	63.98
Ckt2	101.14	47.14	69.41	94.41	50.19	64.78
Ckt3	70.38	31.01	55.06	80.09	42.81	55.48
Ckt4	64.91	18.76	54.32	75.98	38.01	55.16
Ckt5	66.77	35.40	53.13	75.24	39.32	54.50
Ratio	1.000	1.000	1.000	1.081	1.325	0.995

are shown in Table 11.4. Although the placement with non-uniform TSV position has shorter wirelength, better timing, and lower power than the placement with uniform TSV position, its temperature, especially the thermal variation, is worse. Both the non-uniform power density and the non-uniform thermal conductivity, caused by the non-uniform distribution of TSVs in the 3D chip stack, contribute to the problem. In the placement with non-uniform TSV position, we observe that the area with high TSV density has low power density and low temperature, vice versa. These two opposite trends are responsible for high thermal variation.

11.7.2 Comparison with State-of-the-Art

We compare our temperature-aware global placement algorithms with the following recent state-of-the-art temperature-aware placers[2]:

- Goplen and Sapatnekar [3] (force-directed placer): In this work, thermal analysis is performed at the beginning of every global placement iteration. The thermal

[2]This task is challenging due to the discrepancy among the settings and assumptions made in each work. However, we made our best effort to provide fair and meaningful comparison, including in-depth discussions with the authors.

gradient obtained from the analysis is used to compute repulsive force, which moves logic cells from high-temperature area toward low-temperature area. We implement our own version of this work by calling Ansys FLUENT from inside our placer, and combining scaled thermal gradient into density-based force $\mathbf{f}_x^{\text{den}}$.

- Obermeier and Johannes [7] (force-directed placer): Instead of moving logic cells based on placement area density, it moves logic cells based on placement power density. Therefore, logic cells are spread according to their power dissipation, and logic cells with high power dissipation occupy more space than logic cells with low power dissipation, leading to uniform power density and thermal profile across the die. We implement our own version of this work.
- Goplen and Sapatnekar [4] (partitioning-based placer): In this work, logic cells are partitioned into placement area and different dies based on the switching activity and parasitic capacitance of connecting wires and TSVs. We perform global routing to determine the position of TSVs as proposed in [8] after performing global placement using our own implementation of [4].
- Cong et al. [2] (analytical placer): We implement this method by balancing the power density combined across dies in vertical direction against the TSV density and solving the density for potential function. The gradient of potential is used to compute a force to move cells and TSVs to maintain the balance. The force is added to $\mathbf{f}_x^{\text{den}}$ with a user-defined parameter β to provide temperature-wirelength tradeoff similar to the work.

Table 11.5 shows the routed wirelength, delay, power, and temperature comparison based on the GDSII layouts we build using these placers. The wirelength, delay, and power values are normalized to the wirelength-driven non-uniform TSV placement [5] shown in Table 11.3. The temperature values are normalized to the wirelength-driven uniform TSV placement [5] shown in Table 11.4. Recall that non-uniform placer achieves high-quality wirelength, delay, and power results while uniform placer leads to high-quality temperature values.

First, we observe that [3] produces comparable wirelength, delay, and power results to non-uniform TSV placer [5]. In case of temperature, [3] obtains worse result compared with uniform TSV placer [5]. We tried increasing the magnitude of thermal-gradient-based force, and found large increase in wirelength without much additional temperature improvement. Moving cells out of a high-temperature area on a die may not reduce temperature if the high temperature is a result from thermal coupling with other dies. Also, without considering TSV thermal properties during thermal analysis, the thermal gradient does not capture the impact of TSVs on temperature accurately, thereby misguiding the placement. Second, we see that [7] obtains wirelength and delay results that are significantly worse than non-uniform TSV placer. This is mainly because it moves logic cells based only on power density. However, this move helps reduce maximum temperature and thermal variation inside the 3D chip stack significantly. Although it attempts to spread power over placement area, we observe that TSVs obstruct this effort frequently.

Third, the routed wirelength and delay of results from [4] are worse than all other placers. The main reason is that [4] does not consider TSV area during placement.

Table 11.5 Comparison with state-of-the-art temperature-aware placers [2–4, 7]. Our placers are TSA (TSV spread and alignment) and CA (Coupling-aware placement). The routed wirelength, delay, and power values are normalized to the non-uniform TSV placement [5] shown in Table 11.3. The temperature values are normalized to the uniform TSV placement [5] shown in Table 11.4

	Routed wirelength (m)						Longest path delay (ns)					
ckt	[3]	[7]	[4]	[2]	TSA	CA	[3]	[7]	[4]	[2]	TSA	CA
Ckt1	3.04	3.10	3.78	3.24	3.25	3.13	4.93	4.79	5.12	5.06	4.78	4.87
Ckt2	7.74	8.78	14.92	8.34	7.89	8.31	13.67	15.00	15.23	14.41	13.58	14.78
Ckt3	9.34	10.54	16.02	10.70	10.35	10.26	6.56	6.79	7.86	7.27	6.53	6.90
Ckt4	12.48	13.90	19.87	15.23	14.90	14.54	12.51	12.69	16.15	13.60	13.69	13.11
Ckt5	18.86	21.48	27.64	20.12	19.84	19.99	13.93	16.42	15.64	13.67	13.79	14.66
ratio	**1.005**	1.112	1.595	1.120	1.093	1.090	**1.007**	1.066	1.160	1.058	1.015	1.051
	Power consumption (W)						Max junc.-to-amb. temp, $T_{ja,max}$ (°C)					
ckt	[3]	[7]	[4]	[2]	TSA	CA	[3]	[7]	[4]	[2]	TSA	CA
Ckt1	0.729	0.734	0.776	0.736	0.736	0.732	72.48	73.12	82.86	70.69	70.85	70.41
Ckt2	2.463	2.548	2.564	2.521	2.487	2.523	91.70	74.21	101.00	76.89	100.19	73.05
Ckt3	2.290	2.331	2.351	2.318	2.306	2.321	77.74	64.39	69.80	66.34	72.41	65.60
Ckt4	2.640	2.671	2.737	2.682	2.672	2.675	73.79	62.43	80.11	60.14	65.50	59.31
Ckt5	3.127	3.194	3.255	3.166	3.130	3.156	74.86	79.22	76.25	61.95	64.45	61.60
ratio	**1.001**	1.019	1.043	1.015	1.009	1.014	1.056	0.964	1.105	0.909	0.997	**0.895**
	Temp difference, $T_{ja,max} - T_{ja,min}$ (°C)						Average temp, $T_{ja,ave}$ (°C)					
ckt	[3]	[7]	[4]	[2]	TSA	CA	[3]	[7]	[4]	[2]	TSA	CA
Ckt1	16.29	14.94	28.12	14.69	15.55	14.16	63.80	63.70	69.52	63.32	63.27	63.35
Ckt2	46.96	15.15	51.16	22.39	53.87	17.15	64.81	66.84	69.36	66.07	65.14	66.14
Ckt3	39.89	19.68	28.69	23.82	33.65	22.97	55.41	55.49	55.97	54.53	54.14	55.08
Ckt4	35.46	16.69	39.76	15.87	21.83	14.27	55.07	54.35	60.42	53.91	53.63	53.85
Ckt5	38.08	36.39	38.02	23.77	33.07	24.53	54.51	55.08	57.97	53.22	51.91	52.90
Ratio	1.235	0.744	1.360	0.719	1.042	**0.673**	0.994	0.999	1.059	0.984	**0.973**	0.984

Thus, the TSVs inserted during routing affect the placement quality significantly. The maximum temperature, thermal variation, and average temperature are also worse than uniform TSV placer. The router tends to insert TSVs in the middle of the die to minimize wirelength, leaving low thermal conductivity at chip corners, thus high temperature. Fourth, although the wirelength of result from [2] is worse than other placers, temperature improvement is among the best. Because the algorithm considers the impact of TSV on chip area and temperature, it utilizes TSVs more effectively to help improve temperature results.

Fifth, we observe that our TSV spread and alignment method (TSA) achieves comparable delay and power results at the cost of wirelength degradation compared with non-uniform placer. In case of temperature, TSA obtains better average temperature than uniform TSV and comparable maximum temperature and temperature difference. But, the wirelength of TSA method is significantly better than that of uniform TSV placer. These results show that our TSA method is better in reducing wirelength while optimizing temperature compared with uniform TSV placer.

Lastly, our thermal coupling-aware placement (CA) achieves the best temperature results among all placers [2–4, 7], including uniform TSV placer [5]. In particular, our CA method outperforms uniform TSV placer by 10 and 33 % in terms of maximum temperature and temperature difference. CA obtains 9 % worse wirelength and 5 % worse delay results compared with non-uniform TSV placer, but CA is among the best in terms of wirelength and delay among other placers [2–4,7]. The power overhead is negligible. The TSVs in the placement by our CA method are not spread as evenly as our TSA placer and uniform TSV placer, but they are spread only sufficiently to help remove heat from the dies in the stack while maintaining high-quality wirelength. In addition, we observe that high-power logic cells are also placed effectively to dissipate heat using the nearby TSVs that are vertically aligned all the way to the heatsink.

The runtime of all placement algorithms is roughly in the same magnitude. Except for our TSA method, all other placement algorithms require power simulation (and thermal simulation in the case of [3]), resulting in larger runtime than [5].

11.7.3 Power and Thermal Maps

The thermal maps of uniform TSV position [5] and wirelength-driven placement with non-uniform TSV position [5] are shown in Fig. 11.12. In placement with uniform TSV position, power white space (TSVs) is evenly distributed, resulting in lower local power density and thus temperature than placement with non-uniform TSV position.

11.7.4 Temperature versus Wirelength Tradeoff

Our thermal-coupling-aware placement algorithm provides an efficient way to explore temperature-wirelength tradeoff. By increasing the weighting constant α in Eq. (11.24), the placer increases the magnitude of $\mathbf{f}_x^{cond}$ and $\mathbf{f}_x^{pow}$ while decreasing $\mathbf{f}_x^{den}$, i.e., trading wirelength for temperature. The temperature-wirelength tradeoff for Ckt2 is shown in Fig. 11.13. We also implemented the placer from [2] ourselves and show its tradeoff curve in Fig. 11.13. A weighting constant β is used in [2].

With $\alpha = \beta = 0$, both placers perform as a wirelength-driven placer (= leftmost points). As we increase α and β, temperature decreases while wirelength increases. We observe that as α and β increase, our thermal coupling-aware placer outperforms [2]: our placer has shorter routed wirelength at the same temperature, and has lower temperature at the same wirelength. We also observe that [2] shows convergence problem with large β values. When [2] moves a high-power cell into a bin, it moves cells out of other bins in the dies above or below, resulting in potential wirelength increase and convergence problem as discussed in [2]. In addition, [2] does not consider vertical alignment of TSVs so that even if it moves high-power

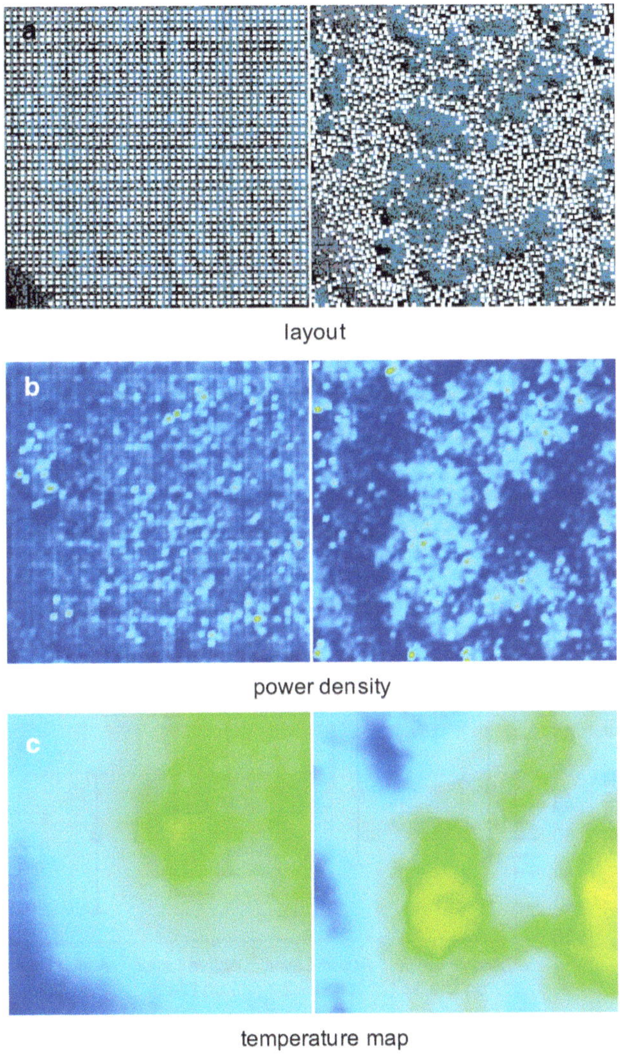

Fig. 11.12 Power and thermal profile of designs with uniform [5] (*left*) and non-uniform [5] (*right*) TSV position (TSVs are in *white* in the layout. Area with low power density or temperature is in *blue*)

cells into a bin with many TSVs, the heat captured in the bin may not be easily dissipated vertically to the heatsink. Our algorithms overcome these limitations.

The temperature of all active layers in various placements is shown in Fig. 11.14. We observe that by simply spreading and aligning TSVs across dies, the thermal variation becomes comparable to that of the uniform TSV placement. The temper-

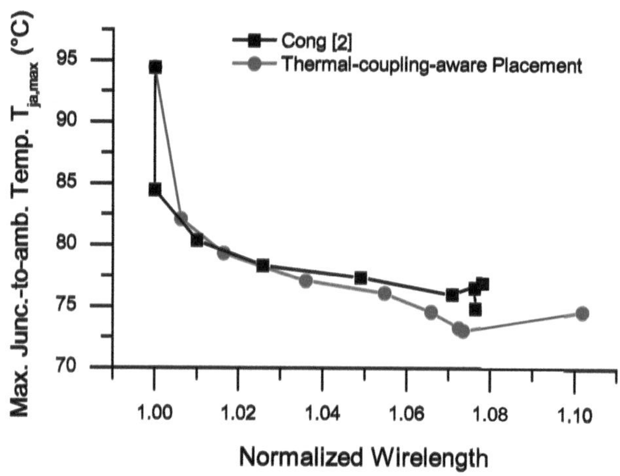

Fig. 11.13 Temperature-Wirelength tradeoff

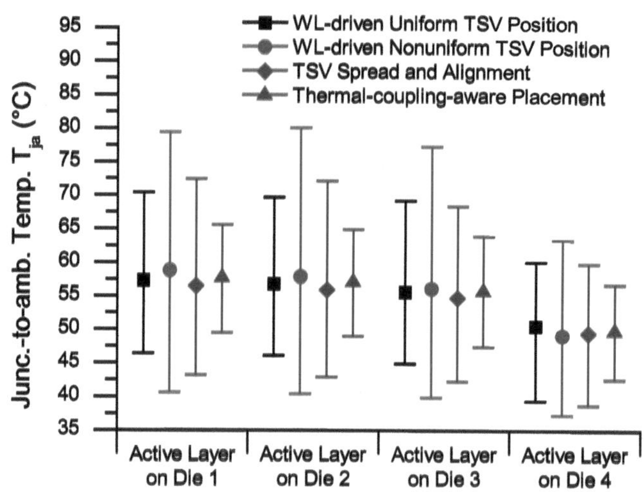

Fig. 11.14 Temperature of Ckt3 placed by different placement algorithms. Die 1 is close to PCB, and Die 4 is close to heatsink

ature of our thermal coupling-aware placement result is also shown in Fig. 11.14. By considering thermal coupling, we obtain even better temperature results than the uniform TSV placement.

11.7.5 Runtime Results

The runtime of wirelength-driven placer with uniform TSV position [5], wirelength-driven placer with non-uniform TSV position [5], state-of-the-art temperature-aware

Table 11.6 Runtime comparison of uniform TSV placement [5], non-uniform TSV placement [5], state-of-the-art temperature-aware placers [2–4, 7] and our placers. Our placers are TSA (TSV spread and alignment) and CA (Coupling-aware placement)

	Runtime (min.)							
Ckt.	[5]	[5]	[3]	[7]	[4]	[2]	TSA	CA
Ckt1	13.04	11.07	19.89	25.78	31.15	21.00	9.06	24.04
Ckt2	62.96	52.38	75.50	96.81	49.49	67.73	52.68	99.56
Ckt3	45.05	42.46	78.09	127.29	65.10	95.70	53.35	102.40
Ckt4	74.88	58.25	102.87	231.77	88.01	262.18	90.59	244.48
Ckt5	169.10	229.08	293.14	388.36	165.47	652.01	168.51	423.03
Total	365.04	393.24	569.50	870.01	399.22	1,098.62	374.18	893.52

placers [2–4, 7], and our placers are shown in Table 11.6. The runtime for [3] includes running power analysis and thermal analysis between iterations. The runtime for [2, 7] and our thermal coupling-aware placer includes running power analysis between iterations.

11.8 Conclusions

In this chapter, we showed that temperature-aware placers must consider TSV thermal properties and die-to-die thermal coupling during placement. We presented two temperature-aware placement algorithms for 3D ICs. TSVs are spread and aligned in the first algorithm. In the second algorithm, logic cells are moved based on the thermal conductivity to the heatsink, and TSVs are moved based on the power density of the neighboring dies. Experimental results showed that our placers achieved the best temperature results among all placers used in our comparison.

References

1. K. Athikulwongse, M. Pathak, S.K. Lim, Exploiting die-to-die thermal coupling in 3D IC placement, in *Proceedings of the ACM Design Automation Conference*, San Francisco, 3–6 June 2012 pp. 741–746
2. J. Cong, G. Luo, Y. Shi, Thermal-aware cell and through-silicon-via co-placement for 3D ICs, in *Proceedings of the ACM Design Automation Conference*, San Diego, 5–9 June 2011, pp. 670–675
3. B. Goplen, S. Sapatnekar, Efficient thermal placement of standard cells in 3D ICs using a force directed approach, in *Proceedings of the IEEE International Conference on Computer-Aided Design*, San Jose, 9–13 Nov 2003, pp. 86–89
4. B. Goplen, S. Sapatnekar, Placement of 3D ICs with thermal and interlayer via considerations, in *Proceedings of the ACM Design Automation Conference*, San Diego, 4–8 June 2007, pp. 626–631

5. D.H. Kim, K. Athikulwongse, S.K. Lim, A study of through-silicon-via impact on the 3D stacked IC layout, in *Proceedings of the IEEE International Conference on Computer-Aided Design*, San Jose, 2–5 Nov 2009, pp. 674–680
6. P. Leduc et al., Challenges for 3D IC integration: bonding quality and thermal management, in *Proceedings of the IEEE International Interconnect Technology Conference*, Burlingame, 4–6 June 2007, pp. 210–212
7. B. Obermeier, F.M. Johannes, Temperature-aware global placement, in *Proceedings of the Asia and South Pacific Design Automation Conference*, Yokohama, 27–30 Jan 2004, pp. 143–148
8. M. Pathak, Y.-J. Lee, T. Moon, S.K. Lim, Through-silicon-via management during 3D physical design: when to add and how many? in *Proceedings of the IEEE International Conference on Computer-Aided Design*, San Jose, 7–11 Nov 2010, pp. 387–394
9. P. Spindler, U. Schlichtmann, F.M. Johannes, Kraftwerk2–A fast force-directed quadratic placement approach using an accurate net model. IEEE Trans. Comput. Aided Des. Integr. Circuit Syst. **27**(8):1398–1411 (2008)
10. G. Van der Plas et al., Design issues and considerations for low-cost 3D TSV IC technology, in *IEEE International Solid-State Circuits Conference on Digest Technical Paper*, San Francisco, 7–11 Feb 2010, pp. 148–149
11. C. Xu et al., Fast 3-D thermal analysis of complex interconnect structures using electrical modeling and simulation methodologies, in *Proceedings of the IEEE International Conference on Computer-Aided Design*, San Jose, 2–5 Nov 2009, pp. 658–665

Chapter 12
3D IC Cooling with Micro-Fluidic Channels

Abstract Heat removal and power delivery have become two major reliability concerns in 3D IC technology. To alleviate thermal problem, two possible solutions have been proposed: thermal-through-silicon-vias (T-TSVs) and micro-fluidic channel (MFC) based cooling. In case of power delivery, a complex power distribution network is required to deliver currents reliably to all parts of the 3D IC while suppressing the power supply noise to an acceptable level. However, these thermal and power networks pose major challenges in signal routability and congestion. This is because signal, power, and thermal interconnects are all competing for routing space, and the related TSVs interfere with gates and wires in each die. In this chapter we study a co-optimization methodology for signal, power, and thermal interconnects in 3D ICs based on design of experiments (DOE) and response surface methodology (RSM). The goal of our holistic approach is to improve signal, thermal, and power noise metrics and to provide fast and accurate design space exploration for early design stage. We also provide an in-depth comparison between T-TSV vs. MFC based cooling method and discuss how to employ DOE and RSM techniques to co-optimize the interconnects. Our DOE-based optimization found the optimal design point with less effort than a gradient search based optimization.

The materials presented in this chapter are based on [16].

12.1 Introduction

The substantially smaller footprint area of 3D ICs inevitably leads to increased power density and chip temperatures. In addition, the thermal conductivity of the material used between dies of 3D ICs is low. Elevated temperatures may lead to inefficiency in performance and power. Furthermore, in 3D ICs the power is fed through TSVs which have significant parasitics. As more dies are stacked together, power noise such as IR-drop becomes more prominent in 3D ICs.

S.K. Lim, *Design for High Performance, Low Power, and Reliable 3D Integrated Circuits*, DOI 10.1007/978-1-4419-9542-1_12,

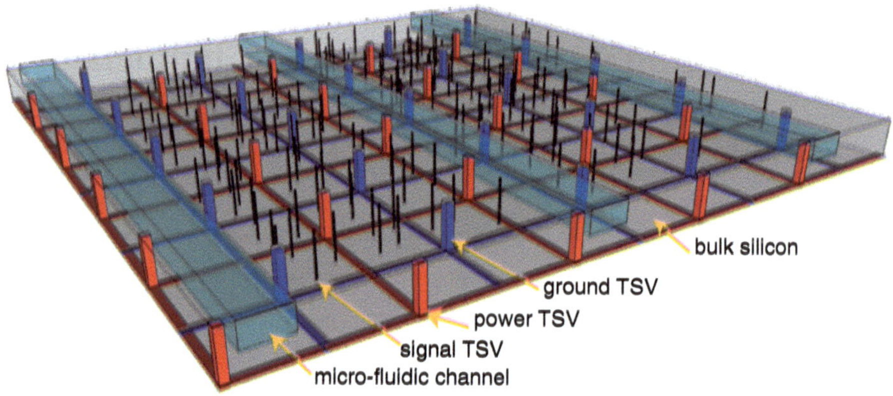

Fig. 12.1 Illustration of a die in a 3D IC with signal TSVs, P/G TSVs, and MFCs. These interconnects all compete for layout space. Transistors and signal wires are not shown for simplicity

Many efforts have been made to solve heat removal and power delivery concerns in the 3D IC technology. Thermal management using thermal-TSVs (T-TSVs) has been proposed as a solution to the heat problem [5]. Also, liquid cooling based on micro-fluidic channels (MFCs) has been proposed as a possible solution to dramatically lower the operating temperatures of 3D ICs with high power densities [24]. Regarding power supply noise, designers use a highly complex hierarchical power distribution network to deliver currents to all parts of 3D ICs while suppressing power noise to an acceptable level [11]. These so called silicon ancillary technologies, however, pose major challenges to routing completion and congestion, because the routing space is shared by these interconnects. As shown in Fig. 12.1, the power and the thermal interconnects are relatively large. Since these interconnects interact in a complex manner, optimizing one interconnect after another may lead to a local optimum. Thus, co-optimization of these interconnects with a holistic approach is highly called for.

In this chapter, we present the co-optimization of signal, power, and thermal interconnects for 3D ICs which is based on design of experiments (DOE) and response surface method (RSM). The goal is to improve signal, power noise, and thermal metrics. In early design stages, this method can provide a quick and accurate design space exploration, and the obtained response models provide insights on the system and are flexible so as to be reused for different optimization goals.

Most of the existing studies on signal, power, and thermal interconnects for 3D ICs are done in isolation, thereby lacking system-level perspective. The following specific topics are covered in this chapter:

- We study 2D IC vs. 3D IC design characteristics of a digital signal processing core and demonstrate the need for more powerful thermal management techniques for 3D IC designs.

- We study the effectiveness of T-TSV and MFC-based cooling for 3D ICs. With signal, power, and thermal analysis results, we discuss the strengths and weaknesses of these two thermal management techniques.
- We study how to co-optimize signal, power, and thermal interconnects using DOE and RSM. We perform optimizations for both the design with MFCs (we call it *MFC case*) and the design with T-TSVs (we call it *T-TSV case*). We discuss strengths and limitations of the co-optimization method. Also, a comparison of design quality with DOE based optimization to that of gradient search based optimization is presented.

12.2 Existing Works

Determining the location and the usage of signal TSVs is related to 3D routing problem. In [28], 3D global routing was performed with thermal optimization, and in [22], the first 3D Steiner routing algorithm with thermal consideration was presented. Regarding power distribution, a compact modeling of power delivery network for 3D ICs was presented in [11].

Several previous works considered T-TSV insertion during floorplanning [26], placement [9], and routing [5]. A pioneering work on MFC [25] developed compact thermal models for MFCs and measured the actual performance of MFCs. It was found that MFCs can handle very high power density ($790\,W/cm^2$). Recently, the performance of MFC based cooling for 3D ICs was thoroughly investigated in [14].

Since its invention [8], DOE has been used for various scientific and engineering applications. DOE has been adopted in VLSI and CAD areas. In [3] the DOE framework for CAD was discussed. A robust interconnect model based on DOE was presented in [27]. In [19] DOE was used to identify performance-critical buses in microarchitectures.

12.3 Routing Resource Modeling

The notations used in this chapter are shown in Table 12.1.

Table 12.1 Notations used in this chapter

DOE	Design of experiments	RSM	Response surface method
ROI	Region of interest	RMSE	Root mean square error
TSV	Through-silicon-via	T-TSV	Thermal-TSV
P/G	Power/ground	MFC	Microfluidic channel

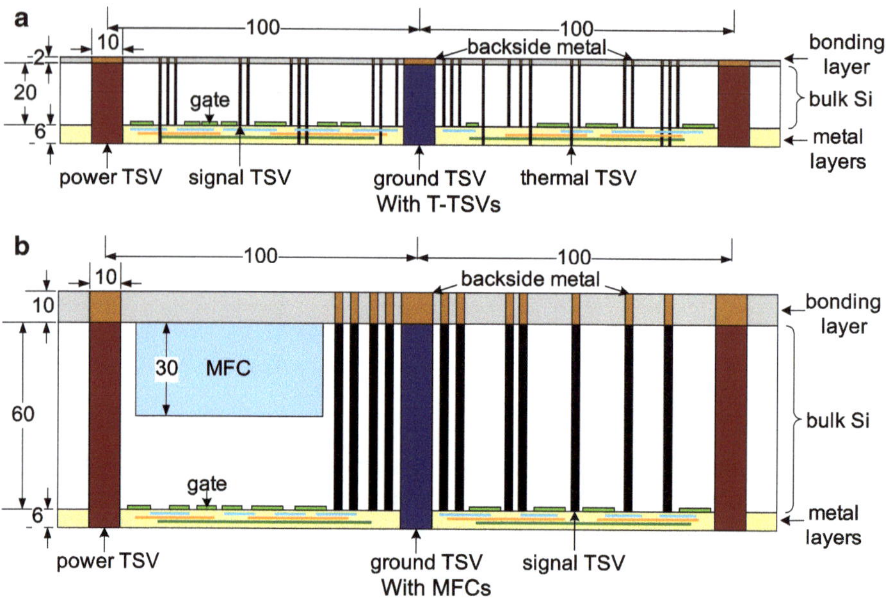

Fig. 12.2 Side view of a die in 3D ICs (**a**) with T-TSVs, and (**b**) with MFCs. In (**b**), bonding layer also seals MFCs and the thickness is larger. Dies are flipped over, and devices are facing down. Shapes are drawn to scale based on our default settings, except for gates. Unit is μm

12.3.1 *Signal Interconnects*

We use the metal interconnect dimensions similar to the ones in the 45 nm technology library from North Carolina State University [20]. Since we do not have industry data and feedback, we use the free technology data with our assumptions and modifications. Out of ten metal layers in [20], we use eight layers. Our assumption for TSV integration scheme is via-first. Via-first TSVs interfere only with device layer and not with metal layers, so they are less intrusive than via-last TSVs. Also we assume that the TSV aspect ratio (= TSV height:TSV diameter) is 10:1 for the baseline case (no T-TSV or MFC) and T-TSV case, and 30:1 for MFC case. The reason for higher TSV aspect ratio with MFC is because dies with MFCs cannot be as much thinned as ones without MFCs and are thicker. If we assume the same TSV aspect ratio for MFC case, the TSV diameter is larger, leading to larger silicon area for signal TSVs. Thus, it is important to assume a high TSV aspect ratio for MFC case. A high TSV aspect ratio of more than 30:1 was demonstrated in [12].

The side views of dies with T-TSVs and MFCs are shown in Fig. 12.2. In both cases, the diameter of signal TSVs is set to a minimum to accommodate as many connections as possible. In contrast, the diameter of P/G TSVs is around 10 μm, because within the same area a big TSV gives lower resistance than a bundle of minimum-sized TSVs. Note, however, that due to manufacturing issues we may have to use a bundle of small TSVs which would increase aggregate resistance.

Each global routing tile has x-, y-, and z-direction routing capacity values. x-and y-direction capacity represents available routing space on metal layers, while z-direction capacity is for signal TSVs. We assume that Metal 1/3/5/7 are for x-direction and Metal 2/4/6/8 are for y-direction. The cell occupancy ratio (*COR*) of a placement tile at (x,y,z) is defined as:

$$COR(x,y,z) = \frac{\sum_{\forall cell \in p\_tile(x,y,z)} S_{cell}}{S_{p\_tile}}$$

where S_{cell} is the area of each cell in the placement tile $p\_tile(x,y,z)$, and S_{p_tile} is the area of a placement tile.

For x- and y-direction, we calculate the default capacity per each metal layer by dividing the routing tile size by the minimum wire pitch of the metal layer. Since Metal 1 is heavily used in standard cells, when a placement tile has $COR = \alpha\%$, we reduce the Metal 1 routing capacity of the corresponding routing tile by $\alpha\%$. Metals 2–6 are dedicated to signal routing. On Metals 7 and 8, due to the P/G nets only part of the space is available for signal nets. When the routing tile is occupied by P/G TSVs, x- and y-direction capacity values are decreased correspondingly. For each routing tile, we calculate the routing capacity of each metal layer then add up to get the total routing capacity of the routing tile.

For z-direction capacity, we check both the available silicon surface area and the MFC area of the routing tile. From the routing tile area we subtract *COR* and the P/G TSV area to get the remaining area. Then the remaining area is further adjusted by the area covered by MFCs. For instance, if the MFC covers 50 % of the routing tile, we multiply 0.5 to the remaining area to get the final remaining area. Then, we divide the area by the area of a minimum-pitch signal TSV to get z-direction capacity.

Figure 12.3 shows the global routing tile objects. We fix the width of our routing tile to 20 μm. Note that some tiles are fully covered by MFCs and thus have zero z-direction capacity. Since the size of a P/G TSV is comparable to that of a routing tile, the tiles that contain P/G TSVs have significantly lower x/y/z-direction routing capacities. Also note that TSV diameter is much larger than the global wire width, making them significant routing obstacles in 3D ICs.

12.3.2 Power Interconnects

In 3D ICs, power is delivered to all devices through a power interconnect hierarchy. The global power distribution network on each die uses grids made of orthogonal interconnects on the top wiring levels. Power is fed from the package through power I/O bumps distributed over the bottom-most die, and travels to the upper dies via P/G TSVs.

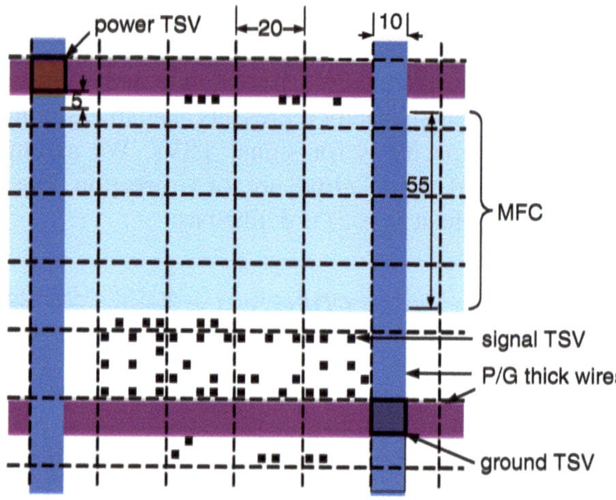

Fig. 12.3 Top view of global placement and routing tiles with MFCs. Only a part of the chip is shown. Objects are drawn to scale based on our default settings. P/G thin wires are not shown for simplicity

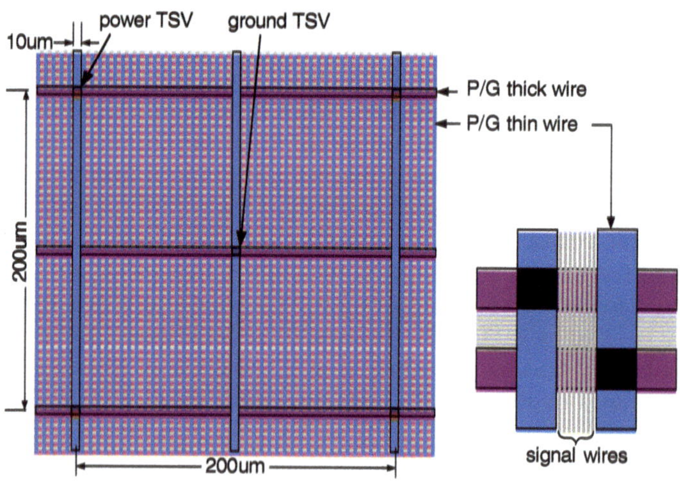

Fig. 12.4 Top view of our P/G network

Figure 12.4 shows the top view of our P/G network. We assume that P/G TSVs are placed regularly in a dual mesh structure, and each P/G TSV has a co-located P/G I/O bump on the bottom side of the chip to reduce the parasitic effects of connecting a P/G TSV to a P/G I/O bump. We predefine the pitch between two power TSVs as 200 μm which is used for all dies. The diameter of P/G TSVs is around 10 μm. P/G wires are globally distributed on metal layers 7 and 8. Thick wires of 10 μm width connect P/G TSVs. Between two thick wires, ten thin wires are placed and the remaining space is used for signal wires. The area ratio between P/G thin wires and signal wires can be varied; if the ratio is 0.4, P/G thin wires occupy 40 % (= 20 % each) of the routing tile area on Metals 7 and 8, and the rest (= 60 %) is for signal routing. Since P/G TSVs provide currents to dies, more TSVs with lower resistance usually decrease power noise. In our 3D IC structure, each P/G TSV pierces through the entire stack for efficient vertical power delivery (see Fig. 12.2). Thus, no gates can be placed and no wires can be routed at the P/G TSV locations. For the global placement tiles with pre-placed P/G TSVs, the placement tile capacity is decreased by a large amount, and the corresponding global routing tile has decreased signal routing capacities in x, y, and z-directions.

12.3.3 Thermal Interconnects

3D ICs bring several challenges in thermal management. By stacking layers, the power consumption per unit footprint area is significantly increased. In addition, the interior layers of 3D ICs are thermally detached from the heat sink. Heat transfer is further restricted by inter-layer dielectric and oxide-based bonding layers with low thermal conductivity. In this chapter, two possible solutions to the thermal problem are discussed: T-TSV and MFC.

12.3.3.1 Thermal-Through-Silicon-Vias

One way of dissipating heat is to insert T-TSVs in the white spaces of 3D ICs. T-TSVs do not provide any electrical functionality. They help decrease the on-chip temperature by lowering the inter-layer thermal resistance, hence providing more thermally conductive paths to the heat sink. Moreover, T-TSVs help distribute the heat more evenly throughout the entire chip, thus reducing the negative impact of high temperature areas (hot spots). T-TSVs go through the entire die (so called via-last TSVs), whereas signal TSVs do not, as shown in Fig. 12.2a. To avoid electrical shorts, signal or P/G nets (wires or TSVs) should not make contact with T-TSVs.

12.3.3.2 Micro-fluidic Channels

Unlike conventional air-cooled heat sinks or T-TSVs, liquid cooling using MFCs offer much larger heat transfer coefficient values and chip-scale cooling solutions. MFCs can be fabricated on the back side of silicon dies, enabling rejection of heat from every layer efficiently. The thermal resistance of the micro-channel heat sink for a single chip was previously measured [1]. When de-ionized water was used as coolant, the junction-to-ambient thermal resistance of the heat sink was 0.24°C/W at a flow-rate of about 65 mL/min without TSVs (impact of copper TSVs on thermal conductivity of the silicon micro-channel wall is negligible), which is significantly better than current state-of-the-art air cooled heat sinks [1]. The smallest resistance possible for air-cooled heat sink is around 0.5°C/W. MFCs capped with a thin polymer coating (Avatrel 2000P, ~30um thickness) were tested up to 2.5 atm pressure with no leakage observed during continuous operation [6].

The on-chip thermal network is composed of fluidic TSVs, manifolds and MFCs. We assume that all the fluidic TSVs and manifolds are located outside the core region in which all gates and metal wires reside. Thus, only MFCs are considered for the analysis. Also a coolant pump and a heat exchanger are assumed to be at off-chip.

The geometries of MFCs – depth, width, and pitch – have impacts on thermal and routability objectives. By increasing MFC depth, the mass flow rate of fluid and thus cooling capability can be improved. However, it also increases die thickness, and for a fixed aspect ratio of TSVs, the diameter of signal TSVs increases proportionally. Since larger TSVs consume more silicon space and have higher parasitics, it is not desirable to have deep MFCs. In contrast, MFC width can be increased without hurting silicon space. However, wide MFCs decrease z-direction routing capacity considerably. Since MFCs should not touch P/G TSVs, MFC pitch should be decided along with P/G TSV pitch. Thus, in this chapter we fix MFC depth and pitch and vary the MFC width only.

12.4 Design and Analysis Flow

12.4.1 Overview of 3D Physical Design

Our design package works with standard cell based circuits and consists of several major steps. The overall design flow with DOE and RSM is summarized in Fig. 12.5. We first define the input factors (or design knobs) and the responses (or assessing

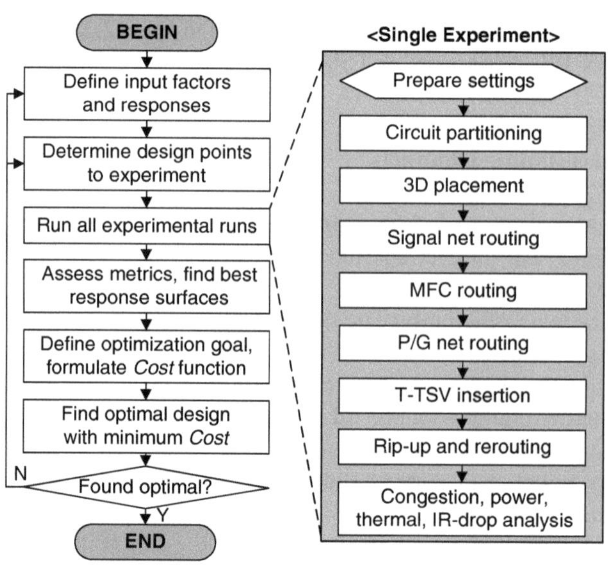

Fig. 12.5 Overall design flow with DOE and RSM

metrics). Our single experimental run is equivalent to performing gate-level global placement and routing. After reading in the input circuit, we perform partitioning in which the input circuit is divided into dies. The partition step not only relates closely to the routing quality but also determines the signal TSV distribution. We use a min-cut based partitioner [4] to minimize signal TSV counts. Note that minimizing signal TSV count does not always lead to optimal design in terms of total wirelength and timing [23].

In the placement step, we perform global placement onto the $N_p \times N_p \times N_{die}$ placement grid. The N_p is the number of placement tiles in x/y-axis, and N_{die} is the number of dies in the 3D IC. We use a force-directed placement algorithm for 3D ICs [13]. Note that the same placement is used for all the experimental runs to limit the solution space change. Even though smaller P/G TSV diameter may allow more gates in the placement tile with a P/G TSV, we observed negligible differences in total wirelength with placements with different P/G TSV diameter settings.

Next, we perform global routing on the $N_r \times N_r \times N_{die}$ routing grid. N_r is the number of routing tiles in x/y-axis. In this chapter, we assumed that the placement and routing tiles are of the same size, 20 μm. The reason for global routing, instead of detailed routing, is to obtain quick but relatively accurate pictures of routing congestion. In our routing flow, we first route signal nets without any MFCs or P/G TSVs. Then we route MFCs (for MFC case), P/G TSVs and wires. After that, T-TSVs may be inserted (for T-TSV case). Since these power and thermal interconnects incur routing congestion, we perform rip-up and reroute for signal nets with routing capacity violations. The reason for performing signal routing before other interconnects is because we use a congestion-aware 3D maze router. If we have other objects, the routing results differ significantly, which disturbs solution space much. This design flow limitation may be lifted if we use a 3D Steiner router [22]. In all experiments we observed at most 1 % nets needed rip-up and reroute. To ensure routability, we check if global routing fails due to insufficient routing capacity.

After all the routings are finished, we generate power map based on the placement and routing results and run power noise and thermal analysis. We evaluate the metrics and complete the experimental run. Once all the experiments are performed, we construct response surfaces and use them to obtain optimal design solutions. The response surface models are only applicable to the chip that the model was built for. Our input factors and the tradeoffs are shown in Table 12.2. Our responses are summarized in Table 12.3. Note that some of the input factors and the responses are only for either MFC case or T-TSV case, while some are for both.

12.4.2 Power Noise Analysis

From the design we build the resistive mesh structure with a current source at each grid node that represents power consumption. Then the IR-drop analysis is performed with a modified nodal analysis technique [10] for faster analysis. We apply a domain decomposition technique [29] that decomposes the circuit into

Table 12.2 Input factors used in this chapter

T-TSV ratio (T-TSV case only)	The maximum T-TSV area ratio per placement tile. This provides tradeoff between thermal and signal
MFC width (MFC case only)	Width of a MFC. All MFCs have the same width. Wider MFC means higher mass flow rate and better cooling capability. This provides tradeoff between thermal and signal, because MFCs block signal TSVs
MFC pressure drop (MFC case only)	The pressure drop between inlet and outlet of a MFC. All MFCs have the same pressure drop. This also affects the mass flow rate and cooling capability, providing tradeoff between thermal and pump power
P/G TSV diameter (both cases)	The diameter of a P/G TSV, which affects the parasitics of P/G TSVs. This provides tradeoff between power noise and signal
P/G thin wire ratio (both cases)	The ratio between P/G thin wires and signal wires on Metals 7 and 8. This also provides tradeoff between power noise and signal

Table 12.3 Responses used in this chapter

Total wirelength (both cases)	Sum of all the wirelength of signal nets. This value represents the quality of signal interconnect
Max. IR-drop (both cases)	The maximum IR-drop of the entire power grid. This represents the quality of power interconnect
Max. Si temp. (both cases)	Maximum silicon temperature of the die stack. This represents the quality of thermal interconnect
Pump power (MFC case only)	The coolant pump power to provide fluid through MFCs. This value may be considered during system power budget planning

several parts and uses a mathematical technique to reduce the time needed for matrix inversion. After the simulation, the IR-drop values of all grid nodes are obtained, of which the maximum is used as the response.

12.4.3 Thermal Analysis for T-TSV Case

For T-TSV case, our thermal analyzer is based on finite element analysis, where the entire 3D IC is mapped onto a 3D thermal mesh structure. To calculate the thermal conductivity of each thermal tile, we check the material composition of the tile. For thermal tiles that corresponds to silicon layer, the z-direction thermal conductivity of the thermal tile is calculated as follows:

$$k_{tile,z} = AR_{TSV} \times k_{Cu} + (1 - AR_{TSV}) \times k_{Si}$$

where AR_{TSV} is the area ratio of total TSVs (signal, P/G, and thermal) in the tile, and k_{Cu} and k_{Si} are thermal conductivities of copper and silicon. Note that k_{Cu} and k_{Si}

are about 400 and 150 W/(m· K), which suggests that inserting copper TSVs may reduce z-direction thermal resistance by about 62 %. The boundary conditions are as follows: The four lateral sides of the chip contacts to ambient air, while the top side has a heatsink which has thermal resistance of 0.25 K/W. The bottom side is assumed to be adiabatic. Then, we solve the following matrix equation: $G \cdot T = P$, where G is the thermal conductance matrix calculated from k_{tile}, T is the temperature vector, and P is the power vector. The temperature distribution can be directly found from T.

Our T-TSV insertion is performed with a predefined maximum T-TSV ratio. The maximum T-TSV ratio is the maximum allowed area ratio of T-TSV per placement tile. For example, if maximum T-TSV ratio is 0.1, up to 10 % of the silicon area of the placement tile may be used for T-TSV. After global routing is completed, we first run thermal analysis to get the temperature distribution without T-TSV. Thermal tiles with higher temperatures are assigned higher temperature severity. Then according to temperature severity, we assign target T-TSV ratio to the routing tile, which is less than or equal to the maximum T-TSV ratio. Target T-TSV area (S_{target_ttsv}) is the T-TSV ratio multiplied by the placement tile area. To see if the target T-TSV area can be accommodated, the white space of the placement tile is calculated as follows:

$$S_{white} = S_{p\_tile} - S_{pgtsv} - S_{gate} \tag{12.1}$$

where S_{white} is the white space of the placement tile, S_{p_tile} is the area of the placement tile, S_{pgtsv} is the P/G TSV area in the placement tile, and S_{gate} is the total gate area placed in the placement tile. The final T-TSV area assigned to the placement tile is the minimum of S_{white} and S_{target_ttsv}. Note that we do not consider signal wire or signal TSV area in (12.1). Instead, we decrease the routing capacity in x/y/z-direction after T-TSV insertion. After T-TSV insertion we rip-up and reroute signal nets that are not routable with updated routing capacity. Then we run thermal analysis again to see the impact of T-TSVs. Note that other thermal insertion algorithms may be used with our DOE and RSM based optimization, as long as it does not change design space characteristics dramatically. As we will discuss in Sect. 12.5, if the design space changes much during the design flow, it may not be possible to find response models.

12.4.4 Thermal Analysis for MFC Case

To analyze the thermal performance of MFCs in 3D ICs, we run numerical simulations. A three-dimensional thermal model [15] is modified to consider the lateral temperature and fluid flow rate distribution caused by non-uniform power/heat flux distribution. Figure 12.6 shows the side view of the 3D IC with MFCs. It is assumed that the temperatures of the fluid and the solid domains are different but uniform at each cross section within each control volume.

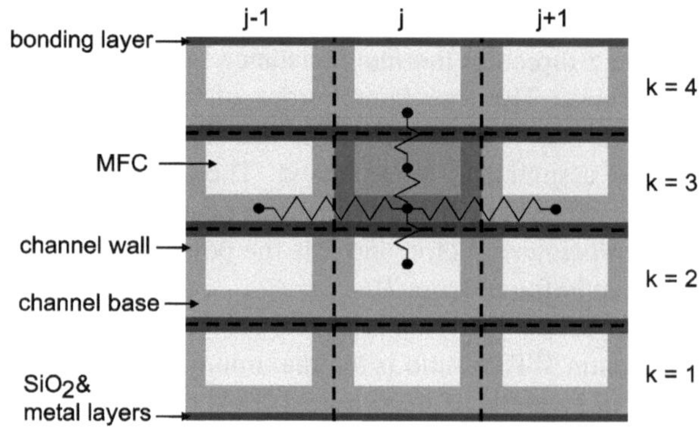

Fig. 12.6 Side view of the thermal grid structure used for 3D IC with MFCs

Thermal and fluid flow in MFCs are described by the following energy and momentum conservation equations:

$$\dot{m}\frac{di}{dz} = \eta_0 h_c \tilde{P}(T_{w,k} - T_{f,k}) + h_c w(T_{w,k+1} - T_{f,k}) \tag{12.2}$$

$$-\frac{dP}{dz} = \frac{2fG^2}{d_h \rho} \tag{12.3}$$

$$\frac{\partial}{\partial x}\left(k\frac{\partial T_w}{\partial x}\right) + \frac{\partial}{\partial y}\left(k\frac{\partial T_w}{\partial y}\right) + \frac{\partial}{\partial z}\left(k\frac{\partial T_w}{\partial z}\right) + \dot{q}_g + \dot{q}_c = 0 \tag{12.4}$$

T_w and T_f represent the temperatures of solid and fluid, respectively, $\dot{m}$, i and h_c are mass flow rate, enthalpy, and convective heat transfer coefficient, respectively. For each MFC, heat is directly supplied only to the channel base, and the channel wall is analyzed as a fin attached to the base (η_0 is the overall surface efficiency for heat transfer, including an array of fins and the base surface). MFC geometry is described by the channel perimeter $\tilde{P}$ and the width w. Equation (12.2) represents the fluid enthalpy change due to the convective heat transfer owing to the temperature difference between the solid and fluid, as well as fluid convective motion. The pressure drop along the MFC is obtained by solving the fluid momentum balance equation, (12.3), wherein P, G and ρ are pressure, mass flux and density of the fluid, respectively, f is the fluid friction factor and d_h is the hydraulic diameter of a MFC. Equation (12.4) is the three-dimensional thermal transport equation for the solid. It has two source/sink terms owing to heat generated ($\dot{q}_g$) from the active and oxide-metal layers and convective heat transfer ($\dot{q}_c$) to the fluid (k denotes the thermal conductivity of solid).

Deionized water is considered as the working fluid, and fluid temperature at the inlet was set to 20°C. The governing equations, (12.2), (12.3), and (12.4), are integrated over a control volume then discretized using the upwind scheme [21]. The resulting system of linear algebraic equations is simultaneously iteratively solved using successive under-relaxation method.

12.5 Design of Experiments

The main goal of design of experiments is to statistically control the experiments so that the output responses can be used for drawing meaningful conclusions on the system. It involves designing the experiments, performing the experiments, and analyzing the responses. After running the experiments and gathering the responses, we find the fitted model per response to understand and optimize the system. This is called response surface methodology [18].

DOE and RSM based optimization is suitable for this work because of the following reasons: (1) The knowledge on the target system, such as the relationship between input factors and responses, is found during the process. (2) Compared with Monte-Carlo or random search method, this method can characterize the system with far less number of experimental runs, providing a faster yet reasonably accurate solution. (3) The response models can be reused if the optimization goal is changed without affecting the whole design settings, which suggests this method is flexible and suitable in early design stages.

12.5.1 Classical DOE

We first perform a classical design: *Box-Behnken* [2]. In this design, each input factor is assigned three levels (minimum, center, maximum). This design can sufficiently fit a quadratic model with less number of experimental runs than a full factorial design.

Since MFCs should not contact with P/G TSVs, we applied the following constraint onto the region of interest (ROI) for MFC case:

$$w_{mfc} + d_{pgtsv} + 2 \cdot msp_{mfc-pgtsv} \leq p_{pgtsv}/2 \tag{12.5}$$

Here, w_{mfc} is MFC width, d_{pgtsv} is P/G TSV diameter, $msp_{mfc-pgtsv}$ is the minimum spacing between MFC and P/G TSV, and p_{pgtsv} is P/G TSV pitch. We divide P/G TSV pitch by 2 to obtain the distance between a power and a ground TSV (see Fig. 12.3). $msp_{mfc-pgtsv}$ is set to 5 μm in this chapter. For the designs that satisfy (12.5), we can place MFCs without touching P/G TSVs.

In addition to the designed data set, a validation data set with four design points was generated per T-TSV case and MFC case to see how the models predict unseen design points. As shown in Sect. 12.6.5, the classical designs did not lead to accurate models.

12.5.2 Advanced DOE

We perform more complex DOE to increase model accuracy. Since our design package does not involve randomized algorithms, there is no random error effects in the experimental results. That is, if we repeat an experiment with the same settings, it will produce the same response. Thus, we do not perform randomization and blocking [17] in our DOE. Since it is hard to theoretically derive optimal design points due to the complex structure of the system, we adopt a space-filling design style. Latin Hypercube sampling distributes N design points at N different levels per each input factor. In this sampling, the number of design points needs to be carefully determined based on the number of input factors and their ranges as well as the response model accuracy. Meanwhile, the same validation data sets as in classical DOE are used to check the model prediction capability.

12.5.3 *Finding Best Response Models*

The accuracy of the response models is important because we use them in the optimization process. Determining the parameters of response models is based on regression analysis. Polynomial response models with n input factors can be expressed in multivariate polynomial equations:

$$y(x) = a_0 + \sum_{l=1}^{n} a_l x_l + \sum_{l<j}^{n} a_{lj} x_l x_j + \sum_{l=1}^{n} a_{li} x_l^2 + \cdots$$

Here, $x_l, x_l^2, \cdots$ are called 'main factors', while $x_l x_j, x_l^2 x_j, \cdots$ are called 'interaction factors'. We estimate the parameters $(a_0, a_l, \cdots)$ such that the response equation fits the data in an optimal way. The goodness-of-fit of the model can be checked with root mean square error (RMSE) and the coefficient of multiple determination (R^2). When R^2 is closer to 1, the model can explain the observed data better.

For the models to predict unseen design points well, it is important to avoid over-fitting problem – the response curve follows not only the underlying truth but also unwanted noise with it. With N design points in the data set, we remove each data point from the data set and the remaining $N-1$ runs are used to fit the prediction model equation, which is the sum of squares of the prediction residuals (PRESS). We can check if the model is over-fitted by comparing PRESS RMSE to RMSE. When PRESS RMSE is much higher than RMSE, we observed over-fitting phenomena.

We also observed that increasing polynomial order of main factors may improve the model accuracy. Increasing model order generally increased R^2 and decreased RMSE. However, PRESS RMSE did not decrease monotonically with increased model order. Each response had a different optimal polynomial order. Additionally, stepwise regression [17] may be performed to determine which polynomial term should be included in the model to minimize PRESS RMSE.

To increase model accuracy further, we tried hybrid radial basis functions (RBFs). A hybrid RBF model has a polynomial model described above and an RBF network model. The polynomial model determines global shape, while the RBF network handles local variations. An RBF network with M distinct RBFs can be described as:

$$\hat{y}(x) = z \sum_{j=1}^{M} \beta_j \Phi(\|x - \mu_j\|)$$

Here, the j-th RBF is centered at μ_j with the weight β_j. The profile functions (Φ) of RBF kernels we tried are as follows:

$$\begin{aligned}
\Phi_{multiquadric}(r) &= \sqrt{r^2 + \beta^2}, \beta > 0 \\
\Phi_{reemultiquadric}(r) &= 1/\sqrt{r^2 + \sigma^2}, \sigma > 0 \\
\Phi_{gaussian}(r) &= e^{-r^2/\sigma^2} \\
\Phi_{thin-plate}(r) &= r^2 ln(r)
\end{aligned}$$

To find the best model, we exhaustively generated candidates of models (different polynomial order per input factor and RBF kernels) then calculated scores for each candidate. The *Score* function is defined as:

$$Score = \frac{R^2}{RMSE + Validation\ RMSE} \cdot e^{-\frac{PRESS\ RMSE}{RMSE}}$$

By this function, we try to minimize RMSE and validation RMSE and maximize R^2, while suppressing the difference between RMSE and PRESS RMSE. Then the models with five top scores are further compared by response surface shape. The model with less needless curvatures on the surface was chosen as the best.

12.5.4 Optimization with Response Surface Models

With multiple responses and design constraints, there can be several optimization scenarios. To consider multiple responses together, each of the responses under consideration is normalized to $[0, 1]$ and forms a partial cost. Then, we combine them into a single desirability function [7] which we call *Cost* function. The *Cost* function can be considered as a new response surface. Then, using optimization algorithms such as nonlinear programming or genetic algorithm, we find the optimal design point with minimum *Cost*. Note that the optimization is fast because it is performed on the *Cost* response surface, not the actual experimental space. That is, no additional experimental run is needed during the optimization process. Since some errors are inevitable in the response models, the actual design point with minimum *Cost* could be different from the optimal design point found by this optimization.

Although we were able to find accurate response models that can be used for optimization, it may not be always possible. Especially, when the design space characteristics changed much with different input factors, we were not able to fit a single response model to the data with a tolerable error.

12.6 Experimental Results

12.6.1 Experimental Settings

We implemented our design package in C++/STL and MATLAB. The simulations were executed on a 64-bit Linux server with two quadcore Intel Xeon 2.5 GHz CPUs and 16 GB main memory. We synthesized the target circuit (*fft*) using Synopsys Design Compiler and the North Carolina State University 45 nm standard cell library [20]. The synthesized circuit has about 370,000 gates and nets. Our technology and default setting parameters are shown in Table 12.4. Note that the technology is solely based on our assumptions. Die size is $700 \times 700\,\mu m$, number of dies is four, die bonding style is face-to-back, clock frequency is 1 GHz, and power supply voltage is 1.1 V, unless stated otherwise. Each signal TSV has a keep-out-zone of 1 μm size around it.

To construct a power map, switching activities of gates were obtained from a commercial tool with proper input stimuli. After routing is finished, dynamic power consumption of each gate is calculated with the parasitic capacitances of the net driven by the gate. Combined with leakage power, the power consumption of each gate is determined, which contributes to the power value of each power map tile. The power map is used in the thermal analyzer and the power noise analyzer. For T-TSV case, the thermal analyzer was written in C++/STL and the runtime was about 3 min. For MFC case, the thermal analyzer was written in MATLAB and the runtime was about 2 min. The runtime of a power noise (IR-drop) simulation was less than 10 seconds.

Table 12.4 Our technology and default setting parameters. Our baseline only uses top-mounted heat-sink, not T-TSVs or MFCs

	Baseline and T-TSV	MFC case
Si layer thickness (μm)	20	60
Metal layer thickness (μm)	6	6
Bonding layer thickness (μm)	2	10
Signal TSV aspect ratio	10:1	30:1
Signal TSV diameter (μm)	2	2
MFC depth (μm)	0	30
MFC pitch (μm)	0	200
P/G TSV pitch (μm)	200	200
Resistance of package pins (mΩ)	3	3

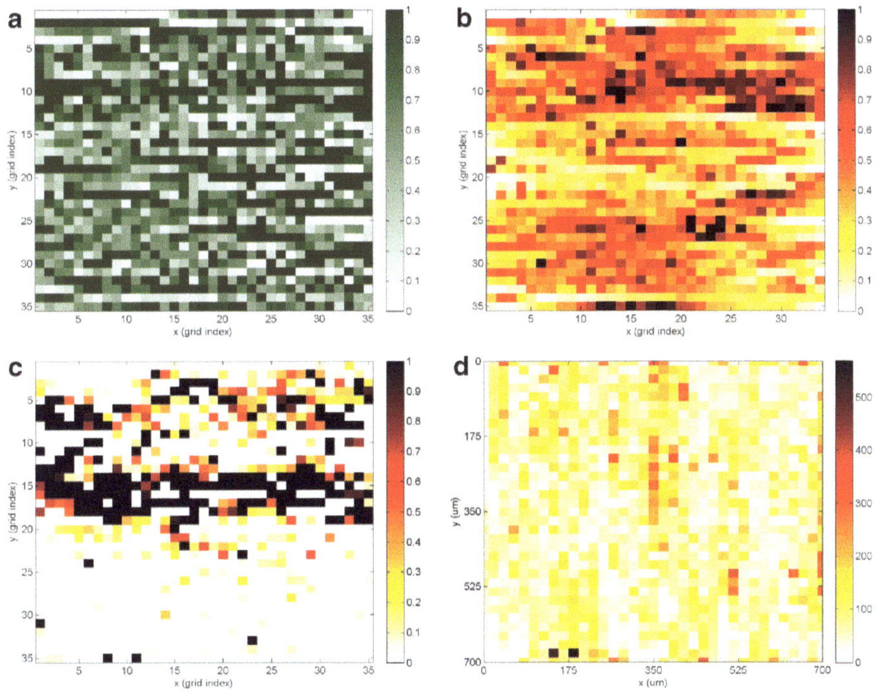

Fig. 12.7 Initial design results with baseline settings. Power density unit is W/cm^2 in power map. (**a**) Placement tile utilization. (**b**) x-direction routing utilization. (**c**) z-direction routing utilization. (**d**) Power map

Figure 12.7 shows the initial design results. Placement result shows that some regions have more cells clustered together. Average utilization is about 62 % including only cells, and about 70 % including signal and P/G TSV area as well. In overall, x- and z-direction congestion is moderate, yet at some regions z-direction congestion is severe. From the power map, we observe that some power hotspots exist.

12.6.2 Comparison of 2D and 3D IC Designs

We first compare the design in 2D and 3D ICs. Table 12.5 shows the comparison of a 2D and two 3D IC designs (two and four dies). We set the die size so that each case has about the same total silicon area. With increased number of dies, total wirelength becomes shorter and total power consumption reduces. Congestion increases mainly because of z-direction congestion. Longest path delay decreases, yet it involves more complex reasons such as how the gates on the path are partitioned into dies and placed. Signal TSVs occupy about 5.38 and 9.34 % silicon area in two and four die cases, respectively. Maximum IR-drop increases with increased number of dies,

Table 12.5 Comparison of 2D and 3D IC designs. Congestion means number of routing edges with 100 % utilization

	2D – 1 die	3D – 2 dies	3D – 4 dies
Footprint (μm^2)	1,960,000	1,000,000	490,000
Total wirelength (μm)	16,543,560	15,410,160	14,609,760
# signal TSVs	0	3,360	8,569
Congestion	39	329	673
Longest path delay (ns)	2.031	1.910	1.796
Power consumption (W)	1.427	1.398	1.304
Max. IR-drop (mV)	4.092	4.330	6.831
Max. Si temp. (°C)	49.681	80.389	131.485

Table 12.6 Comparison of baseline, T-TSV case, and MFC case

	Baseline	T-TSV case	MFC case
Total wirelength (μm)	14,609,760	14,675,180	14,770,840
Congestion	673	793	698
Total power (W)	1.304	1.306	1.366
Max. IR-drop (mV)	6.831	6.838	9.965
Max. Si temp. (°C)	131.485	119.169	34.550

mainly because with smaller footprint fewer P/G TSVs are placed, and the resistance of P/G TSVs contributes to IR-drop.

A big problem with 3D designs is elevated maximum silicon temperatures. The main reasons are that the power density of four die design is very high (about $277 W/cm^2$) and the heat dissipation path, from the bottom die to the heat sink located on the top of the chip, is long. Hence a thermal solution is crucial for the four die design to be practical.

12.6.3 *Comparison of T-TSV and MFC Based Cooling*

Table 12.6 shows the experimental results for baseline, T-TSV case, and MFC case. There are no T-TSVs or MFCs in the baseline case. All three cases have almost the same total wirelength. Congestion in T-TSV case is a little higher than the one in baseline, because T-TSVs consume routing space. Total power consumption is higher in MFC case, because of increased total wirelength and larger signal TSV capacitance. Maximum IR-drop is also higher in MFC case, mainly because P/G TSVs in MFC case are longer and the resistance of them is higher.

Compared with the baseline, T-TSVs decreased the maximum silicon temperature by only 9.4 %, which is small compared with a previous work [9], where silicon-on-insulator technology was assumed. Since the thermal conductivity of the insulator is very low, inserting T-TSVs would dramatically increase the thermal conductivity of insulator layers. However, in this chapter we assumed bulk silicon technology, and the thermal conductivity of silicon is good (about one third of that of

Fig. 12.8 Temperature profiles for (**a**) baseline, (**b**) T-TSV case and (**c**) MFC case. *Dotted lines* in MFC case show MFCs

copper). Thus, inserting T-TSVs did not decrease the maximum silicon temperature so much. In contrast, MFCs reduced the maximum silicon temperature by 74 %. This shows that T-TSV-based cooling is not as efficient as MFC-based cooling.

Temperature profiles for the three cases are shown in Fig. 12.8. The heat sink is attached on top of the chip for baseline and T-TSV case (the top of the chip in MFC case contacts air directly). Hence, the temperatures of lower dies are higher than that of upper dies as they are further away from the heat sink. Comparing T-TSV case profile with baseline profile, we observe that the temperature difference between dies is smaller in T-TSV case. Thus, we can see that T-TSVs helped heat transfer in z-direction. In MFC case, the regions where MFCs are placed show relatively lower temperature than their neighbors, and the temperature along the MFC flow direction increases because the fluid absorbs the heat as it travels.

12.6.4 Varying One Input Factor at a Time

As a preliminary experiment, we vary one input factor at a time to investigate its impact on the responses. By this preliminary experiment, we try to find the ranges of input factors that may satisfy overall target performances and check possible tradeoffs. Each input factor is set to three levels: (−, 0, +). '0' is the default value, '−' is the minimum value, and '+' is the maximum value. While we vary one input factor, the other factors remain at '0' level. For each factor, it is assumed the system shows monotonic responses, and the max/min value will not happen at non-extreme locations of the specified input ranges.

12.6.4.1 MFC Case

For MFC case, our input factor settings are determined as follows (−, 0, +): (1) MFC width (μm) = (30, 55, 80), (2) MFC pressure drop (kPa) = (30, 50, 70), (3) P/G TSV diameter (μm) = (5, 10, 15), (4) P/G thin wire ratio = (0.2, 0.5, 0.8). Figure 12.9 shows the results of preliminary experiments for MFC case. Total

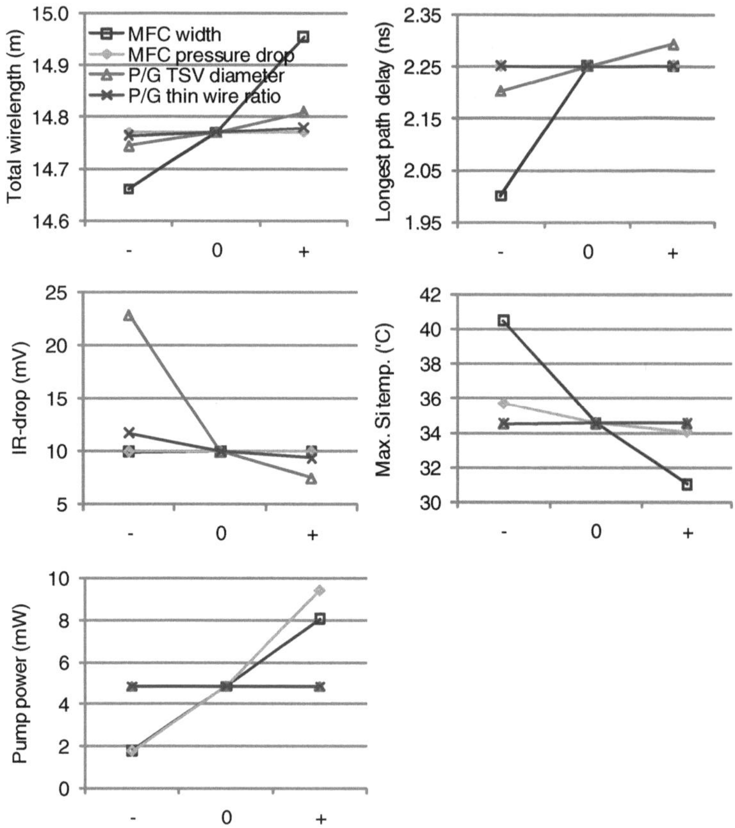

Fig. 12.9 Results of preliminary experiments for MFC case

wirelength and longest path delay are dependent on MFC width and P/G TSV diameter, whereas IR-drop is mostly dependent on P/G TSV diameter and P/G thin wire ratio. Comparing total wirelength and IR-drop graphs, we can see that increasing P/G TSV diameter can decrease IR-drop by a lot at the cost of a little increased total wirelength. Meanwhile, maximum silicon temperature and pump power are mostly dependent on MFC width and MFC pressure drop. For maximum silicon temperature, MFC width affects more than MFC pressure drop, while for pump power MFC pressure drop does more than MFC width.

12.6.4.2 T-TSV Case

For T-TSV case, our input factor settings are determined as follows (−, 0, +): (1) T-TSV ratio = (0, 0.1, 0.2), (2) P/G TSV diameter (μm) = (5, 10, 15), (3) P/G thin wire ratio = (0.2, 0.5, 0.8). Note that we restrict the maximum T-TSV ratio at 20 % or

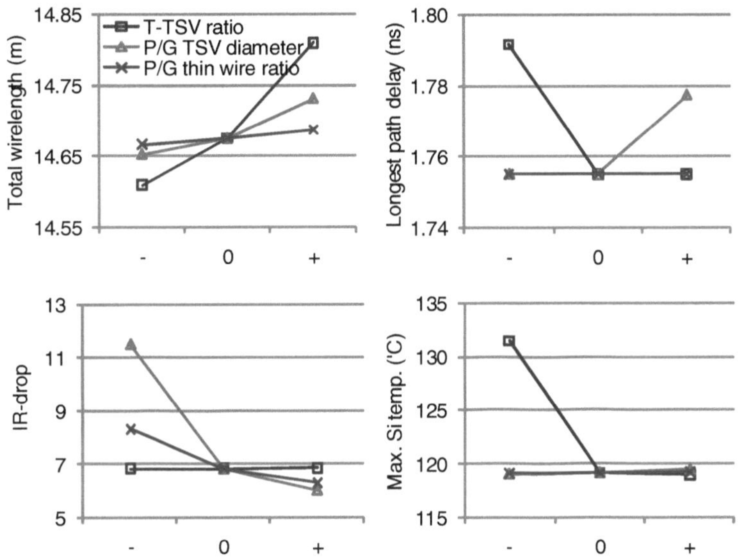

Fig. 12.10 Results of preliminary experiments for T-TSV case

less. Inserting too many T-TSVs may consume too much silicon and routing space and may incur mechanical reliability issues due to thermal expansion coefficient mismatch. Figure 12.10 shows the results of preliminary experiments for T-TSV case. Total wirelength is dependent on all three input factors, of which the most influential one is T-TSV ratio. Longest path delay is affected by T-TSV ratio and P/G TSV diameter. IR-drop is dependent solely on P/G TSV diameter and P/G thin wire ratio. Note that compared to MFC case results, the IR-drop range is smaller. This is mainly because in T-TSV case die thickness is smaller and the resistance of P/G TSVs is lower. Maximum silicon temperature drops sharply when the T-TSV ratio is changed from '−' to '0', however the drop is greatly reduced between '0' and '+'. As we shall see in the later DOE section, this suggests that the response model needs a high order polynomial.

12.6.5 Classical DOE

12.6.5.1 T-TSV Case

For T-TSV case, total 15 design points were generated by Box Behnken design. After running experiments, we gathered the data and built the response models as shown in Table 12.7. Validation RMSE is the RMSE calculated with the validation data set. Compared with the average values (2nd column), RMSE values are very small. However, the PRESS RMSE values are more than twice the RMSE values,

Table 12.7 Summary of models for T-TSV case with classical DOE. The model type is Quadratic, and the number of parameters is 10 for all three responses

Response	Average	RMSE	PRESS RMSE	Validation RMSE	R^2
Total wirelength	14,701,073	8,255.120	19,064.382	13,565.988	0.996
Max. IR-drop	8.120	0.031	0.070	0.476	1.000
Max. Si temp.	122.514	0.394	0.911	5.431	0.998

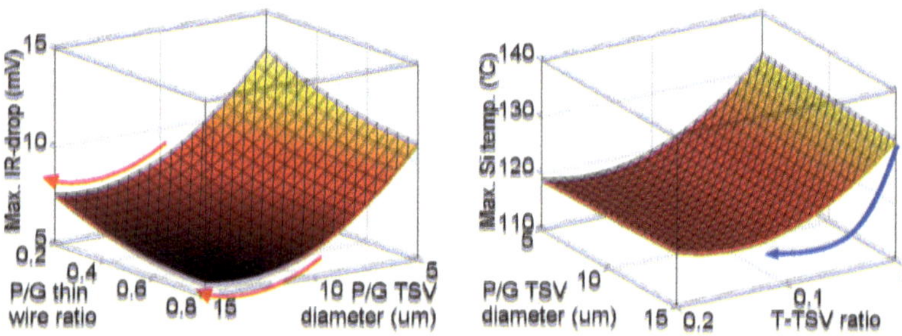

Fig. 12.11 Response surfaces for T-TSV case with classical DOE. Only two significant axes per response are shown

which suggests the response models may not be able to predict unseen design points. It is more evident for maximum silicon temperature model, because validation RMSE of it is very high, about 4.4 % of the average value. For all three response models R^2 is very close to 1. This suggests that we should not rely only on R^2 in assessing overall model quality.

Two response surfaces are shown in Fig. 12.11. We observed that IR-drop model is not monotonic around corners of the ROI, as denoted by red arrows. The maximum silicon temperature model surface did not show enough curvature. As discussed in Sect. 12.6.4, we need a higher order model for this response, as suggested by the blue arrow.

12.6.5.2 MFC Case

For MFC case, total 27 design points were generated by Box Behnken design. Due to the constraint between MFC width and P/G TSV diameter defined by (12.5), we shifted the design points accordingly. The modified Box Behnken design is shown in Fig. 12.12, and the summary of response models are shown in Table 12.8. Compared with the average values, RMSE of pump power is high, about 4.7 %. The difference between RMSE and PRESS RMSE was less than twice. However, validation RMSE reveals that there is a high error in the maximum IR-drop model.

Two response surfaces are shown in Fig. 12.13. Again, we observed needless curvatures on the corners of the ROI for the maximum IR-drop and the maximum silicon temperature models.

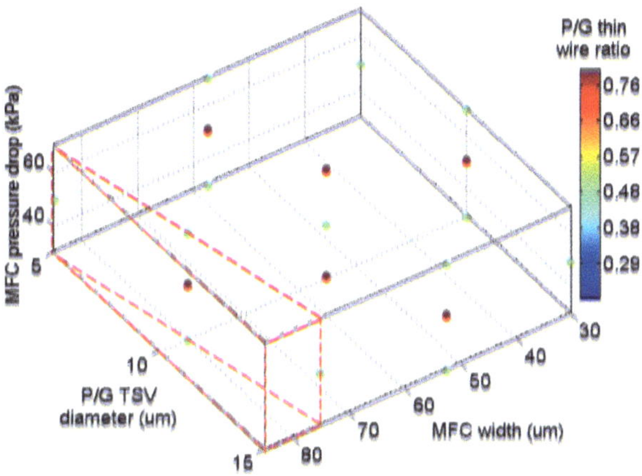

Fig. 12.12 Modified Box Behnken design for MFC case. *Red dotted lines* show the constrained region

Table 12.8 Summary of models for MFC case with classical DOE. The model type is Quadratic, and the number of parameters is 15 for all four responses

Response	Average	RMSE	PRESS RMSE	Validation RMSE	R^2
Total wirelength	14,794,088	2,054.884	3,458.582	8,030.389	1.000
Max. IR-drop	12.549	0.053	0.085	1.379	1.000
Max. Si temp.	35.314	0.307	0.492	0.370	0.996
Pump power	5.202	0.243	0.389	0.256	0.998

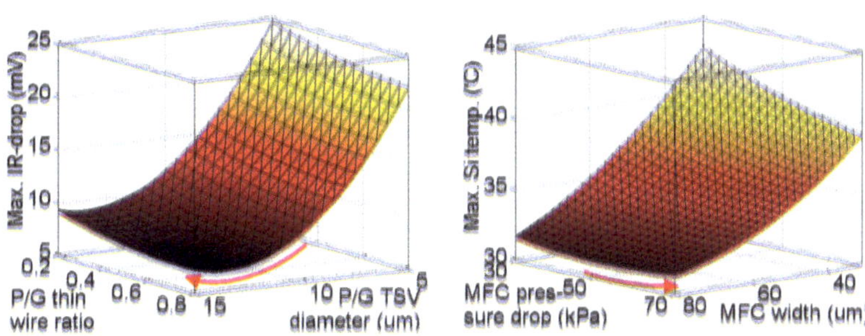

Fig. 12.13 Response surfaces for MFC case with classical DOE. Only two significant axes per response are shown

12.6.6 Advanced DOE: T-TSV Case

To perform advanced DOE for T-TSV case, we generated 32 Latin Hypercube design points, then manually added eight corner design points. The reason for adding corner design points is to reduce the error on the boundaries of the ROI. The input factor ranges are: (1) T-TSV ratio = $[0, 0.2]$, (2) P/G TSV diameter = $[5, 15]$ (μm), and (3) P/G thin wire ratio = $[0.2, 0.8]$. Table 12.9 shows the candidate models for maximum silicon temperature in T-TSV case. We apply the scoring

Table 12.9 Candidate models for maximum silicon temperature in T-TSV case. Only the best five models are shown. The numbers in the parenthesis after Poly means the polynomial order of (T-TSV ratio, P/G TSV diameter, P/G thin wire ratio/interaction), and the name in the parenthesis after RBF means the RBF kernel type. '+stepwise' means stepwise regression was performed

Model type	# para	RMSE	PRESS RMSE	Validation RMSE	R^2	Score
Poly(8,8,8/2) + stepwise	13	0.253	0.338	0.411	0.998	0.3953
Poly(8,8,8/3) + stepwise	23	0.117	0.155	0.463	1.000	0.4565
Poly(10,7,2/3) + stepwise	18	0.133	0.268	0.210	0.999	0.3887
Poly(11,3,3/2) + stepwise	16	0.143	0.278	0.169	0.999	0.4574
Poly(10,5,0/3) + RBF(thin-plate)	27	0.033	0.081	0.154	1.000	0.4660

Table 12.10 Summary of models for T-TSV case with advanced DOE

Response	Model type	# para	Average	RMSE	PRESS RMSE	Validation RMSE	R^2
Total WL	Poly(3,5,2/2) + stepwise	9	14,699,694	4,305.241	4,989.319	9,643.771	0.997
IR-drop	Poly(2,7,7/4) + stepwise	21	7.893	0.003	0.004	0.004	1.000
Max. temp.	Poly(11,3,3/2) + stepwise	16	121.212	0.143	0.278	0.169	0.999

Table 12.11 Parameters for total wirelength model of T-TSV case with advanced DOE. $TTSVrat$, $PGdia$, and $PGthin$ means maximum T-TSV ratio, P/G TSV diameter, and P/G thin wire ratio

Constant	1.467E+07	$PGdia$	4.217E+04
$TTSVrat$	8.926E+04	$PGthin$	1.082E+04
$PGdia^2$	3.158E+04	$PGdia \times TTSVrat$	1.197E+04
$TTSVrat^2$	2.837E+04	$TTSVrat \times PGthin$	4.499E+03
$PGdia^4$	−1.942E+04		

method presented in Sect. 12.5.3. Although the model in the last row had higher Score, the model in the fourth row was chosen as the best, because it had less unwanted curvatures on the response surface.

The best models for T-TSV case are summarized in Table 12.10. Compared with the models in Table 12.7, RMSE, PRESS RMSE and validation RMSE decreased by much. Especially, the validation RMSE of maximum silicon temperature was reduced by 97 %. The difference between RMSE and PRESS RMSE is less than twice, thus over-fitting is unlikely. Validation RMSE is not too far away from PRESS RMSE, suggesting that the models can predict unseen design points well.

The parameters for total wirelength model of T-TSV case are shown in Table 12.11. T-TSV ratio and P/G TSV diameter are strong main factors, and also significant interaction is observed between them.

The response surfaces of all metrics for T-TSV case are shown in Fig. 12.14. For total wirelength and maximum silicon temperature, the most significant input factor was T-TSV ratio. Total wirelength increases with higher maximum T-TSV ratio, due to the congestion incurred by T-TSVs. Also P/G TSV diameter affected the response in the same manner, yet the impact is smaller. In case of maximum IR-drop, P/G

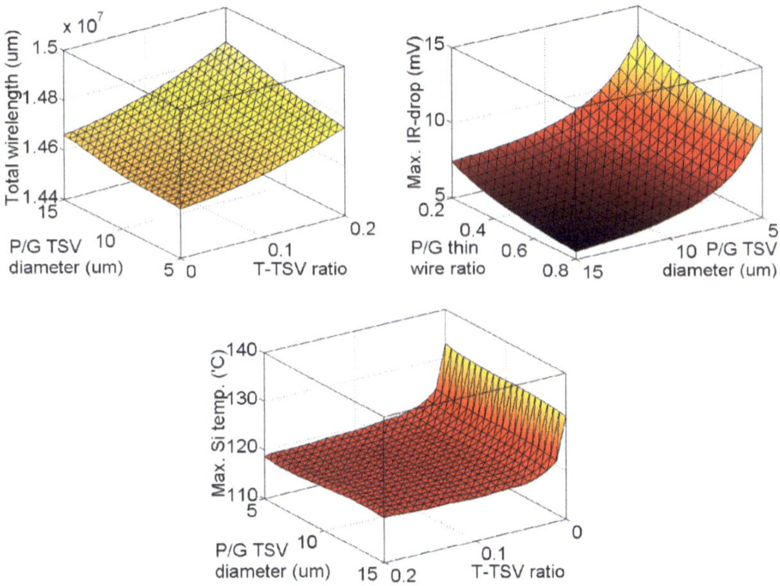

Fig. 12.14 Response surfaces for T-TSV case with advanced DOE. For each metric, we show the two significant input factors

TSV diameter and P/G thin wire ratio were the major factors. Maximum IR-drop decreased with higher P/G thin wire ratio and larger P/G TSV diameter. Maximum silicon temperature drops sharply when maximum T-TSV ratio increases from 0 to 0.05; after 0.05 the slope becomes gentle. Compare the curvature of the maximum silicon temperature to the one in Fig. 12.11. Note that P/G TSV diameter interacts with maximum T-TSV ratio in the maximum silicon temperature model. This is because P/G TSVs occupy silicon space that affect the actual amount of T-TSVs which is different from the maximum T-TSV ratio.

With the response models, we perform optimization. Two optimization scenarios are considered:

- Scenario 1: Minimize total wirelength, maximum IR-drop, and maximum silicon temperature.

  $Cost_1 = \sqrt[3]{Cost^*_{wl} \cdot Cost^*_{ir} \cdot Cost^*_{st}}$

  Here, $Cost^*_{wl}$, $Cost^*_{ir}$, and $Cost^*_{st}$ denote normalized total wirelength, maximum IR-drop, and maximum silicon temperature costs, respectively.
- Scenario 2: Minimize total wirelength and maximum silicon temperature under a maximum IR-drop constraint.

  $Cost_2 = \sqrt{Cost^*_{wl} \cdot Cost^*_{st}}$

  Constraint: (maximum IR-drop) $\leq 10\,\text{mV}$

Table 12.12 Optimization results for Scenarios 1 and 2 in T-TSV case

	Baseline	DOE-pred.	DOE-actual
Scenario 1			
Max. T-TSV ratio	0.1	0.094	0.094
P/G TSV diameter (μm)	10	15	15
P/G thin wire ratio	0.5	0.8	0.8
Total wirelength (μm)	14,675,180	14,736,797	14,741,100
Max. IR-drop (mV)	6.838	5.494	5.497
Max. Si temp. (°C)	119.169	118.940	119.585
$Cost_1$	0.174	0.071	0.089
Scenario 2			
Max. T-TSV ratio	0.1	0.2	0.2
P/G TSV diameter (μm)	10	6.766	6.766
P/G thin wire ratio	0.5	0.388	0.388
Total wirelength (μm)	14,675,180	14,764,854	14,767,680
Max. IR-drop (mV)	6.838	9.065	9.061
Max. Si temp. (°C)	119.169	118.330	118.817
$Cost_2$	0.169	0.076	0.166

Table 12.12 shows the input factor settings and the responses for three cases: **baseline** – This is our baseline settings and the responses from an actual experiment. **DOE-predicted** – This shows the optimal input factor settings found from optimization, and the response values predicted by the response models. **DOE-actual** – With the same optimal settings as DOE-predicted, we run the experiment and get the actual response values. Comparison between DOE-predicted and DOE-actual reveals the accuracy of the response model. Comparing DOE-actual to baseline, we can see that DOE found better solutions with about 49 % less $Cost_1$ for Scenario 1 and about 1.8 % less $Cost_2$ for Scenario 2. Comparing DOE-predicted and DOE-actual, we can see that all the models are quite accurate (error <1 %). In Scenario 1, maximum T-TSV ratio decreased a little because it helped reduce total wirelength without increasing maximum silicon temperature too much. P/G TSV diameter and P/G thin wire ratio increased to maximum because it decreased power noise and did not exacerbate other metrics too much. In Scenario 2, maximum T-TSV ratio reached its maximum to minimize maximum silicon temperature. P/G TSV diameter and P/G thin wire ratio decreased from the baseline, because maximum IR-drop is not minimized but constrained under the target value. The maximum IR-drop value of DOE-actual meets the constraint.

12.6.7 Advanced DOE: MFC Case

For MFC case, we generated 48 Latin Hypercube design points, then manually added 16 corner design points. The input factor ranges are: (1) MFC width $= [30, 85](\mu\text{m})$, (2) MFC pressure drop $= [30, 70](\text{kPa})$, (3) P/G TSV

Table 12.13 Summary of models for MFC case with advanced DOE. In model type column, the numbers in the parenthesis after Poly means the polynomial order of (MFC width, MFC pressure drop, P/G TSV diameter, P/G thin wire ratio/interaction), and the name in the parenthesis after RBF means the RBF kernel type

Response	Model type	# para	Average	RMSE	PRESS RMSE	Validation RMSE	R^2
Total WL	Poly(6,0,6,6/2) + stepwise	12	14,798,019	8,344.520	9,061.613	8,086.943	0.995
IR-drop	Poly(6,0,6,6/3) + stepwise	21	12.656	0.013	0.016	0.021	1.000
Max. temp.	Poly(7,7,0,0/4) + RBF(multiquadric)	30	35.314	0.036	0.045	0.103	1.000
Pump power	Poly(6,6,2,0/4) + stepwise	15	5.188	0.078	0.085	0.092	1.000

diameter = $[5, 15]$(μm), and (4) P/G thin wire ratio = $[0.2, 0.8]$. Again, we apply the input factor constraint defined by (12.5). The best models for MFC case, chosen by the same scoring method as in T-TSV case, are summarized in Table 12.13. Compared with the models in Table 12.8, RMSE, PRESS RMSE and validation RMSE decreased much. The validation RMSE of pump power was reduced by 64 %. PRESS RMSE is close to RMSE, meaning that the models are not overfitted. Validation RMSE is not too far away from PRESS RMSE, except for maximum silicon temperature model. Still, the validation RMSE of maximum silicon temperature model is small compared with the average value.

Figure 12.15 shows the response surfaces of all metrics. Total wirelength was mostly dependent on MFC width and P/G TSV diameter. Wider MFCs incurred z-direction congestion around them, thus some signal nets had to take detours which led to longer wirelength. Maximum IR-drop was mostly dependent on P/G TSV diameter and P/G thin wire ratio, while maximum silicon temperature and pump power were mostly dependent on MFC width and MFC pressure drop. Increasing MFC width lowers maximum silicon temperature throughout its whole input factor range, whereas MFC pressure drop affects maximum silicon temperature more when MFC width is small. That is, smaller MFCs are more sensitive to MFC pressure drop. Wider MFCs with higher pressure drop requires higher pump power.

Three optimization scenarios are considered for MFC case:

- Scenario 1: Minimize total wirelength, maximum IR-drop, maximum silicon temperature, and pump power.

  $$Cost_1 = \sqrt[4]{Cost^*_{wl} \cdot Cost^*_{ir} \cdot Cost^*_{st} \cdot Cost^*_{pp}}$$

  Here, $Cost^*_{wl}$, $Cost^*_{ir}$, $Cost^*_{st}$, and $Cost^*_{pp}$ denote normalized total wirelength, maximum IR-drop, maximum silicon temperature, and pump power costs, respectively.
- Scenario 2: Minimize total wirelength, IR drop, and pump power under a maximum silicon temperature constraint.

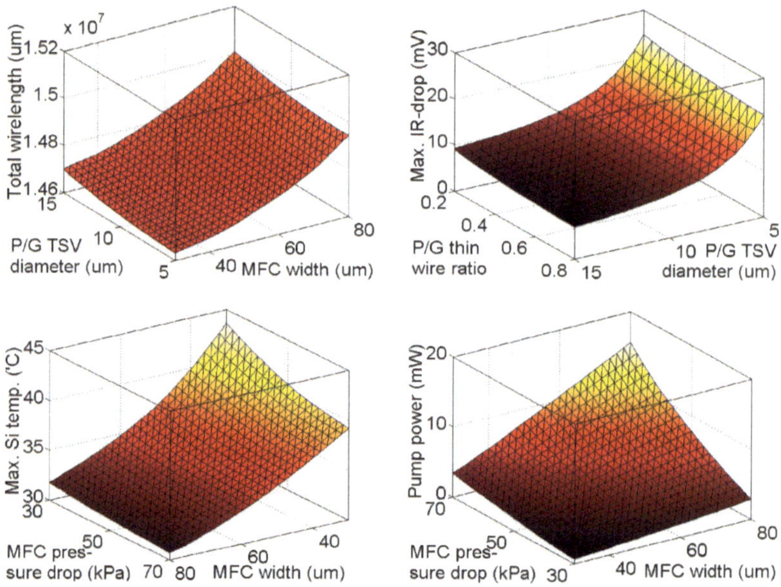

Fig. 12.15 Response surfaces for MFC case with advanced DOE

$$Cost_2 = \sqrt[3]{Cost^*_{wl} \cdot Cost^*_{ir} \cdot Cost^*_{pp}}$$

Constraint: (maximum silicon temperature) $\leq 40\,^{\circ}\mathrm{C}$

- Scenario 3: Minimize total wirelength, IR drop, and maximum silicon temperature under a pump power constraint.

$$Cost_3 = \sqrt[3]{Cost^*_{wl} \cdot Cost^*_{ir} \cdot Cost^*_{st}}$$

Constraint: (pump power) $\leq$ 10 mW

Table 12.14 shows the input factor settings as well as the responses. Comparing DOE-actual with baseline, we can see that DOE consistently found better solutions with about 85 % less $Cost_1$ for Scenario 1, about 85 % less $Cost_2$ for Scenario 2, and about 59 % less $Cost_3$ for Scenario 3. The error between DOE-predicted and DOE-actual for all responses was less than 1 %, except for pump power which had around 5.2 and 3.7 % error in Scenarios 1 and 3. In Scenario 1, MFC width and MFC pressure drop were almost minimized while P/G TSV diameter and P/G thin wire ratio were almost maximized. This is because smaller MFC width and lower MFC pressure drop lowered $Cost^*_{wl}$ and $Cost^*_{pp}$ at the expense of increased $Cost^*_{st}$, which decreased $Cost_1$. Also increasing P/G TSV diameter and P/G thin wire ratio decreased maximum IR-drop without increasing total wirelength too much. In Scenario 2, the optimal design point is similar to the one in Scenario 1, except for MFC width. The MFC width was determined to meet the maximum silicon temperature constraint. The maximum silicon temperature in DOE-actual is close to the constraint value (40 °C). In Scenario 3, compared with baseline, MFC pressure

Table 12.14 Optimization results for Scenarios 1, 2, and 3 in MFC case

	Baseline	DOE-pred.	DOE-actual
Scenario 1			
MFC width (μm)	55	30.009	30.009
MFC pressure drop (kPa)	50	30.063	30.063
P/G TSV diameter (μm)	10	14.993	14.993
P/G thin wire ratio	0.5	0.78	0.78
Total wirelength (μm)	14,770,840	14,712,205	14,714,760
Max. IR-drop (mV)	9.965	6.908	6.950
Max. Si temp. (°C)	34.550	43.618	43.690
Pump power (mW)	4.849	0.729	0.691
$Cost_1$	0.270	0.048	0.040
Scenario 2			
MFC width (μm)	55	39.248	39.248
MFC pressure drop (kPa)	50	30.294	30.294
P/G TSV diameter (μm)	10	14.981	14.981
P/G thin wire ratio	0.5	0.781	0.781
Total wirelength (μm)	14,770,840	14,736,293	14,735,440
Max. IR-drop (mV)	9.965	6.930	6.954
Max. Si temp. (°C)	34.550	39.979	39.954
Pump power (mW)	4.849	1.071	1.066
$Cost_2$	0.254	0.037	0.039
Scenario 3			
MFC width (μm)	55	30.614	30.614
MFC pressure drop (kPa)	50	70	70
P/G TSV diameter (μm)	10	15	15
P/G thin wire ratio	0.5	0.778	0.778
Total wirelength (μm)	14,770,840	14,714,508	14,721,560
Max. IR-drop (mV)	9.965	6.905	6.952
Max. Si temp. (°C)	34.550	39.047	39.171
Pump power (mW)	4.849	3.587	3.453
$Cost_3$	0.269	0.096	0.111

drop increased to maximum to minimize maximum silicon temperature, while MFC width decreased to almost the minimum. The pump power of this solution is not close to the constraint, because the optimum point did not occur along the pump power constraint boundary.

12.6.8 Comparison to Gradient Search

For a comparison, we perform gradient search based optimization for the optimization Scenario 2 in MFC case. To lower the possibility of local minimum, we run total eight search threads independently, starting from different design points. First, we determine the move direction by checking the eight surrounding design points;

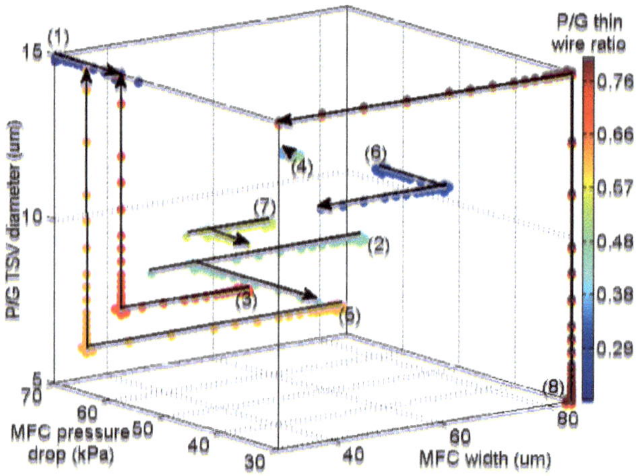

Fig. 12.16 Design space explored by gradient search. Three input factors are mapped to x-, y-, and z-axis, while the fourth input factor is mapped to color. Numbers in parentheses denote search thread numbers, and the *arrows* show directions of search

if the starting design point is (MFC width, MFC pressure drop, P/G TSV diameter, P/G thin wire ratio) = (x_1, x_2, x_3, x_4), then the following design points are tried: $(x_1 \pm \Delta_1, x_2, x_3, x_4)$, $(x_1, x_2 \pm \Delta_2, x_3, x_4)$, $(x_1, x_2, x_3 \pm \Delta_3, x_4)$, $(x_1, x_2, x_3, x_4 \pm \Delta_4)$. If the generated design point lies outside the ROI, it is dropped out. Then we run experiments and find the one with the greatest decrease in $Cost_2$. Next, we perform line search by generating design points along the direction and checking $Cost_2$. If $Cost_2$ keeps decreasing, we increase step size per three experimental runs. If the local minimum along the direction is found or the maximum silicon temperature constraint is violated, we search for a new direction. The process is continued for about 40 experimental runs per each search thread.

The gradient search explored the design space as shown in Fig. 12.16. Search Thread (4) mainly moved in P/G thin wire ratio direction. Although we ran total 311 experimental runs, the gradient search method could not explore much of the design space. The best design point was found by Search Thread (8).

Due to the errors in response models, it is possible that the actual optimal design point is different from the optimal design point on the $Cost_2$ surface. Thus, we further performed a local improvement after the optimization, which we call **DOE+local**. Table 12.15 shows the comparison of optimization results. The CPU time includes running experiments for all design points. Compared with the results of DOE-actual, DOE+local found a design point with a little less $Cost_2$ by increasing P/G thin wire ratio, yet the difference is small. This suggests DOE based optimization indeed found the solution very close to the minimum. Compared with DOE based search, gradient search found a worse solution with about 28 % higher $Cost_2$, even though it spend about 4.8X more CPU time. Although it is possible for the gradient search to reach the optimum point after enough iterations, there

Table 12.15 Comparison of optimization results for Scenario 2 in MFC case. DOE+local means DOE optimization followed by local search

	DOE-actual	DOE+local	Gradient
MFC width (μm)	39.248	39.248	46
MFC pressure drop (kPa)	30.294	30.294	30
P/G TSV diameter (μm)	14.981	14.981	15
P/G thin wire ratio	0.781	0.8	0.8
Total wirelength (μm)	14,735,440	14,736,340	14,782,520
Max. IR-drop (mV)	6.954	6.928	6.929
Max. Si temp. (°C)	39.954	39.955	37.900
Pump power (mW)	1.066	1.066	1.377
$Cost_2$	0.039	0.037	0.050
Experimental runs	64	78	311
CPU time (hours)	5.42	6.60	25.92

is a possibility that gradient search gets stuck at local minimum points. Moreover, the gradient search should be performed from the beginning if the cost function (optimization goal) is changed.

12.6.9 *Discussions*

It is important to find accurate response models to obtain reliable optimization results. When the errors of some of the models were large, the optimization led to suboptimal solutions that showed higher costs than some other design points did. Also, models with too high polynomial order and too many parameters usually exhibited over-fitting problem.

The response models can be reused for multiple optimization scenarios to facilitate early design stage decisions. Response surfaces covering the entire ROI help find the optimal solution in global scope. However, due to the inaccuracy of models, global optimum is not guaranteed.

The congestion of interconnects (signal TSVs, P/G TSVs, T-TSVs and MFCs) affects the amount of tradeoffs the input factors provide. For instance, when the target design was less congested, MFC width or T-TSV ratio did not affect total wirelength much, and the optimization always favored increasing MFC width or T-TSV ratio to the maximum.

12.7 Conclusions

In this chapter, we presented a co-optimization study of signal, power, and thermal interconnects in 3D ICs. The effectiveness of design space exploration based on DOE and RSM was demonstrated for the 3D IC designs with T-TSVs and

MFCs. The benefits as well as limitations of the method were discussed. For high performance circuits with high power density, the proposed optimization method reveals that inserting T-TSVs might not solve the thermal problem effectively. On the other hand, MFCs could bring down die temperature to an acceptable level, however a high TSV aspect ratio is required to avoid chip size increase.

References

1. M. Bakir, B. Dang, J. Meindl, Revolutionary nanosilicon ancillary technologies for ultimate-performance gigascale systems, in *Proceedings of the IEEE Custom Integrated Circuits Conference*, San Jose, 2007, pp. 421–428
2. G. Box, D. Behnken, Some new three level designs for the study of quantitative variables. Technometrics **2**, 455–475 (1960)
3. F. Brglez, R. Drechsler, Design of experiments in CAD: context and new data sets for ISCAS'99. Proc. IEEE Int. Symp. Circuit Syst. **6**, 424–427 (1999)
4. J. Cong, S.K. Lim, Edge separability based circuit clustering with application to circuit partitioning, in *Proceedings of the Asia and South Pacific Design Automation Conference*, Yokohama, 2000, pp. 429–434.
5. J. Cong, Y. Zhang, Thermal-driven multilevel routing for 3-D ICs, in *Proceedings of the Asia and South Pacific Design Automation Conference*, Shanghai, vol. 1, 2005, pp. 121–126
6. B. Dang, M.S. Bakir, J.D. Meindl, Integrated thermal-fluidic I/O interconnect for an on-chip microchannel heat sink. IEEE Electron Device Lett. **27**(2), 117–119 (2006)
7. G. Derringer, R. Suich, Simultaneous optimization of several response variables. J. Qual. Technol. **12**(4), 214–219 (1980)
8. R.A. Fisher, *The Design of Experiments* (Oliver and Boyd, London, 1935)
9. B. Goplen, S. Sapatnekar, Thermal via placement in 3D ICs, in *Proceedings of the International Symposium on Physical Design*, San Francisco, 2005, pp. 167–174
10. C.-W. Ho, A.E. Ruehli, P.A. Brennan, The modified nodal approach to network analysis. IEEE Trans. Circuit Syst. **22**(6), 504–509 (1975)
11. G. Huang et al., Compact physical models for power supply noise and chip/package co-design of gigascale integration, in *Proceedings of IEEE Electronic Components and Technology Conference*, Reno, 2007, pp. 1659–1666.
12. A.J. Joseph et al., Through-silicon vias enable next-generation SiGe power amplifiers for wireless communications. IBM J. Res. Dev. **52**(6), 635–648 (2008)
13. D.H. Kim, K. Athikulwongse, S.K. Lim, A study of through-silicon-via impact on the 3D stacked IC layout, in *Proceedings of the IEEE International Conference on Computer-Aided Design*, San Jose, 2009, pp. 674–680
14. Y.J. Kim, Y.K. Joshi, A.G. Fedorov, Y.-J. Lee, S.-K. Lim, Thermal characterization of interlayer microfluidic cooling of three-dimensional integrated circuits with nonuniform heat flux. J. Heat Transf. **132**(4), 214–219 (2010)
15. J.-M. Koo, S. Im, L. Jiang, K.E. Goodson, Integrated microchannel cooling for three-dimensilonal electronic architecture. J. Heat Transf. **127**, 49–58 (2005)
16. Y.-J. Lee, S.K. Lim, Co-optimization and analysis of signal, power, and thermal interconnects in 3D ICs, *IEEE Transactions on Computer-Aided Design of Integrated Circuits and Syst.* **30**(11), 1635–1648 (2011)
17. R.L. Mason, R.F. Gunst, J.L. Hess, *Statistical Design and Analysis of Experiments With Applications to Engineering and Science*, 2nd edn. (Wiley, New York, 2003)
18. R.H. Myers, D.C. Montgomery. *Response Surface Methodology: Process and Product Optimization Using Designed Experiments* (Wiley, New York, 1995)

19. V. Nookala, Y. Chen, D.J. Lilja, S.S. Sapatnekar, Microarchitecture-aware floorplanning using a statistical design of experiments approach, in *Proceedings of the ACM Design Automation Conference*, Anaheim, 2005, pp. 579–584.
20. North Carolina State University, NCSU FreePDK, available online at http://www.eda.ncsu.edu/wiki/FreePDK
21. S.V. Patankar, *Numerical Heat Transfer and Fluid Flow* (Hemisphere, Washington, DC, 1980).
22. M. Pathak, S.K. Lim, Thermal-aware Steiner Routing for 3D Stacked ICs, in *Proceedings of the IEEE International Conference on Computer-Aided Design*, San Jose, 2007, pp. 205–211
23. M. Pathak, Y.-J. Lee, T. Moon, S.K. Lim, Through silicon via management during 3D physical design: when to add and how many? in *Proceedings of the IEEE International Conference on Computer-Aided Design*, San Jose, 2010, pp. 387–394
24. D. Sekar et al., A 3D-IC technology with integrated microchannel cooling, in *Proceedings of the International Interconnect Technology Conference*, Burlingame, 2008
25. D.B. Tuckerman, R.F.W. Pease, High-performance heat sinking for VLSI. IEEE Electron Device Lett. **2**, 126–129 (1981)
26. E. Wong, S. Lim, 3D floorplanning with thermal vias, in *Proceedings of the Design, Automation and Test in Europe*, Münich, vol. 1, 2006, pp. 1–6
27. Q. Zhang et al., Development of robust interconnect model based on design of experiments and multiobjective optimization. IEEE Trans. Electron Device **48**(9), 1885–1891 (2001)
28. T. Zhang, Y. Zhan, S.S. Sapatnekar, Temperature-aware routing in 3D ICs, in *Proceedings of the Asia and South Pacific Design Automation Conference*, Yokohama, 2006, pp. 309–314
29. Q. Zhou, K. Sun, K. Mohanram, D.C. Sorensen, Large power grid analysis using domain decomposition, in *Proceedings of the Design, Automation and Test in Europe*, Münich, 2006, pp. 1–6

Part IV
Mechanical Reliability in 3D IC Designs

This part of the book covers design-for-mechanical-reliability for 3D ICs. In Chap. 13, we study the full-chip analysis of mechanical stress in 3D IC designs caused by the co-efficient of thermal expansion (CTE) mismatch between TSV and silicon substrate. In Chap. 14, we study the impact of mechanical stress on device mobility and full-chip timing variations in 3D IC. In Chap. 15, we extend the full-chip study in Chap. 13 to investigate the impact of package elements on the mechanical reliability of the entire 3D chip/package system. In Chap. 16, we study the impact of chip/package mechanical stress on device mobility and full-chip path delay variations. In Chap. 17, we study the impact of TSV-induced mechanical stress on crack growth between TSV and its liner (= interfacial crack).

Chapter 13
Mechanical Reliability Analysis and Optimization for 3D ICs

Abstract This chapter studies an efficient and accurate full-chip thermo-mechanical stress and reliability analysis tool and design optimization methodology to alleviate mechanical reliability issues in 3D ICs [5]. First, we analyze the detailed thermo-mechanical stress induced by TSVs in conjunction with various associated structures such as a landing pad and a dielectric liner. Then, we explore and validate the linear superposition principle of stress tensors and demonstrate the accuracy of this method against detailed finite element analysis (FEA) simulations. Next, we apply this linear superposition method to full-chip stress simulation and a reliability metric named the von Mises yield criterion. Finally, we study a design optimization methodology to mitigate the mechanical reliability problems in 3D ICs. Our numerical experimental results demonstrate the effectiveness of the proposed methodology.

The materials presented in this chapter are based on [5].

13.1 Introduction

Due to the coefficients of thermal expansion (CTE) mismatch between TSV fill material such as copper (Cu) and silicon substrate, thermo-mechanical stress is induced during fabrication process and thermal cycling of TSV structures. This thermo-mechanical stress can affect device performance [18] or drive crack growth in 3D interconnects [9, 14]. Most previous works focused on modeling the thermo-mechanical stress and reliability of a single TSV in isolation. These simulations are performed using FEA methods which are computationally expensive or infeasible for full-chip analysis. Furthermore, some works used unrealistic TSV structures such as an extremely large landing pad (LP), mainly because the design context is not considered.

S.K. Lim, *Design for High Performance, Low Power, and Reliable 3D Integrated Circuits*, DOI 10.1007/978-1-4419-9542-1_13,

In this chapter, we study a full-chip TSV thermo-mechanical stress and reliability analysis flow as well as a design optimization methodology to reduce mechanical reliability problems in TSV based 3D ICs. We use von Mises yield criterion as a mechanical reliability metric, and show impacts of design parameters such as TSV size, landing pad size, liner thickness, and keep-out-zone (KOZ) size on the mechanical reliability. The following specific topics are covered in this chapter:

- We study how to model the impact of TSV-induced mechanical stress on substrate reliability. Compared with existing works, we simulate more detailed and realistic TSV structures and study their impact on stress as well as a mechanical reliability metric. We also model the impact of chip operating temperature on stress and reliability.
- We study how to utilize the reliability models to perform full-chip mechanical analysis. We use and validate the principle of linear superposition of stress tensors against FEA simulations, and apply this methodology to generate a stress map and a reliability metric map on a full-chip scale.
- We study how to optimize 3D IC designs to alleviate mechanical reliability problems. We study design methods to reduce von Mises stress, which is a mechanical reliability metric, on full-chip 3D IC designs by tuning design parameters such as landing pad size, liner thickness, KOZ size, and TSV placement.

13.2 Detailed Baseline Modeling

The analytical 2D radial stress model, known as *Lamé* stress solution, was employed to address the TSV thermo-mechanical stress effect on device performance in [18]. This 2D plane solution assumes an infinitely long TSV embedded in an infinite silicon substrate and provides stress distribution in silicon substrate region, which can be expressed as follows [10]:

$$
\begin{aligned}
\sigma_{rr}^{Si} &= -\sigma_{\theta\theta}^{Si} = -\frac{E\Delta\alpha\Delta T}{2}\left(\frac{D_{TSV}}{2r}\right)^2 \\
\sigma_{zz}^{Si} &= \sigma_{rz}^{Si} = \sigma_{\theta z}^{Si} = \sigma_{r\theta}^{Si} = 0
\end{aligned}
\qquad (13.1)
$$

where σ^{Si} is stress in silicon substrate, E is Young's modulus, $\Delta\alpha$ is mismatch in CTE, ΔT is differential thermal load, r is the distance from TSV center, and D_{TSV} is TSV diameter.

Even though this closed-form formula is easy to handle, this 2D solution is only applicable to the structure with TSV and substrate only, hence inappropriate for the realistic TSV structure with a landing pad and a liner. Also, it does not capture the 3D nature of a stress field near the wafer surface around TSVs where devices are located. Moreover, the TSV/substrate interface region near the wafer surface is

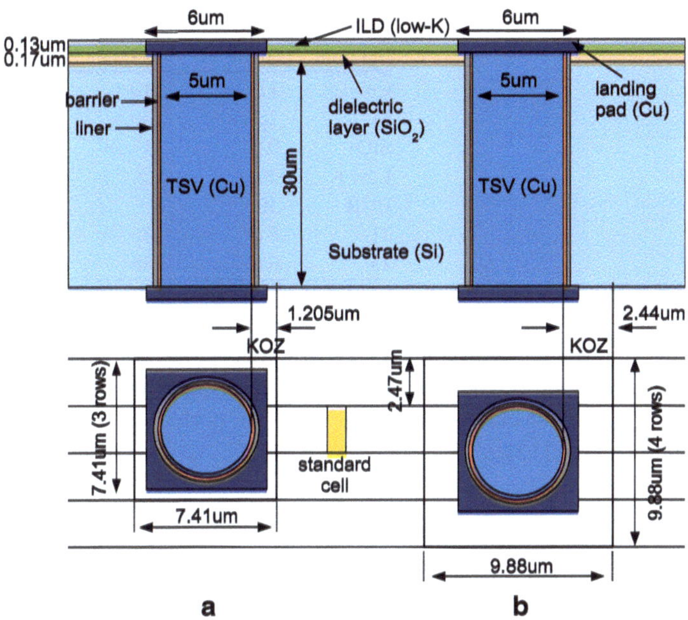

Fig. 13.1 Baseline TSV structure. (**a**) TSV_A cell occupying three standard cell rows (KOZ = 1.205 μm). (**b**) TSV_B cell occupying four standard cell rows (KOZ = 2.44 μm)

known to be a highly problematic area for mechanical reliability [14]. In our study, the wafer surface means the silicon surface right below substrate (Si)/dielectric layer (SiO_2) interface.

Though the authors in [14] proposed a semi-analytic 3D stress model, it is only valid for TSV with a high aspect ratio. Also, their TSV structure only includes TSV and silicon substrate, hence we cannot apply their model to the TSV which contains a landing pad and a dielectric liner because of the change in boundary conditions. Furthermore, since their model is only applicable to a single TSV in isolation, it cannot be directly used to assess mechanical reliability issues in a full-chip scale.

Since there is no known analytical stress model for a realistic TSV structure, 3D FEA models for a TSV structure are created to investigate the stress distribution near wafer surface. To realistically examine the thermo-mechanical stress induced by TSVs, our baseline simulation structure of a TSV is based on the fabricated and the published data [1, 8, 9, 15], as shown in Fig. 13.1 [1, 8, 9, 15]. We construct two TSV cells, i.e. TSV_A and TSV_B, which occupy three and four standard cell rows in NCSU 45 nm technology [3]. We define 1.205 and 2.44 μm from TSV edge as keep-out-zone (KOZ) in which no cell is allowed to be placed for TSV_A and TSV_B cells, respectively. Our KOZ sizes are slightly larger than the one for the single TSV case for digital circuits in [12]. Our baseline TSV diameter, height, Cu diffusion barrier thickness, liner thickness, and landing pad size are 5 μm, 30 μm, 50 nm, 125 nm, and 6 μm, respectively, unless specified, which are close to the data in [1]. We use SiO_2 and Ti as a baseline liner and a Cu diffusion barrier material,

Table 13.1 Material properties

Material	CTE (ppm/K)	Young's modulus (GPa)	Poisson's ratio
Cu	17	110	0.35
Si	2.3	130	0.28
SiO_2	0.5	71	0.16
Low K	20	9.5	0.3
BCB	40	3	0.34
Ti	8.6	116	0.32
Ta	6.8	186	0.34

respectively. Material properties used for our experiments are listed in Table 13.1. We use the FEA simulation tool ABAQUS to perform experiments, and all materials are assumed to be linear elastic and isotropic. Also, perfect adhesion is assumed at all material interfaces [13].

13.2.1 3D FEA Simulation

Before discussing the detailed stress modeling results, we introduce the concept of a stress tensor. Stress at a point in an object can be defined by the nine-component stress tensor:

$$\sigma = \sigma_{ij} = \begin{bmatrix} \sigma_{11} & \sigma_{12} & \sigma_{13} \\ \sigma_{21} & \sigma_{22} & \sigma_{23} \\ \sigma_{31} & \sigma_{32} & \sigma_{33} \end{bmatrix}$$

where, the first index i indicates that the stress acts on a plane normal to the i axis, and the second index j denotes the direction in which the stress acts. If index i and j are same we call this a normal stress, otherwise a shear stress. Since we adopt a cylindrical coordinate system in this modeling for the cylindrical TSV, index 1, 2, and 3 represent r, θ, and z, respectively.

13.2.2 Impact of TSV Liner and Landing Pad

Figure 13.2 shows FEA simulation results of a normal stress components σ_{rr}, $\sigma_{\theta\theta}$, σ_{zz} along an arbitrary radial line from the TSV center at the wafer surface with $\Delta T = -250°C$ of thermal load. That is, we assume TSV structure is annealed at 275°C and cooled down to 25°C to mimic the manufacturing process [6, 10, 14]. We also assume that the entire TSV structure is stress free at the annealing temperature. In our 3D FEA simulations we consider TSV surrounding structures such as dielectric liner and landing pad as well, while the 2D model only considers TSV and substrate which are infinitely long in z-direction. Due to this structural

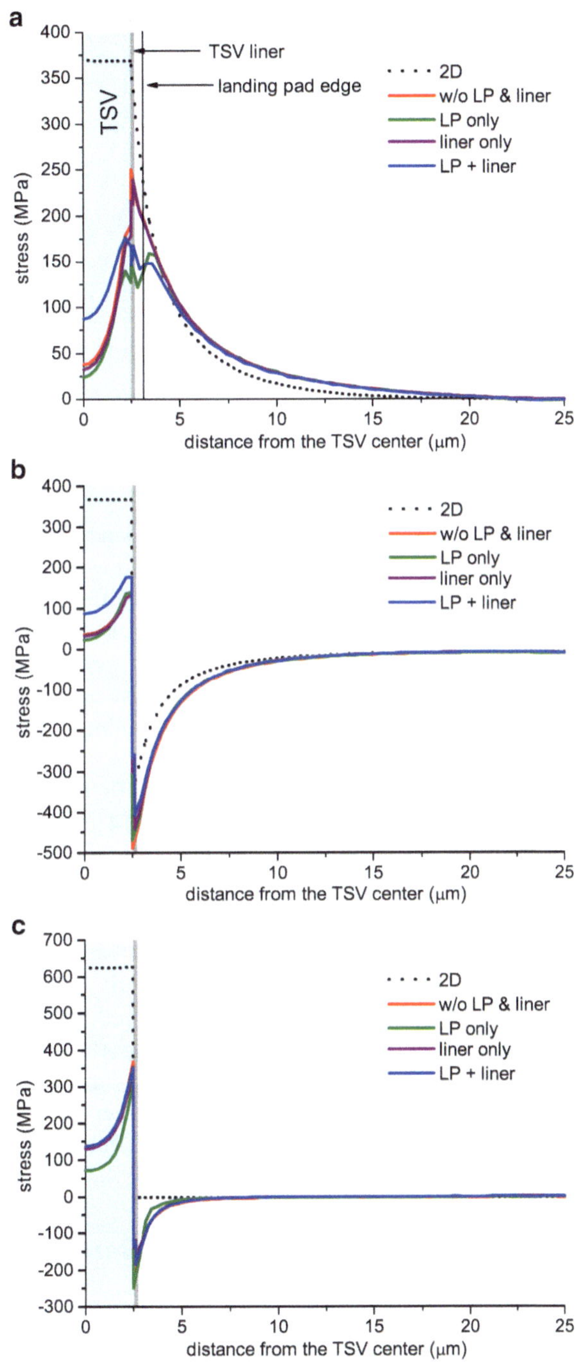

Fig. 13.2 Effect of TSV structures on normal stress components. (**a**) σ_{rr} stress. (**b**) $\sigma_{\theta\theta}$ stress. (**c**) σ_{zz} stress

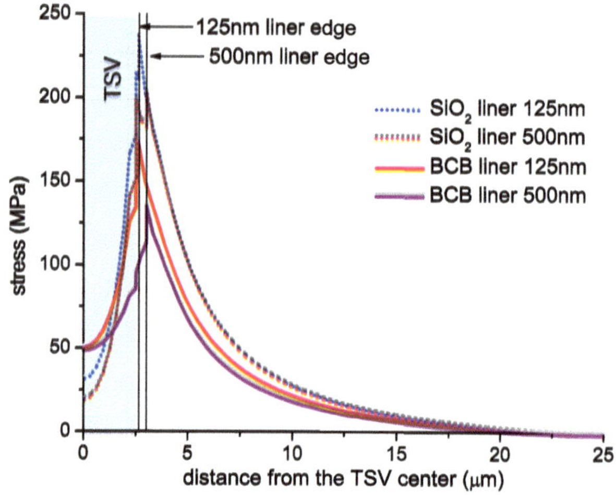

Fig. 13.3 Effect of liner material/thickness on σ_{rr} stress

difference, we observe the huge discrepancy between 2D solution and 3D stress results at the TSV edge. It is widely known that most of mechanical reliability failures occur at the interface between different materials, hence this TSV edge is the critical region for the reliability. Therefore, 2D solution does not predict mechanical failure mechanism for TSVs correctly. Also, the SiO_2 liner, which acts as a stress buffer layer, reduces σ_{rr} stress at the TSV edge by 35 MPa compared with the case without a landing pad and a liner. The landing pad also helps decrease stress magnitude at the TSV edge.

We also employ benzocyclobutene (BCB), a polymer dielectric material, as an alternative TSV liner material [10, 14]. Since Young's modulus, which is a measure of the stiffness of an elastic material, of BCB is much lower than Cu, Si, and SiO_2, this BCB liner can absorb the stress effectively induced by CTE mismatch. Figure 13.3 shows the impact of liner material and its thickness on σ_{rr} stress component. As liner thickness increases, stress magnitude at the TSV edge decreases noticeably, especially for the BCB liner case. It is evident from these experiments that modeling stress distribution considering surrounding structures such as a liner and a landing pad is important to analyze the thermo-mechanical stress around TSVs more accurately. We construct a stress library by varying TSV diameter/height, landing pad size, and liner material/thickness to enable full-chip thermo-mechanical stress and reliability analysis with different TSV structures.

13.2.3 Impact of Cu Diffusion Barrier

For Cu-based interconnects, a barrier layer is needed to prevent Cu diffusion into both dielectrics and Si substrate. This Cu diffusion induces degradation of dielectric layers, hence forms mid-gap defects in Si substrate. These defects serve as a recombination center and reduce the minority carrier lifetime [17]. Therefore, Cu

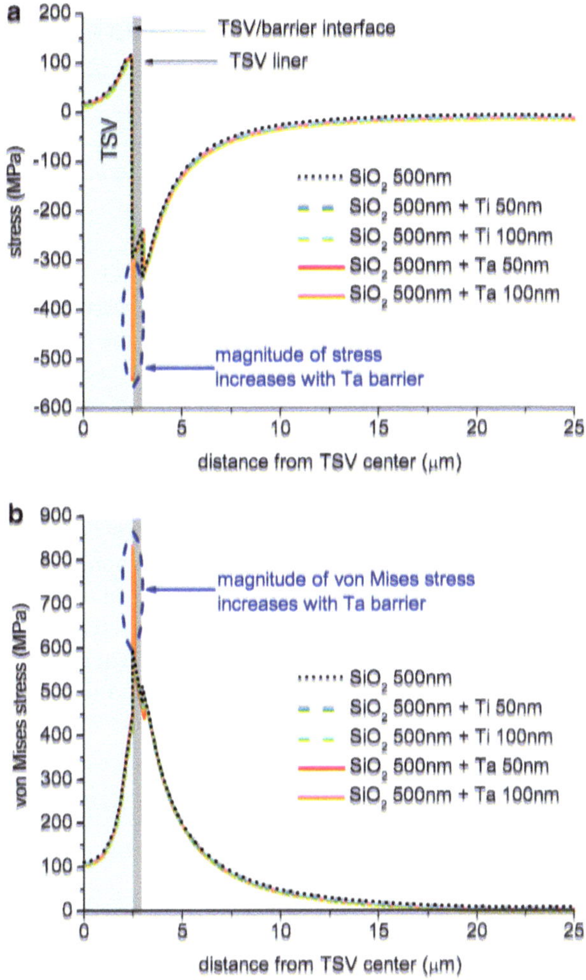

Fig. 13.4 Effect of Cu diffusion barrier on stress. (**a**) $\sigma_{\theta\theta}$ stress. (**b**) Von Mises stress

diffusion barrier is deposited between Cu TSV and dielectric liner. Some previous works ignored Cu diffusion barrier material such as Ti and Ta in FEA simulations, since this barrier thickness is only a small fraction of SiO_2 liner thickness, e.g., one tenth of liner thickness in general. Hence, its impact on stress distribution is negligible [8, 9].

However, thermo-mechanical stress is highly dependent on Young's modulus as well as CTE mismatch. As shown in Table 13.1, CTE of both Ti and Ta are in between Cu and Si, hence it is unlikely that these barrier materials induce additional stress around TSV on top of the stress induced by the CTE mismatch between Cu and Si. However, Ta is the stiffest material used in this chapter. Materials with high Young's modulus cannot absorb stress efficiently. Thus, there is a high chance of stress build-up at this TSV/barrier (Ta) interface. Figure 13.4 shows the impact of

Table 13.2 Effect of Young's modulus of barrier material on von Mises stress at TSV/barrier interface. Young's modulus of Ta changes while CTE and Poisson's ratio remain same

	Young's modulus (GPa)								
	25	50	75	100	125	150	175	200	225
Stress (MPa)	559	572	575	612	674	740	806	875	943

both Ta and Ti barrier material on the stress and mechanical reliability around TSV. In this experiment, we use a 500 nm thick SiO_2 liner, and 50 and 100 nm thick Ta and Ti barrier. We observe a huge increase in stress magnitude in the case of Ta barrier at TSV/barrier interface. For example, in the case of $\sigma_{\theta\theta}$ stress component shown in Fig. 13.4a, we see 241 and 232 MPa increase in compressive stress at TSV/barrier interface for 50 and 100 nm thick Ta barriers respectively, compared with no barrier case. However, there is a negligible change in stress with Ti barrier. This stress increase with Ta barrier also worsens von Mises stress, which is a reliability metric and is discussed in detail in Sect. 13.3.3, as shown in Fig. 13.4b.

To further verify that Young's modulus is the key parameter affecting stress magnitude at TSV/barrier interface, we change Young's modulus of Ta from 25 to 225 GPa with 25 GPa step while CTE and Poisson's ratio are unchanged. As Table 13.2 shows, as Young's modulus increases, maximum von Mises stress at TSV/barrier interface increases as well. Therefore, barrier material for Cu TSV should be chosen carefully to suppress additional mechanical reliability problem on top of existing concerns. Thus, we use Ti as a Cu diffusion barrier material throughout this chapter.

13.2.4 *Stress Influence Zone*

The magnitude of thermo-mechanical stress induced by TSV is highest at the TSV edge. However, as Fig. 13.2 shows, the magnitude of every normal stress component decays fast, and at around 25 μm from the TSV center, stress is almost negligible. For efficient and fast full-chip stress analysis, it is crucial to confine stress analysis to the manageable extent. Thus, we define a stress influence zone as a circle with a radius of 25 μm from the TSV center for our baseline TSV with 5 μm diameter. Beyond the stress influence zone, we neglect stress induced by the TSV under consideration. We further investigate the impact of TSV size on stress influence zone. We use three different TSV diameters with a same aspect ratio of 6; TSV small ($H/D = 15/2.5$ μm), TSV medium ($H/D = 30/5$ μm), and TSV large ($H/D = 60/10$ μm), where H/D is TSV height/diameter. Figure 13.5a shows the magnitude of σ_{rr} stress component from TSV edge, and we see that stress magnitude of smaller TSV decays and reaches zero faster.

Figure 13.5b shows stress magnitude along the normalized distance from TSV center, i.e., r/D_{TSV}, where r is the distance from TSV center, and D_{TSV} is TSV diameter. Even though there are differences in stress magnitude inside TSV and up

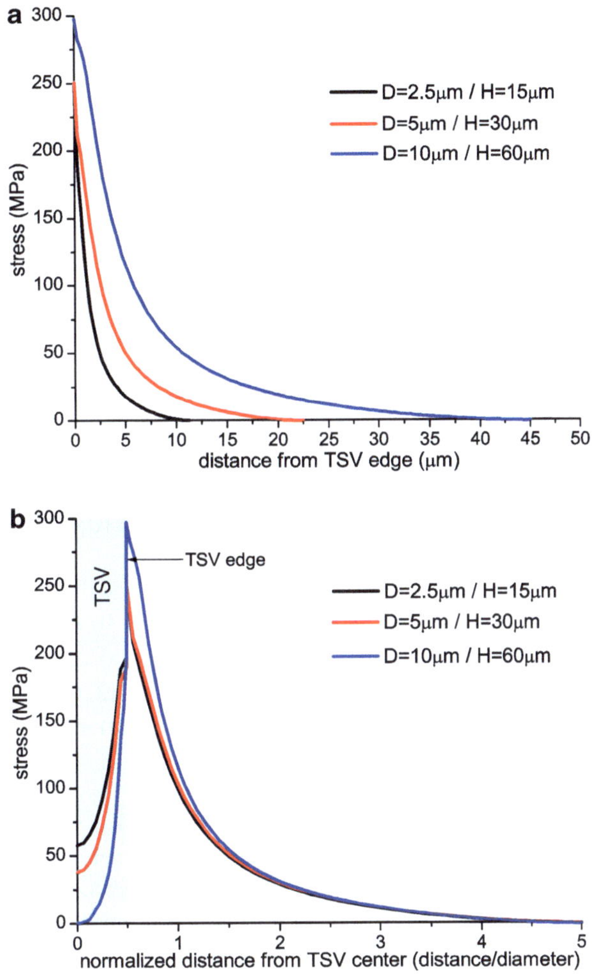

Fig. 13.5 Effect of TSV size on stress influence zone. (**a**) σ_{rr} stress from TSV edge. (**b**) σ_{rr} stress along normalized distance (distance/TSV diameter)

to 1×TSV diameter, stress magnitudes are almost identical beyond that distance. In the 2D solution shown in Eq. (13.1), we see that magnitude of σ_{rr}^{Si} and $\sigma_{\theta\theta}^{Si}$ is proportional to $(D_{TSV}/2r)^2$. That is why we observe similar stress curve along the normalized distance. Most importantly, we observe that stress magnitude becomes negligible at around 5×TSV diameter for all three cases. Therefore, we set stress influence zone as 5×TSV diameter in our study.

13.2.5 Anisotropic Material Property of Silicon

Up to this point, all materials are assumed to be isotropic for simplicity. However, Si is an anisotropic material with elastic behavior that depends on which crystal direction the structure is being stretched. The possible values of Young's modulus (E) for Si range from 130 to 188 GPa, and those for Poisson's ratio (ν) range from 0.048 to 0.4. Thus, the choice of this value can affect analysis results significantly [4]. Also, a recent study showed that TSV-induced stress measurement data matches well with FEA simulation results with anisotropic Si material property [11]. In this section, we examine the impact of anisotropic material property of Si on stress distribution compared with the isotropic material property.

Elasticity is the relationship between stress (σ) and strain (ε). Hooke's law describes this relationship in terms of stiffness C, i.e., $\sigma = C\varepsilon$. For isotropic uniaxial cases, stiffness C can be represented by a single value of Young's modulus E, and the equation takes the form of $\sigma = C\varepsilon$. In an anisotropic material, a fourth rank stiffness tensor with $3^4 = 81$ terms is required to describe the elasticity. Fortunately, due to cubic symmetry of Si, the elastic properties can be expressed in terms of orthotropic material constants. An orthotropic material is one which contains at least two orthogonal planes of symmetry, and Si, with cubic symmetry, can be described this way. The orthotropic elasticity of Si can be expressed with reference axes of a standard (100) Si wafer, which are [110], [$\bar{1}$10], and [001],

$$\begin{bmatrix} \sigma_{xx} \\ \sigma_{yy} \\ \sigma_{yy} \\ \sigma_{yz} \\ \sigma_{zx} \\ \sigma_{xy} \end{bmatrix} = \begin{bmatrix} c1 & c5 & c6 & 0 & 0 & 0 \\ c5 & c1 & c6 & 0 & 0 & 0 \\ c6 & c6 & c2 & 0 & 0 & 0 \\ 0 & 0 & 0 & c3 & 0 & 0 \\ 0 & 0 & 0 & 0 & c3 & 0 \\ 0 & 0 & 0 & 0 & 0 & c4 \end{bmatrix} \begin{bmatrix} \varepsilon_{xx} \\ \varepsilon_{yy} \\ \varepsilon_{yy} \\ \varepsilon_{yz} \\ \varepsilon_{zx} \\ \varepsilon_{xy} \end{bmatrix}$$

where, orientation specific constants c1, c2, c3, c4, c5, c6 are 194.5, 165.7, 79.6, 50.9, 35.7, and 64.1, all in GPa, respectively. This stiffness tensor translates to $E_x = E_y = 169$ GPa, $E_z = 130$ GPa, $\nu_{yz} = 0.36$, $\nu_{zx} = 0.28$, and $\nu_{xy} = 0.064$ [4].

We show stress comparison between anisotropic and isotropic Si in Fig. 13.6 (Young's modulus = 130 GPa and Poisson's ratio = 0.28) material properties. We see that stress magnitude with anisotropic Si is significantly higher than isotropic Si case largely due to higher Young's modulus, especially along X direction as shown in Fig. 13.6a. We also observe the same trend in Y direction (= σ_{yy} stress). However, there is a small difference in Z direction stress component compared with other directions shown in Fig. 13.6b. This can be explained by the same Young's modulus $E_z = 130$ GPa for both anisotropic and isotropic Si. In general, we observe higher stress level by adopting anisotropic Si material property compared with isotropic Si case mostly due to higher Young's modulus. Although FEA simulation with isotropic Si provides a good understanding of stress distribution, proper material properties need to be used to accurately assess thermo-mechanical reliability of TSV-based 3D ICs.

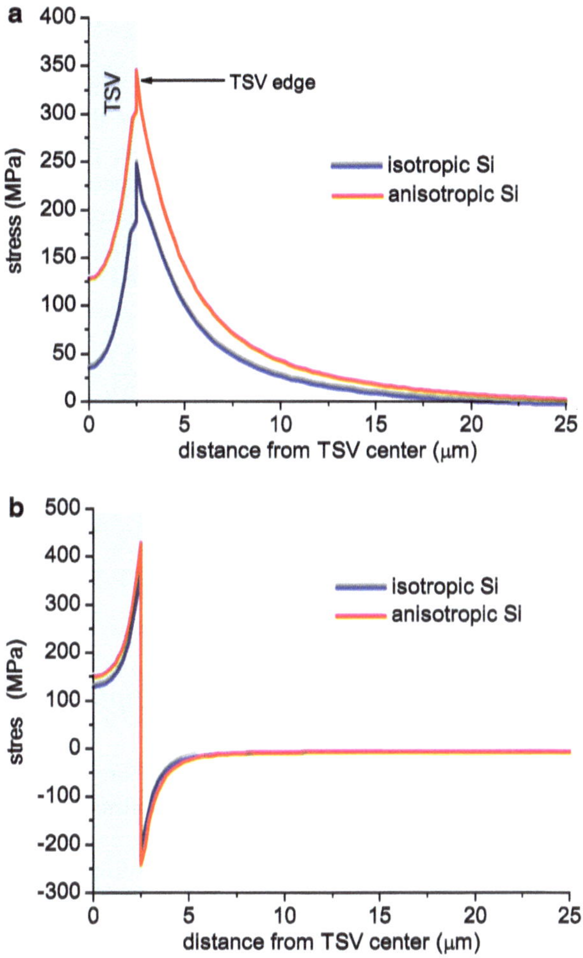

Fig. 13.6 Effect of anisotropic Si material property on stress (**a**) σ_{xx} stress. (**b**) σ_{zz} stress

13.3 Full-Chip Reliability Analysis

FEA simulation involves decomposition of structures into domains of disjoint polygons. These disjoint polygons are used in the formation of the member element functions represented by partial differential equations. They are then assembled and solved by a system of linear equations. FEA simulation of thermo-mechanical stress for multiple TSVs require huge computing resources and time, thus it is not suitable for full-chip analysis. In this section, we present a full-chip thermo-mechanical stress and reliability analysis flow. To enable a full-chip stress analysis, we first explore the principle of linear superposition of stress tensors from individual TSVs. Based on the linear superposition method, we build full-chip stress map and then compute von Mises yield metric to predict mechanical reliability problems in TSV-based 3D ICs.

13.3.1 Linear Superposition Principle

A useful principle in the analysis of linearly elastic structures is that of superposition. The principle states that if the displacements at all points in an elastic body are proportional to the forces producing them, the body is linearly elastic. The effect, i.e., stresses and displacements, of a number of forces acting simultaneously on such a body is the sum of the effects of the forces applied separately. We apply this principle to compute the stress at a point by adding the individual stress tensors at that point caused by each TSV as follows:

$$S = \sum_{i=1}^{n} S_i$$

where, S is the total stress at the point under consideration and S_i is the individual stress tensor at this point due to the ith TSV.

$$\sigma_v = \sqrt{\frac{(\sigma_{xx} - \sigma_{yy})^2 + (\sigma_{yy} - \sigma_{zz})^2 + (\sigma_{zz} - \sigma_{xx})^2 + 6(\sigma_{xy}^2 + \sigma_{yz}^2 + \sigma_{zx}^2)}{2}} \tag{13.2}$$

13.3.2 Stress Analysis with Multiple TSVs

First, based on the observation that the stress field of a single TSV in isolation is radially symmetrical due to the cylindrical shape of a TSV, we obtain stress distribution around a TSV from a set of stress tensors along an arbitrary radial line from the TSV center in a cylindrical coordinate system. To evaluate a stress tensor at a point affected by multiple TSVs, a conversion of a stress tensor to a Cartesian coordinate system is required. This is due to the fact that we extract stress tensors from a TSV whose center is the origin in the cylindrical coordinate system; hence we cannot perform a vector sum of stress tensors at a point from each TSV which has a different center location. That is why we need a universal coordinate system, i.e. Cartesian coordinate system in our case. Then, we compute a stress tensor at the point of interest by adding up stress tensors from TSVs affecting this point. We set a TSV stress influence zone as 25 μm from the center of a TSV with 5 μm diameter, as discussed in Sect. 13.2.4.

Let the stress tensor in Cartesian and cylindrical coordinate system be S_{xyz} and $S_{r\theta z}$, respectively.

$$S_{xyz} = \begin{bmatrix} \sigma_{xx} & \sigma_{xy} & \sigma_{xz} \\ \sigma_{yx} & \sigma_{yy} & \sigma_{yz} \\ \sigma_{zx} & \sigma_{zy} & \sigma_{zz} \end{bmatrix}, S_{r\theta z} = \begin{bmatrix} \sigma_{rr} & \sigma_{r\theta} & \sigma_{rz} \\ \sigma_{\theta r} & \sigma_{\theta\theta} & \sigma_{\theta z} \\ \sigma_{zr} & \sigma_{z\theta} & \sigma_{zz} \end{bmatrix}$$

The transform matrix Q is the form:

$$Q = \begin{bmatrix} \cos\theta & -\sin\theta & 0 \\ \sin\theta & \cos\theta & 0 \\ 0 & 0 & 1 \end{bmatrix}$$

where, θ is the angle between the X axis and a line from the TSV center to the simulation point. A stress tensor in a cylindrical coordinate system can be converted to a Cartesian coordinate system using conversion matrices: $S_{xyz} = QS_{r\theta z}Q^T$.

13.3.3 Mechanical Reliability Analysis

In order to evaluate if computed stresses indicate possible reliability concerns, a critical value for a potential mechanical failure must be chosen. The von Mises yield criterion is known to be one of the most widely used mechanical reliability metric [2, 16, 19]. If the von Mises stress exceeds a yielding strength, material yielding starts. Prior to the yielding strength, the material will deform elastically and will return to its original shape when the applied stress is removed. However, if the von Mises stress exceeds the yield point, some fraction of the deformation will be permanent and non-reversible even if applied stress is removed.

There is a large variation of yield strength of Cu in the literature, from 225 to 600 MPa, and it has been reported to depend upon thickness, grain size, and temperature [19]. We use 600 MPa as a Cu yielding strength in our experiments. The yield strength of silicon is 7,000 MPa, which will not be reliability concerns for the von Mises yield criterion. The von Mises stress is a scalar value at a point that can be computed using components of a stress tensor shown in Eq. (13.2). By evaluating von Mises stress at the interface between TSV and dielectric liner, where highest von Mises stress occurs, we can predict mechanical failures in TSVs.

13.3.4 Validation of Linear Superposition Method

In this section, we validate the linear superposition of stress tensors against FEA simulations by varying the number of TSVs and their arrangement. We set the minimum TSV pitch as 10 μm for all test cases. Stress tensors along a radial line from the TSV center in a single TSV structure (stress tensor list) are obtained through FEA simulation with 0.1 μm interval. In our linear superposition method, simulation area is divided into uniform array style grid with 0.05 μm pitch. If the stress tensor at a grid point under consideration is not obtainable directly from the stress tensor list, we compute stress tensor at the point using linear interpolation with adjacent stress tensors in the list.

Table 13.3 Von Mises stress comparison between FEA simulations and linear superposition method

	FEA		Linear		Max % error	
# TSV	# node	Run time	# sim point	Runtime	Inside TSV	Outside TSV
1	153 K	21 min 35 s	1. M	20.63 s	1.0	−0.4
2	282 K	58 min 11 s	1.2 M	26.21 s	3.3	−0.8
3	358 K	1 h 28 min 24 s	1.44 M	36.43 s	4.8	−1.3
5	546 K	1 h 59 min 05 s	1.68 M	56.02 s	12.7	−1.9
10	1,124 K	4 h 34 min 14 s	2.24 M	65.32 s	13.6	−2.0

Table 13.3 shows some of our comparisons. First, we observe huge run time reduction in our linear superposition method. Note that we perform FEA simulations using four CPUs while only one CPU is used for our linear superposition method. Even though our linear superposition method performs stress analysis on a 2D plane at the wafer surface, whereas FEA simulation is performed on entire 3D structure, we can perform stress analysis for other planes in a similar way if needed. Also, run time in our linear superposition method shows linear dependency on the number of simulation points, which is closely related to the number of TSVs under consideration. Thus, our linear superposition method is highly scalable, hence applicable to full-chip scale stress simulations. More details on scalability is discussed in Sect. 13.3.8.

Most importantly, the error between FEA simulations and the linear superposition method is practically negligible. Results show that our linear superposition method overestimates stress magnitude inside TSV. However, though the maximum % error inside TSV of 10 TSVs case is as high as 13.6%, the stress magnitude difference between FEA and our method is only 5.0 MPa. Also, since most mechanical problem occurs at the interface between different materials, this error inside TSV does not pose a serious impact on our reliability analysis. Figures 13.7 and 13.8 show the stress map of σ_{xx} component and the von Mises stress map for one of test cases which contains 10 TSVs, and it clearly shows our linear superposition method matches well with the FEA simulation result.

13.3.5 Handling Anisotropic Silicon

We also examine the feasibility of linear superposition with anisotropic Si material properties. Even though elastic properties of Si depend on the orientation of the structure, fortunately, Young's modulus along X ([100]) and Y ([$\bar{1}$10]) direction are same, i.e., $E_x = E_y = 169$ GPa. Since our focus is stress on the device layer, which is on the XY-plane, it is possible that the linear superposition method still holds with anisotropic Si elastic properties, though E_z is different from E_x and E_y.

Figure 13.9 shows that the linear superposition method still matches well with the FEA simulation result with anisotropic Si material property. Though, the error

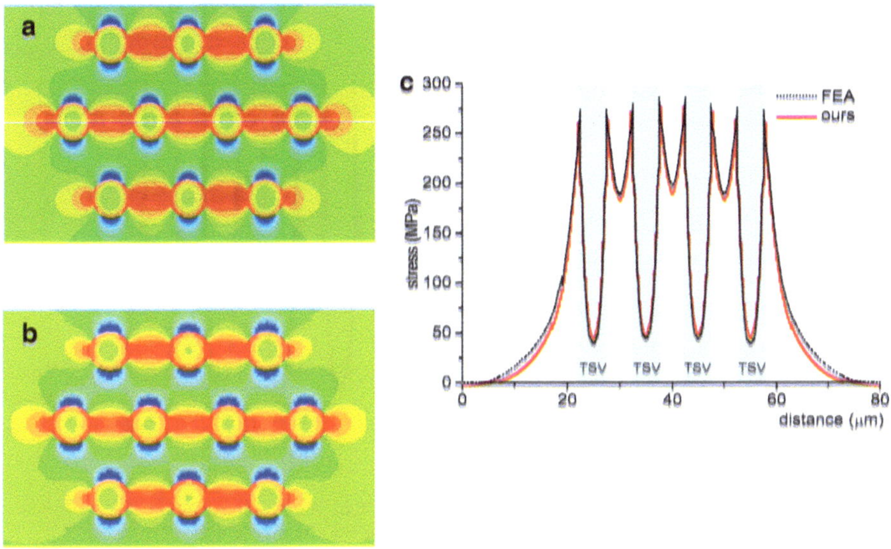

Fig. 13.7 σ_{xx} stress comparison between FEA simulation and linear superposition method. (**a**) FEA, (**b**) linear superposition, and (**c**) FEA vs. linear superposition along the *white line* in (**a**)

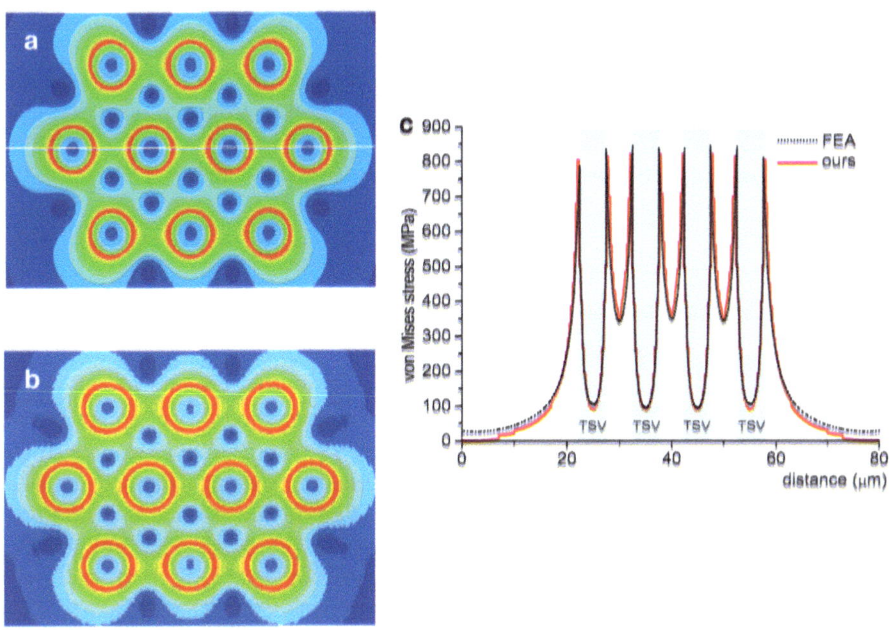

Fig. 13.8 von Mises stress comparison between FEA simulation and linear superposition method. (**a**) FEA and (**b**) linear superposition, (**c**) FEA vs. linear superposition along the *white line* in (**a**)

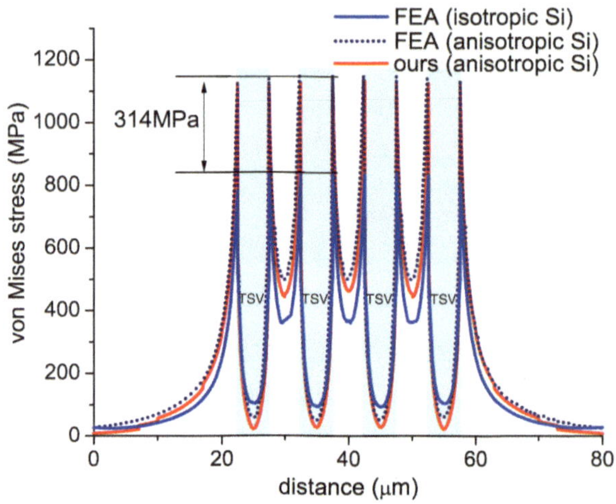

Fig. 13.9 Von Mises stress comparison between isotropic and anisotropic Si. Stress is taken along the *white line* in Fig. 13.8a

Table 13.4 Error in maximum von Mises stress in MPa (and % error) at TSV edge between FEA simulation and linear superposition for both isotropic and anisotropic Si

	# TSV (pitch = 10 μm)			
Si property	2	3	5	10
Isotropic	4.1 (0.5%)	−6.9 (−0.8%)	−5.7 (−0.7%)	−7.0 (−0.9%)
Anisotropic	−12.1 (−1%)	−16.3 (−1.4%)	−16.8 (−1.5%)	−18.3 (−1.6%)

increased compared with the isotropic Si case as listed in Table 13.4, it is still promising to use the linear superposition method for anisotropic Si case. We also observe a huge increase in maximum von Mises stress at TSV edge, 314 MPa in this case, with anisotropic Si case due to higher Young's modulus in the XY-plane. This is expected as discussed in Sect. 13.2.5.

13.3.6 Limitation of Linear Superposition Method

We observe that higher % error occurs inside TSV in Sect. 13.3.4. In this section, we further investigate the limit of our linear superposition method. In our approach, we first obtain stress tensors along a radial line from TSV center to substrate in a single TSV structure through FEA simulation. That is, stress tensors are obtained across at least two different materials. Then, we use linear superposition principle to compute stress tensor at a point with stress tensors that we prepared. This point under consideration could be either inside TSV (Cu) or outside TSV (silicon substrate).

In a full chip, most of simulation points are outside TSV. The stress tensor in the substrate region is computed by adding up stress tensors which are from the substrate region in a single TSV structure. In other words, all the stress tensors in the substrate region are computed by using stress tensors from the substrate region, which is the same material. However, computing stress tensors inside TSV

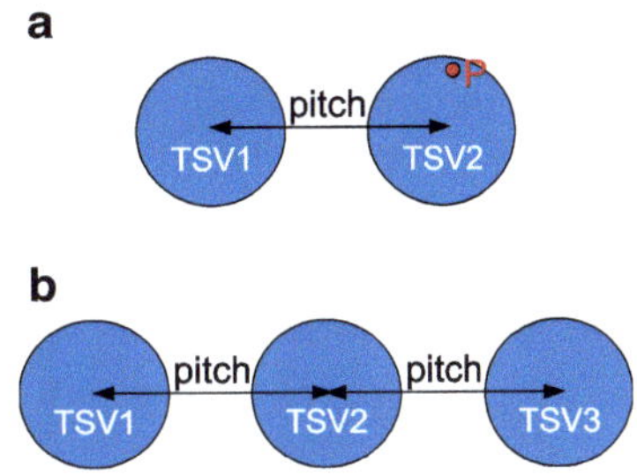

Fig. 13.10 Simulation structures for limit of linear superposition method. (**a**) Two TSVs. (**b**) Three TSVs

Table 13.5 Maximum error (in MPa) in von Mises stress between FEA simulation and linear superposition

		TSV pitch (μm)				
# TSV	Location	7.5	10	15	20	25
2	Inside	16.5	7.8	3.3	1.8	1.7
	Outside	8.9	2.8	0.7	−1.1	−0.7
3	Inside	15.4	6.9	5.5	2.5	1.6
	Outside	7.2	2.3	−1.2	−0.8	−0.5

is different. For example, assume that there is one TSV (TSV1) that affects a point *P* inside another TSV (TSV2) shown in Fig. 13.10a. The stress tensor from TSV1 affects *P* is originated from the silicon substrate region, while that from TSV2 affects *P* is obtained from Cu. Thus, the stress tensor at *P* is calculated by adding stress tensors from two different materials. Even though we use linear elastic models for the whole structure, due to the difference in material properties between Cu and silicon, the stress effect induced by nearby TSVs acting on the TSV under consideration is different between inside and outside the TSV. Therefore, the error from linear superposition is inevitable, especially inside TSV.

Intuitively, if a TSV pitch becomes smaller, the error between FEA simulation and our approach, i.e., stress (ours) – stress (FEA), will be larger due to high stress magnitude. To further explore this, we compare two TSVs structure by varying the TSV pitch from 7.5 to 25 μm shown in Fig. 13.10a. Table 13.5 shows the maximum error between FEA simulation and our linear superposition method for both inside and outside TSV. For example, at 7.5 μm pitch, error is 17.0 MPa inside TSV and 8.8 MPa outside TSV shown in Fig. 13.11a. However, this maximum error inside TSV occurs in the TSV center where the stress magnitude is lowest inside TSV, hence mechanical reliability at this location is not a concern. Also, the highest error outside TSV is found at the center location between TSVs. Nonetheless, the trend of stress magnitude remains same and the error is not significant. Moreover, 7.5 μm pitch with 5 μm diameter TSV is too extreme case for current TSV fabrication process. Therefore, error induced by our linear superposition method is negligible practically. We further examine whether an additional TSV worsens error. We use three TSVs structure shown in Fig. 13.10b. As similar to two TSVs case, the maximum error occurs at TSV2 center. However, as Table 13.5 shows, an additional TSV does not aggravate simulation errors. Thus, our approach is suitable for full-chip analysis with an acceptable error.

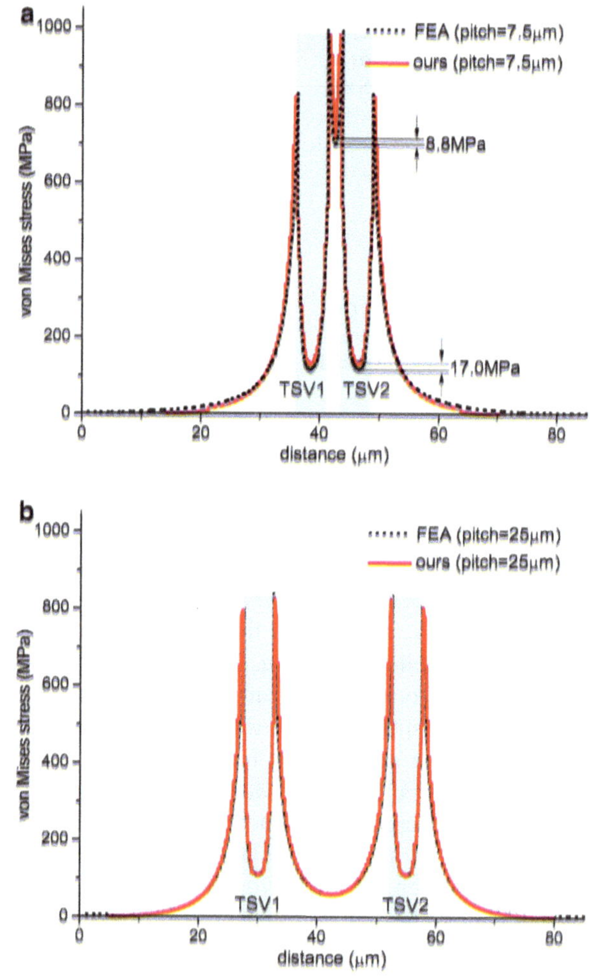

Fig. 13.11 Von Mises stress along the *center line* in Fig. 13.10a with two TSVs with different pitches. (**a**) Two TSVs with pitch of 7.5 μm. (**b**) Two TSVs with pitch of 25 μm

13.3.7 Full-Chip Analysis Flow

Our full-chip thermo-mechanical stress and reliability analysis flow is shown in Fig. 13.12. We first perform a detailed FEA simulation of a single TSV and provide the stress tensors along a radial line from the TSV center as an input to our simulation engine. We also provide the locations of the TSVs from 3D IC layout along with a thermal map to the simulation engine.

The basic algorithm for generating stress and reliability maps is illustrated in Algorithm 9. We first start to find a stress influence zone from each TSV. Then, we associate the points in the influence zone with the affecting TSV. Next, for each simulation point under consideration, we look up the stress tensor from the TSV found in the association step, and use the coordinate conversion matrices to

Fig. 13.12 Simulation flow

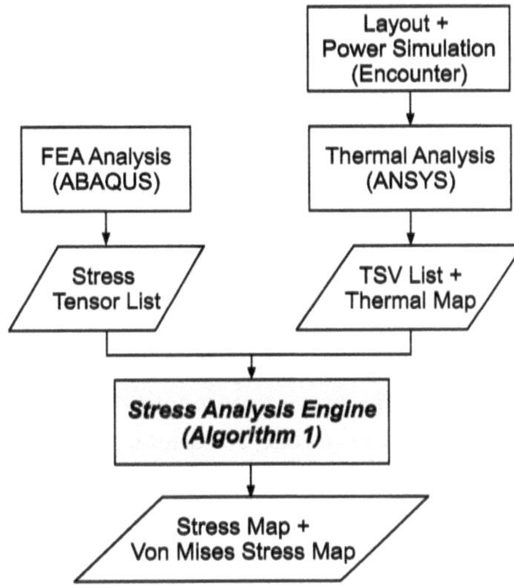


```
input  : TSV list T, stress library, thermal map (optional)
output: stress map, von Mises stress map
for each TSV t in T do
    c ⟵ center of t
    r ⟵ FindStressInfluenceZone(c)
    for each point r' in r do
        r'.TSV ⟵ t
    end
end
for each simulation point p do
    if p.TSV ≠ ∅ then
        for each t ∈ p.TSV do
            d ⟵ distance(t, p)
            S_cyl ⟵ FindStressTensor(d, temperature)
            θ ⟵ FindAngle(line tp, x axis)
            Q ⟵ SetConversionMatrix(θ)
            S_Cart ⟵ Q S_cyl Q^T
            p.S_Cart ⟵ p.S_Cart + S_Cart
        end
    end
    vonMises(p) ⟵ ComputeVonMises(p.S_cart)
end
```

Algorithm 9: Full Chip Stress and Reliability Analysis Flow

Table 13.6 Scalability to full-chip analysis

# TSV	Area (μm × μm)	# sim point	Run time	Time/point
10	60 × 60	90 K	3.6 s	40 μs
100	165 × 165	680 K	25.4 s	37 μs
1,000	495 × 495	6.1 M	5 min 55 s	58 μs
10,000	1,515 × 1,515	57.4 M	1 h 3 min	66 μs
100,000	4,770 × 4,770	568.8 M	10 h 12 min	65 μs

obtain stress tensors in the Cartesian coordinate system. We visit an individual TSV affecting this simulation point and add up their stress contributions. Once we finish the stress computation at a point, we obtain the von Mises stress value using Eq. (13.2). The complexity of this algorithm is $O(n)$, where n is number of simulation points.

13.3.8 Scalability of Algorithm

To verify the scalability of our linear superposition method, we apply our algorithm to 10, 100, 1,000, 10,000 and 100,000 TSVs that are regularly placed across a chip with 15 μm pitch. Note that the entire chip area is divided into uniform array style grid with 0.2 μm pitch in these experiments.

Table 13.6 shows that run time increases almost linearly as the number of simulation points increases. Our algorithm visits each simulation point and calculates stress tensor at this point by linearly superposing stress tensors from TSVs located within a stress influence zone as shown in Algorithm 9. Since the number of TSVs affecting a single point is a small constant in most cases, the number of simulation points dominates entire run time.

We also observe that simulation time per each grid point increases as the number of TSVs increases, and saturates beyond 10,000 TSVs case. At first, the number of TSVs which affect a point under consideration increases as total number of TSVs increases, hence a single point requires more time to compute stress tensor. However, after a certain point, increasing the total number of TSVs does not affect the simulation time per point. This is due to a finite stress influence zone, and hence a finite number of TSVs affects a simulation point.

Note that these test cases in which TSVs are regularly placed across a chip are worst cases in terms of run time. This is because almost every simulation point needs stress computation, since it is highly likely to be within a stress influence zone of a certain TSV. However, in case TSVs are only placed in some part of the chip or irregularly placed, run time can reduce significantly since there could be non-negligible number of simulation points that are outside of the stress influence zone of any TSVs, which are filtered out in the first *for loop* of our Algorithm 9.

Even though run time for 100,000 TSVs case is as long as 10 h, we can decrease it by parallel computing. For this, we first divide entire chip area by a user-defined window. To avoid any mismatch while merging sub-solutions, we need to consider

TSVs that are located outside the current window but affect the point inside the current window. For example, in the case of 5 μm diameter TSV whose influence zone is 25 μm, we need to compute stress from TSVs located within both 25 μm distance from each side of the window in outward direction and the current window. Then, we can compute stress considering all TSVs affecting the window under consideration, and each sub-solution becomes independent of other sub-solutions. Therefore, full chip thermo-mechanical stress and reliability analysis is scalable with our approach.

13.4 Experimental Results

We implement a TSV-aware full-chip stress and reliability analysis flow in JAVA and C++. Four variations of an industrial circuit, with changes in TSV placement style and TSV cell size, are used for our analysis, which are listed in Table 13.7. The number of TSVs and gates are 1,472 and 370 K, respectively, for all cases. These circuits are synthesized using Synopsys Design Compiler with the physical library of 45 nm technology, and final layouts are obtained using Cadence SoC Encounter. All circuits are designed to two-die stacked 3D ICs.

We use our in-house 3D placer for TSV and cell placement, and details of TSV and cell placement algorithms can be found in [7]. In the regular TSV placement scheme, we pre-place TSVs uniformly on each die, and then place cells, while TSVs and cells are placed simultaneously in the irregular TSV placement scheme. The irregular TSV placement shows better wirelength than the regular case [7]. We use a gate-level 3D IC design methodology for these circuits as a baseline, and compare these with block-level designs in Sect. 13.4.8.

We use our in-house 3D placer for TSV and cell placement [7]. In the regular TSV placement scheme, from the given 2D netlist, we perform partitioning and obtain 3D netlist. Then, we insert TSVs into the netlist for 3D nets. Then, we pre-place TSVs in regular array style with user defined pitch. After that, we perform cell placement for each die, and during this stage TSVs are treated as placement obstacles to avoid overlaps between cell and TSVs. With placement finished, we determine which pre-placed TSVs belong to which 3D nets based on the minimum spanning tree based TSV assignment scheme. On the other hand, for the irregular TSV placement scheme, the procedure is same as regular TSV placement until

Table 13.7 Benchmark circuits

Circuit	TSV placement	TSV cell size (μm × μm)	Wirelength (mm)	Area (μm × μm)
$Irreg_A$	Irregular	7.41 × 7.41	9,060	960 × 960
Reg_A	Regular	7.41 × 7.41	9,547	960 × 960
$Irreg_B$	Irregular	9.88 × 9.88	8,884	1,000 × 1,000
Reg_B	Regular	9.88 × 9.88	9,648	1,000 × 1,000

TSV insertion into 3D netlist. Then, we place cells and TSVs simultaneously using force-directed placement algorithm with modifications. In this case, TSV is treated as normal cells. The irregular TSV placement shows better wirelength than the regular case [7]. We use a gate-level 3D IC design methodology for these circuits as a baseline, and compare these with block-level designs in Sect. 13.4.8. In the regular TSV placement scheme, we pre-place TSVs uniformly on each die, and then place cells, while TSVs and cells are placed simultaneously in the irregular TSV placement scheme. The irregular TSV placement shows better wirelength than the regular case [7]. We use a gate-level 3D IC design methodology for these circuits as a baseline, and compare these with block-level designs in Sect. 4.6.

13.4.1 Overall Comparison

In this section, we discuss the impact of TSV structure, TSV placement style, and KOZ size on the thermo-mechanical reliability in 3D ICs. We perform full-chip stress and reliability analysis on our benchmark circuits based on our stress modeling results with different TSV structures. Figure 13.13 shows the maximum von Mises stress in our benchmark circuits. We first observe that designs with irregular TSV placement show worse maximum von Mises stress than those with the regular TSV placement. This is mainly because TSVs can be placed closely in case of the irregular TSV placement scheme to minimize wirelength. Figure 13.14 shows the part of von Mises stress maps of $Irreg_A$ and Reg_A circuits, and we see that most of TSVs in the $Irreg_A$ circuit exceed Cu yielding strength (600 MPa).

Second, as the KOZ size becomes larger stress level reduces significantly for the irregular TSV placement case. By enlarging the KOZ size, i.e., increasing TSV cell size in our design flow, TSV pitch increases accordingly. This in turn reduces stress interference between nearby TSVs, and hence decreases von Mises stress level of TSVs. However, for the regular TSV placement case, since the TSV pitch of Reg_A (23.5 μm) and Reg_B (25 μm) is similar and also interference from nearby TSVs is negligible at this distance, there is no noticeable difference in maximum von Mises stress. Third, these results show the importance of using an accurate TSV stress model to assess the mechanical reliability of 3D ICs. There are significant differences in the von Mises stress depending on the existence of structures surrounding a TSV, such as landing pad or liner. It is possible that we might overestimate the reliability problems by using a simple TSV stress model not considering landing pad or liner. However, most of these test cases violate the von Mises yield criterion for Cu TSV. Section 13.4.6 shows how TSV liners help reduce the violations.

Table 13.8 shows the number of TSVs violating von Mises criterion out of 1,472 TSVs used in our designs. Even though the regular TSV placement scheme shows smaller maximum von Mises stress, all TSVs fail to satisfy von Mises yield criterion unless the BCB liner is used. This is because all TSVs experience similar level of von Mises stress due to the fact that they are evenly spaced.

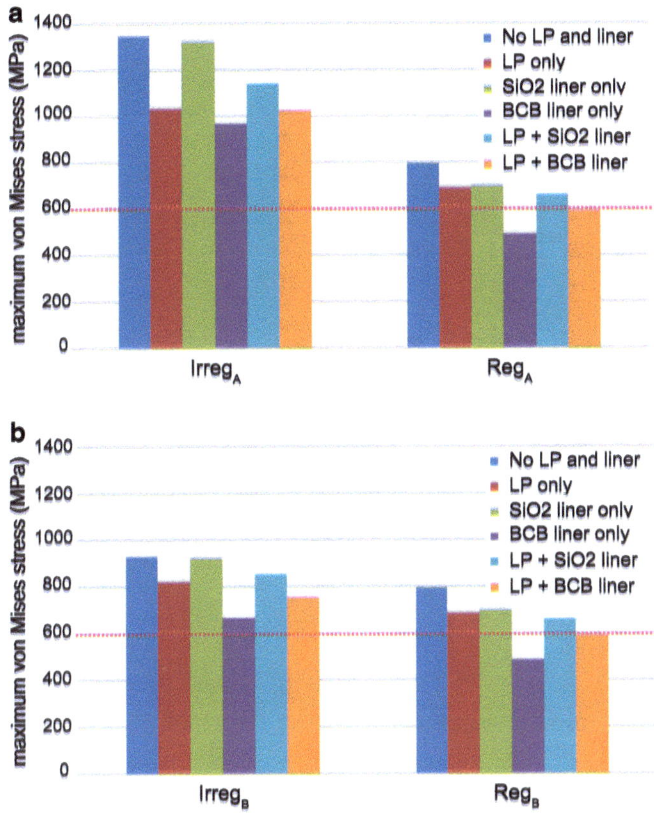

Fig. 13.13 Impact of TSV structure, TSV placement style, and KOZ size on the maximum von Mises stress. (**a**) Designs with TSV_A cell (KOZ $= 1.205\,\mu$m). (**b**) TSV_B cell (KOZ $= 2.44\,\mu$m)

13.4.2 *Impact of TSV Pitch*

The TSV pitch is the key factor that determines stress magnitude in the substrate region between TSVs. In this section, we explore the effect of TSV pitch on von Mises stress. We place TSVs regularly on $1 \times 1\,\text{mm}^2$ chip. We use 1,600, 2,500, 4,356, and 10,000 TSVs whose pitch are 25, 20, 15, and $10\,\mu$m, respectively. We obtain two data sets; one without landing pad, liner, and barrier; and another with $6 \times 6\,\mu\text{m}^2$ landing pad, 125 nm thick BCB liner, and 50 nm thick Ti barrier.

We first observe that von Mises stress magnitude decreases with increasing pitch and starts to saturate at around $15\,\mu$m pitch shown in Fig. 13.15. This is understandable since the stress magnitude induced by a single TSV becomes negligible at the similar pitch. Also, the layout using TSVs with a landing pad and a BCB liner shows a similar trend with lower von Mises stress magnitude than the case without these structures.

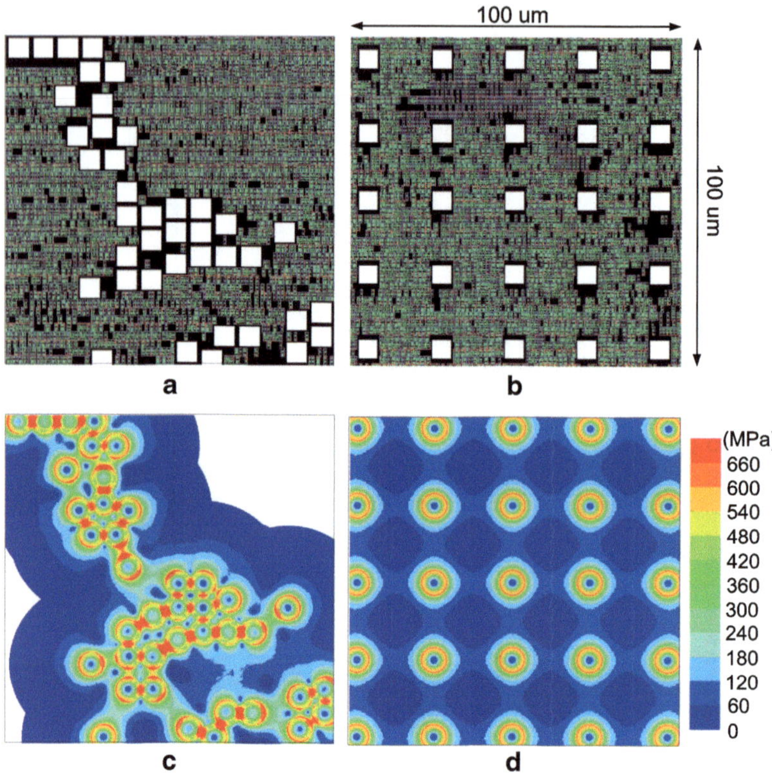

Fig. 13.14 Close-up shots of layouts and von Mises stress maps. (**a**) $Irreg_A$. (**b**) Reg_A. (**c**) Von Mises stress map of $Irreg_A$. (**d**) Von Mises stress map of Reg_A

Table 13.8 Number of TSVs violating von Mises criterion

		# violating TSVs	
TSV placement	TSV structure	TSV 3X cell	TSV 4X cell
Irregular	No LP and liner	1,472	1,472
	LP only	1,472	1,472
	SiO_2 liner only	1,472	1,472
	BCB liner only	1,063	247
	LP + SiO_2 liner	1,467	1,472
	LP + BCB liner	1,407	1,460
Regular	No LP and liner	1,472	1,472
	LP only	1,472	1,472
	SiO_2 liner only	1,472	1,472
	BCB liner only	0	0
	LP + SiO_2 liner	1,472	1,472
	LP + BCB liner	0	0

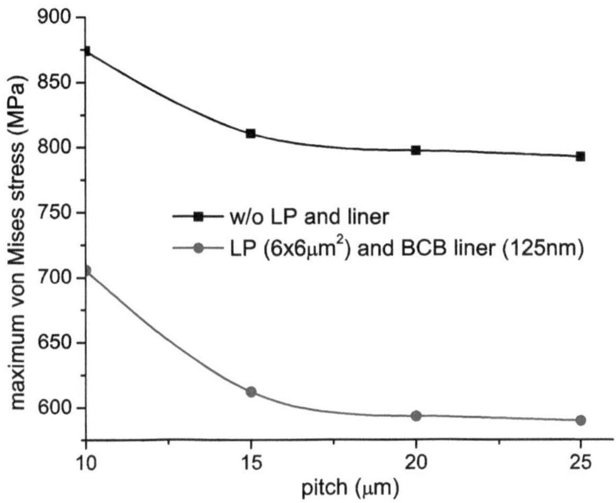

Fig. 13.15 Impact of TSV pitch on maximum von Mises stress

13.4.3 Impact of Relative TSV Orientation

We study the impact of relative orientations of TSVs on mechanical reliability. Figure 13.16 illustrates our simulation structures. We keep TSV1 and TSV2 at fixed locations with 20 μm pitch and rotate TSV3 around TSV1 with a radius of 10 μm, and we monitor maximum von Mises stress in TSV1 shown in Fig. 13.17.

We observe that the maximum von Mises stress in TSV1 occurs when the angle between TSV3, TSV1 and TSV2 is 0° shown in Fig. 13.16a. This is because of the constructive interference of stress components from all three TSVs. Also, since both TSV2 and TSV3 are in the right side of TSV1, net effect of additive stress is maximum at the right side of TSV1 edge. This is why von Mises stress is higher in the case of 0° than that of 180°. The maximum von Mises stress at TSV1 is observed to be minimum when the angle between TSV3, TSV1 and TSV2 is 90° shown in Fig. 13.16c. This is because tensile and compressive stresses of TSV2 and TSV3 at TSV1 are normal to each other and hence destructive interference occurs.

13.4.4 Impact of TSV Size

To investigate the effect of the TSV size, we use three different sizes of TSV with a same aspect ratio of 6; TSV small ($H/D = 15/2.5$ μm and KOZ 1.22 μm), TSV medium ($H/D = 30/5$ μm and KOZ 1.202 μm), and TSV large ($H/D = 60/10$ μm and KOZ 1.175 μm), where H/D is TSV height/diameter. Note that these TSV cells are occupying two, three, and five standard cell rows, respectively, which are selected to minimize the KOZ size difference between them. By setting similar KOZ size, we can focus on the impact of TSV size solely. Additionally, we set the landing pad width is 1 μm larger than the corresponding TSV diameter, and use 125 nm thick SiO_2 liner and 50 nm thick Ti barrier for all cases for fair comparisons.

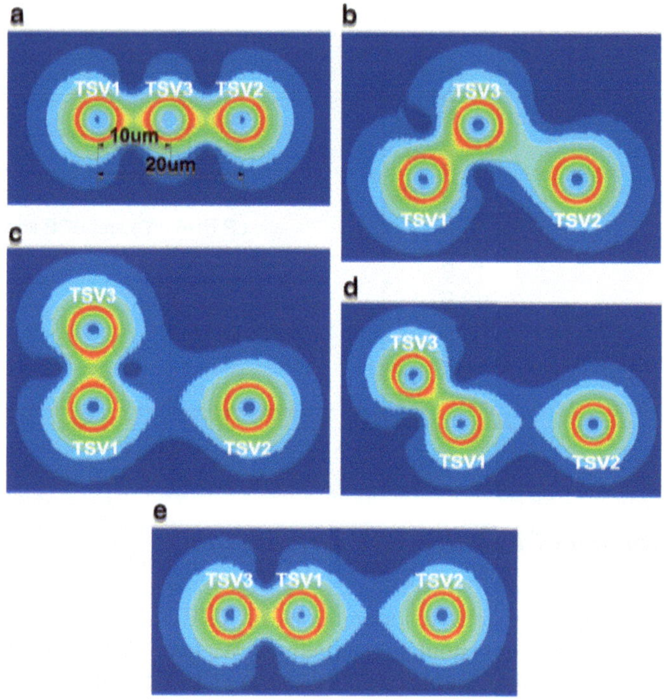

Fig. 13.16 TSV orientation impact on von Mises stress. TSV1 and TSV2 are located in fixed position with 20 μm pitch. TSV3 is rotating around TSV1 with 10 μm pitch. Angle is between TSV1 and TSV3. (**a**) $\theta = 0°$. (**b**) $\theta = 45°$. (**c**) $\theta = 90°$. (**d**) $\theta = 135°$. (**e**) $\theta = 180°$

Fig. 13.17 Impact of TSV orientation on maximum Von Mises stress at TSV1 by rotating TSV3 around TSV1

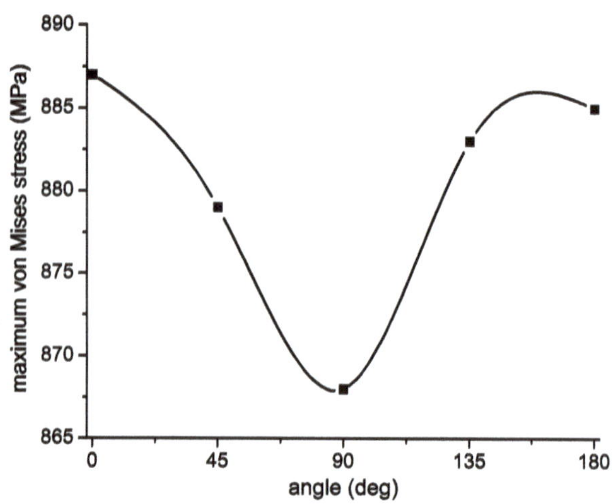

Table 13.9 Impact of TSV size on the maximum von Mises stress. The numbers in parentheses are % reduction compared to TSV large case

	Max von Mises stress (MPa)		
TSV placement	TSV large	TSV medium	TSV small
Irregular	1224.6	1126.4 (8%↓)	902.7 (26%↓)
Regular	749.3	654.6 (13%↓)	449.3 (40%↓)

Table 13.10 Impact of a landing pad size on von Mises criterion. The numbers in parentheses are % reduction compared to LP $6 \times 6\,\mu m^2$ case

	LP $6 \times 6\,\mu m^2$		LP $8 \times 8\,\mu m^2$	
Circuit	Max von Mises stress (MPa)	# violating TSVs	Max von Mises stress (MPa)	# violating TSVs
$Irreg_B$	839.6	1,472	815.3 (3%↓)	1,472 (0%↓)
Reg_B	654.2	1,472	606.7 (7%↓)	1,472 (0%↓)

Table 13.9 shows the maximum von Mises stress. For both irregular and regular TSV placement schemes benefit from smaller TSV diameter significantly. This is mainly because the magnitude of normal stress components decay proportional to $(D/2r)^2$, where r is the distance from the TSV center.

13.4.5 Impact of Landing Pad Size

We now explore the impact of landing pad size, which is normally determined by considering TSV alignment, on reliability issues. Designs with TSV_B cells with a 125 nm thick SiO_2 liner and a 50 nm thick Ti barrier are used. We compare the maximum von Mises stress and the number of violating TSVs with two different landing pad size shown in Table 13.10.

These results show that the regular TSV placement benefits more von Mises stress reduction from the larger landing pad size. This is because the stress reduction in a single TSV directly translates to the overall stress magnitude decrease in a full-chip scale for the regular TSV placement case. However, increasing the landing pad size does not improve von Mises stress significantly. Also we see that all TSVs do not satisfy von Mises criterion for every test case. This is because only the magnitude of σ_{rr} stress component at the wafer surface is reduced due to Cu landing pad, while other stress components hardly changes with a larger landing pad size.

13.4.6 Impact of Liner Thickness

In this section, we examine the impact of liner thickness on von Mises stress. We use designs with both TSV_A cells and TSV_B cells, and set the landing pad size $6 \times 6\,\mu m^2$ and $8 \times 8\,\mu m^2$, respectively. We also use a 50 mn thick Ti barrier for all cases. Figure 13.18 shows the maximum von Mises stress results with liner thickness of 125, 250, and 500 nm.

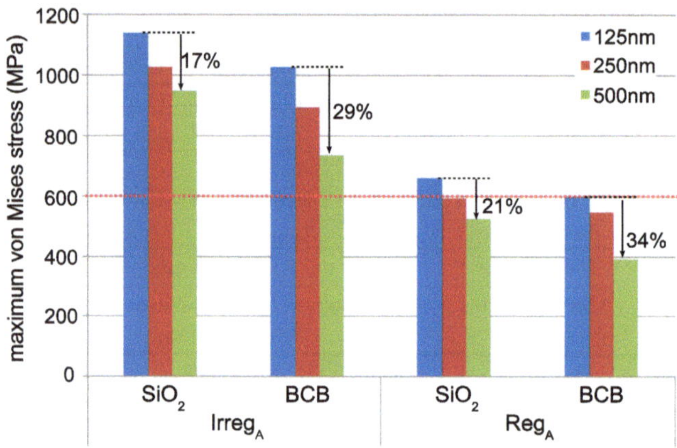

Fig. 13.18 Impact of liner thickness on the maximum von Mises stress of circuits with TSV_A cell

Table 13.11 Impact of liner thickness on the number of TSVs violating von Mises criterion. The numbers in parentheses are % reduction compared to the 125 nm thick liner case

Circuit	Liner material	# violating TSVs 125 nm	250 nm	500 nm
$Irreg_A$	SiO_2	1,462	1,426 (2%↓)	1,281 (12%↓)
	BCB	1,389	1,147 (17%↓)	329 (76%↓)
Reg_A	SiO_2	1,472	0 (100%↓)	0 (100%↓)
	BCB	0	0 (−)	0 (−)
$Irreg_B$	SiO_2	1,472	1,236 (16%↓)	64 (96%↓)
	BCB	974	502 (48%↓)	0 (100%↓)
Reg_B	SiO_2	1,472	0 (100%↓)	0 (100%↓)
	BCB	0	0 (−)	0 (−)

We observe that the liner thickness has a huge impact on the von Mises stress magnitude, since the thicker liner effectively absorbs thermo-mechanical stress at the TSV/liner interface. Especially, the BCB liner shows significant reduction in the maximum von Mises stress compared with the SiO_2 liner due to extremely low Young's modulus shown in Table 13.1. For example, the 500 nm thick BCB liner reduces the maximum von Mises stress by 29% for the $Irreg_A$ and satisfies the von Mises yield criterion for all circuits with a regular TSV placement.

Table 13.11 shows the number of TSVs violating von Mises criterion. Even though there are still many TSVs not satisfying von Mises criterion for the $Irreg_A$ circuit, it is possible to reduce von Mises stress if we place TSVs carefully considering this reliability metric during a placement stage.

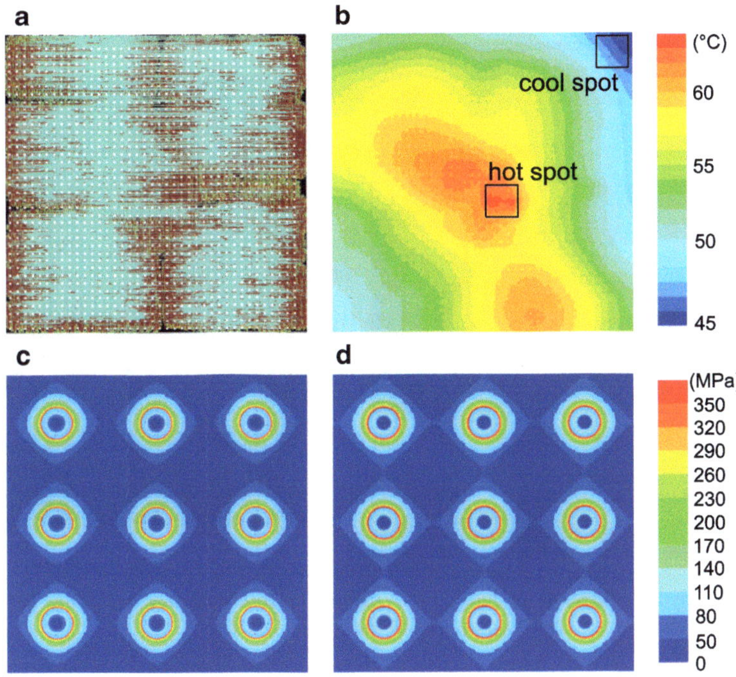

Fig. 13.19 Impact of temperature (°C) during a chip operation on von Mises stress. (**a**) layout of Reg_A. TSV landing pads are *white rectangles*. (**b**) thermal map. (**c**) von Mises stress map in the *hot spot*. (**d**) von Mises stress map in the *cool spot*

13.4.7 Impact of Chip Operation Temperature

Up to this point, we only consider the residual stress caused by the manufacturing process. Now, we examine the reliability problem during the chip operation phase. Our full-chip thermal simulation flow is as follows. We first generate a power map using SoC Encounter, and then feed this to ANSYS Fluent with in-house add-ons which enable a steady-state thermal analysis for GDSII level 3D ICs shown in Fig. 13.19b.

Depending on the temperature distribution across the die area, each TSV experiences a different thermal load. Thus, the significance of mechanical reliability problem of an individual TSV might be different from each other. To support temperature dependent stress analysis, we build stress library with wide range of thermal load.

We use the Reg_A circuit for this experiment since TSVs in the regular TSV placement scheme shows uniform von Mises stress distribution, which enables us to observe the impact of an operating temperature across the die easily. We use a 500 nm thick BCB liner for this experiment.

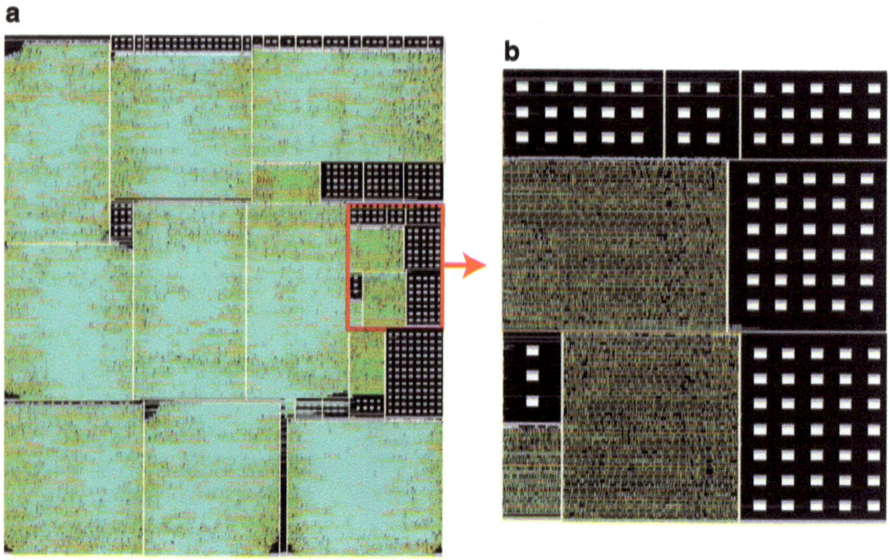

Fig. 13.20 Layout of block-level design (TSV pitch = 15 μm). *White rectangles* are TSV landing pads. (**a**) Full-chip layout. (**b**) Close-up shot of the *red box* in (**a**)

We observe that the cool spot experiences higher von Mises stress, since the temperature difference from the stress free temperature, 275°C in our case, is larger in the cool spot than in the hot spot. However, since the maximum temperature difference across the die is only 20°C, the impact of an operating temperature on the TSV reliability across the die is not significant. In our test case, the difference of the maximum von Mises stress between two spots is 30.4 MPa. Also, both the hot and the cool spot experience less maximum von Mises stress compared to the residual stress case, again due to the reduced thermal load.

However, the reduction of von Mises stress during a chip operation cannot recover the material yielding failure if it already exists, since this is a non-reversible failure mechanism.

13.4.8 Reliability of Block-Level 3D Design

Even though the gate-level 3D design has the potential of highest optimization, the block-level design is attractive in the sense that we can reuse already highly optimized 2D IP blocks. In this section, we study the reliability issues in block-level 3D designs. 3D block-level designs are generated using an in-house 3D floorplanner which treats a group of TSVs as a block shown in Fig. 13.20. We use a 500 nm thick BCB liner for this experiment.

We vary the TSV pitch inside TSV blocks to examine its impact on layout quality as well as reliability issues. Table 13.12 shows that block-level designs use

Table 13.12 Comparison between gate-level and block-level design

Design level	TSV pitch (μm)	# TSV	Wirelength (mm)	Area (μm × μm)	Max stress (MPa)
Gate	Irregular	1,472	9,060	960 × 960	729.3
	23.5	1,472	9,547	960 × 960	386.4
	15	368	8,259	950 × 1,130	414.5
Block	10	394	8,028	1,080 × 1,000	519.9
	7.5	333	7,933	980 × 1,090	716.2

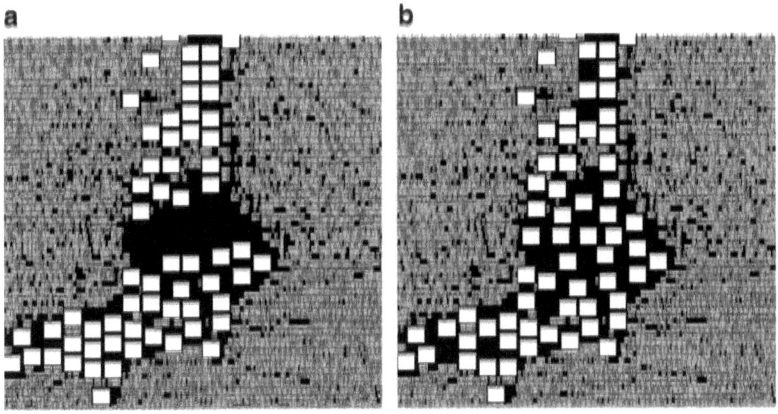

Fig. 13.21 TSV re-placement to reduce von Mises stress. TSV landing pads are *white rectangles*. (**a**) Original layout. (**b**) After TSV re-placement

less number of TSVs, show shorter wirelength, and occupy more area than gate-level designs. Experimental results show that we can control the von Mises stress with area overhead in block-level design, since the TSV pitch in block-level design is controllable. Another benefit of block-level design is that we can localize the thermo-mechanical reliability problems only nearby TSV blocks.

13.4.9 Impact of TSV Re-placement

In this section, we manually optimize TSV locations to show the potential benefit of TSV reliability aware layout optimization while minimizing the change in layout. We use the $Irreg_A$ circuit which shows worst von Mises stress, and employ a 500 nm thick BCB liner for this experiment. Our related study with this BCB liner on the maximum von Mises stress vs. TSV-to-TSV pitch shows that 10 μm pitch is a reasonable choice to reduce von Mises stress considering some safety margin. We reposition densely placed TSVs to nearby white spaces if available to reduce the von Mises stress shown in Fig. 13.21.

Table 13.13 Impact of TSV re-placement on von Mises stress distribution, wirelength, and longest path delay

	von Mises stress (MPa)				WL	LPD
	480–540	540–600	600–660	660–	(mm)	(ns)
Orig	0.100%	0.041%	0.011%	0.002%	9,060	3.607
Re-place	0.092%	0.036%	0.009%	0.0%	9,081	3.636

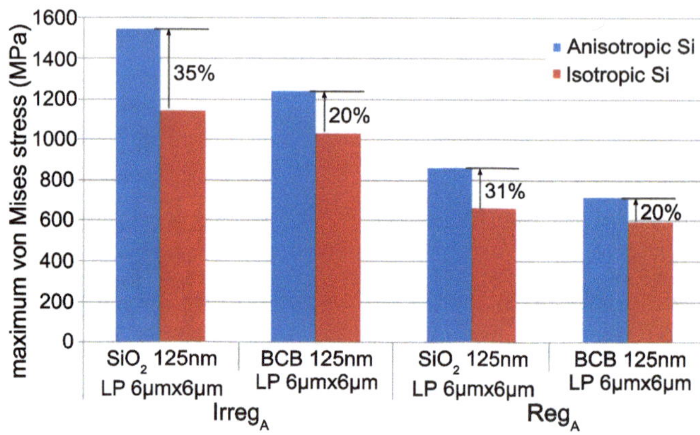

Fig. 13.22 Maximum von Mises stress comparison between isotropic and anisotropic Si

Table 13.13 shows the distribution of von Mises stress higher than 480 MPa across the die, wirelength, and longest path delay before and after the TSV re-placement. We perform 3D STA to analyze timing using Synopsys PrimeTime with TSV parasitic information included. We see the reduction in high von Mises stress region after TSV re-placement. With small perturbations of TSV locations, we could reduce the von Mises stress level and decrease the number of violating TSVs from 329 to 261, which is 21% improvement with only 0.23% wirelength and 0.81% longest path delay increase, respectively. This small test case shows the possibility of a layout optimization without degrading performance too much.

13.4.10 Comparison Between Isotropic and Anisotropic Si

In this section, we compare von Mises stress between isotropic and anisotropic Si material property in full-chip scale. We use a 125 nm thick liner, a 50 nm thick Ti Barrier, and a $6 \times 6\,\mu m^2$ landing pad. Figure 13.22 shows the increased mechanical reliability problems considering the anisotropic Si elastic property; maximum von Mises stress increases more than 30% and around 20% with SiO_2 liner and BCB liner, respectively.

Young's modulus for X and Y direction of anisotropic Si is 30% higher than that of isotropic Si, and this translates to the stress magnitude increase. The BCB liner, again, shows better stress absorption capability even with the anisotropic Si property. However, this clearly shows the importance of using proper material properties to assess the mechanical reliability problems in TSV-based 3D ICs during design stages.

13.5 Conclusions

In this chapter, we studied how TSV surrounding structures such as a landing pad and a dielectric liner affect stress fields and mechanical reliability in 3D ICs. We also studied an accurate and fast full-chip stress and mechanical reliability analysis flow based on the linear superposition principle of stress tensors. This approach can be applicable to mechanical reliability aware placement optimization for 3D ICs. Our results showed that TSV pitch and size, liner material and its thickness, and TSV placement are key design parameters to reduce the mechanical reliability problems in TSV based 3D ICs.

References

1. G.V. der Plas et al., Design issues and considerations for low-cost 3D TSV IC technology, in *IEEE International Solid-State Circuits Conference Digest of Technical Papers*, San Francisco, 2010
2. S. Franssila, *Introduction to Microfabrication* (Wiley, Chichester, 2004)
3. FreePDK45. http://www.eda.ncsu.edu/wiki/FreePDK.
4. M.A. Hopcroft, W.D. Nix, T.W. Kenny, What is the Young's modulus of silicon. J. Microelectromech. Syst. **19**, 229–238 (2010)
5. M. Jung, J. Mitra, D. Pan, S.K. Lim, TSV stress-aware full-chip mechanical reliability analysis and optimization for 3D IC, in *Proceedings of ACM Design Automation Conference*, San Diego, 2011
6. A.P. Karmarkar, X. Xu, V. Moroz, Performance and reliability analysis of 3D-integration structures employing through silicon via (TSV), in *IEEE International Reliability Physics Symposium*, Montreal, 2009
7. D.H. Kim, K. Athikulwongse, S.K. Lim, A study of through-silicon-via impact on the 3D stacked IC layout, in *Proceedings of IEEE International Conference on Computer-Aided Design*, San Jose, 2009
8. J. Lau, X. Zheng, C. Selvanayagam, Failure analyses of 3D Sip (System-in-Package) and WLP (Wafer-Level Package) by finite element methods, in *IEEE International Symposium on Physical and Failure Analysis of Integrated Circuits*, Suzhou, 2009
9. X. Liu, Q. Chen, P. Dixit, R. Chatterjee, R.R. Tummala, S.K. Sitaraman, Failure mechanisms and optimum design for electroplated copper through-silicon vias (TSV), in *IEEE Electronic Components and Technology Conference*, San Diego, 2009
10. K.H. Lu, X. Zhang, S.-K. Ryu, J. Im, R. Huang, P.S. Ho, Thermo-mechanical reliability of 3-D ICs containing through silicon vias, in *IEEE Electronic Components and Technology Conference*, San Diego, 2009
11. C. McDonough, B. Backes, W. Wang, R.E. Geer, Thermal and spatial profiling of TSV-induced stress in 3DICs, in *IEEE International Reliability Physics Symposium*, Monterey, 2011

12. A. Mercha et al., Comprehensive analysis of the impact of single and arrays of through silicon vias induced stress on high-k/metal gate CMOS performance, in *Proceedings IEEE International Electron Devices Meeting*, San Francisco, 2010
13. J.M.G. Ong, A.A.O. Tay, X. Zhang, V. Kripesh, Y.K. Lim, D. Yeo, K.C. Chen, J.B. Tan, L.C. Hsia, D.K. Sohn, Optimization of the thermomechanical reliability of a 65 nm Cu/low-k large-die flip chip package. IEEE Trans. Compon. Packag. Technol. **32**, 838–848 (2009)
14. S.-K. Ryu, K.-H. Lu, X. Zhang, J.-H. Im, P.S. Ho, R. Huang, Impact of near-surface thermal stresses on interfacial reliability of through-silicon-vias for 3-D interconnects. IEEE Trans. Device Mater. Reliab. **11**, 35–43 (2011)
15. J.H. Wu, J.A. del Alamo, Fabrication and characterization of through-substrate interconnects. IEEE Trans. Electron Devices **57**, 1261–1268 (2010)
16. Y. Xiang, X. Chen, J.J. Vlassak, The mechanical properties of electroplated Cu thin films measured by means of the bulge test technique. Proc. Mater. Res. Soc. Symp. **695**, 189–196 (2002)
17. H. Yan, Y.Y. Tay, M.H. Liang, Z. Chen, C.M. Ng, J.S. Pan, H. Xu, C. Liu, V.V. Silberschmidt, Amorphous metallic thin films as copper diffusion barrier for advanced interconnect applications, in *IEEE Electronics Packaging Technology Conference*, Singapore, 2009
18. J.-S. Yang, K. Athikulwongse, Y.-J. Lee, S.K. Lim, D.Z. Pan, TSV stress aware timing analysis with applications to 3D-IC layout optimization, in *Proceedings of ACM Design Automation Conference*, Anaheim, 2010
19. J. Zhang, M.O. Bloomfield, J.-Q. Lu, R.J. Gutmann, T.S. Cale, Modeling thermal stresses in 3-D IC interwafer interconnects. IEEE Trans. Semicond. Manuf. **19**, 437 (2006)

Chapter 14
Impact of Mechanical Stress on Timing Variation for 3D IC

Abstract In this chapter, we study the impact of through-silicon-via (TSV) and shallow trench isolation (STI) stress on the timing variations of 3D IC. We also study systematic TSV-STI-stress-aware timing analysis and show how to optimize layouts for better performance. First, we generate a stress contour map with an analytical radial stress model for TSV. We also develop a stress model for STI from finite element analysis (FEA) results. Then, depending on geometric relation between TSVs, STIs, and transistors, the tensile and compressive stresses are converted to hole and electron mobility variations. Mobility-variation-aware cell library and netlist are generated and incorporated into an industrial engine for timing analysis of 3D IC. We observe that TSV stress and STI stress interact with each other, and rise and fall time react differently to stress and relative locations with respect to both TSVs and STIs. Overall, TSV-STI-stress-induced timing variations can be as much as ±15 % at the cell level. Thus, as an application to layout optimization, we exploit the stress-induced mobility enhancement to improve performance of 3D ICs. We show that stress-aware layout perturbation could reduce cell delay by up to 23.37 % and critical path delay by 6.67 % in our test case.

The materials presented in this chapter are based on [21].

14.1 Introduction

The 3D IC stacking has gained tremendous interest for IC integration in order to reduce wire length and reduce footprint. In addition, several dies manufactured by different process technologies can also be integrated as one chip with 3D integration. Through-silicon-vias (TSVs) are inserted for wafer-to-wafer connection in 3D ICs. Tungsten, poly-silicon, and copper(Cu) have all been considered as fill materials of TSVs. Since copper has low resistivity, it is widely used material for TSV fill.

S.K. Lim, *Design for High Performance, Low Power, and Reliable 3D Integrated Circuits*, DOI 10.1007/978-1-4419-9542-1_14,

However, copper coefficient of thermal expansion (CTE) differs from silicon CTE which can be a source of silicon strain. CTE of copper is $17 \times 10^{-6}\,\mathrm{K}^{-1}$ at 20 °C, while CTE of silicon is $3 \times 10^{-6}\,\mathrm{K}^{-1}$ at 20 °C [2]. The CTE mismatch between copper and silicon causes inevitable stress on silicon for both via-first and via-last approaches. Because the copper electroplating and annealing temperature (250–350 °C) is higher than the operating temperature, tensile stress appears on silicon [9, 13] near TSV at regular operating temperature.

The tensile stress on silicon causes reliability problem such as cracking. In addition, the stress can change mobility of carriers. Therefore, TSV stress induced by CTE mismatch may cause timing violation if the cells on timing critical paths are negatively impacted by the TSV stress. Tensile stress enhances electron mobility. However, hole mobility is either enhanced or degraded depending on stress and transistor channel direction. Longitudinal (with respect to carrier flow) tensile stress reduces hole mobility, while transverse tensile stress increases the mobility [17]. When TSV-induced tensile stress is 100 MPa and the stress works for longitudinal direction, hole mobility degradation can be up to 7.2 %, which makes PMOS transition slow. If the PMOS is on a timing critical path, it can cause unexpected setup and hold time violation, which is not detected with the current timing analysis flow.[1]

Another major stress source in ICs is shallow trench isolation (STI). The CTE of silicon dioxide, the widely used material for STI fill, is $0.5 \times 10^{-6}\,\mathrm{K}^{-1}$ at 20°C. Because it is lower than the CTE of both silicon and copper, STI causes compressive stress on active region it surrounds. Oxidation and oxide densification for STI also take place at much higher temperature than TSV annealing temperature. Therefore, the compressive stress caused by STIs is also not negligible. Longitudinal compressive stress enhances hole mobility, but degrades electron mobility. If an NMOS on a critical path experiences compressive stress, it can cause unexpected setup and hold time violation as well. The following specific topics are covered in this chapter:

- We study compact stress and mobility modeling to consider systematic timing variation caused by TSV stress for 3D ICs. We also develop an STI stress model based on simulation results from FEA simulation. A model for STI-stress-induced mobility variation is also developed. The models allow us to study the impact of interaction between TSV and STI stress on circuit performance.
- We study a stress-aware static timing analysis (SA-STA) flow. The first step for SA-STA is to generate stress map according to TSV and STI positions. Stress calculation is based on analytical model for TSV, a model developed from FEA simulation for STI, and linear superposition. Stress map is used to estimate hole mobility variation and electron mobility variation. We incorporate our stress aware design and library to a commercial STA flow.

[1]The devices in 3D ICs may experience additional timing changes from process variations. The combined effect of stress and process variations on circuit timing is out of the scope of this chapter.

- We show that TSV and STI stresses can change hole mobility quite significantly, e.g., from −20 to +26 %, which was observed to cause more than 15 % variation for single cell delay. Thus, it can deteriorate the overall chip performance, and must be considered during timing analysis and optimization.
- We show that TSV/STI-induced stresses play an important role to optimize performance by adjusting cell locations in order to take advantage of enhanced mobility property due to the combined stresses. Logic cells on critical paths must be placed such that the carrier mobility inside their PMOS/NMOS is not degraded (if not enhanced) by either TSV- or STI-induced stress. Since hole-mobility contour is different from electron mobility contour, PMOS and NMOS should be optimized separately. If a PMOS in a cell is on a critical path, the cell becomes a critical cell for hole mobility optimization. An NMOS critical cell can be optimally placed using the similar procedure.

14.2 Existing Works

Even though several works have been published regarding the impact of TSV stress [21] or STI stress [6] on performance of IC layouts, their impact is studied separately.[2] This is the first work addressing the impact of combined stresses on performance, to our best knowledge. Because TSV/STI stresses are layout dependent, we study a design flow to analyze timing variation by both, and show its implications for layout optimizations during 3D IC design.

A transistor-level STI-stress-aware delay analysis was also proposed in [6]. The method is not suitable for combined stress-aware timing analysis and optimization for two reasons. First, in the work, only mobility variation model is provided after converting stress obtained from TCAD simulation. Because TSV stress and STI stress interact with each other, stress from both structures should be combined before converted to mobility variation. Second, the method is based on SPICE simulation. Because of the amount of time required for simulation, the proposed method is suitable for only a small number of critical paths in a layout, thus limiting its application to the late stages of design flow. During early stages of design flow, e.g. global and detail placement, delay analysis may be required iteratively, and quick gate-level STI-stress-aware timing analysis is desirable.

---

[2] Consideration of other stress enhancement methods such as SiGe is out of the scope of this chapter.

14.3 Preliminaries

14.3.1 TSV/STI-Induced Mechanical Stress

Equation (14.1) shows the relation between stress and strain. E is Young's Modulus, E for silicon is 169 GPa [3]. σ is the applied stress and ε is the deformation rate. For example, 169 MPa stress in silicon results in 0.1 % strain in silicon.

$$\sigma = E \times \varepsilon. \tag{14.1}$$

During 3D IC manufacturing, stress is caused by CTE mismatch between copper TSV and silicon as shown in Fig. 14.1. Investigations [14] show that, at 200 °C, an anneal time of 30–60 min is required in order to achieve reasonable copper layer properties. Since CTE of copper is larger than silicon, at room temperature, copper has less volume compared with that during annealing process because of contraction. Several works have been published to simulate the TSV induced stress [2, 9] using FEA (finite element analysis) simulation. They show that TSV can cause tensile stress of more than 200 MPa.

Strained silicon has been used to enhance I_{on} of a transistor [18]. Unlike TSV stress, its impact on performance, however, is not layout dependent. Several unwanted stress sources are largely layout dependent, and should be considered during the design step. Shallow trench isolation (STI) is one of the unintentional stress sources [6, 12] because silicon dioxide used for STI fill pushes out silicon atoms near STI as shown in Fig. 14.2. Silicon dioxide in STI is generally grown and densified at temperature as high as 1,000 °C [11]. Since CTE of silicon dioxide is smaller than silicon, at room temperature, silicon dioxide contracts slower than silicon when cooling down to room temperature. FEA simulation shows that STI can also cause compressive stress of more than 200 MPa.

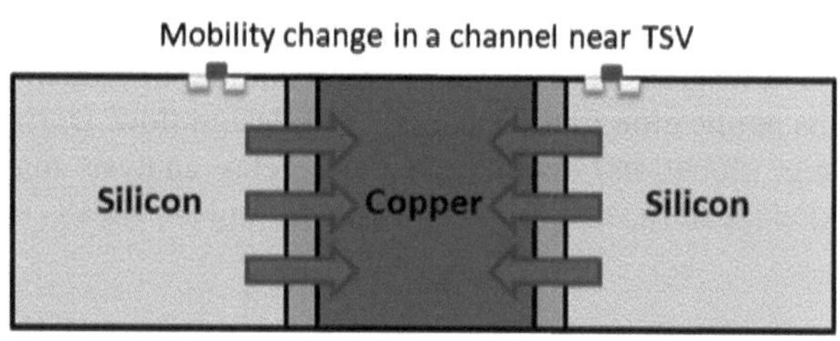

Fig. 14.1 Thermal stress around TSV

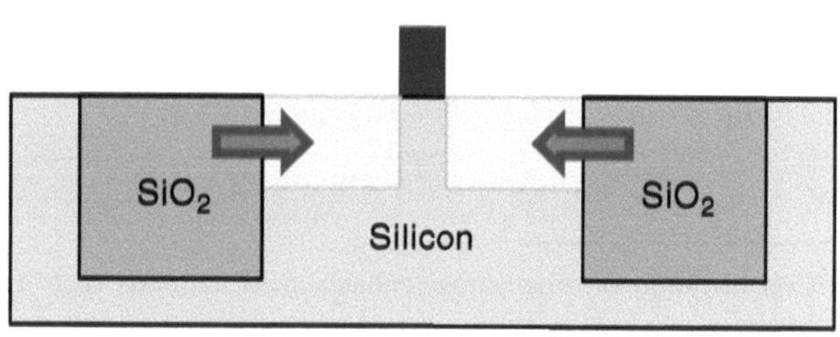

Fig. 14.2 Thermal stress in active region caused by surrounding STIs

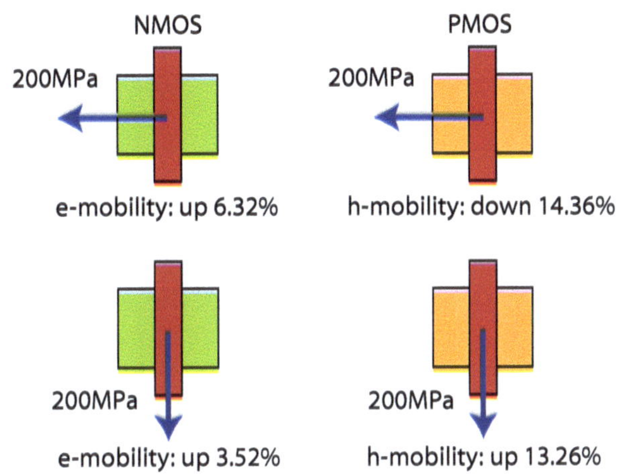

Fig. 14.3 Mobility change due to tensile stress. *Top*: $\Delta\mu/\mu$ for longitudinal tensile stress, *bottom*: $\Delta\mu/\mu$ for transverse tensile stress

14.3.2 Stress Impact on Mobility Variations

Mobility (μ) change as a function of applied stress(σ) has been proposed by the following formula [15]:

$$\frac{\Delta\mu}{\mu} = -\Pi \times \sigma, \tag{14.2}$$

where Π is the tensor of piezo-resistive coefficients for holes and electrons, and σ is the applied stress in silicon. Tensile stress has a positive sign and compressive stress has a negative sign.

Since tensile stress increases mean free path for electron, it enhances NMOS performance. However, longitudinal tensile stress degrades PMOS performance as shown in the top of Fig. 14.3 [4]. With longitudinal stress, piezo-resistive coefficient for electrons is $-3.16 \times 10^{-10}\,\text{Pa}^{-1}$, and the coefficient for holes is $7.18 \times 10^{-10}\,\text{Pa}^{-1}$ for (001) wafer surface and $\langle 110 \rangle$ channel which are the most popular scheme for semiconductor manufacturing [15, 16]. For example, when TSV stress is 200 MPa, $(\Delta\mu/\mu)_e$ is +6.32% for NMOS, and $(\Delta\mu/\mu)_h$ is −14.36 % for PMOS.

However, if TSV is placed perpendicular to a transistor channel, mobility for both holes and electrons is enhanced by adding more space in silicon lattice for carriers to move fast. For transverse stress, piezo-resistive coefficient for electrons is $-1.76 \times 10^{-10}\,\text{Pa}^{-1}$, and the coefficient for holes is -6.63×10^{-10} for (001) surface and $\langle 110 \rangle$ channel. Similarly, we can expect $(\Delta\mu/\mu)_e = +3.52\,\%$, $(\Delta\mu/\mu)_h = +13.26\,\%$ with $\sigma = 200\,\text{MPa}$. Empirically, it is known that $(\Delta I_{on}/I_{on})_{pmos}$ is $0.5 \sim 0.9$ times of $(\Delta\mu/\mu)_h$, and $(\Delta I_{on}/I_{on})_{nmos}$ is $0.4 \sim 0.6$ times of $(\Delta\mu/\mu)_e$ [10, 20] because I_{on} of a transistor is determined by the sum of source, drain, and channel resistance.

Transistor variation due to the stress can change cell timing characteristics. In Fig. 14.4a, buffer rising delay increases because of longitudinal tensile stress.

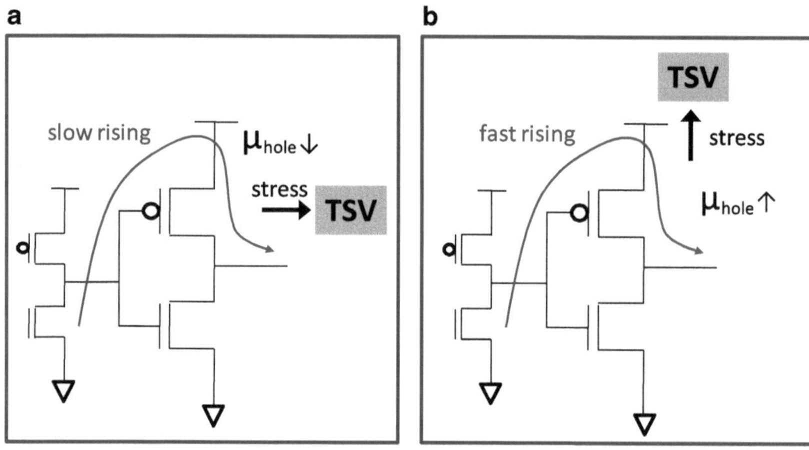

Fig. 14.4 Buffer cell delay change due to TSV induced stress. (**a**) Slower rising delay with longitudinal tensile stress and (**b**) faster rising delay with transverse tensile stress

Even though buffer size is the same in Fig. 14.4b, rising delay decreases in Fig. 14.4b due to transverse tensile stress. Therefore, TSV stress aware timing analysis and layout optimization are essential steps for 3D IC design.

Although stress in the direction perpendicular to a transistor channel can affect performance of the transistor, the major stress variation caused by STI is in horizontal direction. Because of standard cell structure, as suggested by [6, 12], STI stress in horizontal direction is the main STI stress variation that affects mobility. The trend for STI is also different from the trend for TSV because the stress caused by STI is compressive instead of tensile. Longitudinal compressive stress enhances PMOS performance, but degrades NMOS performance.

14.4 Design Methodologies

The overall flow of our 3D IC design methodology is shown in Fig. 14.5. Our stress driven design flow consists of three steps. The first step is to calculate TSV and STI stress and mobility change. Since FEA simulation which provides an accurate solution takes several hours to simulate stress for one TSV, we use the analytical model proposed in [9]. Mobility change can be calculated by extension of the formula (14.2). We will explain the process and device modeling for a single TSV, and extend them to consider multiple TSVs in Sect. 14.5. For STI stress, a model is developed from results obtained from FEA simulations of STI stress. Mobility change caused by STI stress can be calculated in the same way as mobility change caused by TSV stress. We explain the process and device modeling for a single STI, and extend them to consider STIs on both sides of each cell in Sect. 14.6.

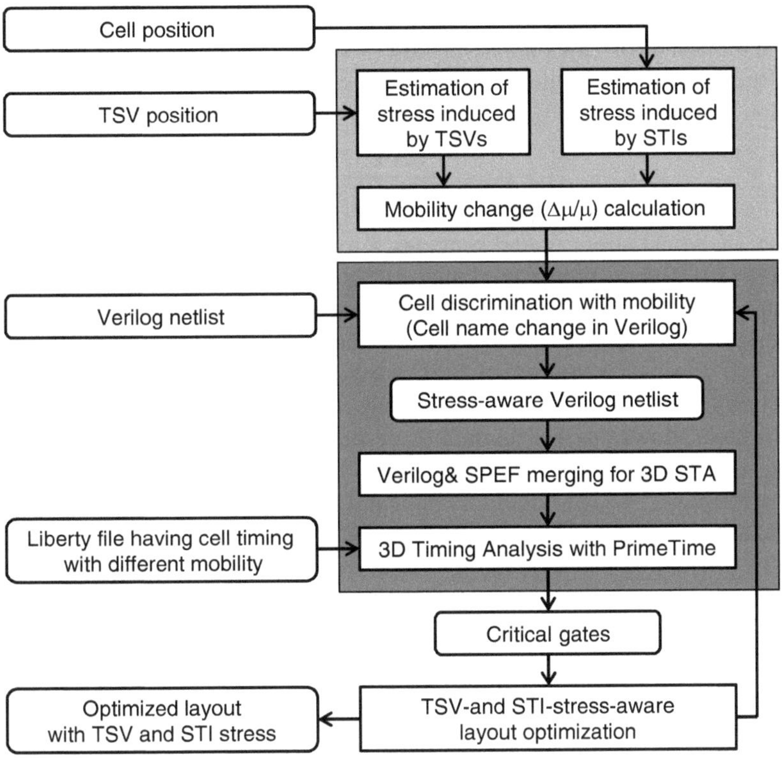

Fig. 14.5 Overall flow for TSV/STI stress modeling and analysis flow

The second step is 3D stress aware static timing analysis (3D SA-STA). We use PrimeTime as an STA (static timing analysis) engine. In Sect. 14.8, we explain how to deal with Verilog netlist and timing library to consider mobility variation. The timing result can be used for layout optimization. Intuitively, if a PMOS in a cell is on a critical path, the cell should be moved to the region of a TSV that has positive $(\Delta\mu/\mu)_h$, or moved in such a way that surrounding STIs cause positive $(\Delta\mu/\mu)_h$. Finally, we can run timing analysis iteratively to verify the layout optimization effect.

14.5 Mobility Variation Under TSV-Induced Stress

14.5.1 Mobility Variation Under Single TSV

In this chapter, we assume that the shape of TSV is a cylindrical type which is widely used for better manufacturability. FEA based TSV simulation has been proposed [2, 9]. The simulation approaches provide an accurate solution with long runtime which is not acceptable for our design flow that should calculate stress for several thousands of TSVs iteratively after each optimization. Assuming 2-D radial plain stress, we use the following analytical solution which is known as *Lamé* stress solution in [9]:

$$\sigma_{rr} = -\frac{B\Delta\alpha\Delta T}{2}\left(\frac{R}{r}\right)^2 \tag{14.3}$$

The analytical stress model provides a relatively accurate solution [9]. In the formula (14.3), B is biaxial modulus, $\Delta\alpha$ is CTE difference between copper and silicon, ΔT is the temperature difference between copper annealing and operating temperature. R is TSV radius, and r is a distance from TSV center. We assume that ΔT is 250 °C which is the case of 25 °C for the room temperature and 275 °C for the copper annealing temperature which is relatively low annealing temperature [14]. The formula shows that the thermal stress near TSV depends on the ratio of TSV radius and a distance from a TSV center.

The formula (14.2) provides an efficient way to calculate mobility variation due to σ_{rr}. As we observed in Sect. 14.3, mobility change depends on not only σ_{rr} but also orientation between applied force and a transistor channel. The empirical value for showing the relation of mobility change and a channel direction has been proposed in [4]. We extend the formula (14.2) to consider stress and channel direction in the formula (14.4) as follows:

$$\frac{\Delta\mu}{\mu}(\theta) = -\Pi \times \sigma_{rr} \times \alpha(\theta)$$

$$\theta = \tan^{-1}\left|\frac{Y_{\mathrm{TSV}} - Y_{\mathrm{poly}}}{X_{\mathrm{TSV}} - X_{\mathrm{poly}}}\right| \tag{14.4}$$

where $\alpha(\theta)$ is an orientation factor as a function of θ which is defined as the degree between the center of TSV and the center of a transistor channel when a transistor is placed vertically as shown in Fig. 14.6a, b. Π is the piezo-resistive coefficient at $\theta = 0$ which works as longitudinal stress.

In Fig. 14.6a, if NMOS is in right side of TSV, θ becomes zero, and $\alpha(0)$ becomes one, which enhances NMOS mobility at its maximum. However, if NMOS is in upper side of TSV, $\alpha(\pi/2)$ is 0.5, which means that NMOS mobility increase is half of the enhancement at $\theta = 0$. PMOS shows opposite trends, which has the best mobility enhancement at $\theta = \pi/2$. If θ is zero, then, PMOS becomes slower than the case of no stress. Figure 14.6c, d show the transistor direction for the best performance. Even though the mixed channel direction is not allowed due to the patterning difficulty, the observation provides a way to optimize layout for 3D ICs.

The top of Fig. 14.7 shows a contour map for hole mobility variation. From the contour, we can see that hole mobility decreases in a horizontal direction, while it increases in a vertical region. Forty-five degrees direction has no hole mobility change. Contour map for electron mobility variation is presented in the bottom of Fig. 14.7. As we see in Fig. 14.6a, horizontal direction has more mobility enhancement zone.

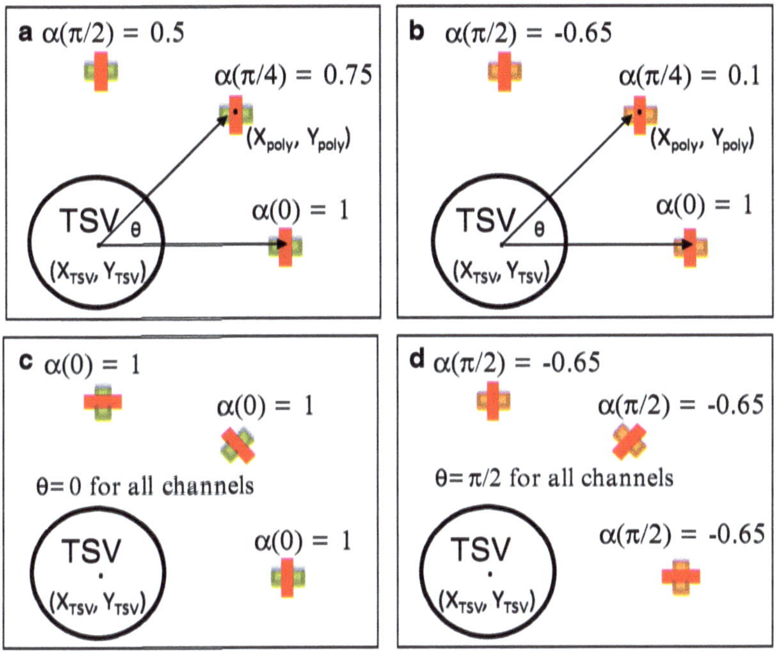

Fig. 14.6 Optimal orientation of MOSFET to maximize the mobility for (001) surface, ⟨110⟩ channel. (**a**) NMOS mobility variation, (**b**) PMOS mobility variation, (**c**) optimal placement for the best NMOS performance, and (**d**) optimal placement for the best PMOS performance

14.5.2 Mobility Variation Under Multiple TSVs

Since we use many TSVs for signaling, power/ground and clock network, we need to consider stress effect for multiple TSVs. Each TSV works as stress source to silicon. When a position in a wafer is strained by multiple stress sources, linear superposition can provide the multiple stress solution [5, 9].

An example of result from simulations of a cell with two TSVs nearby is shown in Fig. 14.8. The contour of stress caused by each TSV is shown in Fig. 14.8a, b, respectively. When we simulate the stress caused by both TSVs, we find that the stress from both TSVs linearly combines.

The horizontal stress caused by both TSVs on a horizontal line across the center of the cell is shown in Fig. 14.9. An additional plot, linear superposition, is created by adding the stress individually caused by each TSV together. We observe that the result from this linear superposition is close to the stress caused by both TSVs. Therefore, it is possible to use linear superposition to estimate stress caused by TSVs nearby a cell.

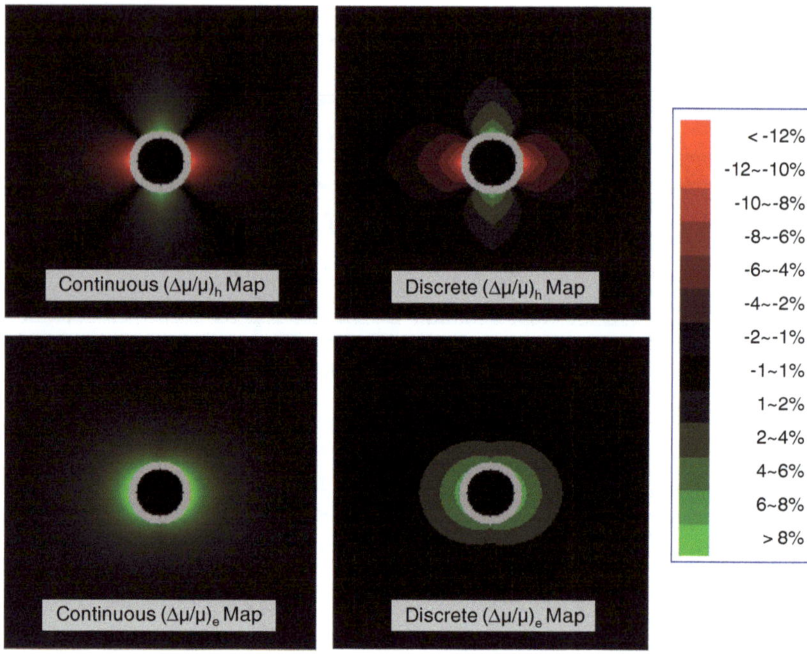

Fig. 14.7 Mobility contour map for a TSV. *Top*: contour map for hole mobility variation, *bottom*: contour map for electron mobility variation

We model the mobility variation for multiple TSVs as follows:

$$\frac{\Delta\mu}{\mu}_{\text{TSV}} = \sum \frac{\Delta\mu}{\mu}(\theta) = -\Pi \sum_{i \in \text{TSVs}} (\sigma_i \times \alpha(\theta_i)) \tag{14.5}$$

where σ_i is the tensile stress caused by ith TSV, $\alpha(\theta_i)$ is the orientation factor of ith TSV. θ_i is the degree between the center of ith TSV and a point that we want to get mobility variation.

Figure 14.10 shows stress and mobility variation contour with linear superposition for four-TSV array. We can see more stress in a region between TSVs. $(\Delta\mu/\mu)_e$ contour has similar trend with stress contour. However, $(\Delta\mu/\mu)_h$ has less variation between TSVs.

In the top of Fig. 14.11, we compare the $(\Delta\mu/\mu)_h$ for two different TSV placement schemes having the same TSV density. Since zigzag TSV placement has compensation effect for positive and negative hole mobility between adjacent rows, the top left of Fig. 14.11 has more stress free zone than the top right of Fig. 14.11 even if the mobility degradation effect within a row remains the same. The bottom of Fig. 14.11 shows electron mobility contour for zigzag and regular TSV placement,

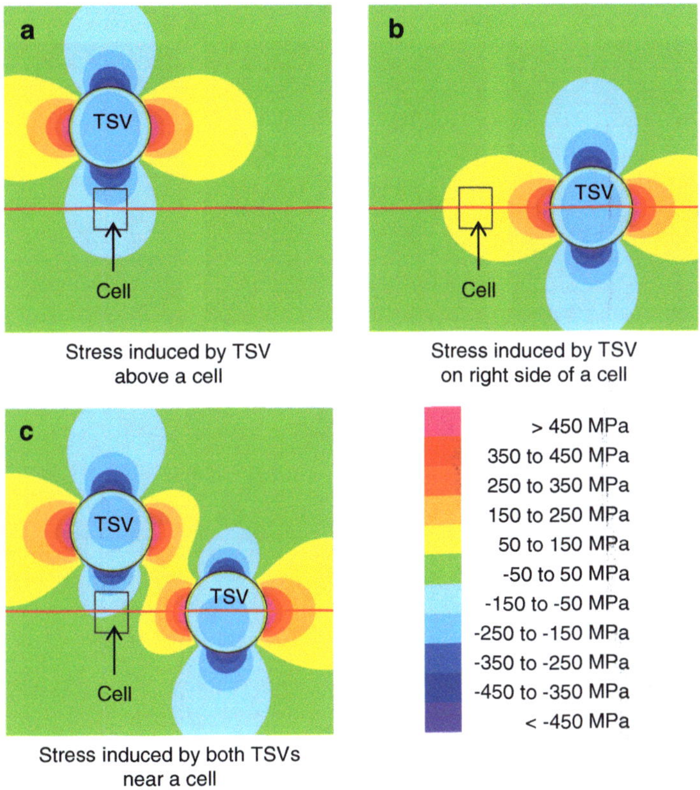

Fig. 14.8 Contour of stress (FEA simulation) caused by TSVs nearby a cell. (**a**) Stress induced by TSV above a cell, (**b**) stress induced by TSV on right side of a cell, and (**c**) stress induced by both TSVs near a cell

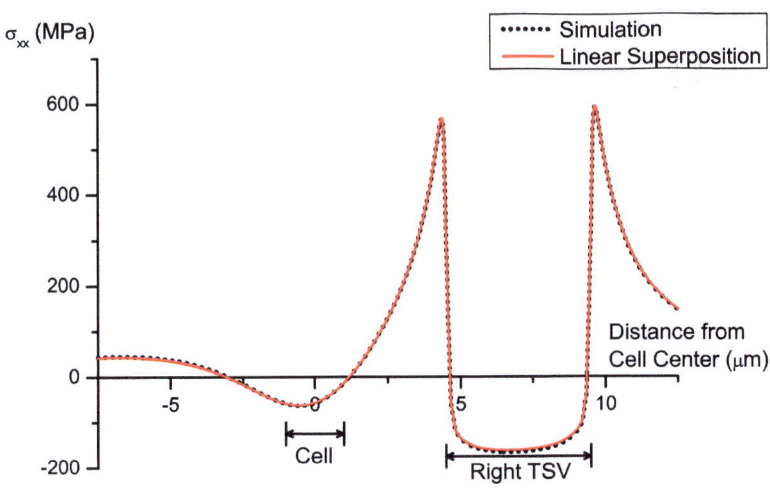

Fig. 14.9 Linear superposition of stress (FEA simulation) caused by TSVs nearby a cell

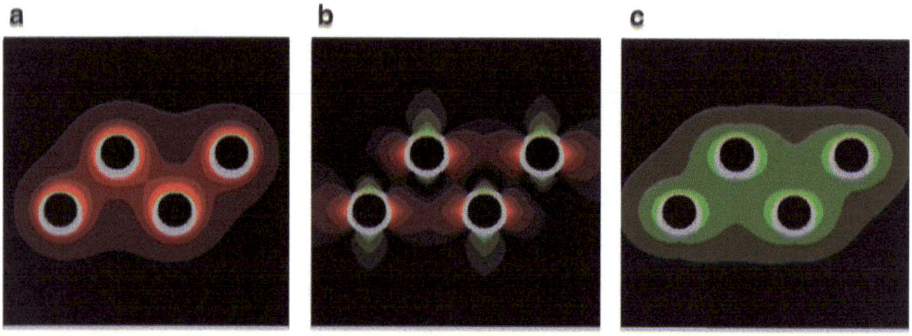

Fig. 14.10 Linear superposition of TSV stress. (a) σ_{total}, (b) $(\Delta\mu/\mu)_{hole}$, and (c) $(\Delta\mu/\mu)_{electron}$

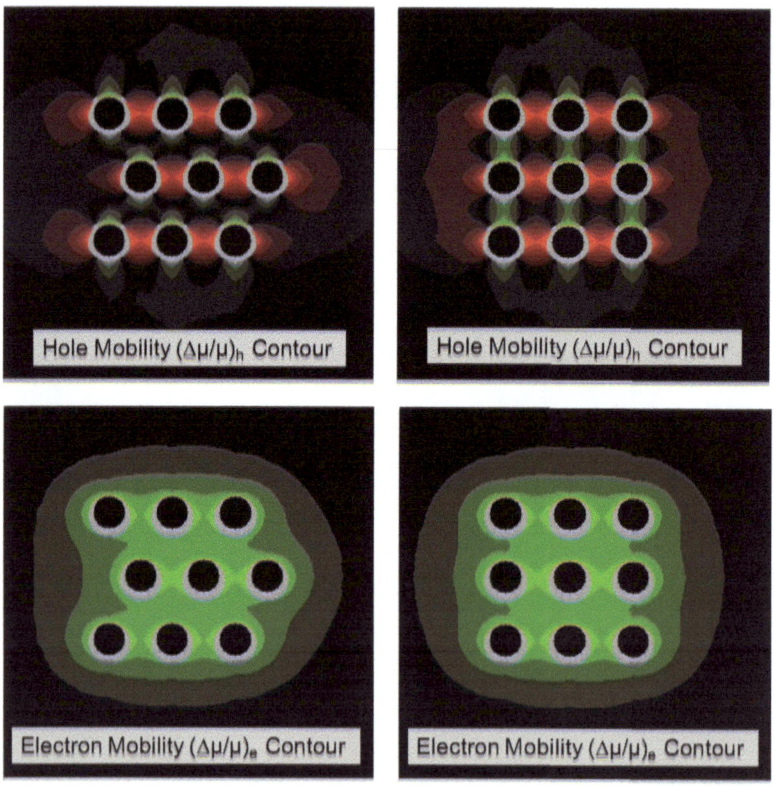

Fig. 14.11 Zigzag TSV placement has less $(\Delta\mu/\mu)_h$ between rows due to compensation

respectively. They do not have compensation effect. From Fig. 14.11, we can see that zigzag TSV placement is preferred for less PMOS variation while regular TSV placement is preferred for more hole mobility enhancement zone.

14.6 Mobility Variation Under STI-Induced Stress

We use ANSYS, a commercial FEA-based simulator, to simulate stress caused by STI in this chapter. An example of simulation result is shown in Fig. 14.12. In the figure, the contour of stress in horizontal direction caused by an STI in a plane of silicon is illustrated. Note that negative stress values represent compressive stress. Compressive stress caused by the STI can be higher than 100 MPa on silicon surface close to the STI, or even higher than 200 MPa on silicon surface adjacent to the STI. We observe that the contour lines in the area close to the STI are parallel to the left and right edges of the STI. Therefore, we approximate that the magnitude of horizontal stress caused by an STI is uniform in vertical direction. This approximation results in some error at a position far from an STI and off its center; however, the actual magnitude of the stress at the position is relatively small, and so is its impact on the mobility variation.

The horizontal stress, caused by the STI, on a horizontal line across the center of the STI is shown in Fig. 14.13. The far left area of the STI is stress free. The stress magnitude increases as the distance from the center of the STI decreases, and rapidly increases in the area adjacent to the STI. Inside STI, the stress is still compressive although its magnitude drops sharply. The trend reverses when we move away from STI center to the right side of the STI.

The simulation setup used to develop a model for STI stress is shown in Fig. 14.14. A patch of STI made of silicon dioxide is deposited on the surface of a silicon plane. STI stress mainly depends on two major parameters [6], distance to STI edge (STID) and width of STI (STIW). The values of these two parameters used for FEA simulations are listed under the figure. Other dimensions of the STI are from NCSU 45-nm cell library. The combinations of these two parameters result in 36 simulations in total. Because STI stress in horizontal direction is the main stress that affects mobility [6, 12], we measure stress along x-axis σ_{xx} at the channel.

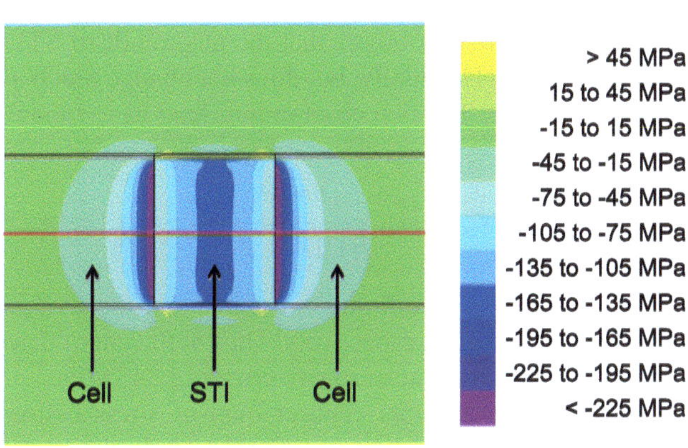

Fig. 14.12 Contour of stress (FEA simulation) caused by STI in *horizontal* direction

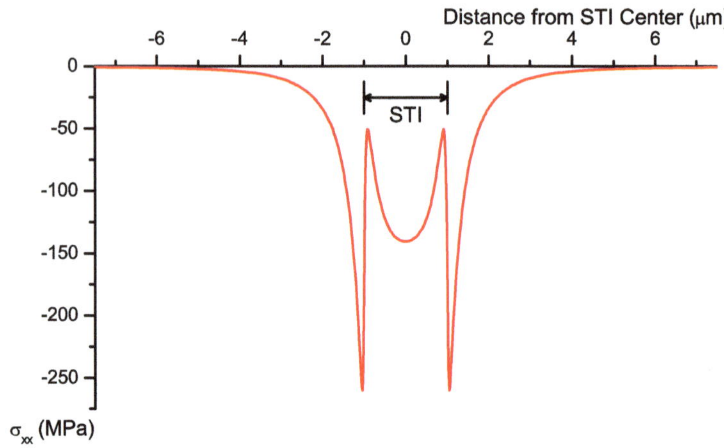

Fig. 14.13 Stress (FEA simulation) on a *horizontal line* across the *center* of the STI in Fig. 14.12

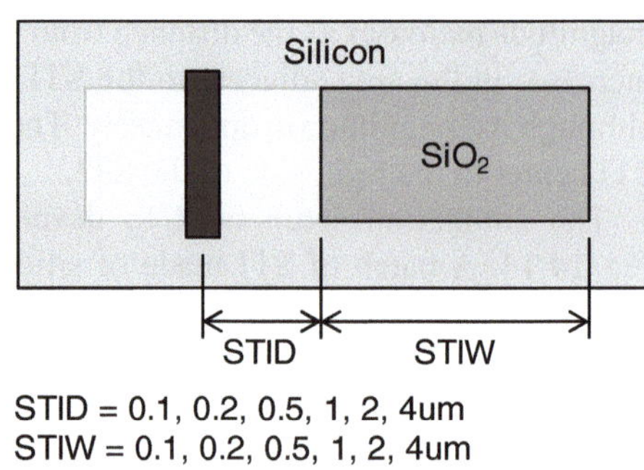

Fig. 14.14 Setup for FEA simulations used to model STI stress

The simulation results are shown in Figs. 14.15 and 14.16. In Fig. 14.15, we observe that the magnitude of STI stress rapidly decreases with the distance from the edge of STI. In Fig. 14.16, we observe that the magnitude of STI stress rapidly increases with the width of STI initially, but does not change much after the width is higher than a certain value. The two observations lead us to model STI stress in the following form:

$$\sigma_{xx} = \frac{\alpha(1 - e^{\beta \cdot \mathrm{STIW}}) + \chi}{\mathrm{STID}^{\delta} + \varepsilon} \tag{14.6}$$

where α, β, χ, δ, and ε are curve-fitting constants, and their value are -37.51, -3.24, 0.8601, 1.594, and 0.1317, respectively. The coefficient of determination for this model is 0.9987, and the root mean square error is 2.843 MPa.

Based on the model, we generate contour map of stress caused by an STI. The contour for a 4 μm-wide STI is show in Fig. 14.17. In the figure, we observe

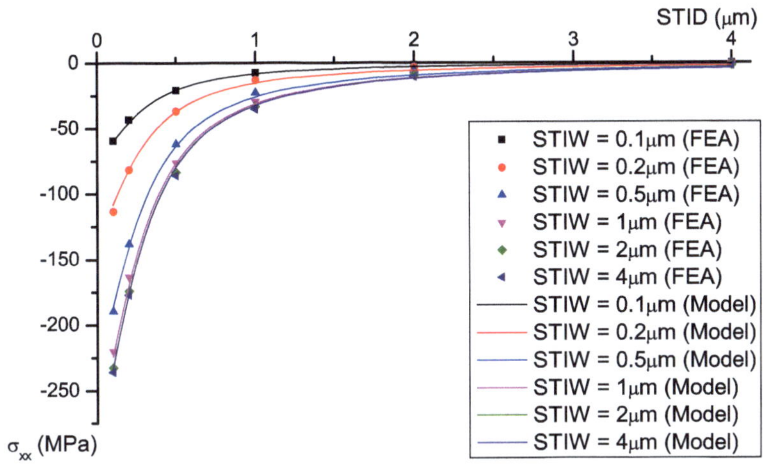

Fig. 14.15 STI stress (FEA simulation and model) at different distances

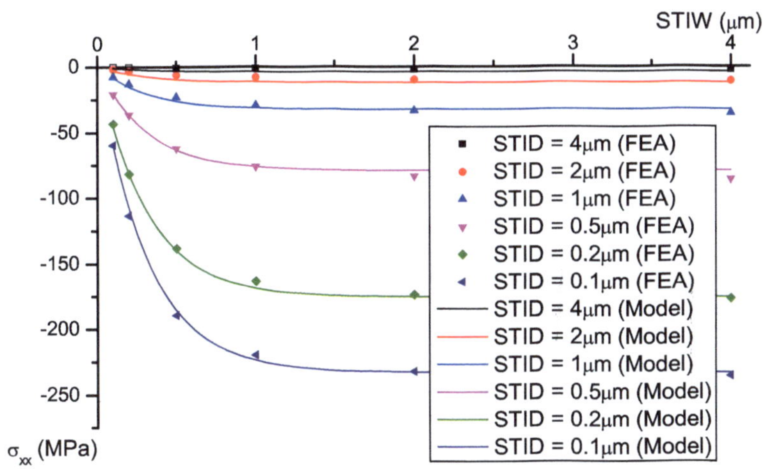

Fig. 14.16 Stress (FEA simulation and model) induced by STI with different widths

that compressive stress of more than 200 MPa is close to the STI edge, but the stress magnitude rapidly drops below 100 MPa in horizontal direction. Note that we show the area occupied by STI in gray in the contour. By using σ_{xx} obtained from Eq. (14.6) in formula (14.2), the contour map for hole and electron mobility variation can be generated, and is shown in Fig. 14.18a, b, respectively. From the contour, we observe that hole mobility is only enhanced by STI stress, whereas electron mobility is only degraded. Both hole mobility enhancement and electron mobility degradation take place on left and right sides of the STI.

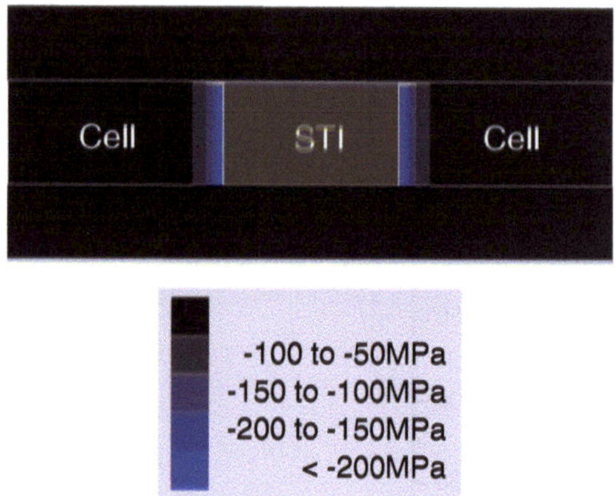

Fig. 14.17 Contour map of stress (model) for a single 4 μm-wide STI

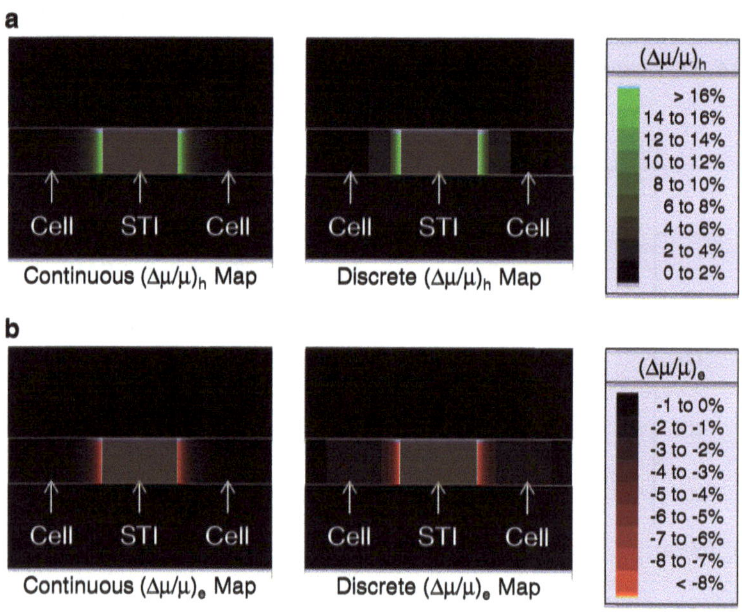

Fig. 14.18 Contour maps of mobility (model) for an STI. (**a**) Hole mobility variation and (**b**) electron mobility variation

Similar to the case for multiple TSVs in Sect. 14.5.2, it is possible to use linear superposition to estimate stress caused by STIs on both left and right sides of a cell. We model the mobility variation for multiple STIs as

$$\frac{\Delta\mu}{\mu}_{\text{STI}} = \sum \frac{\Delta\mu}{\mu} = -\Pi \sum_{i \in \text{STIs}} \sigma_{\text{xx}_i} \tag{14.7}$$

where σ_{xx_i} is the compressive stress caused by ith STI on left or right side of the cell.

14.7 Mobility Variation Under both TSV and STI-Induced Stress

In 3D ICs, stress effect from both TSV and STI must be considered. An example of result from simulations of a cell with a TSV on top and an STI on its right side is shown in Fig. 14.19. The contour of stress caused by the TSV on the top of the cell and the STI on the right side of the cell is shown in Fig. 14.19a, b, respectively. When we simulate the stress caused by both TSV and STI, we find that the stress from both of them interact with each other, and results in change in stress contour in the cell.

The horizontal stress caused by the TSV and STI is shown in Fig. 14.20. An additional plot, linear superposition, is created by adding the stress caused by the TSV and STI together. We observe that the result from this linear superposition is close to the stress caused by the TSV and STI obtained from FEA simulation.

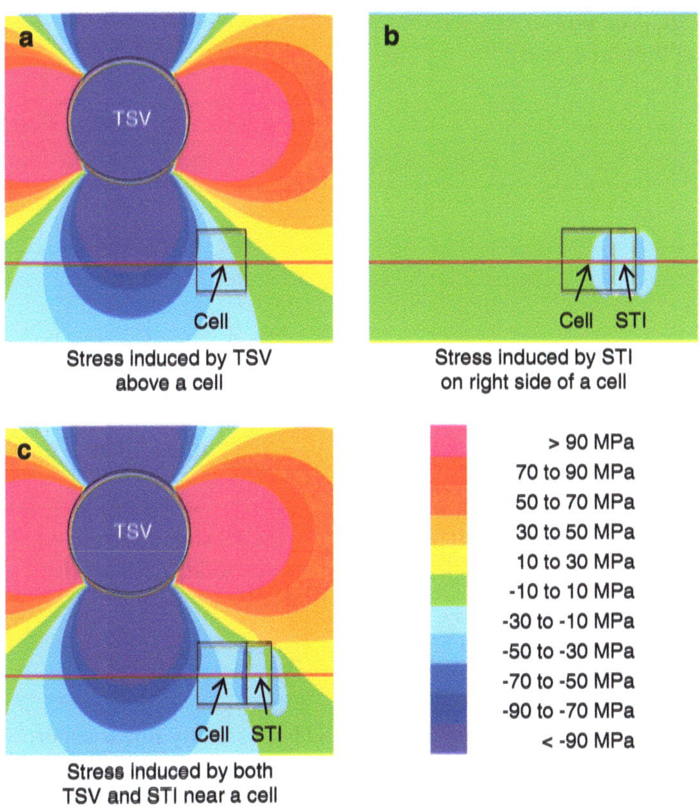

Fig. 14.19 Contour of stress (FEA simulation) caused by TSV on *top* of a cell with an STI on its *right* side. (**a**) Stress induced by TSV above a cell, (**b**) stress induced by STI on *right* side of a cell, and (**c**) stress induced by both TSV and STI near a cell

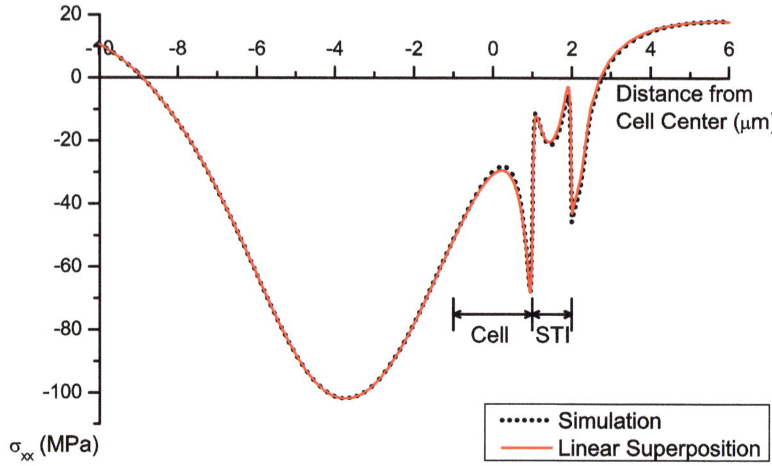

Fig. 14.20 Linear superposition of stress (FEA simulation) caused by a TSV and an STI

Fig. 14.21 Setup for FEA simulation of stress from both TSV and STI

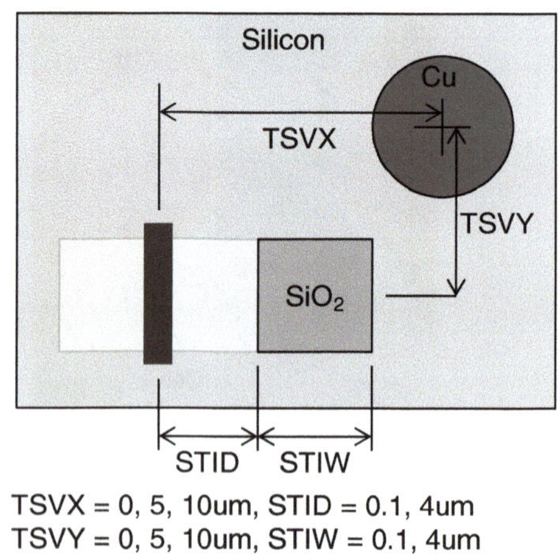

Therefore, it is possible to use linear superposition to estimate stress caused by both TSV and STI during early design stages. Then, we model the mobility variation for both TSV and STI as

$$\frac{\Delta\mu}{\mu}_{\text{total}} = \frac{\Delta\mu}{\mu}_{\text{TSV}} + \frac{\Delta\mu}{\mu}_{\text{STI}} \tag{14.8}$$

To combine the effect of both TSV and STI stresses by linear superposition, we first need to validate by FEA simulations that linear superposition can be used. The simulation setup is shown in Fig. 14.21. A patch of STI deposited on the surface of a silicon plane containing a TSV is simulated. The values of four simulation parameters used for FEA simulations are listed under the figure. Only 24 out of 36 combinations of parameter values are used for simulations because some cases are

Table 14.1 FEA simulation and linear superposition of TSV and STI

TSVX (μm)	TSVY (μm)	STID (μm)	STIW (μm)	FEA simulation $\sigma_{xx_{TSV}}$ (MPa)	FEA simulation $\sigma_{xx_{STI}}$ (MPa)	FEA simulation σ_{xx} (MPa)	Super position (MPa)	Error (%)
0	5	0.1	0.1	−137.03	−15.27	−150.23	−152.30	1.38
0	5	0.1	4	−136.73	−60.58	−159.22	−197.31	23.92
0	5	4	0.1	−136.04	−0.08	−136.10	−136.12	0.01
0	5	4	4	−136.65	−0.59	−137.26	−137.24	−0.02
0	10	0.1	0.1	−29.71	−15.27	−44.65	−44.98	0.75
0	10	0.1	4	−29.73	−60.58	−82.23	−90.32	9.84
0	10	4	0.1	−29.75	−0.08	−29.81	−29.83	0.07
0	10	4	4	−29.76	−0.58	−30.19	−30.35	0.54
5	0	0.1	0.1	152.80	−15.30	133.53	137.50	2.97
5	5	0.1	0.1	8.76	−15.28	−7.12	−6.52	−8.45
5	5	0.1	4	7.13	−60.60	−54.37	−53.46	−1.67
5	5	4	0.1	6.93	−0.08	7.00	6.86	−2.06
5	5	4	4	7.27	−0.59	8.20	6.68	−18.60
5	10	0.1	0.1	−11.63	−15.27	−26.90	−26.90	−0.01
5	10	0.1	4	−11.61	−60.59	−68.94	−72.20	4.73
5	10	4	0.1	−11.60	−0.08	−11.64	−11.68	0.30
5	10	4	4	−11.60	−0.59	−11.82	−12.19	3.10
10	0	0.1	0.1	41.36	−15.28	25.09	26.08	3.94
10	0	0.1	4	41.41	−60.59	−32.39	−19.18	−40.79
10	0	4	0.1	41.41	−0.08	41.23	41.34	0.25
10	5	0.1	0.1	22.22	−15.28	6.33	6.94	9.61
10	5	0.1	4	22.26	−60.58	−45.33	−38.32	−15.47
10	5	4	0.1	22.27	−0.08	22.18	22.20	0.09
10	5	4	4	22.25	−0.60	21.69	21.65	−0.16
Average error								−1.07

illegal (TSV overlaps STI) or unnecessary (TSV is far from STI and cell). For each case, the following three simulations are performed: only TSV, only STI, and both TSV and STI. The resulting σ_{xx} at the channel from the first two simulations are added, and compared with the result from the third simulations. Uniform thermal load of −250 °C is used for this experiment. Usual thermal load of −975 °C for STI only increases stress caused by it, but does not affect applicability of superposition.

The results from the simulations and linear superposition are shown in Table 14.1. Linear superposition overestimates stress in some cases, and underestimates in remaining cases. In most cases, linear superposition provides accurate approximation of stress from both TSV and STI. In few cases, the error of linear superposition is more than 10 %. These cases have STID set to very small value (0.1 μm). These errors are the result of stress singularity [1] mentioned earlier, which is encouraged by short distance to STI edges. Although these cases are included, the average error is still very small at −1.07 %, making linear superposition acceptable for use during early design stages.

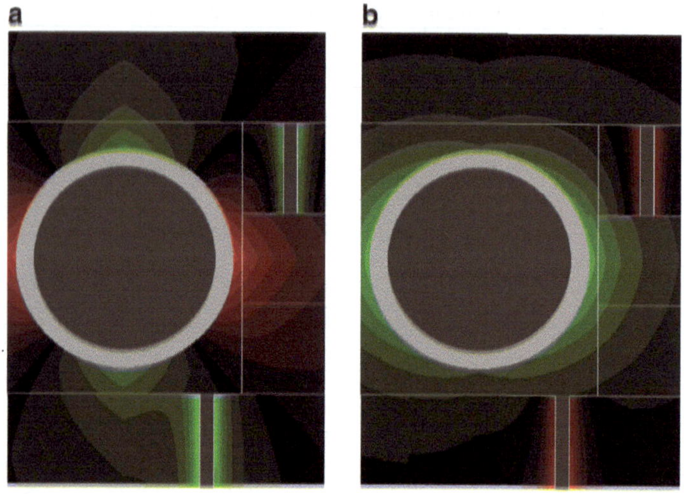

Fig. 14.22 Impact of interaction between TSV and STI stress (model) on mobility variation. (**a**) Hole and (**b**) electron

The model allows us to study the impact of the interaction between TSV and STI stress on circuit performance. The mobility variation contour of a TSV and two STIs is shown in Fig. 14.22. As shown in Fig. 14.22a, the impact of STI stress can compensate the impact of TSV stress on hole mobility variation (the area between the TSV and the right STI), or even increase it (the area under the TSV and on the left of the bottom STI). On the other hand, the impact of STI stress can only reduce the improvement from the impact of TSV stress on electron mobility variation as shown in Fig. 14.22b.

14.8 Full-Chip 3D Timing Analysis Under Mechanical Stress

In this section, we explain how to incorporate the mobility variation into cell level STA flow.

14.8.1 Timing Analysis for 3D ICs

Even though topology of a cell is the same, its timing characteristic will be changed by stress. Figure 14.23 shows an example that cells having the same topology and size can be in different timing corners systematically determined by TSVs (and STIs). When two TSVs are near three inverters, cell characteristics are different in a different position. From the formula (14.8), we can determine $\Delta\mu/\mu$ in any point

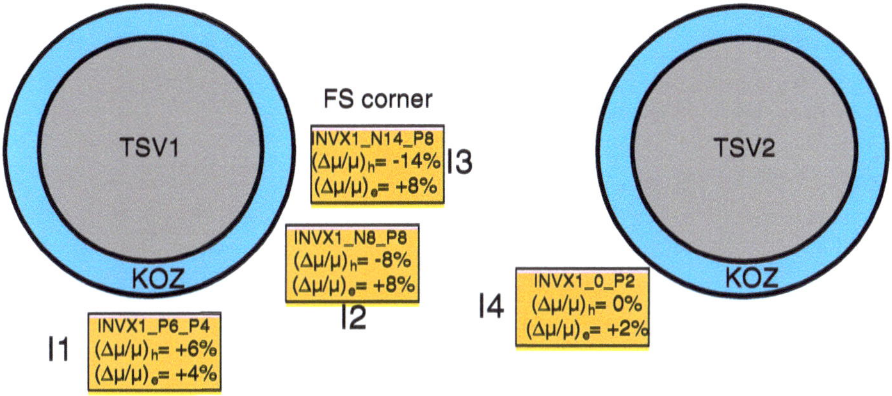

Fig. 14.23 Timing corner determination according to mobility variation

for a given layout. After mobility calculation, our framework renames cells such that mobility variation is included in Verilog netlist. For example, `I2` is renamed to `INVX1_P-8_N+8` which means $-8\,\%$ hole mobility and $+8\,\%$ electron mobility in Fig. 14.23.

We prepare a Verilog netlist and a parasitic extraction file (SPEF) per die. In addition, we make a top level Verilog netlist that instantiates the dies and connects them using wires which correspond to TSV connections. Then we make a top level SPEF file for the TSV connections. With a proper timing constraint file, we can run PrimeTime and get the 3D STA results.

14.8.2 Timing Library Construction Under Mobility Variation

To consider the systematic variation during timing analysis, we characterize a cell with different mobility corners as shown in Fig. 14.24. Hole mobility variation is from -14 to $+8\,\%$, and electron mobility variation is up to $+8\,\%$ to cover stress caused by TSVs in Fig. 14.23. Inverter `I1` in Fig. 14.23 matches the corner near FF corner, whereas `I3` is in FS corner. With mobility-variation-aware library and Verilog netlist having renamed cells, we can run PrimeTime to perform timing analysis with TSV and STI stress.

To cover mobility variation caused by multiple TSVs, we need to extend the mobility variation range ($-20\,\% \leq (\Delta\mu/\mu)_h \leq +8\,\%, 0\,\% \leq (\Delta\mu/\mu)_e \leq +14\,\%$). In addition, to consider both TSV and STI stresses, the mobility variation range needs to be extended even further. The mobility variation ranges needed to be covered for different stress sources are illustrated in Fig. 14.25. Because of their opposite kinds of stress, the mobility variation range needed to be covered for TSV and STI hardly overlaps with each other. The interaction between both TSV and STI stresses requires more than merely adding the covered mobility variation range for both of them.

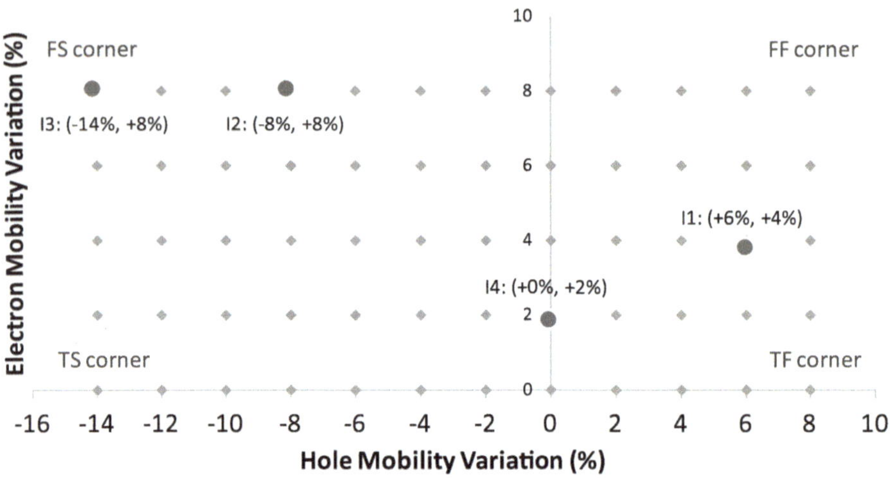

Fig. 14.24 Timing corner with TSV stress

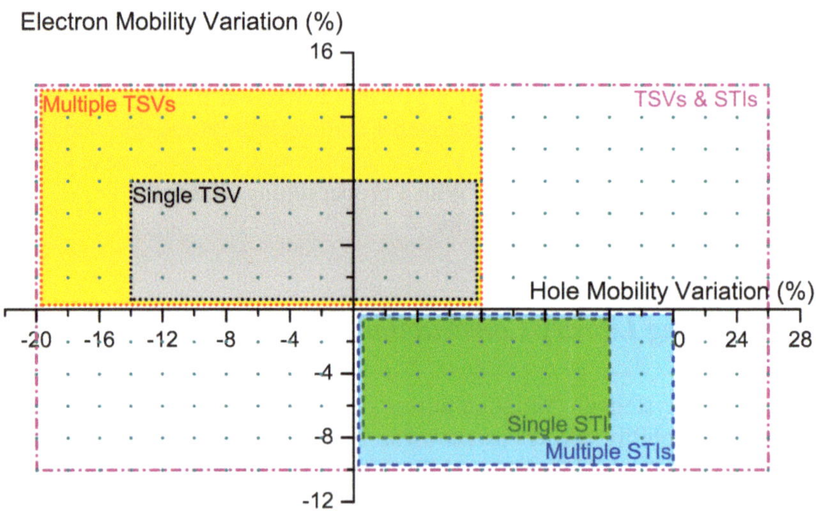

Fig. 14.25 Extended timing corner with both TSV and STI stresses

If mobility step is 2 %, we need to characterize 312 (= 24 × 13) libraries with different mobility values, which is prohibitive. However, we observe that rising delay variation only depends on $(\Delta\mu/\mu)_h$, and falling delay variation only depends on $(\Delta\mu/\mu)_e$ from Fig. 14.26. When we simulate inverter rising delay with mobility variation, electron mobility variation does not contribute to the delay. Similarly, we observe that falling delay only depends on electron mobility variation. In addition, from Fig. 14.26, we observe that hole mobility variation can cause more than 20 % PMOS performance variation depending on device technology, and electron

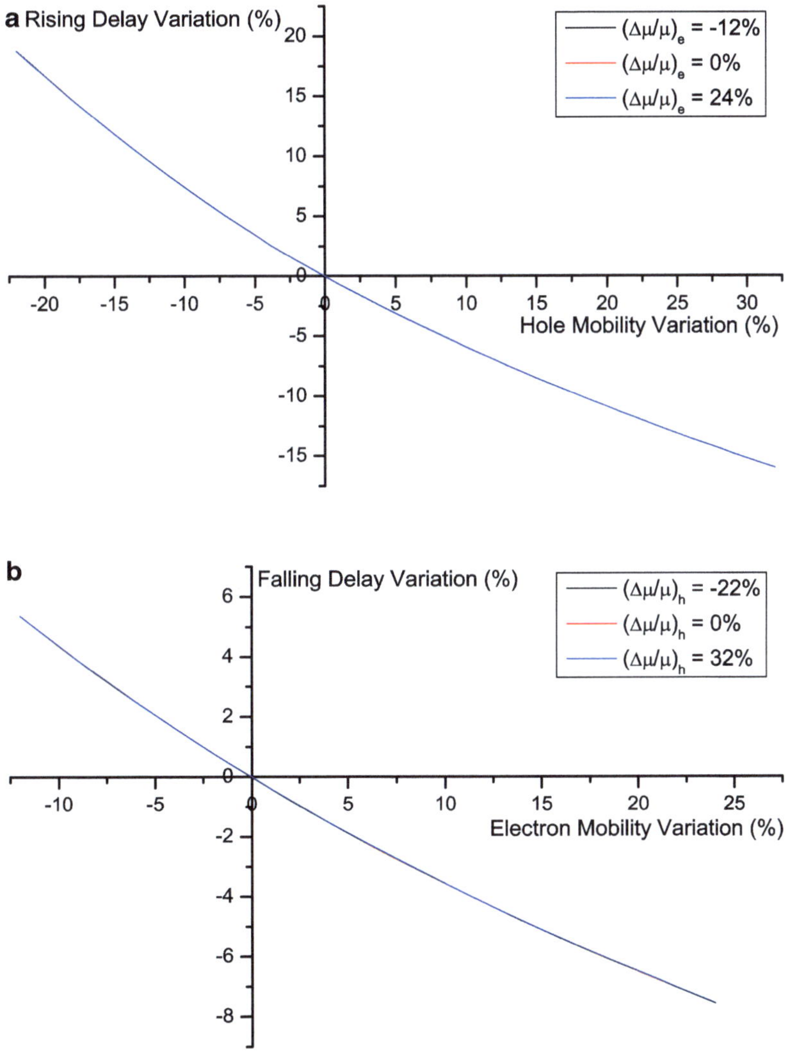

Fig. 14.26 Inverter delay variation with different $(\Delta\mu/\mu)_h$ and $(\Delta\mu/\mu)_e$. (**a**) Rising delay dependency on $(\Delta\mu/\mu)_h$ and (**b**) falling delay dependency on $(\Delta\mu/\mu)_e$

mobility variation can enhance NMOS performance up to 7.5 %. We use inverter in NCSU library and PTM spice model [22] to obtain Fig. 14.26. Therefore, we can fix $(\Delta\mu/\mu)_e$ when we sweep $(\Delta\mu/\mu)_h$. Characterizing 37 (= 24 + 13) libraries will be enough to cover the entire mobility set. If mobility step is 4 %, 20 (= 13 + 7) libraries are required. Since delay variation has semi-linear dependency on mobility variation, we can use interpolation for the mobility value between two libraries.

14.9 Experimental Results

We implemented our 3D SA-STA in C++ and tested on a 3.0 GHz Linux machine with 4 GB RAM. We generated the mobility-aware library based on NCSU 45-nm cell library with 2% mobility step. We used TSV size of 5 μm, TSV parasitic capacitance of 70 fF and resistance of 0.1 Ω. The keep-out-zone (KOZ) size is set to 0.5 μm.

14.9.1 *Full Chip Mobility Variation Map*

First, we show the efficiency of our compact stress and mobility modeling for TSV. When we want to find $\Delta\mu/\mu$ at any point on a die, we can obtain the value promptly. Even though we generate mobility contour in Fig. 14.27 (Die size: $175^2\,\mu m^2$, # TSVs: 462), it takes only 14.9 s. The proposed timing analysis with compact process/device model is fast enough to be used for iterative optimization purpose. Figure 14.27a shows an observation for layout optimization that the leftmost and rightmost sides have wider hole mobility enhanced zone than the middle area because the region has less mobility degradation by horizontally placed neighboring TSVs.

Next, we generate mobility contour as shown in Fig. 14.28 (Die size: $220^2\,\mu m^2$, # TSVs: 600, # cells: 3,422) while considering stress from both TSV and STI. We observe that the contour is noticeably different from that in Fig. 14.27 in two ways. First, because we are not interested in the stress inside STI for this chapter, we shows the area occupied by STI in gray in the contour. Second, the trend of mobility

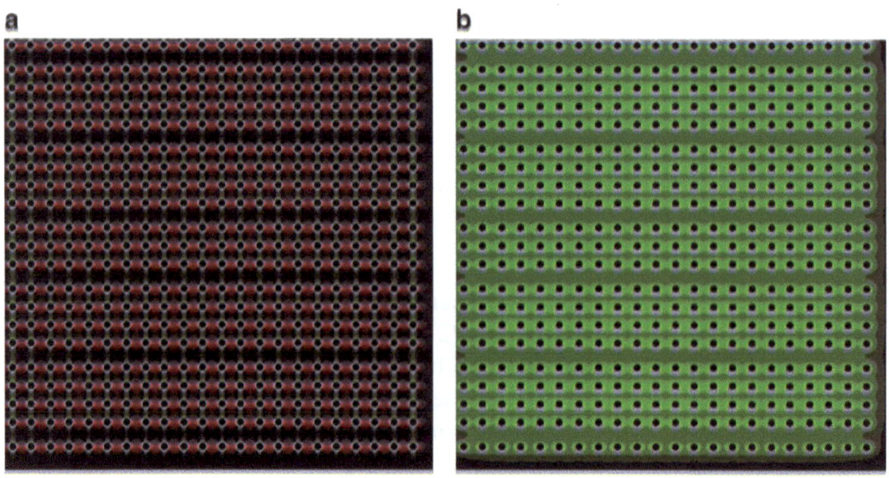

Fig. 14.27 Mobility variation contour map for 22 × 21 TSV array. (**a**) Hole and (**b**) electron

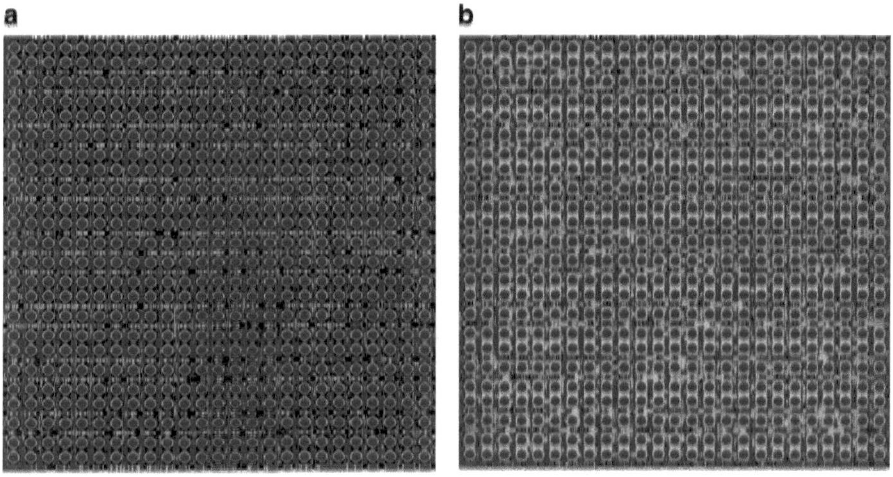

Fig. 14.28 Mobility variation contour maps for a layout considering both TSV and STI stresses. (**a**) Hole and (**b**) electron

Table 14.2 Benchmark circuits

Circuits	# Cells	# Nets	# TSVs	Profile
ex	14,864	15,045	1,483	Execution unit
8051	15,712	15,755	1,575	Micro-controller
8086	19,895	19,909	1,987	Microprocessor
MAC2	29,706	29,980	2,971	Arithmetic unit
Ethernet	77,234	77,381	7,748	Network controller
RISC	88,401	89,154	8,837	Microprocessor
b18	103,711	103,948	10,367	Multiprocessors
des_perf	109,181	109,416	10,916	Data encryption
vga_lcd	126,379	126,484	12,638	Display controller
b19	168,943	169,476	16,869	Multiprocessors

variation changes, i.e., we observe increasing area of hole mobility enhancement, but decreasing area of electron mobility enhancement. This phenomenon is largely due to the inclusion of STI stress.

14.9.2 Full Chip Timing Analysis Results

In this experiment, we compare our SA-STA with no stress case. Ten benchmark circuits used to show the timing variation are listed in Table 14.2. The area utilization of each circuit is around 70 %. All whitespace is occupied by STI. The total amount of STI is in the same range (30–70 %) as in other STI-related works [7, 19]. The benchmark circuits are placed for wire length minimization [8] with neither TSV nor STI stress consideration. The hole and electron mobility variation of each benchmark circuit is shown in Table 14.3. The timing results are shown in Tables 14.4 and 14.5.

When we consider only TSV stress effect, the hole and electron mobility variation of all benchmark circuits are in the same ranges as shown in Table 14.3. Hole mobility variation of cells in each circuit ranges from around −18 to +6 %; whereas, electron mobility variation ranges from 0 to around 13 %. Although the mobility variations of all benchmark circuits are in the same ranges, their timing variation is different. The change of longest path delay of the benchmark circuits has variation from −5.65 to 6.52 %. Some benchmark circuits have timing gain while some benchmark circuits have timing penalty. On average, the impact of TSV stress on timing is 2.82 %. For a random placement, the impact of hole and electron mobility variation may compensate each other, and result in low combined enhancement/degradation in timing for some cases because the average of carrier (both hole and electron) mobility variation is close to zero. In many cases, however, the impact of hole and electron mobility variation is in the same direction, resulting in significant changes (either enhancement or degradation) of longest path delay. If we consider TSV stress effect during cell and TSV placement, we can expect performance improvement for every benchmark circuit. The change of total negative slack (TNS) has variation from −28.48 to +50.43 %, which is bigger than the variation of the change of delay. That motivates the need of TSV-stress-aware layout optimization.

When we consider only STI stress effect, the hole and electron mobility variation of all benchmark circuits are exactly in the same ranges as shown in Table 14.3. Hole mobility variation of cells in each circuit ranges from 0 to +19.72 %; whereas, electron mobility variation ranges from −8.68 to 0 %. The carrier mobility variation of all benchmark circuits are in the exact same range because STI stress depends heavily on the relative size of cell and its adjacent STIs. Wide cells or cells far away from narrow STIs have no carrier mobility variation; whereas, the narrowest cell in the library having wide STIs on both of its side results in the highest carrier mobility variation. The change of longest path delay of the benchmark circuits has variation from −5.92 to +1.92 %, and the average delay variation is 4.11 %. Most benchmark circuits have timing gain because, for a random placement, the average of carrier (both hole and electron) mobility variation is much higher than zero. In addition, STI is pervasive on IC layout. Without considering STI stress, STA only reports pessimistic timing result. Including STI stress, the pessimism in timing results decreases. TNS is significantly reduced by 50.07 % on average because several violating paths become non-violating by STI stress. Wide variation of the change of both delay and TNS suggests the importance of STI-stress-aware layout optimization.

Finally, when we consider both TSV and STI stresses, the range of hole and electron mobility variation of all benchmark circuits shifts from the range when only TSV or only STI stress is considered as shown in Table 14.3. Hole mobility variation of cells in each circuit ranges from around −14 to +25 %; whereas, electron

Table 14.3 Comparison of mobility variation range

	TSV stress		STI stress		TSV + STI stress	
Circuit	Hole mobility var (%)	e-mobility var (%)	Hole mobility var (%)	e-mobility var (%)	Hole mobility var (%)	e-mobility var (%)
ex	−18.6 to +6 .0	0.0 to +13.2	0.0 to +19.7	−8.6 to 0.0	−14.8 to +25.2	−8.6 to +10.8
8051	−18.8 to +6.3	0.0 to +13.4	0.0 to +19.7	−8.6 to 0.0	−13.8 to +24.9	−8.6 to +11.0
8086	−17.8 to +6.3	0.0 to +12.7	0.0 to +19.7	−8.6 to 0.0	−13.7 to +25.5	−8.6 to +11.1
MAC2	−17.4 to +6.3	0.0 to +12.5	0.0 to +19.7	−8.6 to 0.0	−13.3 to +25.8	−8.6 to +10.7
Ethernet	−17.8 to +6.3	0.0 to +12.6	0.0 to +19.7	−8.6 to 0.0	−14.0 to +25.9	−8.6 to +11.0
RISC	−17.9 to +6.4	0.0 to +12.8	0.0 to +19.7	−8.6 to 0.0	−14.2 to +26.0	−8.6 to +11.3
b18	−18.6 to +6.3	0.0 to +13.3	0.0 to +19.7	−8.6 to 0.0	−14.8 to +25.8	−8.6 to +11.4
des_perf	−18.5 to +6.2	0.0 to +13.4	0.0 to +19.7	−8.6 to 0.0	−14.4 to +25.8	−8.6 to +11.4
vga_lcd	−18.6 to +6.3	0.0 to +13.3	0.0 to +19.7	−8.6 to 0.0	−14.3 to +25.9	−8.6 to +11.7
b19	−17.9 to +6.4	0.0 to +13.0	0.0 to +19.7	−8.6 to 0.0	−14.4 to +25.8	−8.6 to +11.4

Table 14.4 Full-chip longest path delay (LPD) and total negative slack (TNS) results for no stress and TSV stress cases. The delay and slack are shown in *ns* (Percentage of change with respect to the no stress case is shown in parenthesis)

	Without stress		With TSV stress	
Ckt	LPD	TNS	LPD	TNS
ex	12.009	−8.815	11.881 (−1.06 %)	−7.215 (−18.15 %)
8051	5.041	−144.035	5.370 (+6.52 %)	−145.363 (+0.92 %)
8086	9.283	−19.317	9.423 (+1.50 %)	−26.779 (+38.63 %)
MAC2	7.797	−87.337	7.905 (+1.38 %)	−93.422 (+6.97 %)
Ethernet	9.294	−474.917	9.484 (+2.05 %)	−463.344 (−2.44 %)
RISC	8.583	−57.101	8.098 (−5.65 %)	−40.840 (−28.48 %)
b18	12.522	−41.301	12.838 (+2.53 %)	−62.128 (+50.43 %)
des_perf	8.467	−40.298	8.720 (+2.99 %)	−45.054 (+11.80 %)
vga_lcd	8.228	−0.991	8.456 (+2.78 %)	−1.191 (+20.25 %)
b19	13.389	−126.528	13.618 (+1.71 %)	−145.533 (+15.02 %)

Table 14.5 Full-chip longest path delay (LPD) and total negative slack (TNS) results for STI stress and STI+TSV stress cases. The delay and slack are shown in *ns* (Percentage of change with respect to the no stress case shown in Table 14.4 is shown in parenthesis)

	With STI stress		With both TSV and STI stresses	
Ckt	LPD	TNS	LPD	TNS
ex	11.686 (−2.69 %)	−5.280 (−40.10 %)	11.577 (−3.59 %)	−4.348 (−50.67 %)
8051	4.768 (−5.42 %)	−60.450 (−58.03 %)	4.761 (−5.56 %)	−61.351 (−57.41 %)
8086	8.734 (−5.92 %)	−3.495 (−81.90 %)	8.888 (−4.26 %)	−7.194 (−62.76 %)
MAC2	7.435 (−4.64 %)	−46.541 (−46.71 %)	7.525 (−3.49 %)	−49.861 (−42.91 %)
Ethernet	9.472 (+1.92 %)	−492.182 (+3.64 %)	9.562 (+2.89 %)	−480.541 (+1.18 %)
RISC	8.434 (−1.73 %)	−27.864 (−51.20 %)	8.387 (−2.29 %)	−17.779 (−68.86 %)
b18	12.013 (−4.06 %)	−22.024 (−46.67 %)	12.308 (−1.71 %)	−30.331 (−26.56 %)
des_perf	8.026 (−5.21 %)	−10.090 (−74.96 %)	8.294 (−2.04 %)	−11.513 (−71.43 %)
vga_lcd	7.835 (−4.77 %)	−0.671 (−32.27 %)	8.078 (−1.82 %)	−0.875 (−11.71 %)
b19	12.760 (−4.70 %)	−43.996 (−65.23 %)	12.821 (−4.25 %)	−48.795 (−61.44 %)

mobility variation ranges from −8.68 to around 11 %. The change of longest path delay of the benchmark circuits has variation from −5.56 to +2.89 %, and the average delay variation is 3.19 %. The changes are in the same direction as the changes considering only STI stress. Compared to the changes considering only STI stress, some benchmark circuits have timing gain while some benchmark circuits have timing penalty. This variation suggests that TSV stress still has significant impact on timing even after STI stress is considered. TNS is significantly reduced by 45.49 % on average. Therefore, both TSV and STI stresses can be exploited together for performance improvement. The potential to exploit them to improve timing is revealed as shown in the next experiment.

Table 14.6 Gate optimization considering only TSV stress on the target path with perturbation

Logic depth	Original gate	Optimized gate	Timing Arc	Original delay (ps)	Optimized delay (ps)	Δ (%)
	Input port	Input port				
1	INVX1_P-6_N+10	INVX1_P-16_N+12	Fall	48.58	43.81	9.82
2	INVX1_P-16_N+12	INVX1_P+4_N+8	Rise	32.77	28.59	12.76
3	INVX1_P-16_N+12	INVX1_P-16_N+12	Fall	160.67	162.23	−0.97
4	INVX1_P-6_N+10	INVX1_P+2_N+8	Rise	469.81	422.60	10.05
5	INVX1_P-6_N+10	INVX1_P-16_N+12	Fall	183.39	168.95	7.87
6	INVX1_P-8_N+10	INVX1_P+4_N+8	Rise	789.49	717.44	9.13
7	INVX1_P+0_N+0	INVX1_P+0_N+0	Fall	896.10	939.16	−4.81
8	INVX1_P-16_N+12	INVX1_P+4_N+8	Rise	1,630.01	1,396.33	14.34
9	INVX1_P-2_N+6	INVX1_P-2_N+6	Fall	327.01	376.35	−15.09
10	INVX1_P+2_N+6	INVX1_P+2_N+6	Rise	221.73	182.14	17.86
11	INVX4_P+2_N+6	INVX4_P+2_N+6	Fall	112.48	95.60	15.01
12	MUX2X1_P+0_N+8	MUX2X1_P-16_N+12	Fall	401.91	406.39	−1.11
13	MUX2X1_P-16_N+12	MUX2X1_P+2_N+8	Rise	922.52	730.53	20.81
14	AOI21X1_P-16_N+12	AOI21X1_P-16_N+12	Fall	528.11	610.77	−15.65
15	NAND3X1_P+2_N+8	NAND3X1_P+2_N+8	Rise	826.82	941.80	−13.91
16	INVX1_P+2_N+8	INVX1_P-16_N+12	Fall	840.11	643.74	23.37
17	NOR2X1_P-6_N+10	NOR2X1_P+4_N+8	Rise	262.29	212.49	18.99
18	INVX1_P+2_N+8	INVX1_P-16_N+12	Fall	49.16	39.85	18.94
19	OAI21X1_P+0_N+8	OAI21X1_P+2_N+8	Rise	17.06	19.20	−12.54
	DFFPOSX1	DFFPOSX1	Rise	0.11	0.12	−9.09
Path delay				8,720.13	8,138.09	6.67

14.9.3 Placement Optimization Results

We manually optimize the critical path in des_perf to present the potential benefit of TSV-stress-aware layout optimization. Before optimization, the path delay is 8.720 ns with TSV-stress-aware timing analysis. However, we could reduce the delay to 8.138 ns with small layout perturbation, which is 6.67 % improvement. It is even less than the path delay without stress, which is 8.467 ns in Table 14.4. Table 14.6 shows the gates on the path. We can see the cell renaming according to the mobility variation. We adjust each cell position with small perturbation to gain timing. The maximum timing gain in a cell is 23.37 % improvement.

In Table 14.6, the timing of some cells is not improved even though the carrier mobility of the cells is enhanced. For example, the delay time of Cell #12 increases by 1.11% while electron mobility increases by 4 % (from +8 to +12 %). The increase in delay is the result from moving a cell that it drives, Cell #13, to improve hole mobility of Cell #13. The impact of increase in wire capacitive load of Cell #12 outweighs the impact of increase in electron mobility of the cell, resulting in delay increase. If the electron mobility of the cell was not increased, the delay increase would be higher than 1.11 %. Although the delay time of some cells on the path increases for this reason, the overall path delay decreases.

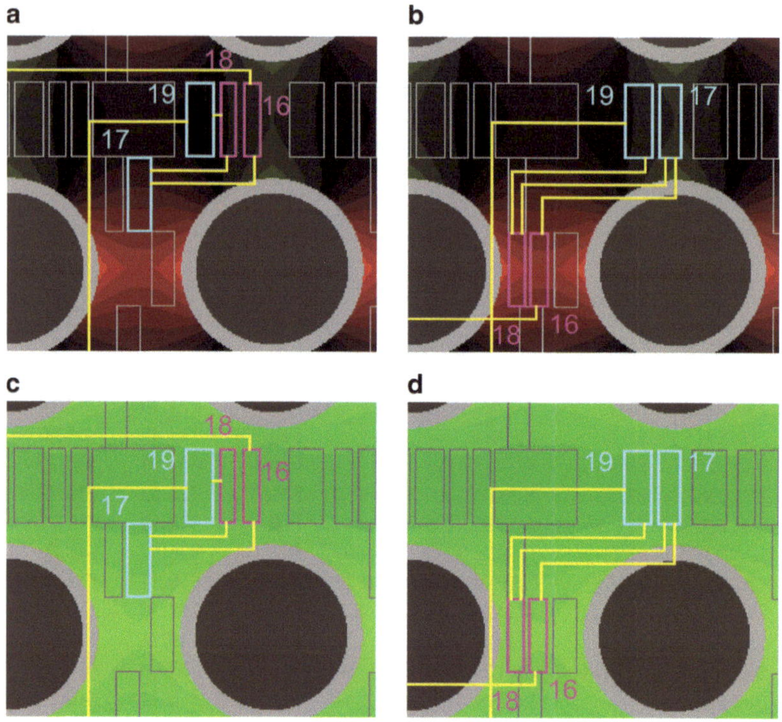

Fig. 14.29 Cell perturbation to take advantage of TSV-stress-induced mobility variation. (**a**) Hole mobility contour with original cell placement, (**b**) hole mobility contour after cell perturbation, (**c**) electron mobility contour with original cell placement, and (**d**) electron mobility contour after cell perturbation

Figure 14.29 shows how cell relocation works for timing optimization. We capture the placement result on a die with TSV-stress-induced mobility variation contours. The cells in logic depth 17 and 19 are hole-mobility critical cells because the timing arc is rising on the path. Therefore, we perturb the cells to be placed close to green area in hole-mobility contour. However, the cells in logic depth 16 and 18 are electron-mobility critical. Therefore, we push the cells to have more mobility enhancement in Fig. 14.29c, d.

Finally, we manually optimize the same critical path in des_perf when considering both TSV and STI stresses to reveal the impact of the interaction between both stresses on performance. Before optimization, the path delay is 8.294 ns with TSV-STI-stress-aware timing analysis. However, we could reduce the delay to 7.867 ns with small layout perturbation which is 5.15 % improvement. Table 14.7 shows the gates on the path. We can see the cell renaming according to the mobility variation. We adjust each cell position with small perturbation to gain timing. The maximum timing gain in a cell is 17.63 % improvement.

Figure 14.30 shows how cell relocation works for this timing optimization. We capture the placement result on a die with TSV-STI-stress-induced mobility

Table 14.7 Gate optimizations considering both TSV and STI stresses on the target path with perturbation

Logic depth	Original gate	Optimized gate	Timing Arc	Original delay (ps)	Optimized delay (ps)	$\Delta(\%)$
	Input port	Input port				
1	INVX1_P+6_N+4	INVX1_P-8_N+10	Fall	49.85	44.62	10.49
2	INVX1_P+4_N+4	INVX1_P+20_N+0	Rise	29.15	24.01	17.63
3	INVX1_P+4_N+4	INVX1_P-6_N+8	Fall	167.51	166.46	0.63
4	INVX1_P+6_N+4	INVX1_P+16_N+2	Rise	428.15	374.05	12.64
5	INVX1_P+6_N+4	INVX1_P-2_N+6	Fall	178.81	163.86	8.36
6	INVX1_P+10_N+2	INVX1_P+18_N+2	Rise	677.35	614.72	9.25
7	INVX1_P+20_N-8	INVX1_P+6_N-2	Fall	914.05	949.16	−3.84
8	INVX1_P+4_N+4	INVX1_P+16_N+2	Rise	1,489.17	1,261.57	15.28
9	INVX1_P+18_N-2	INVX1_P+18_N-2	Fall	352.66	391.58	−11.04
10	INVX1_P+20_N-4	INVX1_P+20_N-4	Rise	175.74	147.00	16.35
11	INVX4_P+18_N-2	INVX4_P+18_N-2	Fall	102.76	89.22	13.18
12	MUX2X1_P+2_N+8	MUX2X1_P-10_N+10	Fall	409.04	410.91	−0.46
13	MUX2X1_P-10_N+10	MUX2X1_P-2_N+8	Rise	874.10	800.69	8.40
14	AOI21X1_P-10_N+8	AOI21X1_P-10_N+8	Fall	538.98	553.49	−2.69
15	NAND3X1_P+10_N+4	NAND3X1_P+12_N+4	Rise	749.84	702.28	6.34
16	INVX1_P+16_N+2	INVX1_P-6_N+8	Fall	852.80	888.45	−4.18
17	NOR2X1_P+4_N+6	NOR2X1_P+12_N+4	Rise	240.80	222.78	7.48
18	INVX1_P+10_N+4	INVX1_P-8_N+8	Fall	46.81	44.56	4.81
19	OAI21X1_P+6_N+6	OAI21X1_P+10_N+4	Rise	16.58	17.53	−5.73
	DFFPOSX1	DFFPOSX1	Rise	0.11	0.12	−9.09
Path delay				8,294.26	7,867.06	5.15

variation contours. Like in previous experiment, the cells in logic depth 17 and 19 are hole-mobility critical cells because the timing arc is rising on the path. Besides moving them to the area that TSVs provide improvement on rise time, surrounding them by STIs improves hole mobility. However, the cells in logic depth 16 and 18 are electron-mobility critical. Therefore, we push the cells to have more mobility enhancement provided by TSVs as shown in Fig. 14.30c, d. Note that manual optimization when considering both TSV and STI stresses is more difficult than when only TSV stress is considered. When only TSV stress is considered, moving cells does not change mobility-variation contour. When both TSV and STI stresses are considered, moving cells to exploit TSV stress changes dimension of STI surrounding it, thus their delay may not be improved as much as expected.

14.9.4 Impact of TSV Diameter on Timing

In this experiment, we change the diameter of TSVs to 10 μm to demonstrate the impact of high TSV stress caused by large TSVs on mobility variation and timing. First, the hole and electron mobility variation of each benchmark circuit

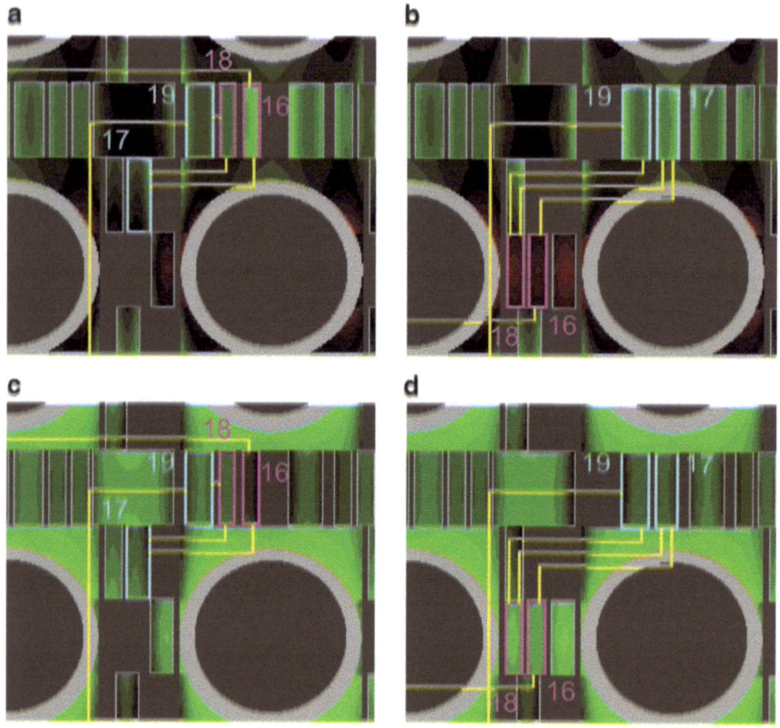

Fig. 14.30 Cell perturbation to take advantage of TSV-STI-stress-induced mobility variation. (**a**) Hole mobility contour with original cell placement, (**b**) hole mobility contour after cell perturbation, (**c**) electron mobility contour with original cell placement, and (**d**) electron mobility contour after cell perturbation

is shown in Table 14.8. We note that individual device mobility variation is larger with 10 μm TSVs compared with 5 μm TSVs. This shows that larger TSVs cause more mechanical stress and thus more variation in mobility of the devices. Full-chip timing results are shown in Tables 14.9 and 14.10. We observe that the delay and slack values are larger than 5 μm TSVs shown in Tables 14.4 and 14.5. This is mainly because of the larger area overhead caused by larger TSVs.

In case of 10μm TSV, the minimum and maximum LPD (longest path delay) change caused by TSV stress are −0.96 % (MAC2) and +7.38 % (b19) as shown in Table 14.9. In case of 5 μm TSV, the minimum and maximum LPD change caused by TSV stress are −1.06 % (ex) and +6.52 % (b19) as shown in Table 14.4. This shows that the larger TSV causes wider range of full-chip timing variations than the smaller TSV. We see similar trend in terms of TNS (total negative slack) metric. Comparison between Tables 14.10 and 14.5 also reveal the same trend in terms of TSV size impact on full-chip variation range. When both TSV and STI stress are considered together, 10 μm TSV causes −6.89 (8086) to +4.06 % (RISC) LPD variation range. In case of 5 μm TSV, the range is from −5.56 (8051) to +2.89 % (ethernet).

Table 14.8 Comparison of mobility variation range for designs with 10 μm TSV

	TSV stress		STI stress		TSV + STI stresses	
Circuit	Hole mobility var (%)	e-mobility var (%)	Hole mobility var (%)	e-mobility var (%)	Hole mobility var (%)	e-mobility var (%)
ex	−26.7 to +10.7	0.0 to +20.6	0.0 to +19.7	−8.6 to 0.0	−25.9 to +29.9	−8.6 to +19.8
8051	−27.9 to +10.5	0.0 to +21.6	0.0 to +19.7	−8.6 to 0.0	−26.6 to +30.2	−8.6 to +21.1
8086	−25.7 to +10.8	0.0 to +20.0	0.0 to +19.7	−8.6 to 0.0	−24.3 to +29.9	−8.6 to +19.3
MAC2	−25.7 to +10.7	0.0 to +19.3	0.0 to +19.7	−8.6 to 0.0	−23.9 to +30.2	−8.6 to +18.8
Ethernet	−26.7 to +10.7	0.0 to +20.8	0.0 to +19.7	−8.6 to 0.0	−26.5 to +30.4	−8.6 to +20.5
RISC	−26.5 to +11.1	0.0 to +20.7	0.0 to +19.7	−8.6 to 0.0	−26.3 to +30.6	−8.6 to +20.5
b18	−27.0 to +11.0	0.0 to +20.8	0.0 to +19.7	−8.6 to 0.0	−26.2 to +30.6	−8.6 to +20.7
des_perf	−26.9 to +10.8	0.0 to +20.8	0.0 to +19.7	−8.6 to 0.0	−26.5 to +30.1	−8.6 to +20.7
vga_lcd	−26.9 to +10.7	0.0 to +20.8	0.0 to +19.7	−8.6 to 0.0	−26.3 to +29.7	−8.6 to +20.7
b19	−26.7 to +11.0	0.0 to +20.8	0.0 to +19.7	−8.6 to 0.0	−25.9 to +30.5	−8.6 to +20.6

Table 14.9 Full-chip longest path delay (LPD) and total negative slack (TNS) results for no stress and TSV stress cases. We use 10 μm TSV for this experiment. The delay and slack are shown in *ns* (Percentage of change with respect to the no stress case is shown in parenthesis)

	Without stress		With TSV stress	
ckt	LPD	TNS	LPD	TNS
ex	12.478	−9.514	12.188 (−2.33 %)	−6.459 (−32.11 %)
8051	5.499	−195.366	5.867 (+6.69 %)	−188.093 (−3.72 %)
8086	9.989	−23.014	9.866 (−1.22 %)	−20.552 (−10.70 %)
MAC2	8.200	−100.314	8.122 (−0.96 %)	−93.859 (−6.43 %)
Ethernet	18.287	−637.391	18.721 (+2.38 %)	−626.982 (−1.63 %)
RISC	13.448	−160.188	14.358 (+6.77 %)	−223.424 (+39.48 %)
b18	13.995	−127.386	14.370 (+2.68 %)	−155.836 (+22.33 %)
des_perf	13.033	−10.807	13.419 (+2.96 %)	−12.466 (+15.35 %)
vga_lcd	14.960	−32.848	14.879 (−0.54 %)	−31.585 (−3.85 %)
b19	16.430	−204.370	17.643 (+7.38 %)	−225.623 (+10.40 %)

Table 14.10 Full-chip longest path delay (LPD) and total negative slack (TNS) results for STI stress and STI + TSV stress cases. We use 10 μm TSV for this experiment. The delay and slack are shown in *ns* (Percentage of change with respect to the no stress case shown in Table 14.9 is shown in parenthesis)

	With STI stress		With both TSV and STI stresses	
ckt	LPD	TNS	LPD	TNS
ex	12.070 (−3.28 %)	−5.065 (−46.76 %)	11.834 (−5.17 %)	−3.406 (−64.20 %)
8051	5.338 (−2.93 %)	−111.730 (−42.81 %)	5.375 (−2.25 %)	−107.988 (−44.73 %)
8086	9.410 (−5.79 %)	−5.060 (−78.01 %)	9.300 (−6.89 %)	−3.812 (−83.43 %)
MAC2	7.812 (−4.73 %)	−50.389 (−49.77 %)	7.727 (−5.77 %)	−41.854 (−58.28 %)
Ethernet	18.345 (+0.32 %)	−622.634 (−2.32 %)	18.719 (+2.36 %)	−618.190 (−3.01 %)
RISC	12.816 (−4.70 %)	−121.124 (−24.39 %)	13.995 (+4.06 %)	−179.671 (+12.16 %)
b18	13.774 (−1.58 %)	−82.244 (−35.44 %)	13.882 (−0.80 %)	−89.664 (−29.61 %)
des_perf	12.591 (−3.39 %)	−7.080 (−34.49 %)	12.969 (−0.49 %)	−8.816 (−18.43 %)
vga_lcd	14.769 (−1.28 %)	−29.360 (−10.62 %)	14.690 (−1.80 %)	−28.165 (−14.26 %)
b19	16.091 (−2.06 %)	−140.633 (−31.19 %)	16.429 (−0.01 %)	−99.188 (−51.47 %)

14.10 Conclusions

The 3D IC stacking requires TSV for interconnection between wafers, and STI is pervasive in nano-meter technology. Cu TSV causes thermal tensile stress in 3D ICs, whereas silicon dioxide in STI causes thermal compressive stress. Both stresses can lead to significant timing variations. Stress, though commonly believed to have negative impact on timing can actually be taken advantage of for timing optimization, since it is a strongly layout-dependent and systematic effect. In this chapter, we studied the first-order compact model for TSV-stress-induced mobility variation and STI-stress-induced mobility variation. We also studied a design methodology to analyze the systematic variation and optimize layout by locating critical cells in a mobility enhanced region of TSVs or changing STIs surrounding them. Our TSV-STI-stress-aware timing analysis framework for 3D ICs also opened the opportunity for stress-aware layout optimizations, such as placement and TSV-STI optimizations.

References

1. D.B. Bogy, Two edge-bonded elastic wedges of different materials and wedge angles under surface tractions. J. Appl. Mech. **38**, 377–386 (1971)
2. T. Dao, D.H. Triyoso, M. Petras, M. Canonico, Through silicon via stress characterization, in *IEEE International Conference on IC Design and Technology* (IEEE, Piscataway, 2009)
3. M.A. Hopcroft, W.D. Nix, T.W. Kenny, What is the Young's modulus of silicon? J. Microelectromech. Syst. **19**(2), 229–238 (2010)
4. H. Irie, K. Kita, K. Kyuno, A. Toriumi, In-plane mobility anisotropy and universality under uni-axial strains in n- and p-MOS inversion layers on (100), (110), and (111) Si, in *IEEE International Electron Devices Meeting* (IEEE, Piscataway, 2004)
5. M. Jung, J. Mitra, D. Pan, S.K. Lim, TSV stress-aware full-chip mechanical reliability analysis and optimization for 3D IC, in *Proceedings of Design Automation Conference* (IEEE, Piscataway, 2011)
6. A.B. Kahng, P. Sharma, R.O. Topaloglu, Exploiting STI stress for perform, in *Proceedings of International Conference on Computer Aided Design*, San Jose, 2007
7. A.B. Kahng, P. Sharma, A. Zelikovsky, Fill for shallow trench isolation CMP, in *Proceedings of International Conference on Computer Aided Design*, San Jose (Association for Computing Machinery, New York, 2006)
8. D.H. Kim, K. Athikulwongse, S.K. Lim, A study of through-silicon-via impact on the 3-D stacked IC layout, in *Proceedings of International Conference on Computer Aided Design*, San Jose, 2009
9. K.H. Lu, X. Zhang, S.-K. Ryu, J. Im, R. Huang, P.S. Ho, Thermo-mechanical reliability of 3-D ICs containing through silicon vias, in *Electronic Components and Technology Conference* (IEEE, Piscataway, 2009)
10. M.S. Lundstrom, On the mobility versus drain current relation for a nanoscale MOSFET. IEEE Electron Device Lett. **22**, 293–295 (2001)
11. M. Miyamoto and other, Impact of reducing STI-induced stress on layout dependence of MOSFET characteristics. IEEE Trans. Electron Devices **51**, 440–443, (2004)
12. V. Moroz, L. Smith, X.-W. Lin, D. Pramanik, G. Rollins, Stress-aware design methodology, in *Proceedings of International Symposium on Quality Electronic Design* (IEEE Computer Society, Los Alamitos, 2006)
13. C.S. Selvanayagam, J.H. Lau, X. Zhang, S. Seah, K. Vaidyanathan, T.C. Chai, Nonlinear thermal stress/strain analysis of copper filled TSV and their flip-chip microbumps, in *Electronic Components and Technology Conference* (IEEE, Piscataway, 2008)
14. N. Serin, T. Serin, S. Horzum, Y. Celik, Annealing effects on the properties of copper oxide thin films prepared by chemical deposition. Electron. J. **20**, 398–401 (2005)
15. C.S. Smith, Piezoresistance effect in germanium and silicon. Phys. Rev. **94**, 42–49 (1954)
16. S. Suthram, J.C. Ziegert, T. Nishida, S.E. Thompson, Piezoresistance coefficients of (100) silicon nMOSFETs measured at low and high channel stress. IEEE Electron Device Lett. **28**, 58–60 (2007)
17. S.E. Thompson, M. Armstrong, C.A. et al., A 90 nm logic technology featuring strained-silicon. IEEE Trans. Electron Devices **51**, 1790–1797 (2004)
18. S.E. Thompson, G. Sun, Y.S. Choi, T. Nishida, Uniaxial-process-induced strained-si: extending the CMOS roadmap. IEEE Trans. Electron Devices **53**, 1010–1020 (2006)

19. R. Tian, X. Tang, M.D.F. Wong, Dummy-feature placement for chemical-mechanical polishing uniformity in a shallow-trench isolation process. IEEE Trans. Comput. Aided Des. Integr. Circuits Syst. **21**, 63–71 (2002)
20. K. Uchida, T. Krishnamohan, K. Saraswat, Y. Nishi, Physical mechanisms of electron mobility enhancement in uniaxial stressed MOSFETs and impact of uniaxial stress engineering in ballistic regime, in *IEEE International Electron Devices Meeting* (IEEE, Piscataway, 2005)
21. J.-S. Yang, K. Athikulwongse, Y.-J. Lee, S.K. Lim, D.Z. Pan, TSV stress aware timing analysis with applications to 3D-IC layout optimization, in *Proceedings Design Automation Conference* (IEEE, Piscataway, 2010)
22. W. Zhao, Y. Cao, New generation of predictive technology model for sub-45 nm early design exploration. IEEE Trans. Electron Devices **53**, 2816–2823. (2006)

Chapter 15
Chip/Package Co-analysis of Mechanical Stress for 3D IC

Abstract In this chapter, we study an efficient chip/package thermo-mechanical stress and reliability co-analysis tool for TSV-based 3D ICs. We also study a design optimization methodology to alleviate mechanical reliability issues in 3D IC. First, we analyze the stress induced by chip/package interconnect elements, i.e., TSV, μ-bump, and package bump. Second, we explore and validate the principle of lateral and vertical linear superposition of stress tensors (LVLS), considering all chip/package elements. This linear superposition principle is utilized to perform full-chip/package-scale stress simulations and reliability analysis. Finally, we study the mechanical reliability issues in practical 3D chip/package designs including wide-I/O and block-level 3D ICs.

The materials presented in this chapter are based on [6].

15.1 Introduction

Most previous works on the thermo-mechanical stress and reliability of TSV-based 3D ICs have been done separately in chip or package domain. The impact of TSV-induced stress due to coefficient of thermal expansion (CTE) mismatch between TSV and substrate materials on device performance [1] and crack growth in TSV [8] were studied in the chip domain. As for the package domain, many works focused on the reliability of package bump (= C4 bump) [10]. Recently, authors in [9] showed the significant impact of package components on the chip domain stress. They proposed a stress exchange file to transfer the boundary conditions from package-level to silicon-level analysis. However, all of these approaches require FEA methods which are computationally expensive or infeasible for full-chip or -package analysis.

S.K. Lim, *Design for High Performance, Low Power, and Reliable 3D Integrated Circuits*, DOI 10.1007/978-1-4419-9542-1_15,

To overcome the limitation of FEA method, linear superposition of stress tensors [5] and response surface method [4] were utilized. However, all of these are limited to the chip domain analysis. In this chapter, we study a full-chip/package-scale thermo-mechanical stress and reliability co-analysis flow as well as a design optimization methodology to reduce the mechanical reliability problems in TSV-based 3D ICs. We show the impact of design parameters such as the size and pitch of chip/package interconnect elements and the number of dies in the stack on thermo-mechanical stress and reliability. The following specific topics are covered in this chapter:

- We study how to model the impact of mechanical stress caused by the CTE mismatch among TSV, substrate, and package on the reliability of chip/package system. Compared with existing works, we simulate more detailed 3D IC structures including both chip and package components and study their interaction and impact on thermo-mechanical stress and reliability.
- We study how to utilize the reliability models to perform full-chip/package mechanical analysis. We use and validate the principle of lateral and vertical linear superposition of stress tensors induced by each chip/package interconnect element such as TSV, μ-bump, and package-bump against FEA simulations. We apply this methodology to generate a stress map and a reliability metric map in full-chip scale.
- We study the mechanical stress and reliability issues in practical 3D chip/package designs including wide-I/O and block-level 3D ICs.

15.2 Motivation

We first examine how various chip/package interconnect components interact and alter the thermo-mechanical stress distribution on the *device layer around TSV* caused by the CTE mismatch between TSV and substrate materials. First, we only consider TSV and substrate which most previous works studied. We employ the same simulation structure used in [5] as shown in Fig. 15.1a. Then, we add a μ-bump and underfill layer *above* the substrate as shown in Fig. 15.1b. All structures undergo $\Delta T = -250^{\circ}C$ of thermal load (annealing/reflow 275°C → room temperature 25°C). As Fig. 15.2 shows, by adding the μ-bump layer (= dotted red line), we see slightly more tensile (= positive) stress than the TSV-only case (= solid black line). This is because ΔCTE of μ-bump and underfill is 24 ppm/K, while that of TSV and substrate is 14.7 ppm/K, hence the deformation of the entire structure is largely determined by the μ-bump and underfill layer. Since the top side of μ-bump layer is free surface, the entire structure easily bends upward as all the elements shrink from the negative thermal load as shown in Fig. 15.1e. Thus, the materials on device layer stretch outward, which results in more tensile stress. Note that if there were no substrate layer *below* the μ-bump layer, the top and bottom side of this μ-bump layer would show symmetrical bending behavior.

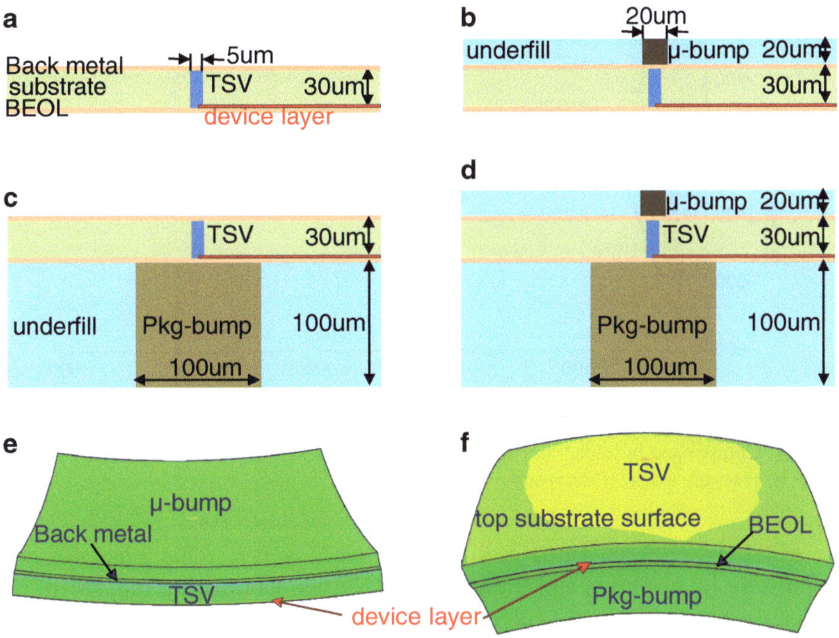

Fig. 15.1 Impact of bumps and underfill on the stress of device layer (= *red line*). (**a**) TSV only [5] (**b**) TSV + μ-bump (**c**) TSV + package-bump (**d**) TSV + μ-bump + package-bump. (**e**) Deformed structure of (**b**). (**f**) Deformed structure of (**c**). Both (**e**) and (**f**) are drawn with 10X deformation scale factor

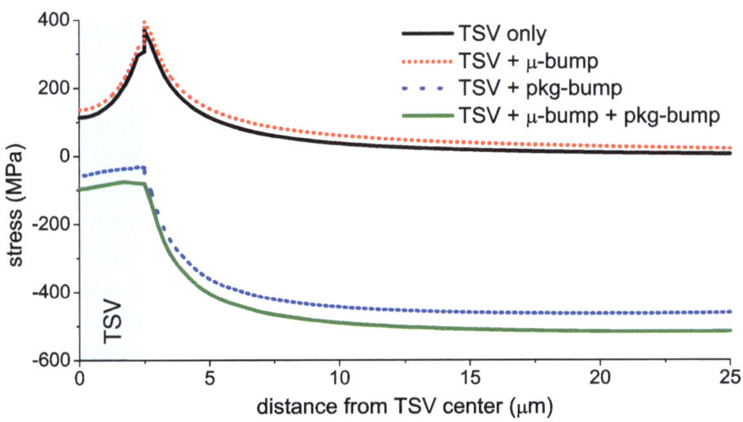

Fig. 15.2 Impact of package components on the stress (σ_{rr}) around TSV on device layer (FEA results)

On the other hand, if we add a package-bump (= C4 bump) layer *below* the substrate as shown in Fig. 15.1c, now the entire structure bends downward as shown in Fig. 15.1f because package elements are shrinking more than chip elements.

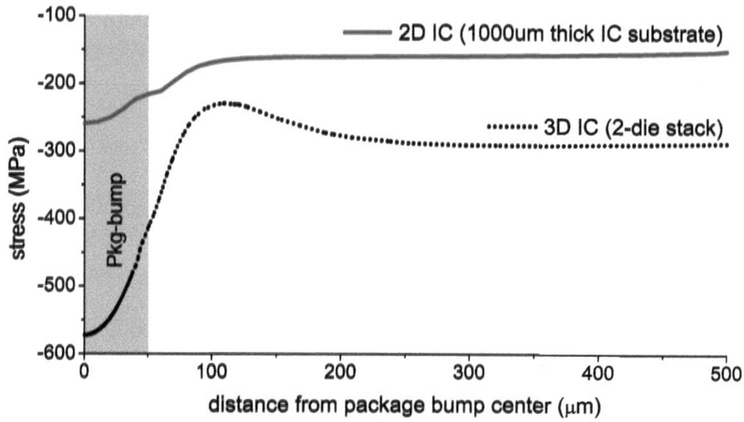

Fig. 15.3 Comparison of impact of package-bump on the device layer stress (σ_{rr}) between 2D IC and 3D IC (two-die stack) (FEA results)

The ΔCTE of package bump and underfill is 22 ppm/K. This generates highly compressive (= negative) stress on the device layer. Comparing Fig. 15.1b, c, we see that the bending direction depends on which layer shrinks more: in both cases, the bump layers shrink more than the silicon substrate.

Lastly, we include both bump layers as shown in Fig. 15.1d. In this case, the ΔCTE is almost the same (24 ppm/K on the top, 22 ppm/K on the bottom). However, the overall structure bends down in a similar fashion as shown in Fig. 15.1f because of the sheer volume of package bump layer (= shrinking more than the μ-bump layer). This in turn causes compressive stress in the device layer. However, the magnitude is slightly *more* (= solid green line in Fig. 15.2) than the package-bump layer only case (= dotted blue line). One might expect the overall compressive stress would be less because the μ-bump layer tries to bend upward while the package-bump layer tries to bend downward (= canceling effect). However, this additive effect is because the μ-bump layer eventually bends down and adds more compressive stress to the device layer. Remember that the bending direction of the μ-bump layer is affected by adjacent layers. Since now the deformation of the entire structure is dominated by the package-bump layer, the flexible underfill material in the μ-bump layer easily bends downward. These basic simulations clearly show the importance of considering package element impact on the chip-domain stress distribution.

Figure 15.3 shows the stress contributions of package bump and underfill layer to the chips (2D vs. 3D) mounted on it. For the 3D IC/package structure, we build a two-die stack chip/package structure similar to Fig. 15.4a excluding TSV and μ-bump. This was to examine the impact of package-bump solely. The bottom die (= die0) is thinned, and we examine the device layer of this thin die. One 2D IC/package structure is also created, where we use a single un-thinned die of 1,000 μm thickness. We examine the device layer of this un-thinned die. We observe in Fig. 15.3 that the 3D IC experiences more severe compressive stress than the 2D

IC case. The main reason is the thickness and the flexibility of the die that we are monitoring. Even though the thickness of the entire structure is thicker in 3D IC, the thin die (30 μm thick) and the underfill material above the thin die is much more flexible than the un-thinned substrate in 2D IC. Thus, this thin die is highly affected by the package-bump underneath it. This indicates that the impact of package-bump is more significant in 3D IC.

15.3 3D IC/Package Stress Modeling

We use the von Mises yield criterion [11] as a mechanical reliability metric for TSVs. However, we do not use a specific threshold value for the von Mises criterion in this chapter, since it is greatly affected by fabrication process.

15.3.1 Stress Tensor and Von Mises Criterion

To help understand stress modeling results, we introduce the concept of a stress tensor. Stress at a point in a body can be described by the nine-component stress tensor:

$$\sigma = \sigma_{ij} = \begin{bmatrix} \sigma_{11} & \sigma_{12} & \sigma_{13} \\ \sigma_{21} & \sigma_{22} & \sigma_{23} \\ \sigma_{31} & \sigma_{32} & \sigma_{33} \end{bmatrix}$$

where, the first index i indicates that the stress acts on a plane normal to the i axis, and the second index j denotes the direction in which the stress acts. If index i and j are same we call this a normal stress, otherwise a shear stress. Since we adopt a cylindrical coordinate system in this modeling for the cylindrical TSV, μ-bump, and package-bump, index 1, 2, and 3 represent r, θ, and z, respectively.

In order to evaluate if computed stresses indicate possible reliability concerns, a critical value for a potential mechanical failure must be chosen. The von Mises yield criterion is known to be one of the most widely used mechanical reliability metric [11]. If the von Mises stress exceeds a yielding strength, material yielding starts. Prior to the yielding strength, the material will deform elastically and will return to its original shape when the applied stress is removed. However, if the von Mises stress exceeds the yield point, some fraction of the deformation will be permanent and non-reversible even if applied stress is removed [5].

There is a large variation of yield strength of Cu in the literature, from 225 MPa to 3.09 GPa, and it has been reported to depend upon thickness, grain size, and temperature [11]. In this chapter, rather than selecting a specific value of yield stress for Cu TSV, we show how von Mises stress level changes under various circumstances. The yield strength of silicon is 7,000 MPa, which will not be reliability concerns for the von Mises yield criterion.

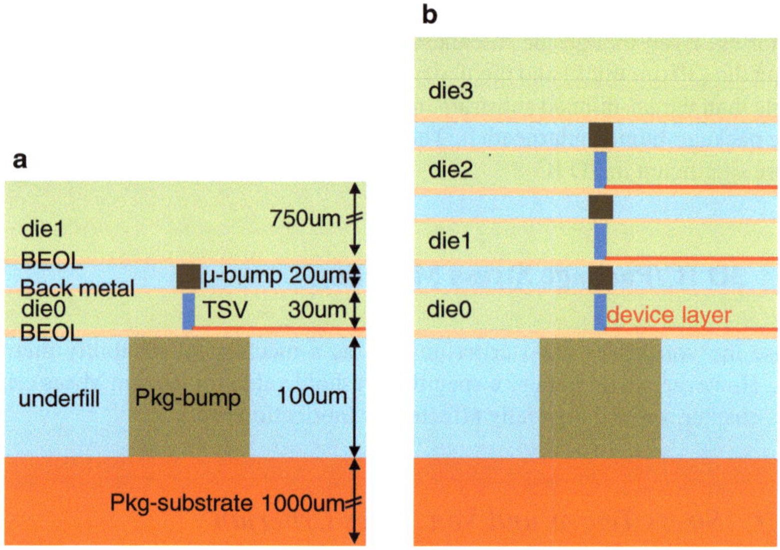

Fig. 15.4 Side view of baseline chip/package simulation structures. (**a**) Two-die stack. (**b**) Four-die stack

The von Mises stress is a scalar value at a point that can be computed using components of a stress tensor shown in Eq. 15.1. By evaluating von Mises stress at the interface between TSV and dielectric liner, where highest von Mises stress occurs, we can predict mechanical failures in TSVs.

$$\sigma_v = \sqrt{\frac{(\sigma_{xx}-\sigma_{yy})^2+(\sigma_{yy}-\sigma_{zz})^2+(\sigma_{zz}-\sigma_{xx})^2+6(\sigma_{xy}^2+\sigma_{yz}^2+\sigma_{zx}^2)}{2}} \qquad (15.1)$$

15.3.2 3D IC/Package Simulation Structure

Figure 15.4 shows our simulation structure, where the dimensions of our baseline simulation structures are based on the fabricated and/or published data [2, 9]. In this chapter, we specifically examine the stress distribution on device layer for each die shown in red lines in Fig. 15.4. Our baseline TSV diameter, height, landing pad size, Cu diffusion barrier thickness, and dielectric liner thickness are 5 µm, 30 µm, 6 µm, 50 nm, and 125 nm, respectively. We use Ti and SiO_2 as Cu diffusion barrier and liner materials. Also, diameter/height of µ-bump and package-bump are 20 and 100 µm, respectively, unless otherwise specified. Material properties used for our experiments are as follows: CTE (ppm/K) / Young's modulus (GPa) for Cu = (17/110), Si = (2.3/188), SiO_2 = (0.5/71), Ti = (8.6/116), package-bump (SnCu) = (22/44.4), μ-bump ($Sn_{97}Ag_3$) = (20/26.2), underfill = (44/5.6), package substrate (FR-4) = (17.6/19.7).

We use a FEA simulation tool ABAQUS to perform experiments, and all materials are assumed to be linear elastic and isotropic. We also discuss the impact of the anisotropic Si material property on the thermo-mechanical stress and reliability in Sect. 15.3.6. The entire structure undergoes $\Delta T = -250°C$ of thermal load (annealing/reflow 275°C → room temperature 25°C) to represent a fabrication process. In addition, all materials are assumed to be stress free at the annealing/reflow temperature.

15.3.3 Impact of Die Stacking

Previous works on the full-chip thermo-mechanical stress analysis used the same stress pattern for different dies in a multiple-die stack [1, 5]. In this section, for the first time, we examine how the thermo-mechanical stress distribution on the device layer around TSV differs across strata. We employ the four-die stack structure for this purpose. Also, we use only one TSV, μ-bump, and package-bump for each die or layer, respectively, and their center locations are aligned as shown in Fig. 15.4.

First of all, the stress level, the extent of compression or tension, differs significantly across dies as shown in Fig. 15.5a. The overall stress trend remains similar: the stress is highest at the TSV edge and decays then saturates as distance increases from the TSV center. However, the bottom-most die (= die0, solid red line), which is closest to the package-bump layer, shows the most compressive stress among three dies containing TSV. This is because the impact of package-bump is most significant in die0 due to their proximity.

Also, as we go to the upper dies, the stress level becomes closer to the case considering only TSV and substrate. We also see that the stress curve of die0 is very close to the case of TSV + μ-bump + package-bump (= dotted purple line), which does not contain the package substrate and un-thinned top die shown in Fig. 15.1d. This also indicates that the stress level in die0 is mostly determined by package-bump. The stress distribution in die3 (un-thinned top die without TSVs) is almost flat (-110 ± 5 MPa). Since die3 does not contain any TSVs, there is no local von Mises stress peak (= dangerous region) caused by TSVs. Thus, we only consider the dies containing TSVs in this chapter.

Moreover, we observe that the mechanical reliability problem is most severe in die0 shown in Fig. 15.5b. The maximum von Mises stress at TSV edge in die0 is about 110 MPa higher than the upper two dies. This is again mostly due to the package-bump that induces large deformation at the nearest die.

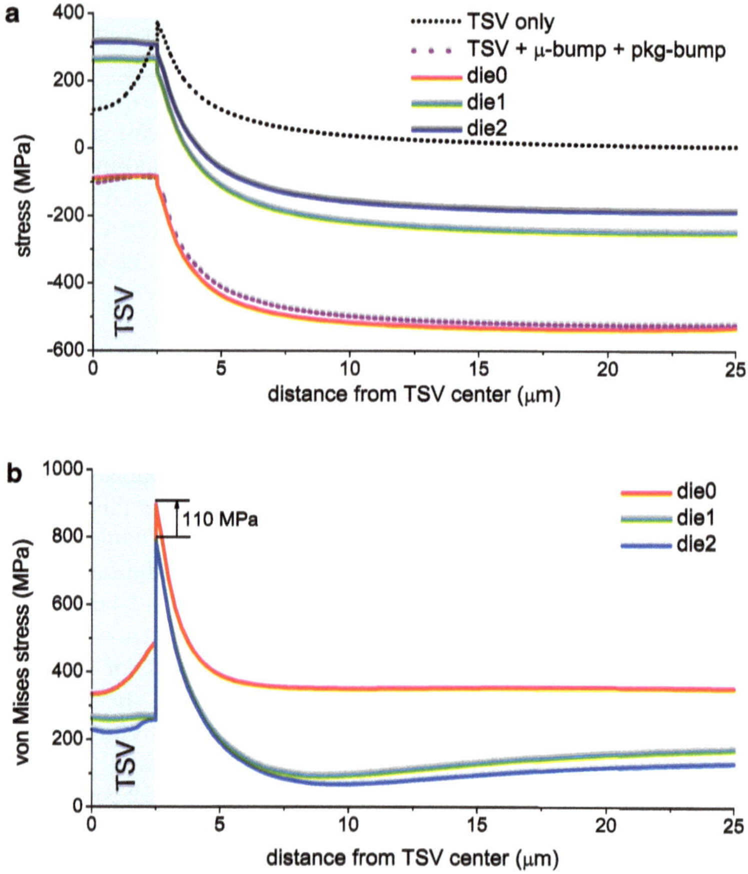

Fig. 15.5 Impact of die stacking on device layer stress (FEA results). (**a**) σ_{rr} stress on device layer in each die in four-die stack. (**b**) Von Mises stress in each die in a four-die stack

15.3.4 Impact of Thickness of Substrate

In this section, we study the impact of the thickness of package substrate and un-thinned top silicon substrate on the thermo-mechanical stress. We use a 1 mm thick package substrate and a 750 μm thick un-thinned top die as a baseline structure.

We first vary the package substrate thickness from 0.75 to 3 mm, and monitor the stress around TSV on device layer in die0 in a four-die stack structure. We observe that the stress becomes more compressive as thickness increases, but the difference is almost indistinctive shown in Fig. 15.6a. This is mainly because this package substrate is already much thicker than other layers, hence its impact on device layer is negligible.

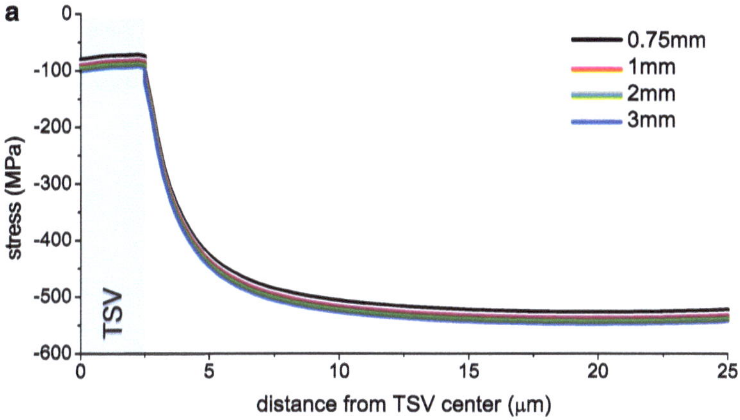

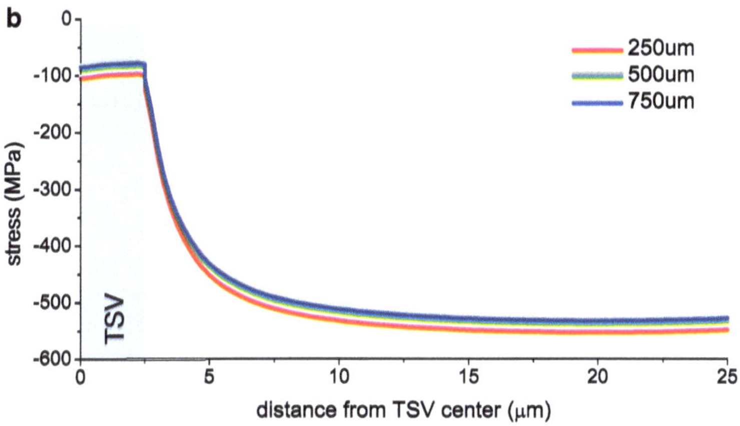

Fig. 15.6 Impact of package substrate and un-thinned *top die* thickness on stress (FEA results). (**a**) Impact of package substrate thickness. (**b**) Impact of un-thinned *top die* thickness

We also change the thickness of the un-thinned top die from 250 to 750 μm, and observe that thinner die induces more compressive stress. This is because thinner die is more flexible as we see in the thin die case, and hence helps the entire structure bend more easily. However, still the difference is not significant. Thus, we use the baseline 1 mm thick package substrate and 750 μm thick un-thinned top die in our experiments.

15.3.5 Impact of Multiple Die Stacking

We now examine the stress magnitude in each die with a different number of die stacking. Figure 15.7 shows stress distributions in die0 with a two-die, a three-die, and a four-die stack. As more dies are stacked, more compressive stress occurs

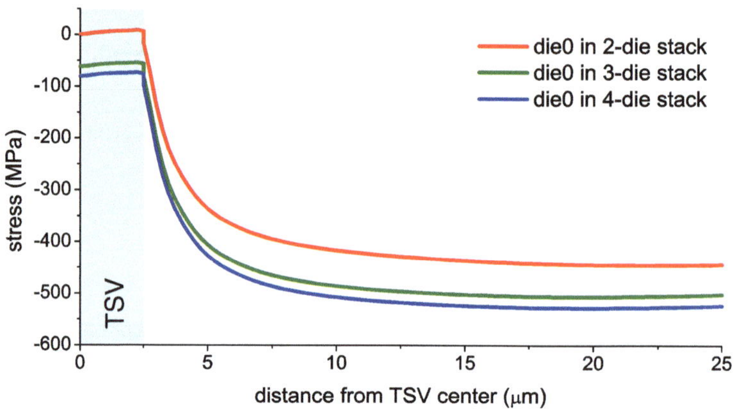

Fig. 15.7 σ_{rr} stress on die0 with a different number of die stacking

in die0 due to the additional stress from dies above. However, we see that this difference becomes smaller as we go to the upper dies, e.g., die1 stress in a three-die and a four-die stack.

15.3.6 Isotropic Versus Anisotropic Si Property

Up to this point, all materials are assumed to be isotropic for simplicity. However, Si is an anisotropic material with elastic behavior that depends on which crystal direction the structure is being stretched. The possible values of Young's modulus (E), which is a measure of stiffness of a material, for Si range from 130 to 188 GPa, and those for Poisson's ratio (ν) range from 0.048 to 0.4. Thus, the choice of this value can affect analysis results significantly [3]. In this section, we examine the impact of anisotropic material property of Si on the stress distribution compared with the isotropic Si material property.

Elasticity is the relationship between stress (σ) and strain (ε). Hooke's law describes this relationship in terms of stiffness C, i.e., $\sigma = C\varepsilon$. For isotropic uniaxial cases, stiffness C can be represented by a single value of Young's modulus E, and the equation takes the form of $\sigma = C\varepsilon$. In an anisotropic material, a fourth rank stiffness tensor with $3^4 = 81$ terms is required to describe the elasticity. Fortunately, due to the cubic symmetry of Si, the elastic properties can be expressed in terms of orthotropic material constants. An orthotropic material is one which contains at least two orthogonal planes of symmetry, and Si, with cubic symmetry, can be described this way. The orthotropic elasticity of Si can be expressed with reference axes of a standard (100) Si wafer, which are [110], $[\bar{1}10]$, and [001],

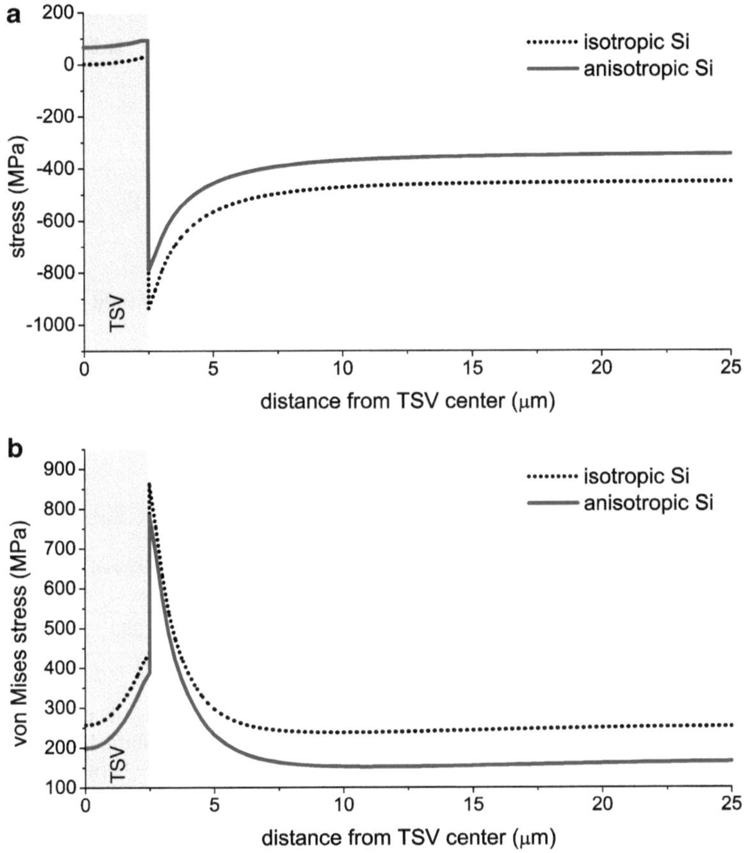

Fig. 15.8 Impact of silicon material property on stress (FEA results). (**a**) $\sigma_{\theta\theta}$ stress. (**b**) Von Mises stress

$$\begin{bmatrix} \sigma_{xx} \\ \sigma_{yy} \\ \sigma_{yy} \\ \sigma_{yz} \\ \sigma_{zx} \\ \sigma_{xy} \end{bmatrix} = \begin{bmatrix} c1 & c5 & c6 & 0 & 0 & 0 \\ c5 & c1 & c6 & 0 & 0 & 0 \\ c6 & c6 & c2 & 0 & 0 & 0 \\ 0 & 0 & 0 & c3 & 0 & 0 \\ 0 & 0 & 0 & 0 & c3 & 0 \\ 0 & 0 & 0 & 0 & 0 & c4 \end{bmatrix} \begin{bmatrix} \varepsilon_{xx} \\ \varepsilon_{yy} \\ \varepsilon_{yy} \\ \varepsilon_{yz} \\ \varepsilon_{zx} \\ \varepsilon_{xy} \end{bmatrix}$$

where, orientation specific constants c1, c2, c3, c4, c5, c6 are 194.5, 165.7, 79.6, 50.9, 35.7, and 64.1, all in GPa, respectively. This stiffness tensor translates to $E_x = E_y = 169\,\text{GPa}$, $E_z = 130\,\text{GPa}$, $\nu_{yz} = 0.36$, $\nu_{zx} = 0.28$, and $\nu_{xy} = 0.064$ [3].

Figure 15.8 shows the stress comparison between anisotropic and isotropic Si (Young's modulus = 188 GPa for all directions) material properties. We see that the normal stress component becomes less compressive and the von Mises stress is

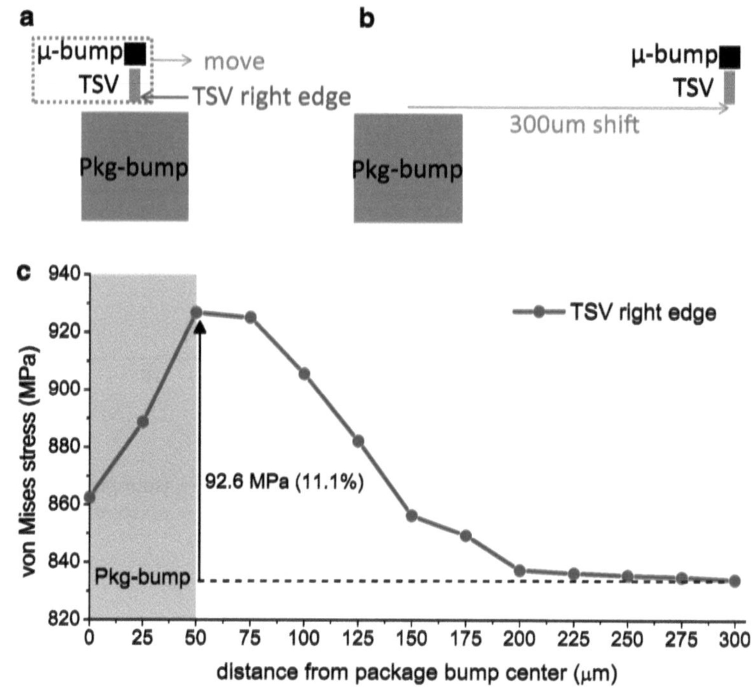

Fig. 15.9 Impact of relative position between TSV/μ-bump and package-bump on von Mises stress. (**a**) Initial position. (**b**) Final position where TSV/μ-bump are shifted by 300 μm from package bump center. (**c**) Von Mises stress at TSV edge along the distance between TSV/μ-bump and package-bump (FEA results)

lower with the anisotropic Si compared with the isotropic Si case. This is largely due to the fact that we use the maximum Young's modulus for the isotropic Si case. With higher Young's modulus Si substrate becomes stiffer, hence higher stress builds up at the TSV/substrate interface. In this chapter, even though anisotropic Si property is more realistic, we use the isotropic Si property as a worst case scenario.

15.3.7 Impact of TSV and Bump Alignment

In this section, we explore the impact of alignment between TSV, μ-bump, and package-bump on the mechanical reliability of TSVs. We first examine the impact of relative position between TSV/μ-bump and package-bump. We use a two-die stack structure in which center locations of TSV, μ-bump, and package-bump are aligned as shown in Fig. 15.9a. Then we shift both TSV and μ-bump together from the package-bump center with a 25 μm step and monitor the von Mises stress at the right edge of TSV.

Figure 15.9c shows that the von Mises stress is maximum around package-bump edge region and then decreases and saturates as distance increases. The difference between minimum and maximum is as high as 11.1 %. As Fig. 15.3 shows, the highest stress gradient occurs around package-bump edge which results in the highest deformation of the structure near this region. Hence, this higher deformation causes more severe mechanical reliability problem in TSV.

We also see the decrease in von Mises stress near the package-bump center. This is because the material around this area is the same (= package-bump material), hence its deformation is relatively smaller than the edge which is the interface between two different materials.

We also examine whether relative position between μ-bump and TSV/package-bump affects the mechanical reliability of TSV. We fix the location of TSV and package-bump whose centers are aligned, then move μ-bump only with a 5 μm step up to 30 μm and monitor the von Mises stress at TSV edges. We observe the similar trend as before. However, the difference between minimum and maximum is only 6.5 MPa (0.8 %), which is negligible. Thus, we identify that the relative position between TSV and package-bump is a critical factor that affects the mechanical reliability of TSV.

15.4 Handling Full-Chip/Package Co-analysis

FEA simulations for multiple TSVs, μ-bumps, and package-bumps require huge computing resources and time, thus it is not feasible for a full-system-scale analysis. In this section, we present a chip/package thermo-mechanical stress co-analysis flow in full-chip/package scale. We use the principle of lateral and vertical linear superposition of stress tensors from individual TSVs, μ-bumps, and package-bumps to enable a full-system-level analysis. We validate our approach against FEA simulation results. Based on the linear superposition method, we build full-chip stress maps and then compute the von Mises yield metric to assess the mechanical reliability problems in TSV-based 3D ICs.

15.4.1 Lateral and Vertical Linear Superposition

In [5], authors used the principle of linear superposition of stress tensors to perform a full-chip stress and reliability analysis considering many TSVs. In that case, all stress contributors (= TSVs) are on the same layer, hence we call this lateral linear superposition. However, as we consider the impact of μ-bump and package-bump, which are not in the same layer where TSVs are located, this lateral linear superposition cannot be used alone. Fortunately, the principle of linear superposition is not limited to 2D plane, but applicable to any linearly elastic structures including 3D structures.

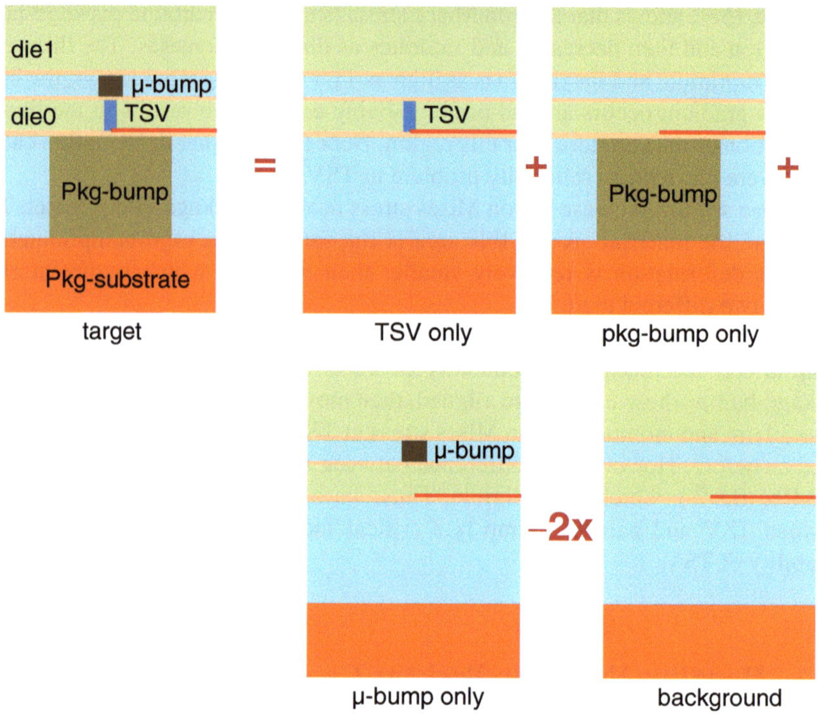

Fig. 15.10 Illustration of *vertical* linear superposition with a two-die stack structure. Stress is extracted along the *red line* on device layer from each structure using FEA tool

Figure 15.10 illustrates our vertical linear superposition method, which enables us to consider the stress induced by elements which are not in the same layer. We first decompose the target structure into four separate structures: TSV only, package-bump only, μ-bump only, and background which does not contain TSV and bumps. Next, we obtain stress tensors along the red line on device layer affected by each interconnect element separately from FEA simulations. Then, we add up the stress tensors from TSV only, package-bump only, and μ-bump only structures, and subtract twice the magnitude of the background stress tensors since this background stress is already included in previous three structures. If the point under consideration is affected by n components, then we need to subtract $n-1$ times the background stress.

Figure 15.11 shows the stress distributions from each structure as well as the stress obtained by the vertical linear superposition. We see that μ-bump induces more tensile stress than background and package-bump generates much more compressive stress than background, which is discussed in Sect. 15.2. We also observe that even without interconnect elements (= background) device layer is in compression due to the fast shrinking of the underfill material which has the highest CTE (= 44 ppm/K) among all materials in the simulation structure.

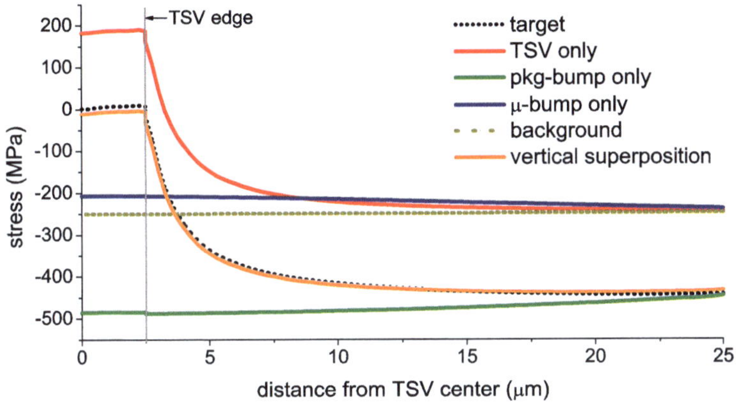

Fig. 15.11 *Vertical* linear superposition of σ_{rr} stress in a two-die stack shown in Fig. 15.10

Most importantly, our vertical linear superposition method matches well with the target stress distribution. Although we see the maximum error (11 MPa) occurs inside TSV, this is inevitable since we ignore the direct interaction between TSV, μ-bump, and package-bump by decomposing the structure. Nonetheless, this error is acceptable for a fast full-system-scale analysis.

To obtain the stress tensor at a point affected by multiple TSVs, μ-bumps, and package-bumps, we apply both lateral and vertical linear superposition (LVLS) as follows:

$$S = \sum_{i=1}^{n_{TSV}} S_{TSV\,i} + \sum_{j=1}^{n_{\mu B}} S_{\mu B\,j} + \sum_{k=1}^{n_{pkgB}} S_{pkgB_k} - (n_{TSV} + n_{\mu B} + n_{pkgB} - 1) \times S_{bg} \quad (15.2)$$

where, S is the total stress at the point under consideration and $S_{TSV\,i}$, $S_{\mu B\,j}$, and S_{pkgB_k} are individual stress tensor at this point due to *i*th TSV, *j*th μ-bump, and *k*th package-bump, respectively. S_{bg} indicates the background stress at that point.

15.4.2 Full-Chip/Package Stress Analysis Flow

We briefly explain how we perform a full-chip/package stress analysis based on the LVLS method. We first build a stress library from FEA simulations. This library contains stress tensors along an arbitrary radial line on the device layer induced by each interconnect element, i.e., TSV, μ-bump, and package-bump, separately. Given locations of TSVs, μ-bumps, and package-bumps, we find a stress influence zone for each element. Beyond this stress influence zone of each interconnect element, the stress induced by the element under consideration is negligible [5]. In our work, we use five times the diameter of each component as a stress influence zone, which

Table 15.1 Von Mises stress comparison between FEA and LVLS for a four-die stack structure (die0). Error = LVLS − FEA (At TSV edge, typical von Mises stress level is around 900 MPa)

# TSV	FEA		LVLS		Max error (MPa)		
/μ-B /pkg-B	# node	Run time	# grid	Run time	Inside TSV	TSV edge	Outside TSV
1/1/1	754 K	1 day 2 h	1 M	23 s	−11.4	−12.6	7.9
2/2/1	812 K	1 day 2 h	1 M	26 s	−12.7	−13.2	7.3
5/5/2	902 K	1 day 6 h	6 M	2 min 43 s	−14.1	−15.3	8.2
10/10/4	1.3 M	1 day 20 h	9 M	6 min 44 s	−23.1	−19.8	9.4
10/10/9	1.4 M	2 day 0 h	16.8 M	11 min 11 s	−22.5	−20.5	11.9

is determined by FEA simulations. Then, we associate each grid point with all the interconnect elements whose stress influence zone overlaps with the point. Next, we apply the LVLS method at the point under consideration to obtain the stress tensor induced by every component found in the association step. Finally, we compute the von Mises stress value using Eq. 15.1 to assess the mechanical reliability problem in TSVs. More details of our algorithm is discussed in Sect. 15.4.4.

15.4.3 Validation of LVLS

In this section, we validate our LVLS method against FEA simulations by varying the number of TSVs, μ-bumps, and package-bumps as well as their arrangement. We set the minimum pitch of TSV, μ-bump, and package-bump as 10, 20, and 200 μm for all test cases. Stress tensors along the radial line on device layer induced by each interconnect element (stress tensor library) are obtained through FEA simulation with 0.25 μm interval. In our linear superposition method, simulation area is divided into uniform array style grid with 0.1 μm pitch. If the stress tensor at the grid point under consideration is not obtainable directly from the stress library, we compute the stress tensor using linear interpolation with adjacent stress tensors in the library.

Table 15.1 shows some of our comparisons in die0 in a four-die stack, which shows the largest errors among three dies containing TSVs due to its proximity to package-bumps. Also, we only list the cases with the minimum pitches for each component, which again shows maximum errors. First, we observe a huge run time reduction in our LVLS method. Note that we perform FEA simulations using eight CPUs while only one CPU is used for our linear superposition method. Even though the LVLS method performs stress analysis on a 2D plane (= device layer), whereas FEA simulation is performed on the entire 3D structure, we can perform stress analysis for other planes in a similar way if needed.

The error between FEA simulations and LVLS is very small. Results show that our LVLS method underestimates stress magnitude inside TSV and TSV edge, and overestimates outside TSV, as shown in Fig. 15.11. In general, the most critical region for the mechanical reliability is the interface between different materials,

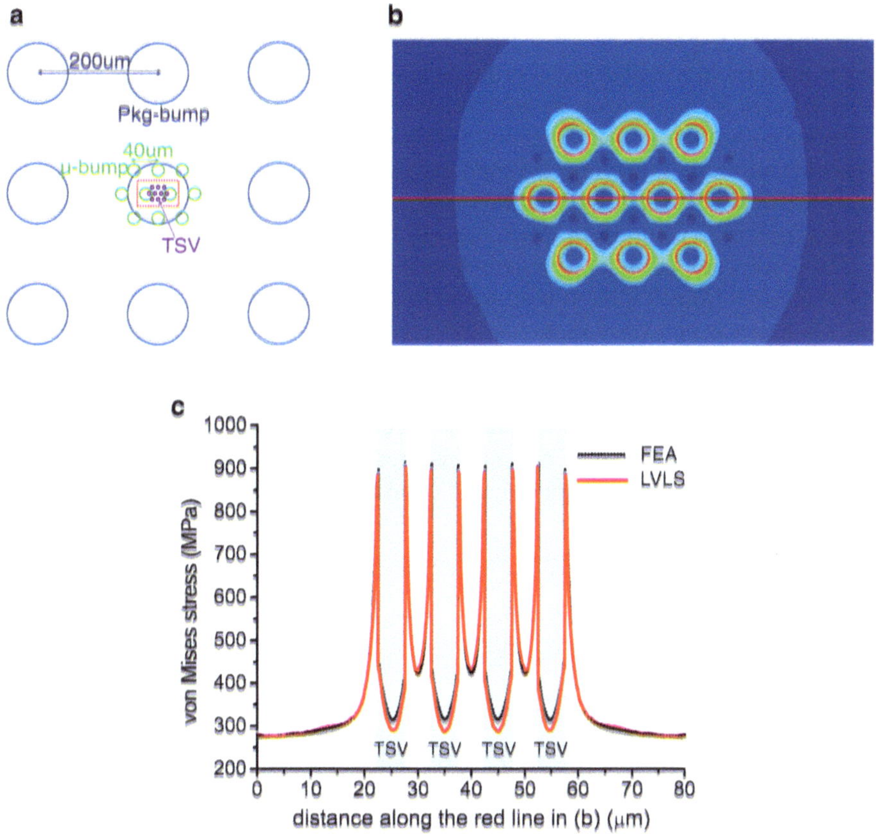

Fig. 15.12 Sample stress comparison between FEA and LVLS. (**a**) Test structure. (**b**) Close-up shot of von Mises stress map (using LVLS) taken from the *red box* in (**a**) on the device layer in die0 in a four-die stack. (**c**) FEA vs. LVLS along the *red line* in (**b**)

hence TSV edge is most important in our case. Even though the maximum error at TSV edge is as high as −20.5 MPa, its % error is only −2.24 %. Figure 15.12 shows one test case comparison of von Mises stress between FEA and LVLS. The structure has 10 TSVs (5 μm diameter and 10 μm pitch), 10 μ-bumps (20 μm diameter and 40 μm pitch), and 9 package-bumps (100 μm diameter and 200 μm pitch). It clearly shows our LVLS method matches well with the FEA simulation result.

15.4.4 Full-Chip/Package Analysis Algorithm

In this section, we discuss details of our full-system-scale thermo-mechanical stress and reliability analysis flow. First, based on the observation that the stress field induced by a single TSV, a μ-bump, and a package-bump in isolation is radially symmetrical due to their cylindrical shape, we obtain stress tensors for each interconnect component along an arbitrary radial line on the device layer from their center location in a cylindrical coordinate system. To evaluate the stress tensor at a point affected by multiple interconnect elements, a conversion of a stress tensor to a Cartesian coordinate system is required. This is due to the fact that we extract stress tensors from these interconnect components whose center position is the origin in the cylindrical coordinate system; hence we cannot perform a vector sum of stress tensors from each component which has a different center location.

Then, we compute the stress tensor at the point of interest by adding up the stress tensors from TSVs, μ-bumps, and package-bumps that affect this point. We set a stress influence zone of TSV, μ-bump, and package-bump 25, 100, 500 μm from the center of each component, which is five times the diameter of each component, respectively. This is because the magnitude of each stress tensor component saturates well before this distance, hence there is a negligible impact from the interconnect element beyond this stress influence zone.

Let the stress tensor in Cartesian and cylindrical coordinate system be S_{xyz} and $S_{r\theta z}$, respectively.

$$S_{xyz} = \begin{bmatrix} \sigma_{xx} & \sigma_{xy} & \sigma_{xz} \\ \sigma_{yx} & \sigma_{yy} & \sigma_{yz} \\ \sigma_{zx} & \sigma_{zy} & \sigma_{zz} \end{bmatrix}, \quad S_{r\theta z} = \begin{bmatrix} \sigma_{rr} & \sigma_{r\theta} & \sigma_{rz} \\ \sigma_{\theta r} & \sigma_{\theta\theta} & \sigma_{\theta z} \\ \sigma_{zr} & \sigma_{z\theta} & \sigma_{zz} \end{bmatrix}$$

The transform matrix Q is the form:

$$Q = \begin{bmatrix} \cos\theta & -\sin\theta & 0 \\ \sin\theta & \cos\theta & 0 \\ 0 & 0 & 1 \end{bmatrix}$$

where, θ is the angle between the x-axis and a line from the center of each interconnect element to the simulation grid point. A stress tensor in a cylindrical coordinate system can be converted to a Cartesian coordinate system using conversion matrices: $S_{xyz} = QS_{r\theta z}Q^T$ [5].

Our full-system-scale thermo-mechanical stress and reliability analysis flow is shown in Algorithm 10. We first start to find a stress influence zone from each TSV, μ-bump, and package-bump. Then, we associate the points in the influence zone with the affecting interconnect elements. Next, for each grid point under consideration, we look up the stress tensors from each interconnect component found in the association step, and subtract background stress from the stress tensor. Then, we use the coordinate conversion matrices to obtain stress tensors in the Cartesian coordinate system. We visit an individual TSV, μ-bump, and package-bump affecting this simulation point and add up their stress contributions. After visiting all the components effecting this point, we add one background stress back. Once we finish the stress computation at the point, we obtain the von Mises stress value using Eq. 15.1.

```
Algorithm: Lateral and Vertical Linear Superposition Method
input  : TSV list T, pkg-bump list P, μ-bump list M, stress library
output: stress map, von Mises stress map
for each TSV t, pkg-bump p, and μ-bump m in T, P, and M do
    (it,ip,im) ⟵ FindStressInfluenceZone(t,p,m);
    for each point it', ip', and im' in it, ip, and im do
        it'.TSV ⟵ it;
        ip'.pkg-bump ⟵ ip;
        im'.μ-bump ⟵ im;
    end
end
for each simulation point r do
    if r.TSV ≠ ∅ —— r.pkg-bump ≠ ∅ —— r.μ-bump ≠ ∅ then
        for each (t,p,m) ∈ (r.TSV,r.pkg-bump,r.μ-bump) do
            (dt,dp,dm) ⟵ distance(t,p,m,r);
            S_cyl(t,p,m) ⟵ GetStressTensor(dt,dp,dm);
            S_cyl(t,p,m) ⟵ S_cyl(t,p,m) − BGstress;
            θ(t,p,m) ⟵ GetAngle(linetr, pr, mr, x-axis);
            Q(t,p,m) ⟵ SetConversionMatrix(θ_t,θ_p,θ_m);
            S_Cart(t,p,m) ⟵ Q(t,p,m)S_cyl(t,p,m)Q(t,p,m)^T;
            r.S_Cart ⟵ r.S_Cart + S_Cart(t,p,m);
        end
    end
    r.S_Cart ⟵ r.S_Cart + BGstress;
    vonMises(r) ⟵ ComputeVonMises(r.S_cart);
end
```

Algorithm 10: Lateral and vertical linear superposition method (LVLS) for full-chip/package stress and reliability analysis

15.5 Experimental Results

We implement a chip/package thermo-mechanical stress and reliability co-analysis flow based on LVLS in C++/STL. More details can be found in Sect. 15.4.4. We explore the impact of package-bump and μ-bump on the reliability in full-system scale. Also, we examine the reliability concerns in wide-I/O DRAM and block-level 3D IC designs.

In our experiments, we adopt a regular TSV placement style in which TSVs are placed uniformly across each die or inside TSV blocks with pre-defined pitch. In all cases, the pair of TSV and μ-bump is vertically aligned. Default diameter/height (μm) of TSV, μ-bump, and package-bump are 5/30, 10/10, and 100/100, respectively, unless otherwise specified.

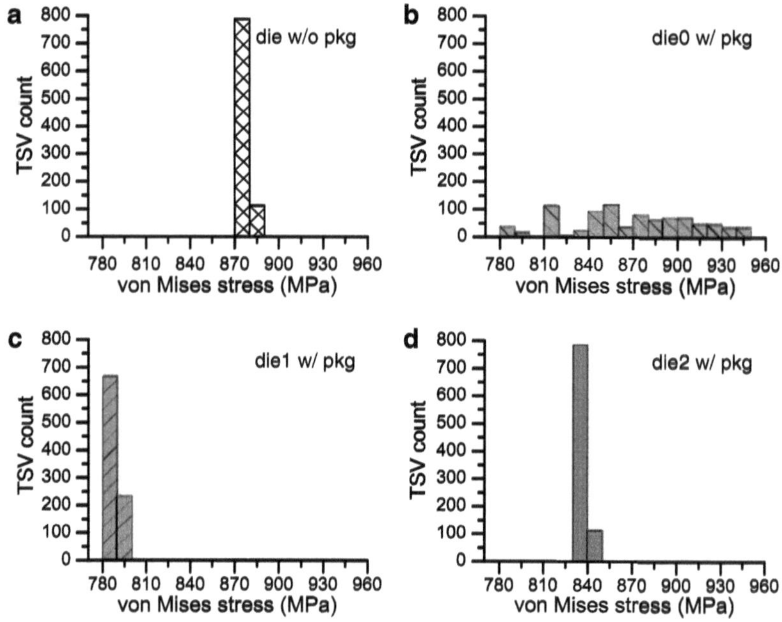

Fig. 15.13 Impact of package components and die stacking on the mechanical reliability of TSVs (900 TSVs in each die)

15.5.1 *Impact of Package-Bump and Micro-Bump*

We first study the impact of package-bump and μ-bump on the mechanical reliability of different dies in a four-die stack. We also compare this to the case without these components as in the previous work [5] as shown in Fig. 15.1a. In this experiment, the pitch of TSV/μ-bump and package-bump are 20 and 100 μm, respectively; the total number of TSV/μ-bump and package-bump are 900 and 16, respectively, as shown in Fig. 15.14a.

We first observe that unlike the die without package-bumps and μ-bumps (Fig. 15.13a) and the upper dies with package components (Fig. 15.13c, d), TSVs in die0 (Fig. 15.13b) experience large variations of von Mises stress across the die. This is because die0 is highly affected by package-bumps underneath it, and hence depending on the relative position between TSVs in die0 and package-bumps the von Mises stresses of TSVs change noticeably.[1]

We also identify that higher von Mises stress occurs around package-bump edge and in between package-bumps due to constructive stress interference shown in

[1]Note that we see higher von Mises stress level in (Fig. 15.13a than the previous work [5] even with the same simulation structure. This is because we use the Young's modulus of 188 GPa for Si instead of 130 GPa in [5] as a worst case scenario. More details are discussed in Sect. 15.3.6.

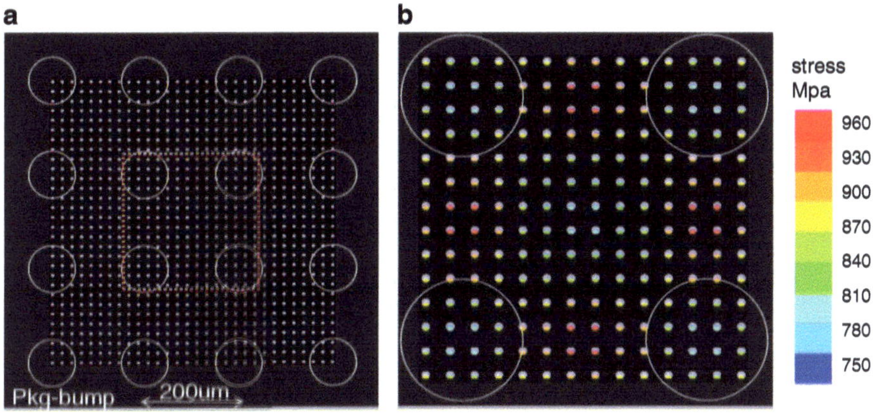

Fig. 15.14 Von Mises stress map for TSVs (die0 in a four-die stack). *Colored dots* are TSVs and *white circles* are package-bumps. (**a**) Test structure. (**b**) Close-up shot of *red box* in (**a**)

Fig. 15.14b. However, as we see in the center of Fig. 15.14b, if the distance between TSV and package-bumps is long enough, the von Mises stress of TSV becomes low.

Interestingly, die1 shows lowest von Mises stress level among all cases even though die2 is farthest from package-bumps. This is due to the fact that die2 is affected by the rigid un-thinned top silicon substrate above it. Since die0 is most problematic in terms of the mechanical reliability, we only consider die0 in a four-die stack in the subsequent experiments.

In addition, we identify that the von Mises stress distribution in die2 is similar to the die without package components case as the impact of package-bump decreases. Interestingly, die1 shows lowest von Mises stress among all cases. This is due to the fact that die2 is affected by the rigid un-thinned top silicon substrate.

15.5.2 Impact of Bump Size

In this section, we study the impact of package-bump and μ-bump size on the reliability. First, we vary the package-bump diameter/height from 100 to 300 μm, while fixing the package-bump pitch and the TSV/μ-bump count and pitch as 400, 1,600, and 20 μm, respectively. Table 15.2 shows that the number of TSVs experiencing higher von Mises stress increases with larger package-bumps due to the larger deformation of a stack and the increased package-bump circumference where the highest von Mises stress occurs. However, in the 300 μm package-bump case, there are more TSVs with lower von Mises stress (780–870 MPa) than the

Table 15.2 Maximum von Mises stress distribution of TSVs with different size of package-bump and μ-bump (die0 in four-die stack with 1,600 TSVs)

Von Mises	pkg-bump size (μm)			μ-bump size (μm)		
stress (MPa)	100 (%)	200 (%)	300 (%)	10 (%)	20 (%)	30 (%)
780–810	31.2	17	17.8	4.2	5.6	0
810–840	33.8	18	27.8	6.9	6.3	6.9
840–870	19	14	17.2	22.9	22.2	15.3
870–900	14	28.5	12.2	21.5	22.9	18.1
900–930	2	20.5	13.5	27.1	22.9	22.9
930–960	0	2	10	17.4	20.1	36.8
960–	0	0	1.5	0	0	0
Median (MPa)	824.6	871.7	848.2	893.3	890.1	908.0

200 μm package-bump case. As discussed in Sect. 15.3.7, TSVs located near the package-bump center region show lower von Mises stress than those around package-bump edge. Hence, with larger package-bumps more TSVs reside near the package-bump center, which results in lower von Mises stress level for these TSVs.

We now vary the μ-bump size, and use a 100 μm package-bump with a 200 μm pitch. Note that since we align center locations of TSV and μ-bump, we set the TSV pitch as 35 μm to accommodate the largest μ-bump diameter of 30 μm. We observe that larger μ-bump causes more TSVs to experience higher von Mises stress. However, this μ-bump size impact is less significant than the package-bump size.

15.5.3 Impact of TSV Size

In general, package-bumps and μ-bumps generate global stress distribution, while TSVs create local stress distribution. Therefore, TSV size and pitch are still critical factors that affect the mechanical reliability problem in TSVs even with the presence of other interconnect elements. In this section, we study the effect of TSV size. We use three different sizes of TSV with the same aspect ratio of 6; TSV small ($H/D = 15/2.5$ μm), TSV medium ($H/D = 30/5$ μm), and TSV large ($H/D = 60/10$ μm), where H/D is TSV height/diameter. We set the pitch of TSV and package-bump as 25 and 200 μm, respectively for all cases.

Figure 15.15 shows that smaller TSVs reduce the von Mises stress level significantly. This is mainly because larger TSV induces higher stress level at TSV edge due to the sheer volume of TSV. Also, the magnitude of normal stress components decay proportional to $(D/2r)^2$, where r is the distance from the TSV center. In other words, larger TSV affects larger area, hence increases stress level around it more than smaller TSV.

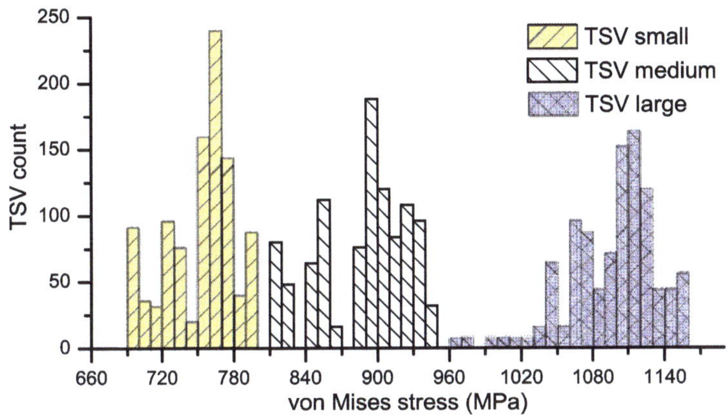

Fig. 15.15 Impact of TSV size on von Mises stress distribution of TSVs (die0 in four-die stack with 1,024 TSVs)

Table 15.3 Impact of package-bump and TSV/μ-bump pitch on von Mises stress (die0 in four-die stack with 900 TSVs)

von Mises	pkg-bump pitch (μm)			TSV/μ-bump pitch (μm)		
stress (MPa)	200 (%)	250 (%)	300 (%)	15 (%)	25 (%)	35 (%)
780–810	4.7	6.3	19.5	0.6	4.6	7.1
810–840	4.7	21.5	27.0	3.1	4.6	6.9
840–870	21.9	27	31.6	19.1	21.6	22.9
870–900	19.5	33.2	20.3	23.5	20.9	21.5
900–930	24.2	12.1	1.6	26.4	23.7	17.5
930–960	25.0	0	0	23.4	24.6	24.1
960–	0	0	0	3.9	0	0
Median (MPa)	897.9	863.9	844.1	901.8	897.9	893.2

15.5.4 Impact of Pitch

In this section, we explore the effect of package-bump and TSV/μ-bump pitch on the reliability. We employ a 100 μm package-bump and change its pitch from 200 to 300 μm. The pitch of TSV/μ-bump is set to 25 μm. Table 15.3 shows that larger package-bump pitch reduces the von Mises stress level noticeably by reducing constructive stress interference between package-bumps. However, we cannot arbitrarily increase the package-bump pitch considering the package size increase given the required number of pins.

We also examine the impact of TSV pitch on the von Mises stress. In this case, we set the package-bump pitch as 200 μm. In Table 15.3, we see that larger TSV pitch reduces von Mises stress level. However, there is not much difference between 25 and 35 μm pitch cases. This is because the stress influence zone of a 5 μm diameter TSV is 25 μm, hence there is a negligible difference between these two cases in terms of the stress induced by TSVs solely. Thus, in this case, the von

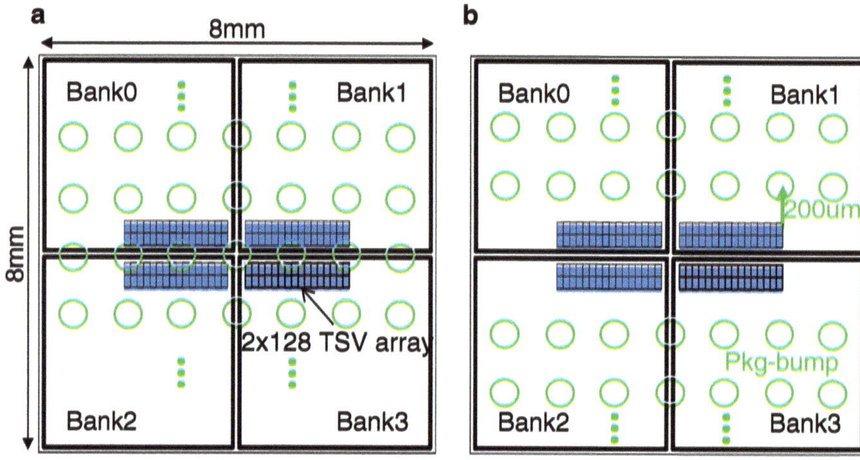

Fig. 15.16 Mechanical reliability in wide I/O DRAM. Thousand twenty four TSVs are placed in the *middle* of a chip. (**a**) Package-bumps are placed underneath TSV arrays. (**b**) Package-bumps are placed 200 μm apart from TSV arrays (not drawn in scale)

Mises stresses of TSVs are largely determined by relative position between TSVs and package-bumps. Therefore, the proper TSV placement considering the locations of package-bumps is a key design knob to mitigate the reliability concerns in TSV-based 3D ICs.

15.5.5 Case Study I: Wide-I/O DRAM

Wide-I/O based 3D DRAM is fast becoming the first mainstream product that utilizes TSV in 3D ICs, mainly targeting mobile computing applications such as smart phones which need lower power consumption and high data bandwidth. In this section, we evaluate the reliability concerns of TSVs in wide-I/O DRAM. We follow the TSV placement style similar to the work in [7], where TSV arrays are placed in the middle of a chip. We assume that 2×128 TSV array (per memory bank) is placed in the middle of a chip shown in Fig. 15.16. We employ four memory banks and 1,024 TSVs in total. We set the pitch of TSV/μ-bump and package-bump as 15 and 200 μm, respectively. We compare two cases; (a) Package-bumps are placed right underneath TSV arrays; (b) Package-bumps are placed with 200 μm spacing from TSV arrays. This 200 μm distance is chosen since we see that the effect of package-bump on the TSV reliability is negligible beyond 200 μm in case of the 100 μm diameter package-bump shown in Fig. 15.9.

Table 15.4 clearly shows that the chip/package co-design can greatly reduce the mechanical reliability concerns in TSV-based 3D ICs. With a safe margin of

Table 15.4 Reliability in wide-I/O DRAM

	von Mises stress distribution (MPa)					Median
Case	780–810	810–840	840–870	870–900	900–930	(MPa)
(a)	30	114	52	220	608	944.8
(b)	182	842	0	0	0	856.2

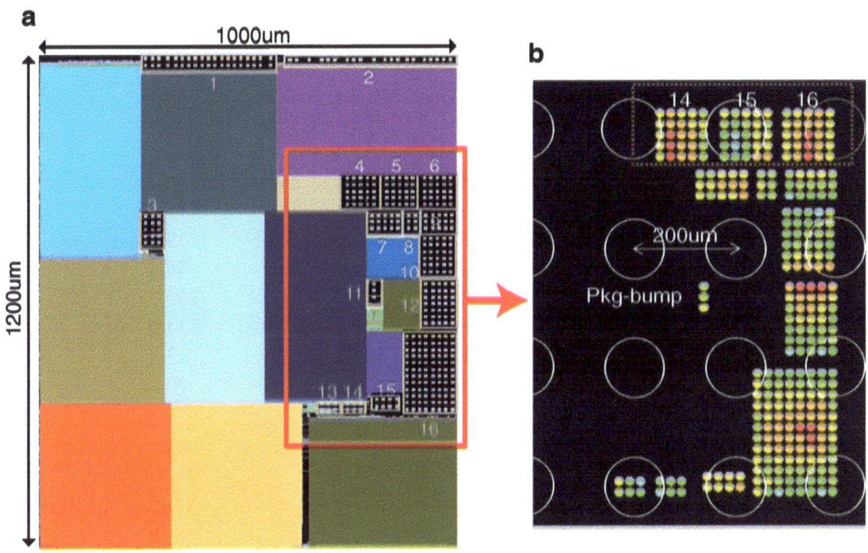

Fig. 15.17 Mechanical reliability in block-level 3D IC. (**a**) Sample layout of block-level design. (**b**) Von Mises stress map for TSVs in *red box* in (**a**)

200 μm (= case(b)), von Mises stress magnitude reduces significantly. Thus, given the TSV placement, we can find safe locations for package-bumps without affecting the package design much, or vice versa.

15.5.6 Case Study II: Block-Level 3D IC

In this section, we study the reliability issues in block-level 3D designs. 3D block-level designs are generated using an in-house 3D floorplanner which treats a group of TSVs as a block shown in Fig. 15.17. Total 16 TSV blocks (368 TSVs) are used and the TSV pitch is 15 μm. Package-bumps are regularly placed with 200 μm pitch.

Table 15.5 shows von Mises stress level in the TSV blocks. We first observe that larger TSV blocks experience more variation of von Mises stress within the TSV block. This is because the distance between each TSV in the block and package-bumps can vary more than small TSV blocks, which is a key factor that affects the

Table 15.5 Mechanical reliability in block-level 3D IC. TSV blocks are shown in Fig. 15.17

TSV block #	# TSV	Von Mises stress (MPa) Max	Min	Avg	Std dev	blk-bump dist (μm)
1	2×19	909.0	798.5	839.6	34.0	96.4
2	1×20	921.9	805.2	846.5	35.2	97.9
3	5×3	901.0	811.1	859.5	26.0	96.4
4	5×5	939.6	853.5	902.6	24.0	67.6
5	5×5	908.6	816.0	858.7	33.3	24.1
6	5×5	942.3	874.4	910.4	22.0	91.4
7	3×5	915.2	855.3	891.3	16.6	61.0
8	3×2	887.2	854.3	865.4	11.2	78.4
9	3×5	889.3	802.5	856.3	24.6	106.0
10	6×5	933.6	812.7	857.8	36.0	111.2
11	3×1	896.6	855.9	871.0	18.2	39.3
12	7×5	952.7	797.1	871.3	43.9	98.8
13	2×3	879.4	807.4	836.9	24.4	100.7
14	2×3	834.7	800.7	820.4	10.9	114.8
15	2×4	909.6	888.5	895.3	7.1	73.9
16	12×8	943.7	806.0	877.2	33.6	90.7

reliability of TSVs. We also see that TSV blocks with the same size can show quite different characteristics depending on the distance to the nearest package-bump. For example, although TSV block 4, 5, and 6 are all 5×5 TSV blocks and are located side-by-side, TSV block five shows the lowest von Mises stress level. However, its standard deviation of von Mises stress is highest among three blocks. We observe lower von Mises stress if TSV is placed near the package-bump center or far away from it; however, we see higher stress in TSV located around package-bump edge shown in Fig. 15.9 in Sect. 15.3.7. In case of TSV block 5, most TSVs are near the package-bump center, which lowers von Mises stress level. However, at the same time a few TSVs are around the package-bump edge, which increases the standard deviation of von Mises stress inside the TSV block.

From this experiment, we observe two possible ways to reduce the mechanical reliability problems in block-level 3D designs: (1) Assign TSV blocks right above package-bump center locations if possible. (2) Place package-bumps outside the TSV block locations with a safe margin such as outside the red box in Fig. 15.17a. However, other design constraints such as package area and the required number of pins should be carefully considered as well.

15.6 Conclusions

In this chapter, we showed how package elements affect the stress field and the mechanical reliability on top of the TSV-induced stress in 3D ICs. We observed that the mechanical reliability of TSVs in the bottom-most die in the stack are highly affected by packaging elements, and that effect decreases as we go to the upper dies. We also studied an accurate and fast full-chip/package stress and mechanical reliability co-analysis flow based on the principle of lateral and vertical linear superposition of stress tensors (LVLS), considering all chip/package elements.

References

1. K. Athikulwongse, A. Chakraborty, J.-S. Yang, D.Z. Pan, S.K. Lim, Stress-driven 3D-IC placement with TSV keep-out zone and regularity study, in *Proceedings of IEEE International Conference on Computer-Aided Design*, San Jose, 2010
2. G.V. der Plas et al., Design issues and considerations for low-cost 3D TSV IC technology, in *IEEE International Solid-State Circuits Conference Digest of Technical Papers*, San Francisco, 2010
3. M.A. Hopcroft, W.D. Nix, T.W. Kenny, What is the Young's modulus of silicon. J. Microelectromech. Syst. **19**, 229–238 (2010)
4. M. Jung, X. Liu, S.K. Sitaraman, D.Z. Pan, S.K. Lim, Full-chip through-silicon-via interfacial crack analysis and optimization for 3D IC, in *Proceedings of IEEE International Conference on Computer-Aided Design*, San Jose, 2011
5. M. Jung, J. Mitra, D.Z. Pan, S.K. Lim, TSV stress-aware full-chip mechanical reliability analysis and optimization for 3D IC, in *Proceedings of ACM Design Automation Conference*, San Diego, 2011
6. M. Jung, D. Pan, S.K. Lim, Chip/package co-analysis of thermo-mechanical stress and reliability in TSV-based 3D ICs, in *Proceedings of ACM Design Automation Conference*, San Francisco, 2012
7. J.-S. Kim et al., A 1.2 V 12.8 GB/s 2 Gb mobile wide-I/O DRAM with 4x128 I/O using TSV-based stacking, in *IEEE International Solid-State Circuits Conference Digest of Technical Papers*, San Francisco, 2011
8. K.H. Lu, S.-K. Ryu, J. Im, R. Huang, P.S. Ho, Thermomechanical reliability of through-silicon vias in 3D interconnects, in *IEEE International Reliability Physics Symposium*, Monterey, 2011
9. M. Nakamoto et al., Simulation methodology and flow integration for 3D IC stress management, in *Proceedings of IEEE Custom Integrated Circuits Conference*, San Jose, 2010
10. S.R. Vempati et al., Development of 3-D silicon die stacked package using flip chip technology with micro bump interconnects, in *IEEE Electronic Components and Technology Conference*, San Diego, 2009
11. J. Zhang et al., Modeling thermal stresses in 3-D IC interwafer interconnects. IEEE Trans. Semicond. Manuf. **19**, 437 (2006)

Chapter 16
3D Chip/Package Co-analysis of Stress-Induced Timing Variations

Abstract In this chapter, we study a chip/package stress-aware timing co-analysis methodology for TSV-based 3D ICs. While previous works ignore the stress and mobility variation due to die-stacking and package components, we address these impacts on full-stack 3D IC timing. First, we build hole and electron mobility variation maps based on the chip/package stress co-analysis that considers on/off-chip elements such as TSVs, μ-bumps, and package-bumps (= C4 bumps). Second, we compare our approach with conventional TSV stress aware timing analysis methods that ignore packaging impacts. Our major finding is that we observe different mobility variation behavior across the stack when we consider both chip and package components. In addition, we observe significant mobility variations in the die closest to the package-bump layer due to the highly compressive stress caused by package-bumps and underfill. Based on these findings, we develop a full-stack 3D static timing analysis engine and provide results for practical 3D IC designs including wide-I/O and block-level 3D ICs.

16.1 Introduction

Thermo-mechanical reliability issues caused by the coefficient of thermal expansion (CTE) mismatch among the materials used in TSV-based 3D ICs has emerged as one of the most critical concerns in 3D ICs. Most previous studies on the thermo-mechanical stress of TSV-based 3D ICs have been done separately in either chip or package domain. The impact of TSV-induced stress due to the CTE mismatch between TSV and substrate materials on the mobility variation and timing performance was studied in the chip domain [1, 9, 11]. These works used a 2D analytical stress model to obtain full-chip stress and mobility maps. However, the non-negligible discrepancy between the 2D stress model and the 3D stress from finite element analysis (FEA) was identified [4], which may affect the mobility variation and hence timing analysis results. The following specific topics are covered in this chapter:

S.K. Lim, *Design for High Performance, Low Power, and Reliable 3D Integrated Circuits*, DOI 10.1007/978-1-4419-9542-1_16,

- We model and simulate more detailed 3D IC structures including both chip and package components and study their impact on hole and electron mobility variations as well as cell path delay. In addition, we provide a theoretical background on why 2D stress and 3D stress models generate different mobility variations.
- We develop a full-stack static timing analysis (STA) flow considering the stress induced by chip/package interconnect elements such as TSV, μ-bump, and package-bump. We also compare this with the 2D stress model [11] and the 3D stress without package components [4].
- We demonstrate the chip/package stress impact on timing using practical 3D chip/package designs including wide-I/O and block-level 3D ICs.

16.2 Existing Works

Recently, authors in [10] showed the significant impact of package components on the chip domain stress. They proposed a stress exchange file to transfer the boundary conditions from package-level to silicon-level analysis. However, this requires FEA methods which are computationally expensive or infeasible for full-chip or -package analysis. To overcome the limitation of FEA method and to consider both chip and package components in full-chip/package scale, the principle of lateral and vertical linear superposition of stress tensors was proposed to consider each chip/package interconnect element such as TSV, μ-bump, and package-bump [5]. However, none of these works address the mobility and full-stack timing variations caused by the CTE mismatch among the materials in full-chip/package scale.

16.3 Stress and Mobility Variation Model

16.3.1 Need for True 3D Chip/Package Stress Model

Stress at a point in an object can be defined by the nine-component stress tensor:

$$\sigma = \sigma_{ij} = \begin{bmatrix} \sigma_{11} & \sigma_{12} & \sigma_{13} \\ \sigma_{21} & \sigma_{22} & \sigma_{23} \\ \sigma_{31} & \sigma_{32} & \sigma_{33} \end{bmatrix}$$

where the first index i indicates that the stress acts on a plane normal to the i axis, and the second index j denotes the direction in which the stress acts. If index i and j are same we call this a normal stress, otherwise a shear stress. Since we adopt a cylindrical coordinate system for the cylindrical TSV, μ-bump, and package-bump, index 1, 2, and 3 represent r, θ, and z, respectively.

The analytical 2D radial stress model, known as *Lamé* stress solution, was employed to address the TSV thermo-mechanical stress. This 2D plane solution assumes an infinitely long TSV embedded in an infinite silicon substrate and provides stress distribution in silicon substrate region, which can be expressed as follows [8]:

$$\sigma_{rr}^{Si} = -\sigma_{\theta\theta}^{Si} = -\frac{E\Delta\alpha\Delta T}{2}\left(\frac{D_{TSV}}{2r}\right)^2$$
$$\sigma_{zz}^{Si} = \sigma_{rz}^{Si} = \sigma_{\theta z}^{Si} = \sigma_{r\theta}^{Si} = 0 \qquad (16.1)$$

where σ^{Si} is stress in silicon substrate, E is Young's modulus, $\Delta\alpha$ is mismatch in CTE, ΔT is differential thermal load, r is the distance from TSV center, and D_{TSV} is TSV diameter.

Authors in work [11] used this 2D analytical solution to assess the impact of TSV-induced stress on the mobility variation and full-chip timing. However, in [11] only σ_{rr} stress term was considered while all other eight stress tensor elements were set to zero. When only one normal stress component is considered, we call this uniaxial stress. However, stress is biaxial in nature in an elastic object as Eq. (16.1) indicates: there exist two non-zero normal stress components, i.e., σ_{rr} and $\sigma_{\theta\theta}$. Since the mobility variation depends on the piezo-resistive effect due to stress, the mobility variation pattern may change depending on the choice of stress mode.

Although this closed-form formula is easy to handle, this 2D solution is only applicable to the structure with TSV and substrate only, hence it is inappropriate for the realistic TSV structure with Cu diffusion barrier and dielectric liner. Also, a huge stress magnitude discrepancy was observed around TSV edge on the device layer between the 2D stress model [11] and the 3D FEA simulations [4]. This is simply because a 3D TSV structure cannot be correctly modeled by the 2D plane solution due to the change in boundary conditions, especially near the top and bottom of the structure. Moreover, packaging elements and die-stacking affect stress distribution on each device layer differently [5], which will be explained in Sect. 16.4. Therefore, if we consider the 3D stress tensors, i.e., non-zero nine stress components, as well as packaging elements, the mobility variation pattern can be significantly different from 2D stress cases.

To obtain stress tensors induced by TSV, μ-bump, and package-bump from 3D simulation structures, we use a FEA tool ABAQUS. Dimensions of our baseline simulation structures are based on the fabricated and/or published data [2, 10] shown in Fig. 16.1. In this chapter, we specifically examine the stress distribution on the device layer for each die shown in red lines in Fig. 16.1 to assess the mobility variation. Our baseline TSV diameter, height, Cu diffusion barrier thickness, and dielectric liner thickness are 5 μm, 30 μm, 50 nm, and 125 nm, respectively. We use Ti and SiO_2 as Cu diffusion barrier and liner materials. Also, diameter/height of μ-bump and package-bump are 10 and 100 μm, respectively, unless otherwise specified.

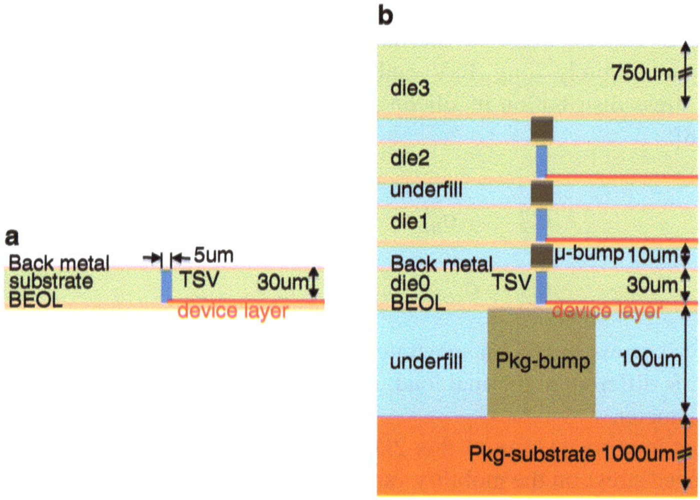

Fig. 16.1 Comparison of simulation structures (side view). (**a**) Die without package components [4]. (**b**) Four-die stack with package components used in this chapter

Material properties used for our experiments are as follows: CTE (ppm/K)/ Young's modulus (GPa) for Cu = (17/110), Si = (2.3/188), SiO_2 = (0.5/71), Ti = (8.6/116), package-bump (SnCu) = (22/44.4), μ-bump ($Sn_{97}Ag_3$) = (20/26.2), underfill = (44/5.6), package substrate (FR-4) = (17.6/19.7). All materials are assumed to be linear elastic and isotropic. The entire structure undergoes $\Delta T = -250°C$ of thermal load (annealing/reflow 275°C → room temperature 25°C) to represent a fabrication process. In addition, all materials are assumed to be stress free at the annealing/reflow temperature.

16.3.2 Piezo-Resistivity

In semiconductors, changes in inter-atomic spacing resulting from strain affect the bandgaps, making it easier or harder for electrons – depending on the material and strain – to be raised into the conduction band. This results in a change in resistivity of the semiconductor, which also can be translated to a change in mobility as follows [3]:

$$\begin{aligned}\frac{\Delta R}{R} = -\frac{\Delta\mu}{\mu} &= \left[\pi'_{11}\sigma_{xx} + \pi'_{12}\sigma_{yy}\right] cos^2\phi \\ &+ \left[\pi'_{11}\sigma_{xx} + \pi'_{12}\sigma_{yy}\right] sin^2\phi \\ &+ \pi_{12}\sigma_{zz} + \pi'_{44}\sigma_{xy} sin2\phi \end{aligned} \tag{16.2}$$

where σ_{ij} is the stress in the silicon substrate in Cartesian coordinate system, and ϕ is an angle between the wafer orientation and the transistor channel.

Table 16.1 Piezo-resistivity coefficient (TPa^{-1}) in (100) Si wafer [3]

Type	π_{11}	π_{12}	π_{44}	π'_{11}	π'_{12}	π'_{44}
N-type Si	−1,022	534	−136	−312	−176	−1,556
P-type Si	66	−11	1,381	718	−663	77

In this chapter, we assume the (100) Si wafer with reference axes of [110], [$\bar{1}$10], and [001]. We also assume that the transistor channel direction and the x-axis ([110]) are identical. In this setup, π'_{ij} is the piezo-resistivity coefficient defined along the reference axes of (100) Si wafer listed in Table 16.1:

$$\pi'_{11} = \frac{\pi_{11} + \pi_{12} + \pi_{44}}{2}$$

$$\pi'_{12} = \frac{\pi_{11} + \pi_{12} - \pi_{44}}{2}$$

$$\pi'_{44} = \pi_{11} - \pi_{12}$$

16.3.3 Mobility Variation: 2D Versus 3D Stress

In this section, we examine the impact of different stress cases on the mobility variation around a single TSV. To utilize Eq. (16.2), we first need to convert stress tensors from cylindrical coordinate system (S_{xyz}) to Cartesian coordinate system ($S_{r\theta z}$).

$$S_{xyz} = \begin{bmatrix} \sigma_{xx} & \sigma_{xy} & \sigma_{xz} \\ \sigma_{yx} & \sigma_{yy} & \sigma_{yz} \\ \sigma_{zx} & \sigma_{zy} & \sigma_{zz} \end{bmatrix} \quad S_{r\theta z} = \begin{bmatrix} \sigma_{rr} & \sigma_{r\theta} & \sigma_{rz} \\ \sigma_{\theta r} & \sigma_{\theta\theta} & \sigma_{\theta z} \\ \sigma_{zr} & \sigma_{z\theta} & \sigma_{zz} \end{bmatrix}$$

The transform matrix Q is the form:

$$Q = \begin{bmatrix} \cos\theta & -\sin\theta & 0 \\ \sin\theta & \cos\theta & 0 \\ 0 & 0 & 1 \end{bmatrix}$$

where θ is the angle between the x-axis and a line from the origin to the center of a transistor channel. A stress tensor in a cylindrical coordinate system can be converted to a Cartesian coordinate system using conversion matrices: $S_{xyz} = QS_{r\theta z}Q^T$.

Now we examine how different stress cases affect the mobility variation pattern. We first show stress tensor components in Cartesian coordinate system converted from cylindrical coordinate system shown in Eq. (16.2). Then, we derive the mobility variation formula for each case. We assume that the x-axis and the transistor channel direction are identical ($\phi = 0$).

- **2D uniaxial stress**: $\sigma_{rr} \neq 0$, all other stress terms = 0

$$\sigma_{xx} = \sigma_{rr}cos^2\theta, \sigma_{yy} = \sigma_{rr}sin^2\theta, \sigma_{zz} = 0$$
$$-\Delta\mu/\mu = \pi'_{11}\sigma_{rr}cos^2\theta + \pi'_{12}\sigma_{rr}sin^2\theta \quad (16.3)$$

- **2D biaxial stress**: $\sigma_{rr} = -\sigma_{\theta\theta} \neq 0$, all other stress terms = 0

$$\sigma_{xx} = -\sigma_{yy} = \sigma_{rr}cos2\theta, \sigma_{zz} = 0$$
$$-\Delta\mu/\mu = \pi'_{11}\sigma_{rr}cos2\theta - \pi'_{12}\sigma_{rr}cos2\theta = \pi_{44}\sigma_{rr}cos2\theta \quad (16.4)$$

- **3D stress**: all stress tensor components $\neq 0$

$$\sigma_{xx} = \sigma_{rr}cos^2\theta + \sigma_{\theta\theta}sin^2\theta - \sigma_{r\theta}sin2\theta$$
$$\sigma_{yy} = \sigma_{rr}sin^2\theta + \sigma_{\theta\theta}cos^2\theta + \sigma_{r\theta}sin2\theta$$
$$\sigma_{zz} \neq 0$$
$$-\Delta\mu/\mu = \pi'_{11}\sigma_{xx} + \pi'_{12}\sigma_{yy} + \pi_{12}\sigma_{zz} \quad (16.5)$$

It is clear from above expressions that the trend of mobility variation is different between these stress cases. Mobility variation maps around a single TSV for the 2D biaxial stress (*2D biaxial*) and the 3D stress with package components (*3D wPkg*) are shown in Fig. 16.2. We see a significant difference in the electron mobility variation maps, which will be discussed in detail in Sect. 16.4.

Comparing both 2D stress cases, we observe that the electron mobility in the 2D uniaxial stress (*2D uniaxial*) improves regardless of angle θ, since both π'_{11} and π'_{12} are negative for N-type silicon and $\sigma_{rr}cos^2\theta$ and $\sigma_{\theta\theta}sin^2\theta$ terms are non-negative. On the other hand, the sign of electron mobility variation in the *2D biaxial* case depends on θ, which is shown in Fig. 16.3b. We also observe that the *2D uniaxial* case underestimates the hole mobility variation range compared with the *2D biaxial* case. Thus, using *2D uniaxial* model in [11] may result in erroneous results.

As for the 3D stress without package components case (*3D woPkg*) shown in Fig. 16.3a, b, the hole mobility variation range is larger than the *2D biaxial* case. Also, the electron mobility variation is not symmetric along the x-axis and the y-axis from TSV center unlike the *2D biaxial* case. This is largely due to the non-zero σ_{zz} term. Note that in cases of *2D uniaxial*, *2D biaxial*, and *3D woPkg*, stress tensors are assumed to be identical across tiers, hence there is no difference in mobility variations in different dies in the 3D stack.

As we include package components, the electron mobility variation differs across the stack as shown in Fig. 16.3d. This is mainly due to the large compressive stress generated by the package-bump. This effect is most significant in die0, which is closest to the package-bump layer shown in Fig. 16.1b. We will discuss more details in Sect. 16.4.

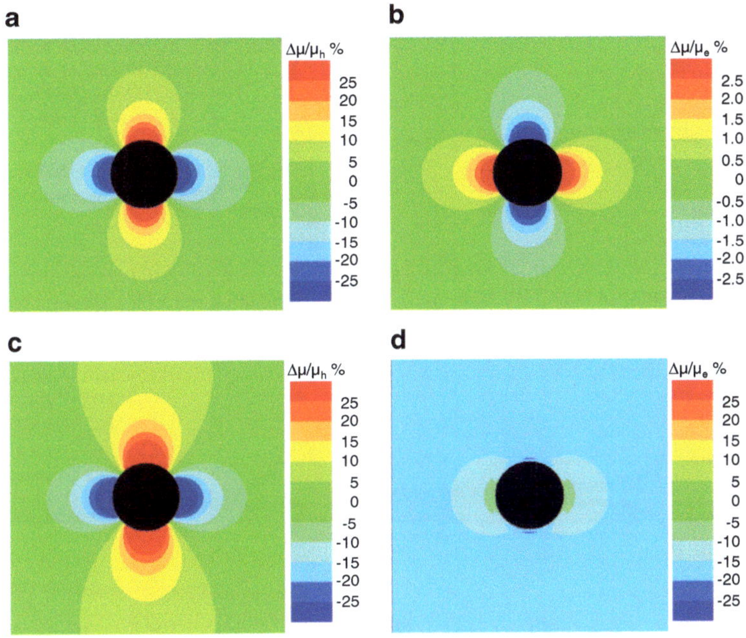

Fig. 16.2 Mobility variation map around a single TSV. (**a**) Hole mobility (2D biaxial stress). (**b**) Electron mobility (2D biaxial stress). (**c**) Hole mobility in die0 in four-die stack (3D stress with package components). (**d**) Electron mobility in die0 in four-die stack (3D stress with package components). For both (**c**) and (**d**) TSV, μ-bump, and package-bump are vertically aligned

16.4 Chip/Package Stress Impact on Mobility Variation

16.4.1 Linear Superposition Principle

In this section, we discuss the impact of both chip and package elements on the mobility variation in 3D ICs in detail. Before discussing the mobility variation, we briefly explain how to obtain the stress at a point considering the effect of both chip and package components. The method of lateral and vertical linear superposition of stress tensors was proposed in [5] to handle multiple TSVs, μ-bumps, and package-bumps that are not in the same layer. We adopt this method in our work.

First, the target structure which contains all components shown in Fig. 16.1b is decomposed into four separate structures: TSV only, package-bump only, μ-bump only, and the background which does not contain TSV and bumps. Next, stress tensors along the red line on device layers affected by each interconnect element are obtained separately from FEA simulations. To compute the stress tensor at a point affected by these interconnect elements, the stress tensors from TSV only, μ-bump only, and package-bump only structures are added up. Then, we subtract twice the magnitude of the background stress tensor since this is already included in previous

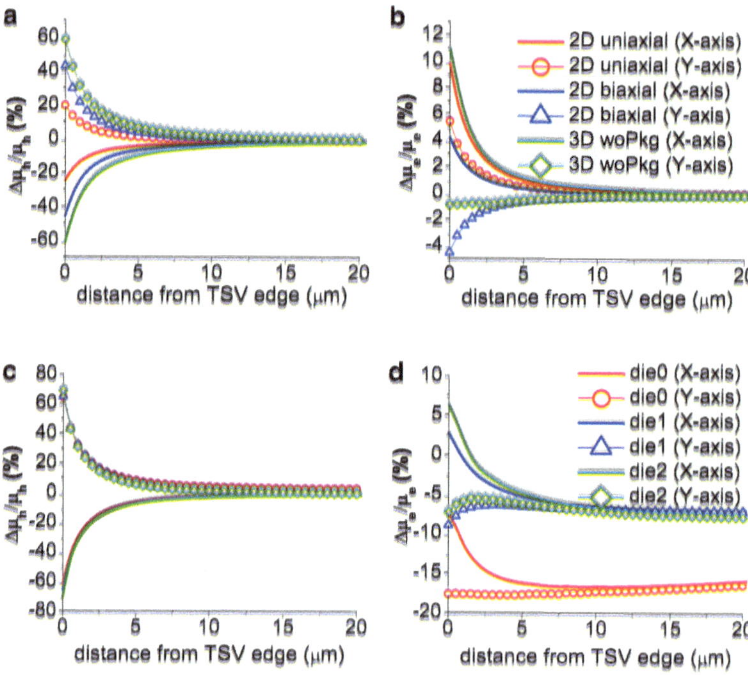

Fig. 16.3 Mobility variation range of a single TSV with different stress cases. Mobility variation numbers are collected along the x-axis and the y-axis from a TSV center on device layers. (**a**) Hole mobility under 2D and 3D stress without package components. (**b**) Electron mobility under 2D and 3D stress without package components. (**c**) Hole mobility under 3D stress with package components in four-die stack. (**d**) Electron mobility under 3D stress with package components in four-die stack

three structures. If the point under consideration is affected by n components, then the background stress need to be subtracted $n-1$ times. This method can be expressed as follows [5]:

$$S = \sum_{i=1}^{n_{TSV}} S_{TSV\,i} + \sum_{j=1}^{n_{\mu B}} S_{\mu B\,j} + \sum_{k=1}^{n_{pkgB}} S_{pkgB\,k} - (n_{TSV} + n_{\mu B} + n_{pkgB} - 1) \times S_{bg} \tag{16.6}$$

where S is the total stress at the point under consideration and $S_{TSV\,i}$, $S_{\mu B\,j}$, and $S_{pkgB\,k}$ are individual stress tensor at this point due to ith TSV, jth μ-bump, and kth package-bump, respectively. S_{bg} indicates the background stress at that point. Based on the stress analysis results, we obtain both hole and electron mobility variation maps using Eq. (16.5).

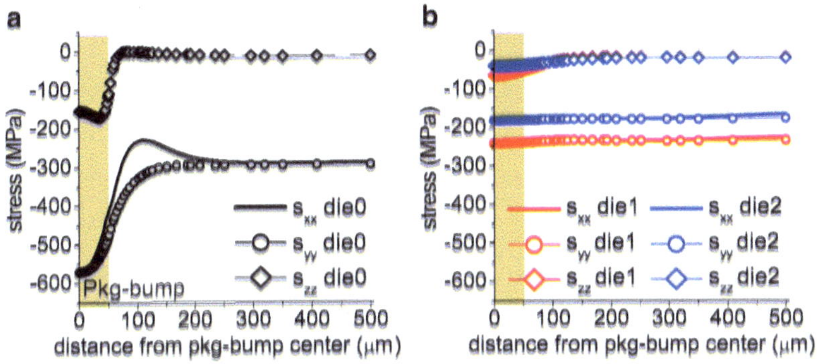

Fig. 16.4 Normal stress components induced by package-bump on device layers. Stress tensors are collected along the x-direction from the package-bump center. (**a**) Stress in die0. (**b**) Stress in die1 and die2

16.4.2 Mobility Variations from Chip/Package Elements

From FEA simulations, a highly compressive stress is observed on device layers due to package-bumps, which is induced by the CTE mismatch between package-bumps and underfill. As Fig. 16.4 shows, die0 (=closest to package-bump layer) experiences the most compressive stress due to their proximity. The stress becomes less compressive as we go to upper dies. The stress distribution (σ_{xx} and σ_{yy}) in die3 (un-thinned top die) is almost flat (-110 ± 5 MPa), since die3 does not contain any TSVs. Thus, we only compute the stress in the dies containing TSVs in this chapter.

In Eq. (16.5), the electron mobility variation is approximately proportional to the sum of σ_{xx} and σ_{yy} due to the same sign (= negative) of π'_{11} and π'_{12}, while the hole mobility variation is roughly proportional to the difference between σ_{xx} and σ_{yy} due to the opposite sign of π'_{11} and π'_{12}. Figure 16.4 shows the stress distribution on device layers induced by package-bump only. Although there is a noticeable difference between σ_{xx} and σ_{yy} near the package-bump edge in die0, their difference is almost negligible in other regions. Thus, this package-bump induced stress will not alter the hole mobility variation significantly except near the package-bump edge area. On the other hand, the electron mobility will be degraded under the influence of the package-bump since both σ_{xx} and σ_{yy} are compressive (= negative), which is shown in Fig. 16.3d. Furthermore, the level of electron mobility degradation is most severe in die0.

Figure 16.5 shows hole and electron mobility variation maps in a four-die stack with 441 TSVs/μ-bumps with 20 μm pitch and nine package-bumps with 200 μm pitch. Both hole and electron mobility variation range is largest in die0 due to the direct impact of package-bump-induced stress. Especially, the hole mobility degrades in between package-bumps along the x-direction and improves along the y-direction. This is because of the difference between σ_{xx} and σ_{yy} stress components near package-bump edge area shown in Fig. 16.4a: along the x-direction σ_{xx} is

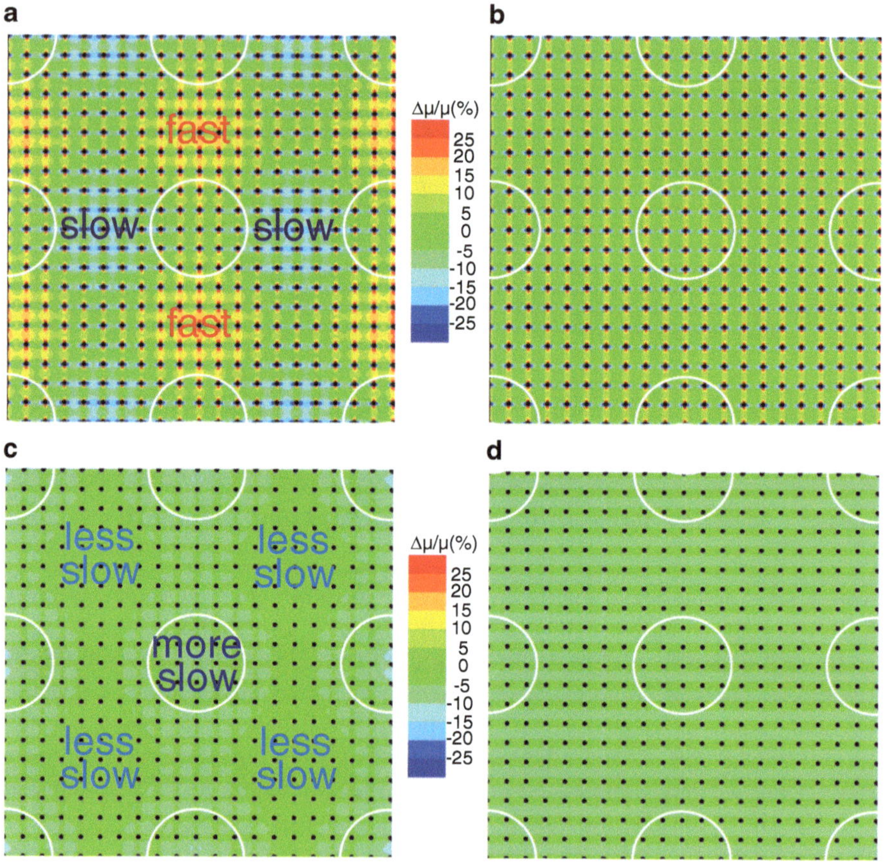

Fig. 16.5 Mobility variation map with 441 TSVs/μ-bumps (*black dots*) and nine C4 bumps (*white circles*). (**a**) Hole mobility variation map in die0. (**b**) Hole mobility variation map in die2. (**c**) Electron mobility variation map in die0. (**d**) Electron mobility variation map in die2

higher than σ_{yy}, while along the y-direction σ_{yy} is higher than σ_{xx}. The electron mobility in die0 degrades in most cases, and the worst spot is inside the package-bump area since the most compressive stress occurs in this region as shown in Fig. 16.4.

In addition, Fig. 16.5 shows that the stress induced by package-bumps affects the mobility variation of a large number of cells, while TSVs generate the mobility variation pattern only for the cells nearby these TSVs. We also observe that as we go to upper dies, mobility variations due to package-bumps are almost negligible, hence the mobility variation pattern is mostly determined by TSVs.

16.5 Chip/Package Stress-Aware Timing Analysis

In this section, we present our stress-aware static timing analysis (STA) flow. First, we build a Verilog netlist and a parasitic extraction file (SPEF) for each die from 3D IC layouts. Each instance name in the netlists are replaced by corresponding hole and electron mobility variation based on our stress and mobility analysis results. For example, INV_X1 with +4% hole mobility and −8% electron mobility variation becomes INV_X1_Hp4_Em8. Then, we create a top-level Verilog netlist that instantiates each die design and connects the 3D nets using TSV. We also create a top-level SPEF file that contains parasitic models of the TSVs. Lastly, we run Synopsys PrimeTime to perform 3D STA.

For this stress-aware STA, we build a timing library to capture the mobility variation impact on cell delay. We first obtain both hole and electron mobility variation range affected by multiple TSVs, μ-bumps, and package-bumps. Since this range is different across the stack and also affected by the alignment and the pitch of TSVs, μ-bumps, and package-bumps, we generate several test cases by varying these knobs. Figure 16.5 is one of the test cases. We find that the hole mobility varies from −76 to 76% and the electron mobility ranges from −24 to 12% without any TSV keep-out-zone (KOZ), where devices cannot be placed. Actual mobility variation range is reduced by introducing KOZ. We characterize cell timing with the mobility variation using Cadence Encounter Library Characterizer with 4% mobility step size.

Figure 16.6 shows the FO4 delay of INV_X1 and NAND_X1 gates with mobility variations. We see that the delay variation range is similar for both gates with given mobility variations. Note that the rising delay is not affected by the electron mobility variation and the falling delay is not much influenced by the hole mobility change. Thus, we can fix $\Delta\mu_e/\mu_e$ when we sweep $\Delta\mu_h/\mu_h$, and vice versa. This is useful to reduce the number of library characterization. Instead of characterizing 390 (=39×10 with 4% step size) libraries, we need to prepare 49 (=39 + 10) libraries [11].

16.6 Experimental Results

We implement our chip/package thermo-mechanical stress and mobility analysis flow in C++/STL. In our experiments, we build four-die stack 3D IC designs using Cadence Encounter with Nangate 45 nm cell library. We adopt a regular TSV placement style in which TSVs are placed uniformly across each die or inside TSV blocks with a pre-defined pitch. In all cases, a pair of TSV and μ-bump is always vertically aligned. The default diameter/height (μm) of TSV, μ-bump, and package-bump are 5/30, 10/10, and 100/100, respectively, unless otherwise specified. The package-bump pitch is assumed to be 200 μm for all cases.

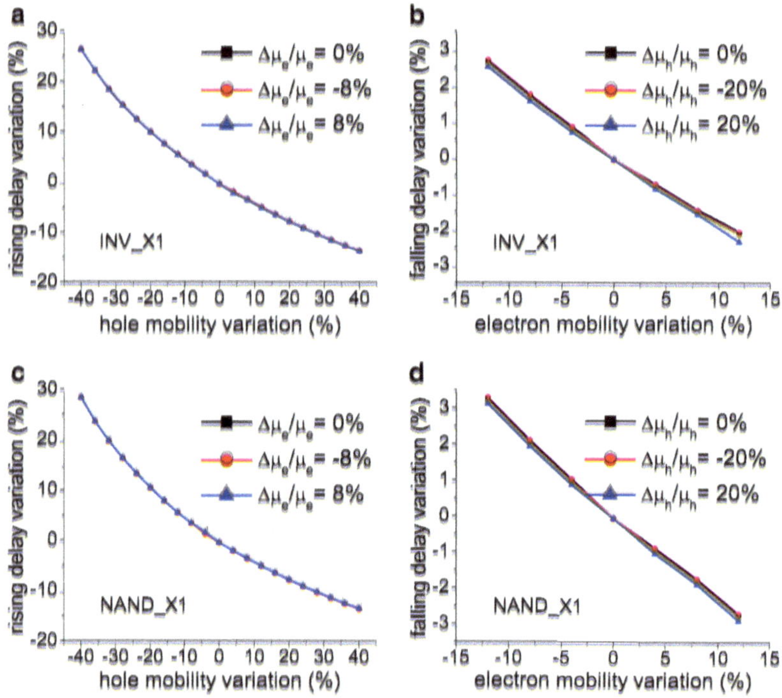

Fig. 16.6 Mobility variation impact on cell FO4 delay. (**a**) Rising delay dependency on hole mobility variation (INV_X1). (**b**) Falling delay dependency on electron mobility variation (INV_X1). (**c**) Rising delay dependency on hole mobility variation (NAND_X1). (**d**) Falling delay dependency on electron mobility variation (NAND_X1)

Table 16.2 Benchmark circuits with TSV KOZ = 1.0 μm

Circuit	# cell	Area (μm × μm)	WL (mm)	# TSV	TSV pitch (μm)
ckt1	53 K	300 × 300	1,250	1,075	15
ckt2	586 K	790 × 790	15,565	2,116	25
ckt3	1.29 M	1,140 × 1,140	36,676	6,414	20

16.6.1 2D Versus 3D Stress Impact on Mobility and Timing

In this section, we examine the impact of different stress cases, i.e., 2D stress (*2D uniaxial* and *2D biaxial*) and 3D stress (*3D woPkg* and *3D wPkg*), on the full-stack timing and mobility variation. We use three circuits in Table 16.2 with the TSV KOZ size of 1 μm. Note that all benchmark circuits are designed with the timing optimization objective, but the stress impact is not considered in design stages.

Figure 16.7 shows the cell mobility distribution in die0 in a four-die stack of Ckt2. We first observe that the electron mobility is highly concentrated within $0 \sim 2\,\%$ range for both 2D stress cases and the *3D woPkg*. Note that the *2D uniaxial*

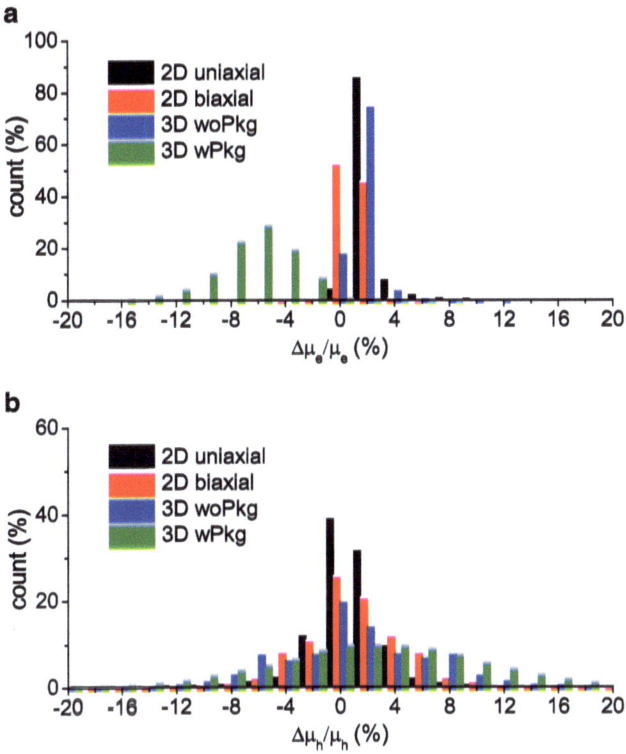

Fig. 16.7 Cell mobility variation histogram in die0 in four-die stack (ckt2). (**a**) Electron mobility. (**b**) Hole mobility

case always improves the electron mobility, while the *2D biaxial* case can degrade the electron mobility as well. Most importantly, the electron mobility variation with package components shows quite a different behavior: the mobility variation range is wider than other cases and most of cells in die0 experience the electron mobility degradation. The degradation is mainly due to the compressive stress from package-bumps. Also, the wider distribution originates from the relative positions between cells, TSVs/μ-bumps, and package-bumps.

As for the hole mobility distribution, all cases show wider distribution than the electron mobility case, which is expected from Fig. 16.3. However, still the *3D wPkg* case generates the largest variation, which is clear as shown in Fig. 16.5a. Note that as we go to upper dies, the hole mobility distribution of the *3D wPkg* becomes comparable to the *3D woPkg* case.

Figure 16.8 shows stress-aware 3D STA results. We show the longest path delay (LPD) and total negative slack (TNS) for different stress cases. First, we observe that the *2D uniaxial* case always underestimates the LPD compared with the *2D biaxial* case. Interestingly, the LPD of Ckt2 in the *3D wPkg* case shows better timing than other stress cases shown in Fig. 16.8a. This can be explained with Figs. 16.9

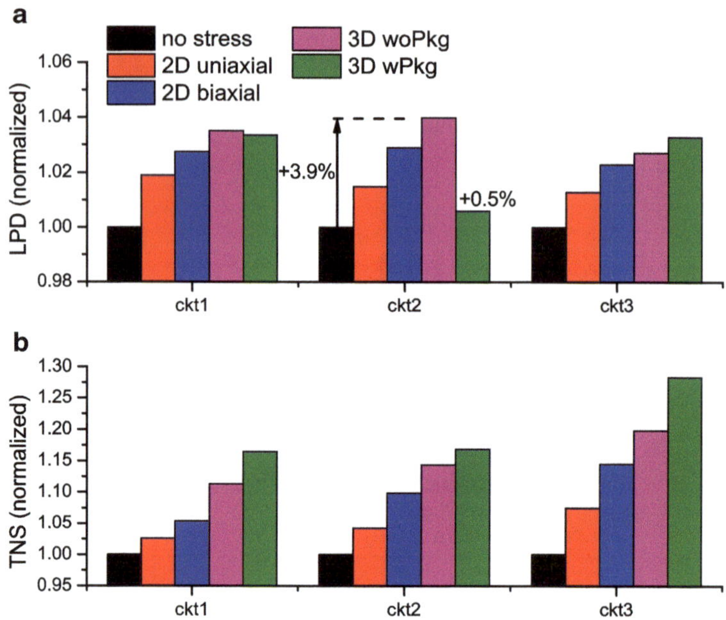

Fig. 16.8 Impact of 2D and 3D stress cases on the longest path delay (LPD) and total negative slack (TNS). Timing numbers are normalized to the no-stress case. TSV KOZ is 1 μm for all cases. (**a**) LPD variation. (**b**) TNS variation

and 16.10. As Fig. 16.9a shows, the cells in the critical path are located in between package-bumps in y-direction. In this case, the hole mobility improves as shown in Fig. 16.5a. Moreover, the hole mobility further improves when cells are placed in between TSVs in y-direction as shown in Fig. 16.9d.

The opposite case can also happen as shown in Fig. 16.10a, where the cells in the critical path are placed in between package-bumps in x-direction. In this case, the LPD degrades by 5.6% in the *3D wPkg* case compared with the *no-stress* case, while the *3D woPkg* case degrades the LPD by 2.5%.

The impact of package-bump stress on the mobility in die0 is clear if we compare Figs. 16.9c and 16.10c. Although the relative positions between TSV and cells are similar, the hole mobility variation is significantly different depending on package-bump locations.

The stress impact on timing is more evident in TNS. In the *3D wPkg* case, TNS is larger than the *no-stress* case up to 28% as shown in Fig. 16.8b. This is because most cells in the design are affected by the stress induced by TSVs, μ-bumps, and package-bumps, and thus undergo mobility variations. We also show the slack histogram in Fig. 16.11. The slack distributions of the *no-stress* and the *2D uniaxial* case are similar, while the *3D wPkg* case shows more paths with higher negative slack as well as the shift in positive slack distribution. Since designers use this

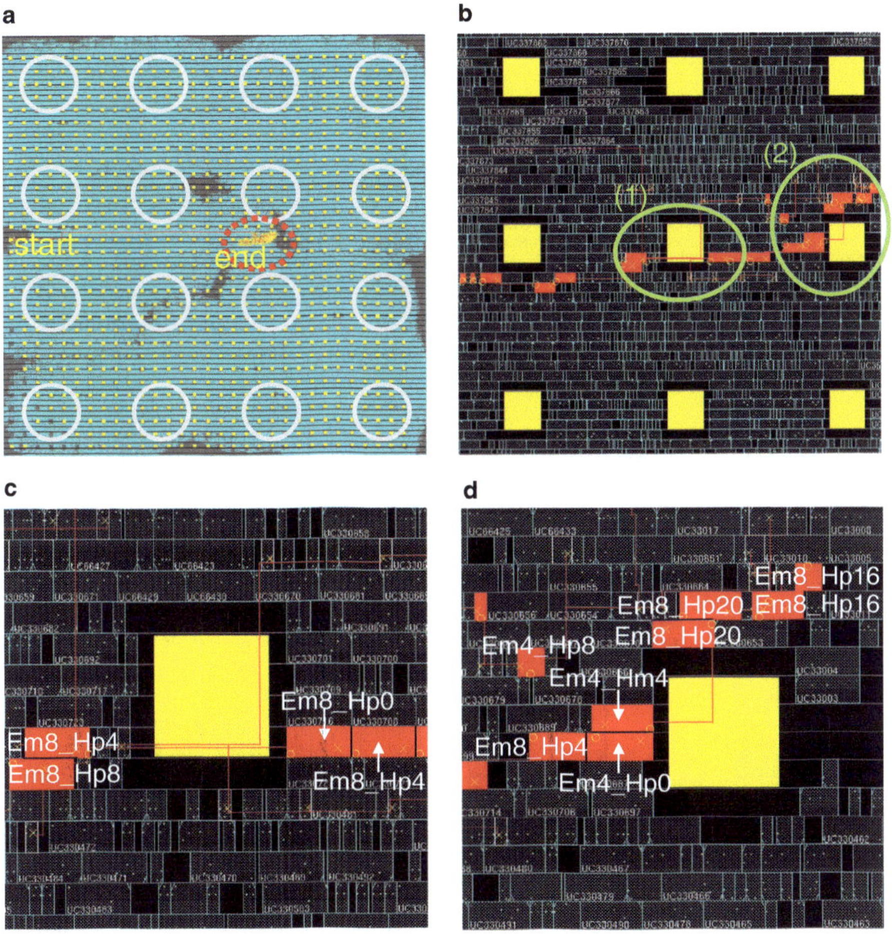

Fig. 16.9 Full-chip layout (die0 in four-die stack) with the highlighted longest path. *Yellow squares* are TSVs (cell mobility naming convention: e.g., Em8_Hp4 = electron mobility minus 8% and hole mobility plus 4%). (**a**) Layout of Ckt2 (KOZ = 1.0 μm). (**b**) Cells in *red circle* in (**a**). (**c**) Close-up shot of *green circle* (*1*) in (**b**). (**d**) Close-up shot of *green circle* (*2*) in (**b**)

positive slack for power minimization, a correct assessment of slack considering the stress impact is important.

16.6.2 Impact of KOZ Size

In this section, we explore the impact of KOZ size on the timing with 3D stress cases. Table 16.3 shows cell mobility variation ranges across the stack with five different KOZ sizes (0.3 ~ 3.1 μm) in Ckt3. First, we observe that the mobility

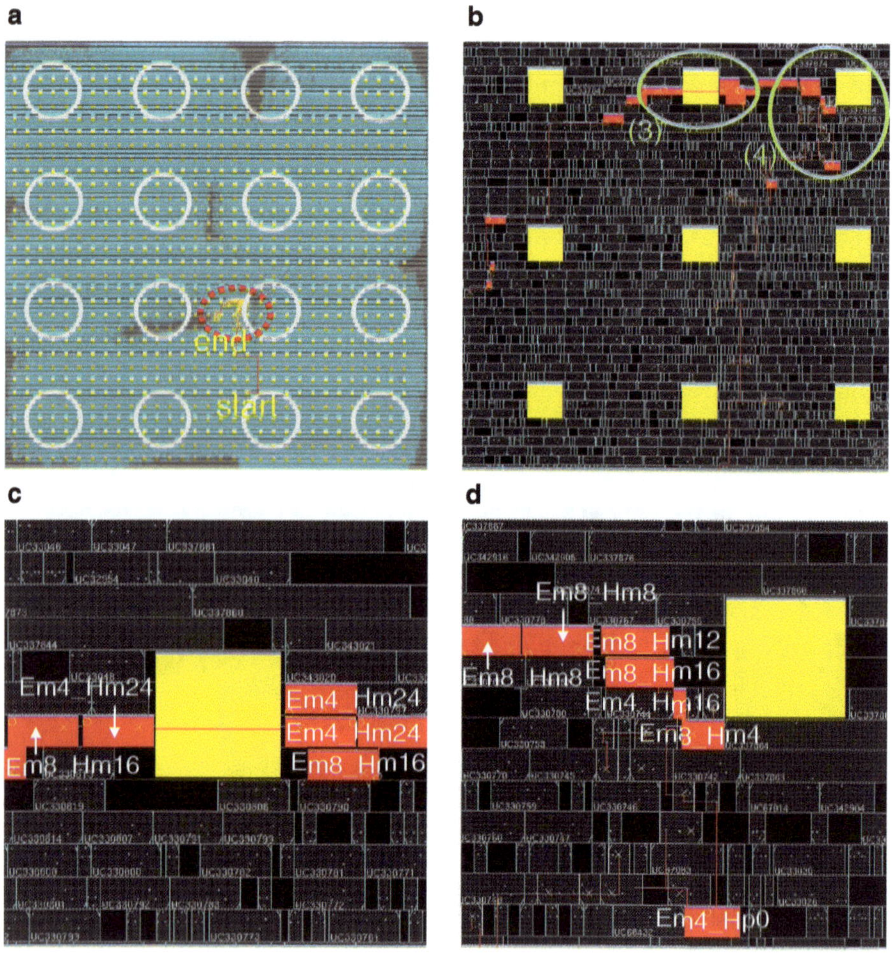

Fig. 16.10 Full-chip layout (die0 in four-die stack) with the highlighted longest path. *Yellow squares* are TSVs (cell mobility naming convention: e.g., Em8_Hp4 = electron mobility minus 8% and hole mobility plus 4%). (**a**) Layout of Ckt2 (KOZ = 0.3 μm). (**b**) Cells in *red circle* in (**a**). (**c**) Close-up shot of *green circle* (*3*) in (**b**). (**d**) Close-up shot of *green circle* (*4*) in (**b**)

variation range as well as the standard deviation for both electron and hole decrease as KOZ size increases for all dies. Interestingly, the average electron mobility of cells slightly degrades as KOZ size increases. As discussed in Sect. 16.4, the electron mobility variation largely depends on the sum of σ_{xx} and σ_{yy} components. As KOZ size increases, the stress level of σ_{xx} and σ_{yy} induced by TSVs that cells experience also decreases. This is because the magnitude of these normal stress components decreases proportional to $(1/d)^2$, where d is the distance from TSV center, and this d is effectively increasing as KOZ size increases.

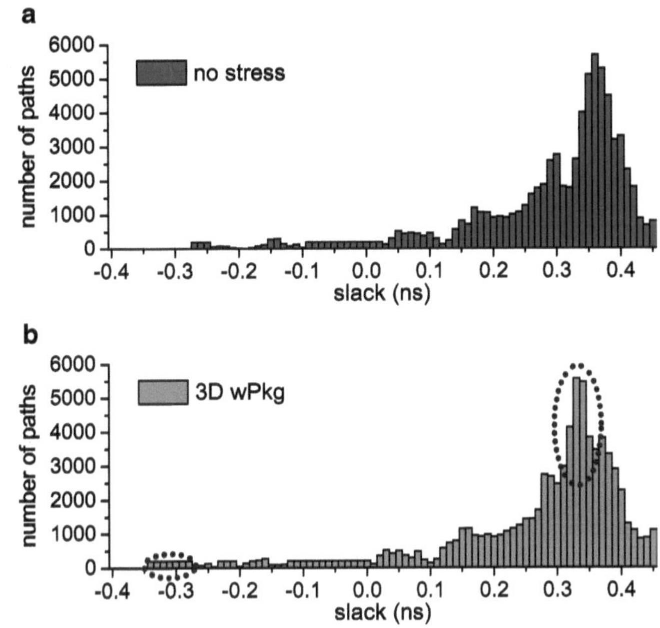

Fig. 16.11 Impact of 2D and 3D stress cases on the slack distribution of Ckt2. (**a**) No-stress case. (**b**) 3D stress with package components case

Table 16.3 Impact of die-stack and KOZ on cell mobility variation (ckt3). TSV/μ-bump pitch is 25 μm and package-bump pitch is 200 μm for all cases

	KOZ (μm)	Mobility variation (%) Electron Range	Avg	Stddev	Hole Range	Avg	Stddev
Die0	0.3	−17.8 ∼ 6.0	−7.53	2.92	−64.1 ∼ 57.5	1.02	9.22
	1.0	−17.6 ∼ 2.8	−7.78	2.78	−43.6 ∼ 31.8	0.55	7.96
	1.7	−17.0 ∼ 0.0	−8.06	2.63	−31.2 ∼ 31.7	0.75	7.45
	2.4	−17.6 ∼ −2.3	−8.29	2.60	−26.7 ∼ 25.3	1.02	6.96
	3.1	−17.2 ∼ −2.6	−8.45	2.55	−23.3 ∼ 25.7	1.20	6.78
Die1	0.3	−10.0 ∼ 4.6	−7.51	1.34	−62.7 ∼ 47.5	1.08	4.12
	1.0	−10.1 ∼ 2.4	−7.66	1.20	−34.2 ∼ 22.8	1.01	2.82
	1.7	−10.2 ∼ −1.3	−7.79	1.16	−22.6 ∼ 22.9	1.13	2.49
	2.4	−10.4 ∼ −2.3	−7.96	1.12	−16.1 ∼ 13.4	1.07	1.87
	3.1	−10.4 ∼ −3.1	−8.10	1.09	−12.2 ∼ 13.2	1.13	1.64
Die2	0.3	−7.7 ∼ 7.8	−5.64	1.03	−67.1 ∼ 47.4	0.86	6.63
	1.0	−7.2 ∼ 0.9	−5.71	0.91	−37.7 ∼ 22.5	0.45	4.97
	1.7	−7.3 ∼ −1.0	−5.85	0.76	−25.1 ∼ 22.5	0.83	4.32
	2.4	−7.5 ∼ −2.3	−5.97	0.69	−17.4 ∼ 13.2	0.70	3.48
	3.1	−7.6 ∼ −3.1	−6.08	0.62	−12.9 ∼ 13.1	0.91	3.17

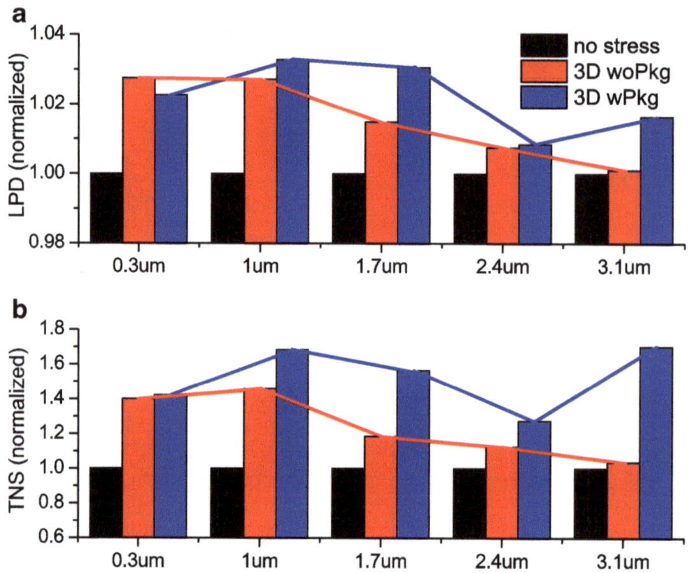

Fig. 16.12 Impact of KOZ size on timing variation in Ckt3. (**a**) LPD variation. (**b**) TNS variation

Table 16.4 Block-level and wide-I/O style 3D IC designs. Package-bump pitch is 200 μm

Circuit	# cell	Area (μm × μm)	WL (mm)	# TSV	TSV pitch (μm)
ckt2_block	570 K	832 × 918	16,055	1,769	15
ckt2_wideIO	570 K	800 × 800	15,404	2,116	10

Figure 16.12 shows full-stack timing results with different KOZ sizes. In the *3D woPkg* case, the deviation from the *no-stress* case for both LPD and TNS decreases as the KOZ size increases, which agrees with [1]. However, in the *3D wPkg* case, there is no obvious trend with the KOZ size. This is because the cell mobility variation is affected by both TSVs and package-bumps. Depending on the relative position between cells, TSVs, and package-bumps each path delay variation can be positive or negative.

16.6.3 Case Study: Block-Level 3D Designs

In this section, we study the chip/package stress impact on the full-stack timing in practical 3D IC designs, i.e., block-level and wide-I/O style designs listed in Table 16.4. The block-level 3D IC design is attractive because existing 2D IP blocks can be reused. We use a 3D floorplanner in [7] to generate layouts using Ckt2 shown in Fig. 16.13a. TSVs are inserted into the whitespace and form TSV arrays.

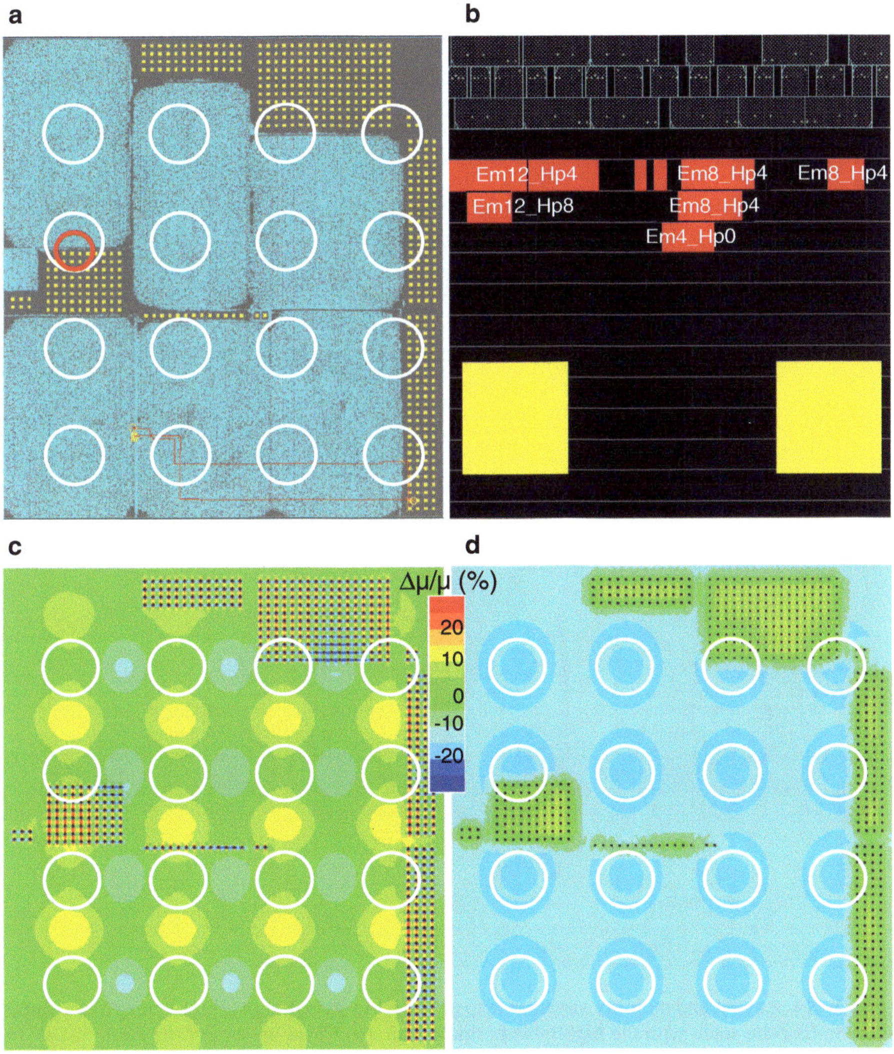

Fig. 16.13 Block-level 3D IC (ckt2_block). (**a**) Layout of die0 in four-die stack with the critical path highlighted. TSVs are *yellow squares* and package-bumps are *white circles*. (**b**) Close-up shot of *red circle* in (**a**). (**c**) Hole mobility variation map. (**d**) Electron mobility variation map

Figure 16.13c, d show hole and electron mobility maps, and we observe that the high mobility variation regions are limited to nearby TSV blocks. Although the global mobility variation pattern is largely determined by package-bumps, the local mobility minima and maxima are mostly caused by TSVs. Thus, most of the cells inside functional blocks do not experience high mobility variations.

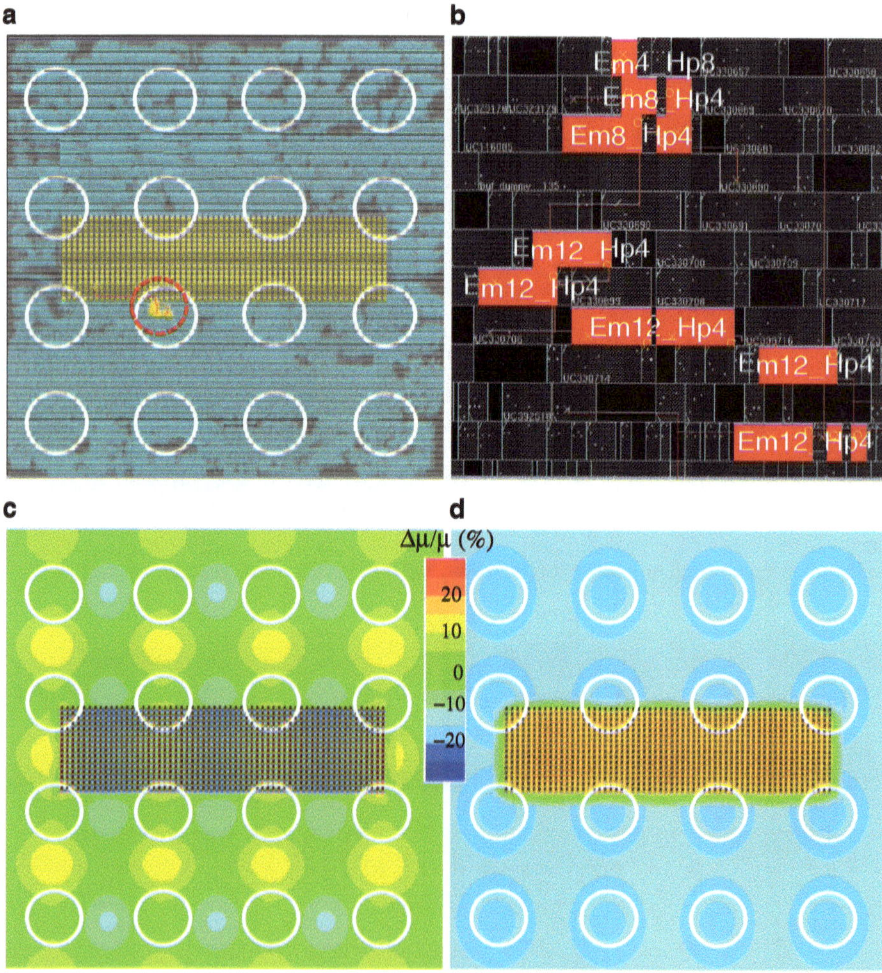

Fig. 16.14 Layout and mobility variation map of wide-I/O style design (ckt2_wide). (**a**) Layout of die0 in four-die stack with the highlighted critical path. (**b**) Close-up shot of *red circle* in (**a**). (**c**) Hole mobility variation map. (**d**) Electron mobility variation map

16.6.4 Case Study: Wide-I/O Style 3D Designs

In case of the wide-I/O style design, we follow the TSV placement style similar to the work in [6], where TSV arrays are placed in the middle of a chip. We assume that 8×30 TSV array (per memory bank) is placed in the middle of a chip. We also assume that there are four memory banks, hence the total 960 TSVs are employed in die0 as shown in Fig. 16.14. The hole and electron mobility maps in Fig. 16.14c, d

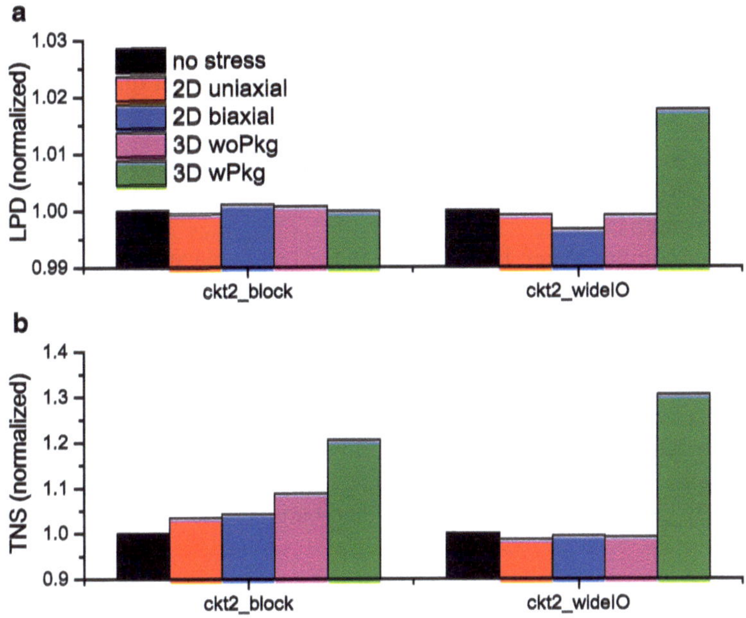

Fig. 16.15 Impact of 2D and 3D stress cases on the longest path delay (LPD) and total negative slack (TNS) in block-level and wide-I/O style 3D IC designs. TSV KOZ is 1.7 μm for all cases. (**a**) LPD variation. (**b**) TNS variation

clearly show that high mobility variation region is confined to inside and nearby the TSV array. Thus, majority of cells are not affected by the TSV stress similar to the block-level design.

Figure 16.15 shows 3D STA results for the block-level and the wide-I/O style designs. As for the LPD, we observe that there is an almost negligible impact from all stress cases for both block-level and wide-I/O style designs, since most cells are not affected by the TSV stress. One exception is the *3D wPkg* case in the wide-I/O style design. This is because the cells in the critical path are placed nearby TSV array and right above a package-bump as shown in Fig. 16.14a. Cells that are placed in the y-direction with respect to TSVs experience electron mobility degradation and hole mobility improvement. However, the electron mobility further decreases inside package-bump area as shown in Fig. 16.5c, hence the net effect is timing degradation.

We also observe more TNS variation in the block-level design than that in the wide-I/O style design for *2D uniaxial*, *2D biaxial*, and *3D woPkg* cases. The block-level design contains more TSV blocks than the wide-I/O style design, hence the number of cells nearby these TSV blocks also increases. Thus, more paths are affected by the TSV stress than the wide-I/O style design. However, as we include the impact of package-bumps, all cells in these designs are affected by package-bumps, hence we observe non-negligible variations in TNS for both design styles.

16.6.5 Key Findings and Design Guides

We summarize our work and offer design guidelines for better timing improvement in 3D ICs.

1. Both hole and electron mobility variation is largest in the die nearest to a package-bump layer due to the highly compressive stress induced by the package-bump layer.
2. The hole mobility of cell improves when the cell is placed in between package-bumps in y-direction as well as in between TSVs in y-direction.
3. The electron mobility of cell degrades when the cell is placed inside package-bump area. Also, when the cell is in between TSVs in x-direction, the electron mobility improves.
4. It is recommended to place cells in the critical path in between package-bumps in y-direction due to the high hole mobility boost.
5. Proper gate sizing is necessary to compensate the electron mobility degradation especially in the die nearest to the package-bump layer.

16.7 Conclusions

In this chapter, we demonstrated how chip and package components affect the mobility and full-stack timing variations in 3D ICs. We also studied a chip/package stress-aware timing analysis method, which is applicable to stress-aware full-stack timing optimization for 3D ICs. Our results showed that the mobility variation pattern is different across the stack, and especially the die nearest to the package-bump layer experiences the largest hole and electron mobility variations.

References

1. K. Athikulwongse, A. Chakraborty, J.-S. Yang, D.Z. Pan, S.K. Lim, Stress-driven 3D-IC placement with TSV keep-out zone and regularity study, in *Proceedings of IEEE International Conference on Computer-Aided Design*, San Jose, 2010
2. G.V. der Plas et al., Design issues and considerations for low-cost 3D TSV IC technology, in *IEEE International Solid-State Circuits Conference Digest of Technical Papers*, San Francisco, 2010
3. R.C. Jaeger, J.C. Suhling, R. Ramani, A.T. Bradley, J. Xu, CMOS stress sensors on (100) silicon. IEEE J. Solid State Circuits **35**, 85–95 (2000)
4. M. Jung, J. Mitra, D.Z. Pan, S.K. Lim, TSV stress-aware full-chip mechanical reliability analysis and optimization for 3D IC. In *Proceedings of ACM Design Automation Conference*, San Diego, 2011
5. M. Jung, D. Pan, S.K. Lim, Chip/package co-analysis of thermo-mechanical stress and reliability in TSV-based 3D ICs, in *Proceedings of ACM Design Automation Conference*, San Francisco, 2012

6. J.-S. Kim et al., A 1.2 V 12.8 GB/s 2 Gb mobile wide-I/O DRAM with 4x128 I/O using TSV-based stacking, in *IEEE International Solid-State Circuits Conference Digest Technical Papers*, San Francisco, 2011
7. D.H. Kim, R. Topaloglu, S.K. Lim, Block-level 3D IC design with through-silicon-via planning, in *Proceedings of Asia and South Pacific Design Automation Conference*, Sydney, 2012
8. K.H. Lu, X. Zhang, S.-K. Ryu, J. Im, R. Huang, P.S. Ho, Thermo-mechanical reliability of 3-D ICs containing through silicon vias, in *IEEE Electronic Components and Technology Conference*, San Diego, 2009
9. A. Mercha et al., Comprehensive analysis of the impact of single and arrays of through silicon vias induced stress on high-k/metal gate CMOS performance, in *Proceedings IEEE International Electron Devices Meeting*, San Francisco, 2010
10. M. Nakamoto et al., Simulation methodology and flow integration for 3D IC stress management, in *Proceedings of IEEE Custom Integrated Circuits Conference*, San Jose, 2010
11. J.-S. Yang, K. Athikulwongse, Y.-J. Lee, S.K. Lim, D.Z. Pan, TSV stress aware timing analysis with applications to 3D-IC layout optimization, in *Proceedings of ACM Design Automation Conference*, Anaheim, 2010

Chapter 17
TSV Interfacial Crack Analysis and Optimization

Abstract In this chapter, we study an efficient and accurate full-chip through-silicon-via (TSV) interfacial crack analysis flow and design optimization methodology to alleviate TSV interfacial crack problems in 3D ICs. First, we analyze the TSV interfacial crack at the TSV/dielectric liner interface caused by TSV-induced thermo-mechanical stress. Then, we explore the impact of TSV placement in conjunction with various associated structures such as a landing pad and a dielectric liner on the TSV interfacial crack. Next, we study a full-chip TSV interfacial crack analysis methodology based on design of experiments (DOE) and response surface method (RSM). Finally, we study design optimization methodologies to mitigate the mechanical reliability problems in 3D ICs.

The materials presented in this chapter are based on [2].

17.1 Introduction

Due to the coefficients of thermal expansion (CTE) mismatch between a TSV fill material such as copper (Cu) and silicon (Si) substrate, thermo-mechanical stress is induced during fabrication process and thermal cycling of TSV structures. This stress can affect device performance [10] and drive crack growth in 3D interconnects [6, 8, 9]. Most previous works focused on modeling the thermo-mechanical stress and reliability of a single TSV in isolation. These simulations are performed using finite element analysis (FEA) method which is computationally expensive or infeasible for full-chip analysis. Furthermore, some works used unrealistic TSV structures such as an extremely large landing pad (LP), mainly because the design context is not considered.

Even though there are several works on thermo-mechanical reliability issues induced by TSV stress, this is the first work addressing TSV interfacial crack in a full-chip scale to the best of our knowledge. In this chapter, we study a fast and

S.K. Lim, *Design for High Performance, Low Power, and Reliable 3D Integrated Circuits*, DOI 10.1007/978-1-4419-9542-1_17,

efficient full-chip TSV interfacial crack analysis flow based on DOE and RSM. We use energy release rate (ERR) as a mechanical reliability metric, and show the impact of TSV placement style on ERR. The following specific topics are covered in this chapter:

- We study more detailed and realistic TSV structures and study their impact on the TSV interfacial crack.
- We study DOE and RSM methods to generate ERR models of TSV interfacial crack for full-chip analysis. We validate our methodology against FEA simulations.
- We study design methods to reduce ERR for TSV based 3D IC designs.

17.2 Preliminaries

17.2.1 TSV Interfacial Crack

TSVs pose a significant challenge to thermo-mechanical reliability of 3D ICs. In particular, the CTE mismatch between the conducting metal in TSV and silicon substrate can generate thermal stress inside and around TSVs. Such stress can induce a cohesive crack in the silicon substrate [7] and drive an interfacial crack between TSV and dielectric liner [6, 8, 9].

It is widely known that most of the mechanical reliability failures occur at the interface between different materials. Thus, in this chapter, we focus on the TSV interfacial crack at the TSV/dielectric liner interface. This TSV interfacial crack can cause not only mechanical reliability problems, but functional failures due to leakage.

However, it is hard to obtain realistic crack structures and crack growth behavior models after crack initiation without measurement data. Even with the same initial crack, every crack can grow in a different manner depending on the surrounding environment. Therefore, we adopt a crack structure well studied from previous works [8, 9]. Figure 17.1 shows our TSV interfacial crack structure. This crack initiates around the circumference of the TSV near the wafer surface and grows vertically downward. In our study, wafer surface means the dielectric layer surface right below the dielectric layer (SiO_2)/ILD (low K) interface shown in Fig. 17.2. Also, we assume that the crack front propagates uniformly to simplify crack modeling.

17.2.2 Energy Release Rate

Energy release rate (ERR) is defined as the energy dissipated during fracture, i.e. crack, per newly created fracture surface area. According to the law of conservation of energy, energy supplied to a crack front for it to grow must be balanced by the

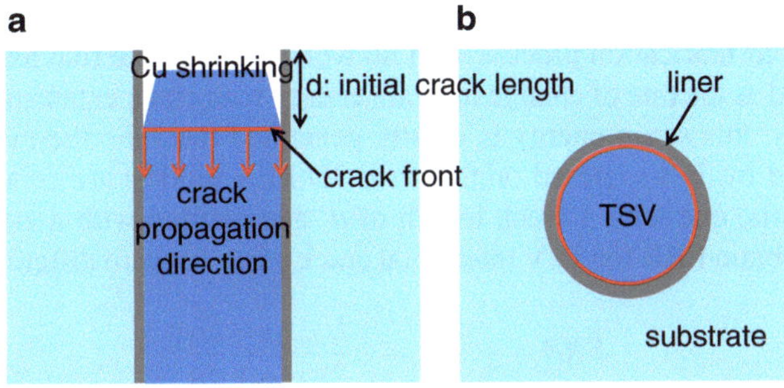

Fig. 17.1 TSV interfacial crack structure under negative thermal load. (**a**) Side view with initial crack length of d. (**b**) Top view

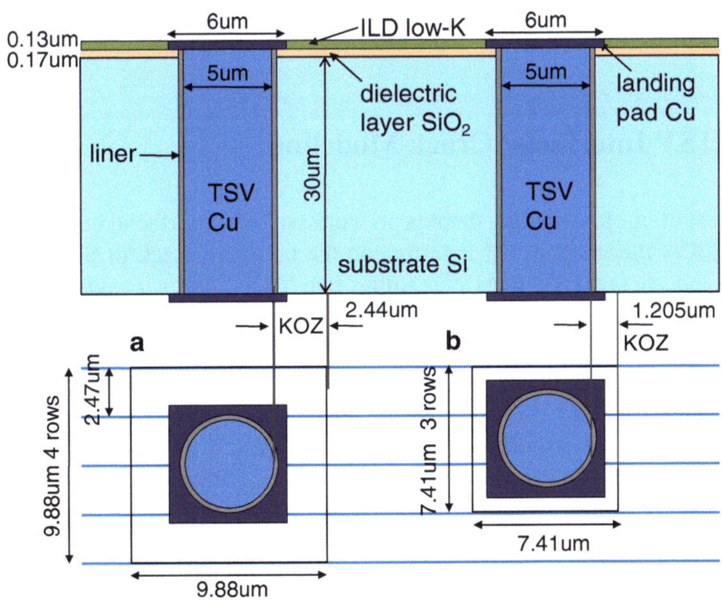

Fig. 17.2 Baseline TSV structure. (**a**) TSV_A cell occupying four standard cell rows (KOZ = 2.44 μm). (**b**) TSV_B cell (KOZ = 1.205 μm)

amount of energy dissipated due to the formation of new surfaces. In other words, ERR is the measure of the amount of energy available for fracture. If high energy is available around crack front, then there is a high chance of crack growth.

However, even with an initial crack, if ERR of the crack under consideration is lower than a threshold value, the crack does not grow further and stays in a stable state. The debonding energy between different materials is this threshold, and its value is material and fabrication process specific. For example, the debonding energy of Cu/SiO_2 interface ranges from 0.7 to $10\,J/m^2$ depending on fabrication process [6].

Since the loading in our simulation structure is solely due to the thermal expansion from fabrication process with no work done by external loads, ERR can be determined as the rate of change in strain energy with crack extension [6]. In TSV based 3D ICs, this strain energy is mostly generated from the thermo-mechanical stress induced by TSVs. Based on this, two 3D FEA models are created for strain energy analysis, one with a crack length of d, and another with a crack length of $d+\Delta d$. We obtain ERR for TSV interfacial crack using forward difference approach as follows:

$$ERR = -\frac{\partial U}{\partial A} = -\frac{U_{d+\Delta d} - U_d}{2\pi r_{TSV} \cdot \Delta d}$$

where, U is a strain energy, A is area, d is an initial crack length, Δd is a crack increment, and r_{TSV} is TSV radius. In this chapter, we set d as 1 μm and vary Δd from 0.1 to 0.5 μm to capture the crack initiation from the wafer surface as well as the near surface thermal stress impact [9].

17.3 TSV Interfacial Crack Modeling

To the best of our knowledge, there is no work on TSV interfacial crack considering nearby TSVs the work in [7] investigated the cohesive crack in Si substrate with straight and zigzag TSV lines containing five TSVs, and showed that the zigzag type is a better choice to mitigate crack driving force than the straight line. However, this work was performed based on a 2D stress model. Thus, it does not capture the 3D nature of a stress field near a wafer surface around TSVs where devices are located [9]. Also, they did not consider dielectric liner and landing pad in TSV structures, which are essential components for TSV.

Although authors in [9] proposed a semi-analytic ERR model for TSV interfacial crack, it is only valid for an infinitely long TSV. Also, their model is only applicable to a single TSV in isolation and their TSV structure includes only TSV and silicon substrate. Thus, it cannot be directly used to assess TSV interfacial crack considering multiple TSVs as well as a TSV which contains a landing pad and a dielectric liner because of the change in boundary conditions.

Before discussing detailed crack modeling, we introduce two terminologies: (1) **Victim TSV**: TSV with an interfacial crack. (2) **Aggressor TSV**: TSV located nearby a victim TSV and affecting crack growth of the victim TSV.

17.3.1 3D FEA Simulation

Since there is no known analytical ERR model for a realistic TSV structure, 3D FEA models for a TSV interfacial crack analysis are created to investigate the impact of aggressor TSVs on the interfacial crack of a victim TSV. To realistically examine the interfacial crack, our baseline simulation structure of a TSV is based on the fabricated and the published data [1], as shown in Fig. 17.2.

Table 17.1 Material properties

Material	CTE (ppm/K)	Young's modulus (GPa)	Poisson's ratio
Cu	17	110	0.35
Si	2.3	130	0.28
SiO_2	0.5	71	0.16
Low K	20	9.5	0.3
BCB	40	3	0.34

We construct two TSV cells, i.e., TSV_A and TSV_B, which occupy four and three standard cell rows in 45 nm technology. We define 2.44 and 1.205 μm from TSV edge as keep-out-zone (KOZ) in which no cell is allowed to be placed for TSV_A and TSV_B cells, respectively. Our baseline TSV diameter, height, landing pad width, and liner thickness are 5 μm, 30 μm, 6 μm, and 125 nm, respectively, unless specified, which are close to the data in [1]. We use SiO_2 as a baseline liner material, and ignore Cu diffusion barrier material such as Ta and Ti in these experiments. In general, this barrier thickness is negligible compared with SiO_2 liner, hence its impact on stress distribution is negligible.

Material properties used for our experiments are listed in Table 17.1. We use the FEA simulation tool ABAQUS to perform experiments. We apply $\Delta T = -250\,^{\circ}C$ of thermal load for the entire simulation structures. That is, we assume TSV structure is annealed at 275 °C and cooled down to 25 °C to represent the manufacturing process [3, 7, 9]. We also assume that the entire TSV structure is stress free at the annealing temperature.

17.3.2 Impact of TSV Liner and Landing Pad

We first explore the impact of surrounding structures such as a liner and a landing pad. For this experiment, we use simulation structures without aggressor TSVs. Figure 17.3 shows ERR of a victim TSV with four different configurations. First, we observe that a 125 nm thick SiO_2 liner, which acts as a stress buffer layer, reduces ERR by 6.5 % compared with the case without landing pad and liner. The landing pad also helps decrease ERR by preventing the TSV/liner interface from separating. Finally, when both a SiO_2 liner and a landing pad are considered, ERR decreases by 18.9 %.

We also employ benzocyclobutene (BCB), a polymer dielectric material, as an alternative TSV liner material [7, 9]. Since Young's modulus, which is a measure of the stiffness of an isotropic elastic material, of BCB is much lower than Cu, Si, and SiO_2, this BCB liner can absorb the stress effectively caused by the CTE mismatch. Figure 17.4 shows the impact of liner material and thickness on ERR. For this experiment, we set the landing pad width as 6 μm for all cases. As liner thickness increases, ERR decreases noticeably for both liner materials. Also, BCB liner outperforms SiO_2 on reducing ERR.

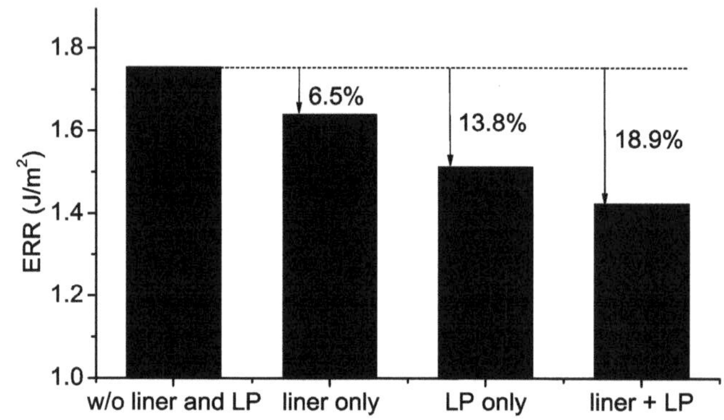

Fig. 17.3 Impact of TSV surrounding structures on ERR

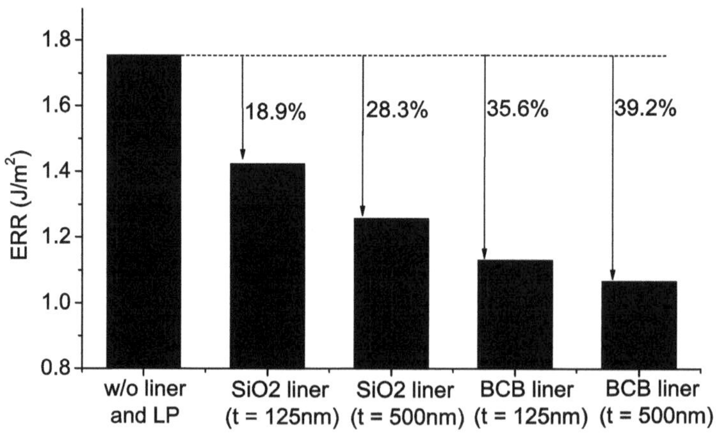

Fig. 17.4 Impact of TSV liner on ERR. Landing pad width of 6 μm is used with each liner case

We examine the impact of landing pad size on the TSV interfacial crack as well. We use four landing pad widths; 6, 8, 10, and 12 μm. We observe that ERR is lower with a landing pad than ERR without a landing pad case for all landing pad sizes. We also see that for the landing pad width up to 10 μm, ERR increases and then saturates. This is because the magnitude of all normal stress components at the TSV/liner interface underneath a landing pad increases due to increased Cu volume with enlarged landing pad area, which is an additional CTE mismatch source. Thus, for the TSV interfacial crack, larger landing pad size is not beneficial. It is thus evident from these experiments that modeling TSV interfacial crack considering surrounding structures such as a liner and a landing pad is important to analyze the thermo-mechanical reliability in 3D ICs more accurately.

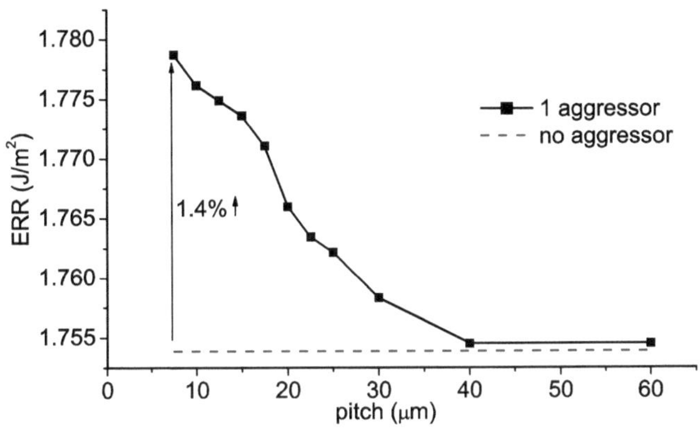

Fig. 17.5 Impact of TSV pitch on ERR

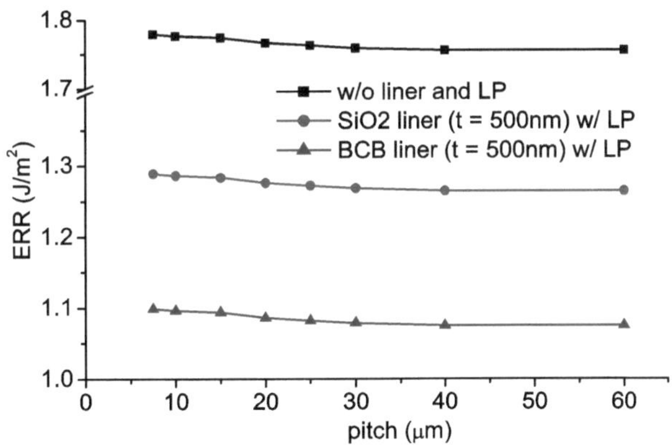

Fig. 17.6 Impact of TSV pitch with liner and landing pad (6 μm) on ERR

17.3.3 Impact of Pitch and Angle among TSVs

In this section, we study the impact of TSV pitch on the TSV interfacial crack. With fixed victim TSV location, we vary the TSV pitch between a victim and an aggressor from 7.5 to 60 μm. As Fig. 17.5 shows, ERR decreases monotonically as the pitch increases and approaches to the level when there is no aggressor at around 40 μm pitch. However, when only one aggressor is considered, ERR increase at the minimum pitch compared with the maximum pitch is only 1.4 %, which is negligible. Figure 17.6 shows ERR curve with two 500 nm thick liner materials and a 6 μm wide landing pad. The magnitude of ERR decreases when we use a liner and a landing pad, but overall ERR trend remains similar.

As we introduce additional TSVs, both distance and angle between TSVs become important to the TSV interfacial crack. Stress at a point can be computed by adding individual stress tensors induced by each TSV at this point. Depending on relative angle between TSVs, even with the same pitch, stress at the point can be either added up or canceled out. Since this stress directly affects strain energy of a TSV structure, ERR value also varies over different angles.

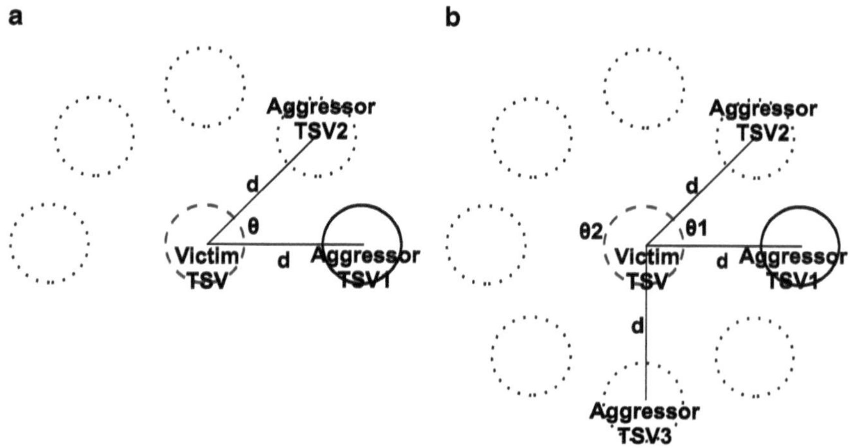

Fig. 17.7 Simulation structure for angular dependency. Distance from victim to all aggressors is d. (**a**) Two aggressor TSVs. Aggressor 1 is fixed and aggressor 2 rotates. (**b**) Three aggressor TSVs. Aggressor 1 is fixed and aggressor 2 and 3 rotate

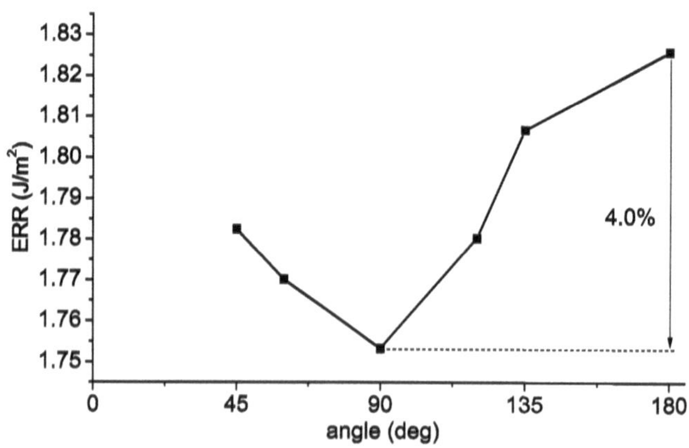

Fig. 17.8 Impact of angle between victim and two aggressor TSVs on ERR

Now we perform experiments with two aggressors which are located at 10 μm distance from the victim TSV. Then, we vary angle among three TSVs from 45° to 180° to explore the impact of angle on ERR of the victim TSV shown in Fig. 17.7a. As Fig. 17.8 shows, ERR is minimum when three TSVs form 90° angle, and increases as the angle approaches to either 45° or 180°. If we plot σ_{xx} stress component in Cartesian coordinate system, TSV structure with negative thermal load creates tensile stress along the x-axis and compressive stress along the y-axis. Thus, if a victim and aggressors form 90°, tensile and compressive stress from each aggressor TSV cancels out at the victim TSV location, that is why we see the lowest ERR at 90°.

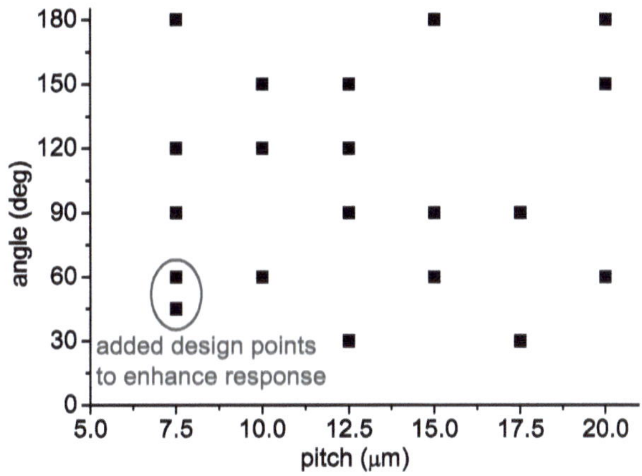

Fig. 17.9 DOE based simulation points of two aggressor TSVs

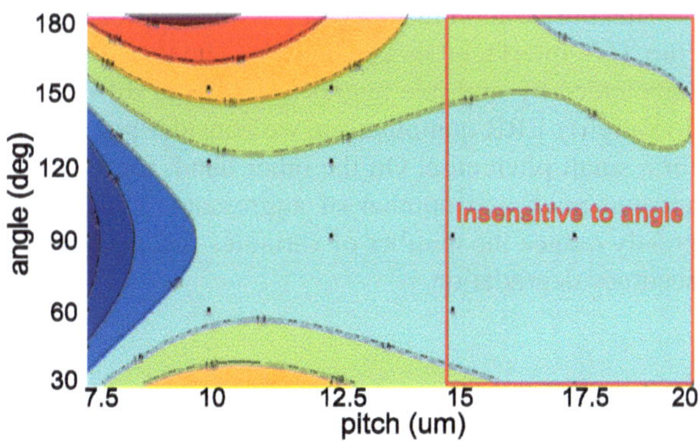

Fig. 17.10 ERR contour map of pitch and angle

17.3.4 Relative Importance of Pitch Over Angle

In this section, we explore the combined effect of TSV pitch and angle. We first use two aggressor TSVs, and change both pitch and angle. We also employ DOE, which will be discussed in detail in Sect. 17.4, to produce simulation points shown in Fig. 17.9. Figure 17.10 shows an ERR contour map for different pitch and angle. We observe high angular dependency in a small pitch region. However, as the pitch exceeds 15 μm, impact of angle is almost negligible.

To further investigate the relative importance between pitch and angle, we now use three aggressor TSVs shown in Fig. 17.7b. As Fig. 17.11 shows, angular dependency is dominant for small pitches. However, as the pitch increases, even though there are still some fluctuations along angle axis, angular dependency of ERR is not significant. We also perform simulations with more aggressors up to eight aggressors. We find that angular dependency is almost not noticeable beyond 10 μm pitch, and the number of aggressor TSVs as well as TSV pitch mostly determine the ERR value of the victim TSV.

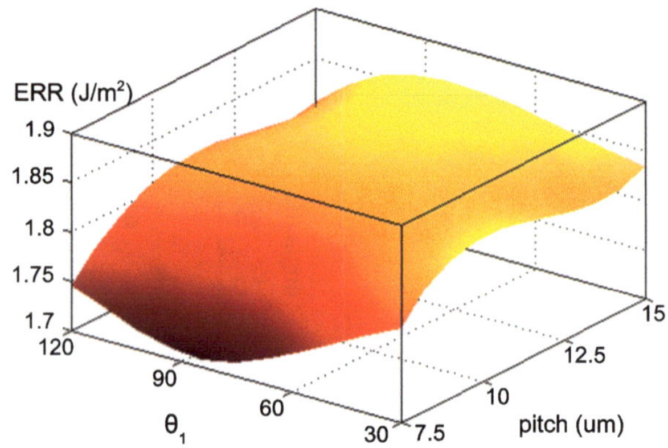

Fig. 17.11 ERR surface map of pitch and θ_1 in three aggressor TSVs case

This observation indicates a possible approach to generate ERR model for full-chip TSV interfacial crack analysis. The victim TSV under consideration could be surrounded by multiple aggressor TSVs at different distances and relative angles. To simplify ERR computation, we consider pitch, angle, and number of aggressors for a small pitch case. On the other hand, for a large pitch case, we only take into account pitch and number of aggressors. This simplification of ERR model can greatly reduce the number of variables that needs to be considered with acceptable accuracy degradation.

17.4 DOE-Based Full-Chip TSV Interfacial Crack Modeling

FEA simulation of TSV interfacial crack with multiple TSVs requires huge computing resources and time. In our simulations, depending on the number of TSVs and mesh structure, single FEA simulation takes about 1–12 h using four CPUs. Thus, it is not feasible for full-chip analysis.

Meanwhile, DOE has been used for many science and engineering applications. Recently, DOE was even used for co-optimization of power network, thermal TSV, and micro-fluidic channel in 3D ICs [5]. It has been proven to be an effective technique when analysis is desired for complex systems with multiple input factors. It provides a well-organized way of performing experiments so that we can use the experimental results to find meaningful relations between input factors and responses of the system. In this section, we present a design of experiments (DOE) and response surface method (RSM) based full-chip TSV interfacial crack analysis flow.

In general, TSV placement style is largely divided into two categories: (1) Regular TSV placement. (2) Irregular TSV placement. In the regular TSV placement scheme, we pre-place TSVs uniformly on each die, and then place cells. In this case, the pitch between TSVs is the most critical factor to predict ERR of a victim TSV. On the other hand, TSVs and cells are placed simultaneously in the irregular TSV

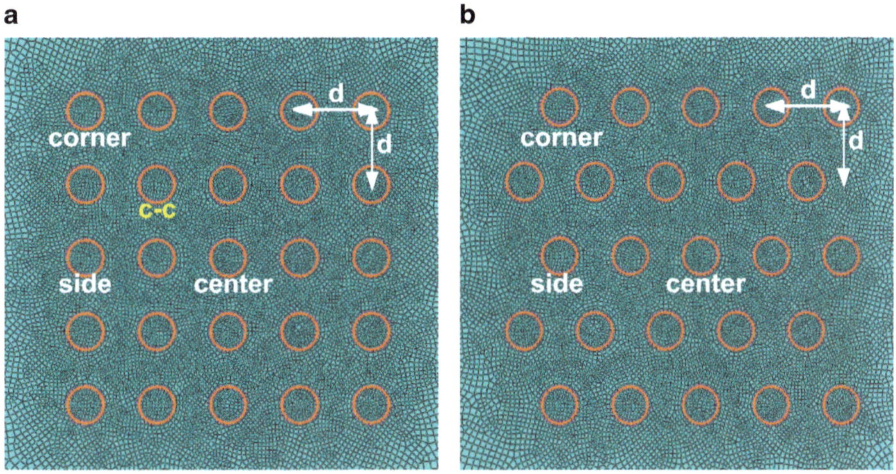

Fig. 17.12 Top view of meshed simulation structures for 5 × 5 TSV block. *Orange circles* are TSVs. (**a**) Array type. (**b**) Staggered type

placement scheme the irregular TSV placement shows shorter wirelength than the regular case [4]. However, in this irregular TSV placement case, possible positions of aggressor TSVs around a victim TSV are innumerable. Thus, it is infeasible to examine all possible TSV arrangements to assess ERR of a victim TSV.

In the following sections, we discuss DOE and RSM based full-chip TSV interfacial crack modeling for both regular and irregular TSV placement styles, and validate our model against FEA simulations. We use *Model-Based Calibration Toolbox* in MATLAB to design experiments and obtain response surface model.

17.4.1 Designing Experiments

To use DOE and RSM, we first need to define design knobs (= input factors) and metrics (= responses). We use ERR as our metric to assess TSV interfacial crack in a full-chip scale. However, input factors are different for regular and irregular TSV placement style.

Figure 17.12 shows two possible regular TSV placement of 5 × 5 TSV block with a same pitch. In the array type, TSVs are aligned both in horizontal and vertical direction, whereas in the staggered type, TSVs in every other row are shifted by half pitch. For both of these regular TSV placement schemes, the most important factor that determines ERR of a victim TSV is the pitch and the position of the victim TSV inside the block such as center, side, or corner shown in Fig. 17.12. We set TSV pitch as the only input factor and find ERR models for aforementioned critical victim TSV locations separately for regular TSV placement scheme.

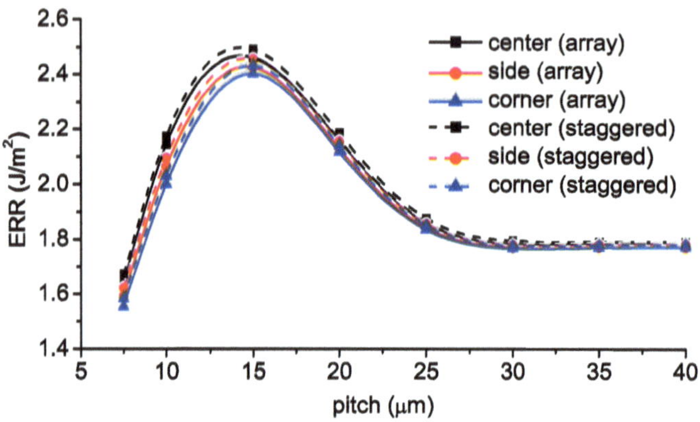

Fig. 17.13 ERR vs. pitch for array and staggered type

Unlike the regular TSV placement style, there are countless possible combinations of TSV arrangement in the irregular TSV placement. However, simulating all these possibilities is impossible. In Sect. 17.3.4, we identified that the relative angle between victim and aggressor TSV is important only when the pitch is small. Also, as shown in Figs. 17.5 and 17.6, gradient of ERR along pitch is not steep. From these observations, we simplify ERR model for the irregular TSV placement as follows: If the distance between a victim and aggressors is less than 10 μm, we consider both the number of aggressors and the angle between them. If the distance exceeds 10 μm, we only count the number of aggressors at each distance bin with 5 μm interval shown in Fig. 17.16. In this way, we remove the distance from input factors, and use the number of aggressors at each bin and the angle of nearest aggressors to design experiments.

We generate design points using Stratified Latin Hypercube from space filling design styles. Based on the design points, we create FEA simulation structures and obtain ERR data from each simulation. With these ERR values, we build the response surface and obtain the analytical ERR model for full-chip TSV interfacial crack analysis.

17.4.2 ERR Model for Regular TSV Placement

First, we monitor ERR values of victim TSVs in center, side, and corner location shown in Fig. 17.12. As Fig. 17.13 shows, ERR is highest at the center, and decreases as victim TSV location moves to side. The lowest ERR occurs at corner due to decreased number of aggressors surrounding the victim TSV. We also observe that ERR of the victim TSVs in array type is always lower than the counterpart in staggered type, even though the difference is not significant. This is because large numbers of nearby aggressors are forming 90° angle in array type, hence reduces stress magnitude at the victim TSV location.

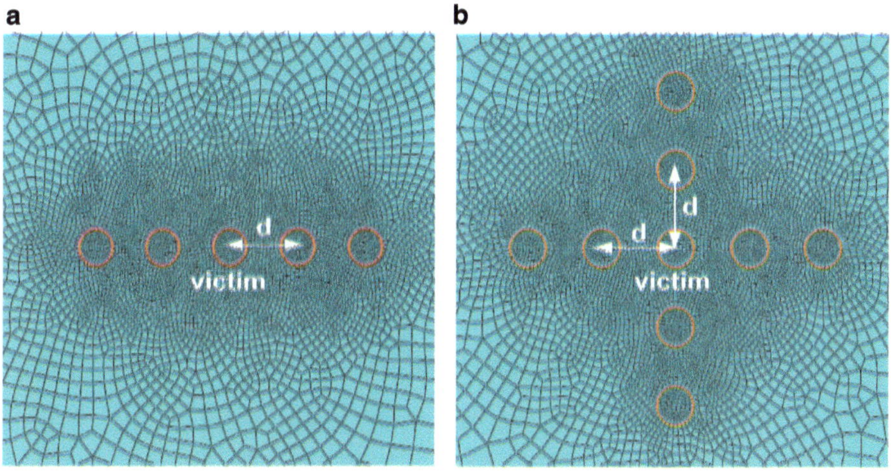

Fig. 17.14 Top view of meshed simulation structures for (**a**) line style, (**b**) cross style

Interestingly and counter-intuitively, ERR is minimum in the smallest pitch, and increases up to 15 μm, then decreases and finally saturates at around 30 μm pitch. To verify why minimum ERR occurs at the smallest pitch, we build two simulation structures shown in Fig. 17.14. In the line style, the victim TSV is only affected by constructive stress interference from aggressors, whereas the victim TSV experiences both destructive and constructive stress interference from aggressors in the cross style. As Fig. 17.15 shows, even though there are four more aggressors in the cross structure, ERR is always higher in the line style for simulated pitches. Most importantly, ERR is minimum at the smallest pitch in the cross structure due to the highest destructive stress interference. This observation indicates that it is always better to build TSV blocks in the array type rather than in the line style with a given number of TSVs to help suppress TSV interfacial crack growth. We also observe that ERR difference among three victim TSV locations, i.e., center, side, and corner, is higher in smaller pitches. This is again due to higher stress interference in smaller pitches, which results in larger stress magnitude difference among different victim TSV locations.

We decide to use the array type for the regular TSV placement ERR model, which shows the lowest ERR. We generate eight design points for the regular TSV placement and build RSM model based on FEA simulation results. RSM model can be expressed as a multivariate polynomial equation. In our case, the ERR model of the regular TSV placement is expressed as a fourth order polynomial with one variable (= pitch) as follows:

$$ERR_{reg} = c_1 + c_2 \cdot d + c_3 \cdot d^2 + c_4 \cdot d^3 + c_5 \cdot d^4$$

where, d is pitch and $c_1 - c_5$ are TSV dimension dependent coefficients. We build ERR models for center, side, and corner locations separately. We observe that

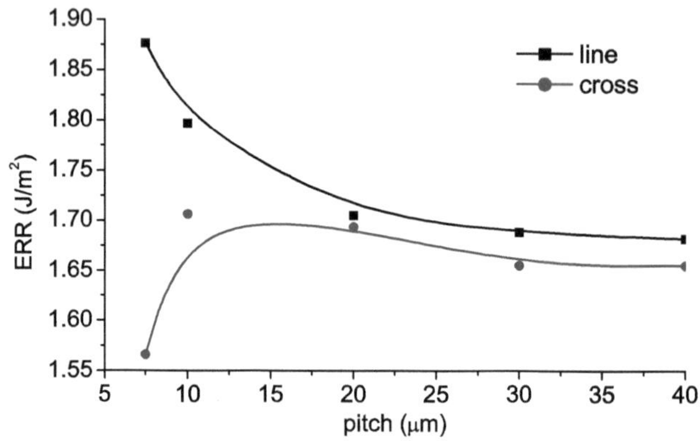

Fig. 17.15 ERR vs. pitch for *line* and *cross* type

ERR of an intermediate point such as c-c shown in Fig. 17.12a can be obtained by averaging ERR values of victim TSVs in center and corner locations with a negligible error. We also generate ERR models for different TSV array blocks, such as 3×3 and 7×7 array.

17.4.3 ERR Model for Irregular TSV Placement

As we discussed, there are innumerable scenarios for the irregular TSV placement. We reduce the number of input factors by distance binning and considering angular dependency within 10 μm distance range from victim TSV shown in Fig. 17.16. In addition, the number of aggressors at each distance bin cannot be arbitrarily large. Also, as the pitch becomes smaller, possible number of aggressors at that bin is also smaller due to reduced bin area. We generate 50 design points with this constraint. Since we use 13 input factors, 8 for angle (8 aggressors are maximum possible number that can be placed in the bin 1) and 5 for number of aggressors at each distance bin, 50 design points are not enough to obtain high quality RSM model. However, based on the observation that if we rotate entire simulation structure by same angle θ around victim TSV, ERR will remain same since relative positions of TSVs are unchanged, we generate 885 data points from 50 simulations for better response surface fitting.

Figure 17.17 shows predicted ERR (RSM model) and data points. There are six outliers which occur when there are no aggressors at bin 1, hence eight input factors for angle are not exercised, which causes deviation from the predicted model. This ERR model can be enhanced by simulating more design points on this particular case. However, the proportion of the case with no aggressors at bin 1 is less than 6 % (87 out of 1,472 TSVs) in the worst case in our benchmark circuits. Also, since the ERR model fits well with data points in general, we use the model without

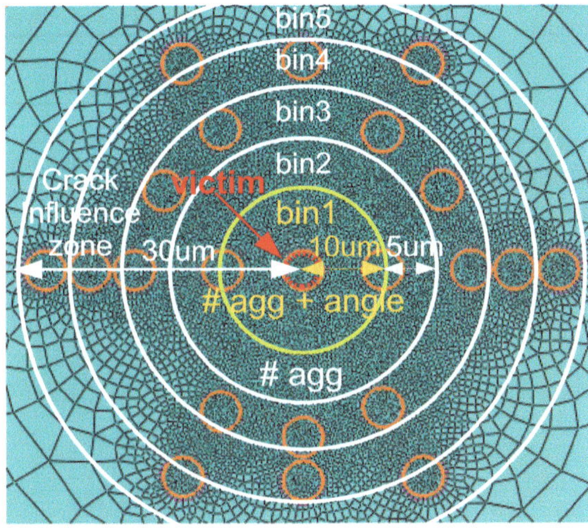

Fig. 17.16 Top view of meshed simulation structure for irregular TSV placement with 21 aggressor TSVs

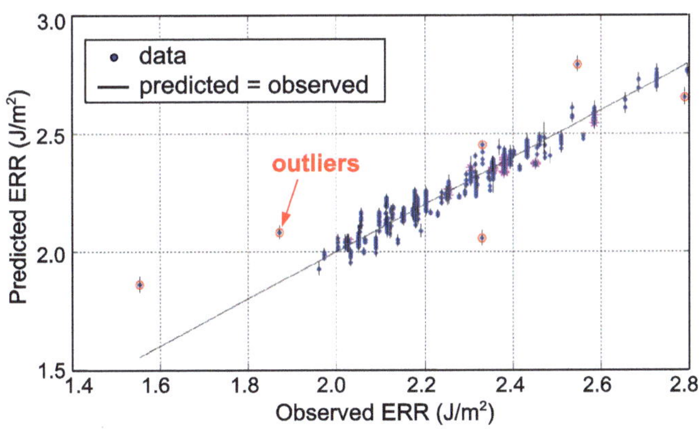

Fig. 17.17 Predicted ERR using DOE and RSM vs. observed ERR

further simulations. The ERR model of the irregular TSV placement is expressed as a second order polynomial with 13 variables. The details of our ERR model is not shown due to space limit.

17.4.4 Quality of ERR Model

The goodness-of-fit of a model can be tested with statistics such as coefficient of determination (R^2), root mean square error (RMSE), and prediction error sum of squares RMSE (PRESS RMSE), which is evaluated by excluding one data point at a time, building a new RSM model, and computing RMSE [5]. Table 17.2 shows

Table 17.2 Quality of ERR model

Placement type	R^2	RMSE	PRESS RMSE
Regular	0.993	0.034	0.086
Irregular	0.956	0.044	0.098

Table 17.3 Validation of ERR model. Simulation case shows pitch for regular TSV placement and number of aggressors for irregular TSV placement

Placement type	Simulation case	ERR (model)	ERR (simulation)	Validation RMSE
Regular	9 μm	1.996	1.985	0.033
	12.5 μm	2.401	2.371	
	17.5 μm	2.355	2.335	
	22.5 μm	2.015	1.988	
	27.5 μm	1.789	1.778	
Irregular	10 agg	1.901	1.971	0.055
	21 agg	2.229	2.324	
	28 agg	2.320	2.305	
	36 agg	2.394	2.371	
	43 agg	2.572	2.547	

that R^2 values of our ERR models are close to 1, and both RMSE and PRESS RMSE is less than 0.1. Considering the fact that ERR values from our simulations range from 1.5 to 3.0 in general, quality of fitting is acceptable.

Even though our models match well with simulation data, it is essential to validate whether our ERR models predict unseen data points correctly. We design five new simulation structures to validate ERR models for both regular and irregular TSV placement cases. Table 17.3 shows predicted ERR from our model and ERR from simulations. Since the regular TSV placement type uses only one input factor, i.e., pitch, the validation RMSE is lower than the irregular TSV placement case to the model RMSE value. The Validation RMSE of the irregular TSV placement type is also acceptable compared with the model RMSE value.

17.4.5 Full-Chip Analysis Flow

In this section, we briefly summarize our full-chip TSV interfacial crack analysis flow. Our full-chip TSV interfacial crack analysis flow is shown in Algorithm 11. We assume that each TSV can be a candidate for victim TSV. Thus, while we visit each TSV, we set this TSV as a victim TSV and other TSVs within influence zone as aggressors. In our simulation, we use 30 μm as a crack influence zone, since at around this pitch ERR saturates shown in Fig. 17.13. Then, we find angle and distance between aggressor and victim TSV, and insert aggressor into corresponding distance bin for the irregular TSV placement or find a pitch for the regular TSV placement. Once we prepare this information, we compute ERR using analytical models based on DOE and RSM.

Input: TSV list T, TSV placement style S, ERR model M
Output: ERR map
for *each TSV t in T* **do**
    $c \longleftarrow$ center of t
    $r \longleftarrow$ FindCrackInfluenceZone(c)
    $r.Agg \longleftarrow$ FindAggressorTSVs(r)
    for *each aggressor a in r.Agg* **do**
        $t.\theta \longleftarrow angle(t,a)$
        $t.d \longleftarrow distance(t,a)$
        if $S = Iregular$ **then**
            $t.n \longleftarrow$ AddAggressorAtDistanceBin($t.d$)
        end
        else if S = *Regular orBlock* **then**
            $p \longleftarrow$ FindPitch(T)
        end
    end
    $t.ERR \longleftarrow$ ComputeERR($t.\theta, t.d, t.n, p, S, M$)
end

Algorithm 11: Full-chip TSV interfacial crack analysis flow

Table 17.4 Benchmark circuits

Circuit	TSV placement	TSV cell size (μm × μm)	Wirelength (mm)	Area (μm × μm)	Pitch (μm)
Irreg_A	Irregular	9.88×9.88	8,884	$1{,}000 \times 1{,}000$	–
Reg_A	Regular	9.88×9.88	9,648	$1{,}000 \times 1{,}000$	25
Irreg_B	Irregular	7.41×7.41	9,060	960×960	–
Reg_B	Regular	7.41×7.41	9,547	960×960	22

17.5 Experimental Results

We implement a full-chip TSV interfacial crack analysis flow in C++. Four variations of a gate-level 3D circuit, with changes in TSV placement style and TSV cell size, are used for our analysis, which are listed in Table 17.4. The number of TSVs and gates are 1,472 and 370 K, respectively, for all cases to compare impact of placement style on ERR fairly. These circuits are synthesized using Synopsys Design Compiler with the physical library of 45 nm technology, and designed using Cadence SoC Encounter to two-die stacked 3D ICs.

17.5.1 Impact of KOZ

We first investigate the impact of KOZ size on ERR of both regular and irregular TSV placement style. Figure 17.18 shows histogram of number of TSVs for

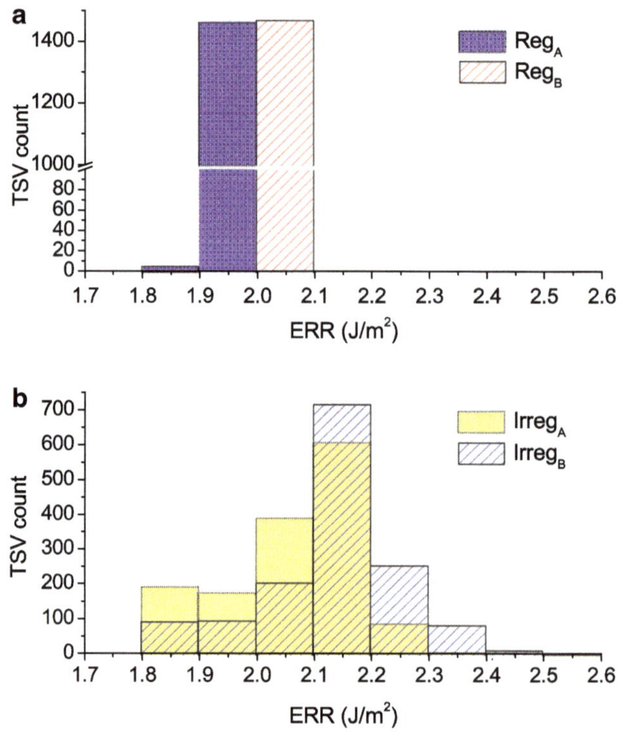

Fig. 17.18 Impact of keep-out-zone on ERR. TSV_A cell (KOZ = 2.44 μm) and TSV_B cell (KOZ = 1.205 μm). (**a**) Regular TSV placement. (**b**) Irregular TSV placement

observed ERR ranges. We first observe that ERR values are highly concentrated in a small range in the case of regular TSV placement. Even though there is a difference of ERR between center and corner locations in TSV array for example, that is negligible in the pitch of 22 μm (Reg_B) and 25 μm (Reg_A). Also, KOZ size impact on ERR is not significant for this regular TSV placement case, since both TSV pitches are already close to crack influence zone (30 μm), and their difference is only 3 μm.

On the other hand, the irregular TSV placement case shows larger variations of ERR and large number of TSVs experience higher ERR than the regular TSV placement case. This is mainly because TSVs can be placed either densely or sparsely to minimize wirelength in the case of irregular TSV placement scheme. Thus, ERR of victim TSV can vary noticeably depending on the placement of nearby aggressor TSVs. Furthermore, since there are regions where group of TSVs are closely placed as shown in Fig. 17.19a, higher ERR values are observed in the irregular TSV placement style. We also see that the KOZ size affects ERR values significantly in the irregular TSV placement. This is because the number of aggressors at each distance bin decreases due to increased KOZ size, hence reduces stress magnitude at victim TSV.

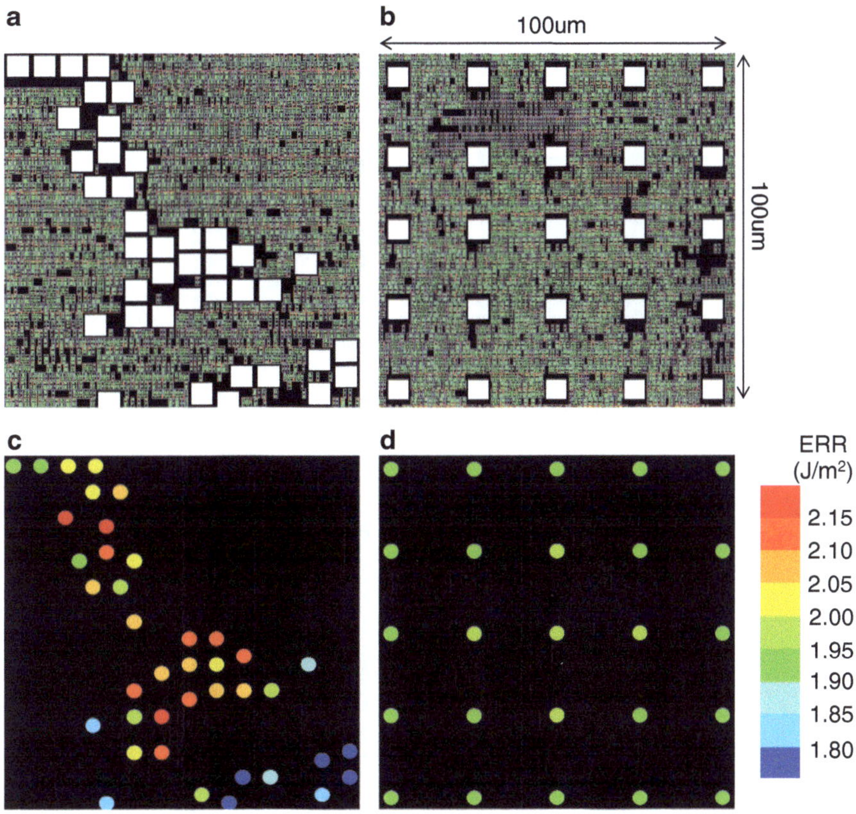

Fig. 17.19 Close-up shots of layouts and ERR maps. (**a**) $Irreg_B$. (**b**) Reg_B. (**c**) ERR map of $Irreg_B$. (**d**) ERR map of Reg_B

17.5.2 Impact of Liner

We identify that ERR is highly dependent on the liner material and its thickness in Sect. 17.3.2. In this section, we study the impact of liner on ERR in a full-chip scale. We use a $6 \times 6\,\mu m^2$ landing pad for all cases. Figure 17.20a shows that ERR values of both irregular and regular TSV placement schemes reduce significantly with use of liner. We also observe that liner thickness has a huge impact on the maximum ERR magnitude, since the thicker liner effectively absorbs thermo-mechanical stress at the TSV/liner interface. Especially, the BCB liner shows significant reduction in the maximum ERR compared with the SiO_2 liner due to extremely low Young's modulus shown in Sect. 17.3.1. Furthermore, we see that ERR decrease is higher in the irregular TSV placement case compared with the regular TSV placement since the stress buffer effect of liner is more effective where aggressor TSVs are close to victim TSV.

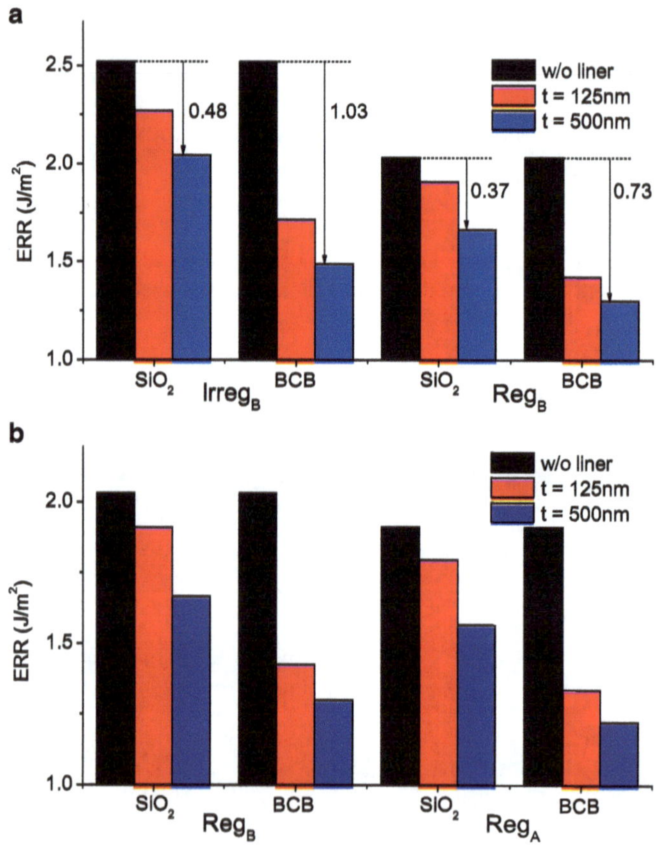

Fig. 17.20 Impact of liner material and thickness on maximum ERR. (**a**) Irreg_B vs. Reg_B. (**b**) Reg_B vs. Reg_A

Figure 17.20b shows that differences of ERR values between Reg_B and Reg_A circuits are not significant with different liner materials and their thickness. This is again because the pitch difference between Reg_B and Reg_A circuits is negligible and the pitch itself is already close to the crack influence zone range.

17.5.3 Reliability of Block-Level 3D Design

Even though the gate-level 3D design has the potential of highest optimization, the block-level design is attractive in the sense that we can reuse highly optimized 2D IP blocks. In this section, we examine the TSV interfacial crack in block-level 3D designs. 3D block-level designs are generated using an in-house 3D floorplanner which treats a group of TSVs as a block shown in Fig. 17.21. We use a 500 nm thick

Table 17.5 Comparison between gate-level and block-level design

Level	TSV pitch (μm)	# TSV	WL (mm)	Area (μm × μm)	Max ERR (J/m^2)	Std' dev'
Gate	Irregular	1,472	9,060	960 × 960	1.489	0.081
	22	1,472	9,547	960 × 960	1.300	0.003
Block	7.5	333	7,933	980 × 1,090	1.232	0.129
	10	394	8,028	1,080 × 1,000	1.500	0.160
	15	368	8,259	950 × 1,130	1.805	0.270

BCB liner and a 6 ×6 μm^2 landing pad for all cases. We vary the TSV pitch inside TSV blocks to examine its impact on layout quality as well as reliability issues. Note that the pitch inside TSV block is smaller than regular TSV placement case, in general.

Table 17.5 shows that block-level designs use less number of TSVs, show shorter wirelength, and occupy more area than gate-level designs. Experimental results show that the block-level design with 7.5 μm pitch shows smallest ERR among all cases. This is observed in Fig. 17.13 due to strong TSV-to-TSV stress interference in small pitches. However, it comes with larger variation of ERR across TSVs compared with both irregular and regular TSV placement cases. This is mainly due to the small TSV pitch and different types of TSV blocks used in block-level design such as the line type shown in top-right part of Fig. 17.21a. Also, it is possible that decreased TSV pitch could worsen signal integrity due to high TSV-to-TSV coupling. Therefore, TSV pitch in block-level designs should be carefully determined considering both mechanical and electrical issues.

17.5.4 Summary and Key Findings

In this section, we summarize our work and briefly discuss key findings.

1. We present an efficient full-chip TSV interfacial crack analysis flow based on DOE and RSM. Note that our full-chip TSV interfacial crack analysis methodology is general enough to be applied to other types of crack structures.
2. TSV interfacial crack is affected by TSV placement style, KOZ size, and TSV surrounding structures such as a liner and a landing pad.
3. Irregular TSV placement suffers from higher ERR and larger ERR variation compared with regular TSV placement, hence hard to control TSV interfacial crack problems across a chip.
4. It is recommended to use array type TSV blocks wherever possible for regular TSV placement and block-level design to benefit from ERR reduction by highest destructive stress interference. In addition, TSV pitch inside TSV blocks should be carefully determined by considering both mechanical and electrical issues.

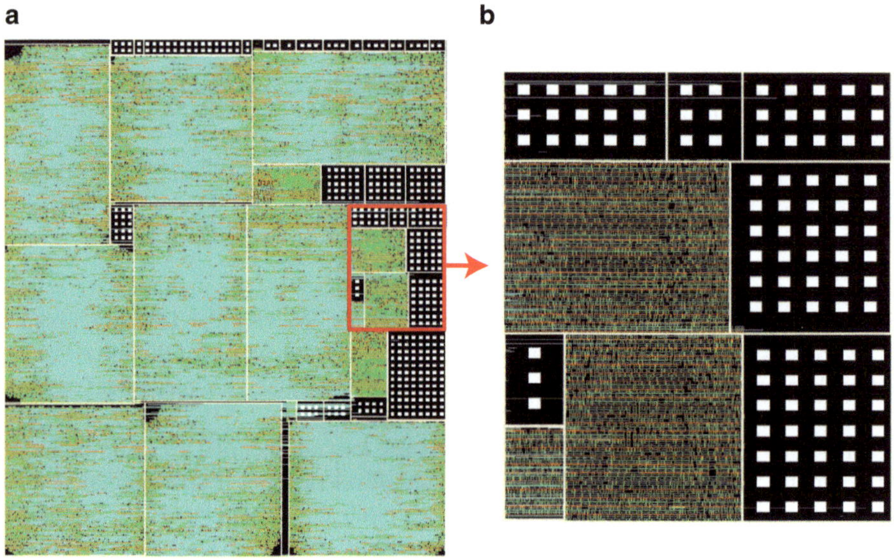

Fig. 17.21 Layout of block-level design (TSV pitch = 15 μm). *White rectangles* are TSV landing pads. (**a**) Full-chip layout. (**b**) Close-up shot of the *red box* in (**a**)

5. Larger KOZ relieves TSV interfacial crack due to increased spacing among TSVs and decreased number of aggressors affecting victim TSV inside crack influence zone. However, since it comes with larger footprint area, a careful design of KOZ is required.
6. Liner material and thickness are key design knobs to alleviate TSV interfacial crack problem by reducing stress magnitude at TSV/liner interface.

17.6 Conclusions

In this chapter, we studied how TSV placement as well as TSV surrounding structures such as a liner and a landing pad affect TSV interfacial crack in 3D ICs. We also studied a DOE and RSM based accurate and fast full-chip TSV interfacial crack analysis flow, which is applicable to placement optimization for 3D ICs. Our results showed that KOZ size, liner material/thickness, and TSV placement are key design parameters to reduce the TSV interfacial crack problems in TSV based 3D ICs.

References

1. G.V. der Plas et al., Design issues and considerations for low-cost 3D TSV IC technology, in *IEEE International Solid-State Circuits Conference Digest Technical Papers*, San Francisco, (IEEE, Piscataway, 2010)
2. M. Jung, X. Liu, S.K. Sitaraman, D.Z. Pan, S.K. Lim, Full-chip through-silicon-via interfacial crack analysis and optimization for 3D IC, in *Proceedings of IEEE International Conference on Computer-Aided Design*, San Jose, 2011
3. A.P. Karmarkar, X. Xu, V. Moroz, Performance and reliability analysis of 3D-integration structures employing through silicon via (TSV), in *IEEE International Reliability Physics Symposium*, Montreal, (IEEE, New York/Piscataway, 2009)
4. D.H. Kim, K. Athikulwongse, S.K. Lim, A study of through-silicon-via impact on the 3D stacked IC layout, in *Proceedings of IEEE International Conference on Computer-Aided Design*, San Jose, (ACM, New York, 2009)
5. Y.-J. Lee, R. Goel, S.K. Lim, Multi-functional interconnect co-optimization for fast and reliable 3D stacked ICs, in *Proceedings of IEEE International Conference on Computer-Aided Design*, San Jose, (ACM, New York, 2009)
6. X. Liu, Q. Chen, P. Dixit, R. Chatterjee, R.R. Tummala, S.K. Sitaraman, Failure mechanisms and optimum design for electroplated copper through-silicon vias (TSV), in *IEEE Electronic Components and Technology Conference*, San Diego, (IEEE, Piscataway, 2009)
7. K.H. Lu, X. Zhang, S.-K. Ryu, J. Im, R. Huang, P.S. Ho, Thermo-mechanical reliability of 3-D ICs containing through silicon vias, in *IEEE Electronic Components and Technology Conference*, San Diego, (IEEE, Piscataway, 2009)
8. K.H. Lu, S.-K. Ryu, Q. Zhao, X. Zhang, J. Im, R. Huang, P.S. Ho, Thermal stress induced delamination of through silicon vias in 3-D interconnects, in *IEEE Electronic Components and Technology Conference*, Las Vegas, (IEEE, Piscataway, 2010)
9. S.-K. Ryu, K.-H. Lu, X. Zhang, J.-H. Im, P.S. Ho, R. Huang, Impact of near-surface thermal stresses on interfacial reliability of through-silicon-vias for 3-D interconnects. IEEE Trans. Device Mater. Reliab. **11**, 35–43 (2011)
10. J.-S. Yang, K. Athikulwongse, Y.-J. Lee, S.K. Lim, D.Z. Pan, TSV stress aware timing analysis with applications to 3D-IC layout optimization, in *Proceedings of ACM Design Automation Conference*, anaheim, (ACM, New York, 2010)

Part V
Other Topics

This part of the book covers other topics in 3D ID design. In Chap. 18, we study the density, performance, and power benefit of monolithic 3D integration, where NMOS and PMOS are placed in two different tiers and connected with extremely small monolithic inter-tier vias (MIVs). In Chap. 19, we study the impact of TSV scaling on the area, wirelength, timing, and power quality of 3D designs done at the current and future technology nodes. In Chap. 20, we study the design, manufacturing, and testing of the 3D-MAPS (massively parallel processor with stacked memory), where one tier of 64 general-purpose cores and another tier of SRAM memory are face-to-face bonded for core-to-memory communication and utilize TSVs to communicate with the package.

Chapter 18
Ultra High Density Logic Designs Using Monolithic 3D Integration

Abstract Recent innovations in monolithic 3D technology enable much higher-density vertical connections than today's through-silicon-via (TSV)-based technology. In this chapter, we study the benefits and challenges of monolithic 3D integration technology for ultra high-density logic designs. Based on our layout experiments, we compare important design metrics such as area, wirelength, timing, and power consumption of monolithic 3D designs with the traditional 2D designs. We also explore various interconnect options for monolithic 3D ICs that improve design density and quality. Depending on the interconnect settings of monolithic 3D ICs and the benchmark circuit characteristics, we observe that our two-tier monolithic 3D design provides up to 40 % reduced footprint, 27.8 % shorter wirelength, 39.7 % faster operation, and 9.2 % lower power consumption over the 2D counterpart.

The materials presented in this chapter are based on [7].

18.1 Introduction

Monolithic 3D IC is a vertical integration technology that builds up two or more tiers of devices sequentially, rather than bonding two fabricated dies together using bumps and/or TSVs. Compared with other existing 3D integration technologies (wire bonding, interposer, through-silicon-via, etc.), monolithic 3D integration truly allows ultra fine-grained vertical integration of devices and interconnects, thanks to the extremely small size of inter-tier vias (typically 50 nm in diameter). The manufacturing technology has matured to allow very high alignment precision [3] and extremely thinned die. A side view of a typical monolithic 3D IC is shown in Fig. 18.1. Since the dimension of monolithic inter-tier-via (MIV) is very small (about local via size), and the parasitic capacitance of an MIV is almost negligible (<1 fF), many more vertical interconnects can be utilized than TSV-based 3D ICs to improve design quality.

S.K. Lim, *Design for High Performance, Low Power, and Reliable 3D Integrated Circuits*, DOI 10.1007/978-1-4419-9542-1_18,

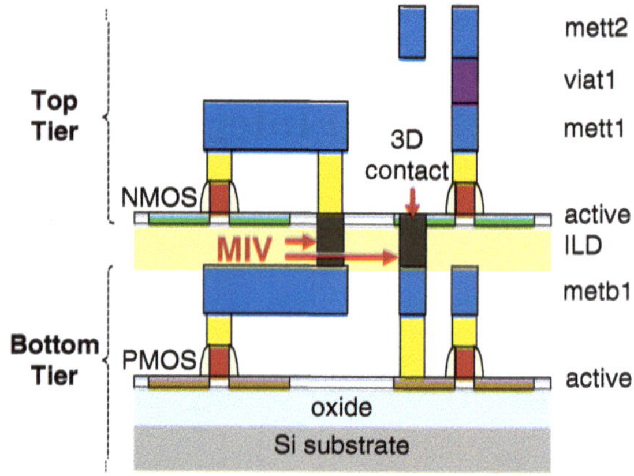

Fig. 18.1 Side view of a two-tier monolithic 3D IC. MIV and ILD stand for monolithic inter-tier via and inter-layer dielectric. On the *top tier*, only the first two metal layers *(mett1, mett2)* are shown. Objects are drawn to scale

In monolithic 3D IC designs, we may place the conventional 2D standard cells in different layers and connect them vertically using MIVs, which is called gate-level monolithic 3D. A finer-grained 3D placement method is to split NMOS and PMOS transistors in each standard cell and place them in different layers, which is called transistor-level monolithic 3D (we call it MI-T in this chapter). In this study, we focus on MI-T technology because of the following reasons: First, TSV-based 3D IC designs or gate-level monolithic 3D IC designs require 3D-aware physical design engines (partitioner, placer, etc.). However commercial electronic design automation (EDA) tools with native 3D support are not mature yet. In contrast, MI-T designs can be handled by existing EDA tools for 2D ICs with proper modifications. Second, since PMOS and NMOS transistors are separately fabricated on different dies, die manufacturing process can be further optimized. Third, manufacturing cost per die can be reduced because each die requires less number of masks than when both PMOS and NMOS are on the same die.

In this chapter, we study the benefits and challenges of monolithic 3D IC technologies for high-density logic designs. The following specific topics are covered in this chapter:

- We study how the standard cells for MI-T are designed for high-density integration. Various practical layout and design techniques for density and performance are discussed.
- We study the routing congestion problem in MI-T and explore various interconnect options to overcome it. Several metal layer structures and their dimensions are investigated. With layout-based experiments, we provide detailed analysis on wirelength, timing, and power metrics with several benchmark circuits.
- We study how the benefit of monolithic 3D IC increases or diminishes based on the future technology scaling trends. We analyze the impact of faster and more power-efficient cells and higher interconnect resistance/capacitance (RC) on the quality of monolithic 3D designs.

18.2 Existing Works

Currently, there are very few related works on the design of monolithic 3D ICs. Jung et al. demonstrated the single-crystal thin-film-based process for their SRAM design [6], which reduced the SRAM cell footprint by 46.4 %. Recently, Golshani et al. demonstrated the monolithic 3D integration of SRAM and image sensor [5]. Also, Naito et al. demonstrated the first 3D FPGA design implementation based on a monolithic 3D IC technology [9]. Batude et al. enabled high-quality top silicon layer using a molecular bonding technique and a low thermal budget process [3]. Recently, monolithic 3D standard cell libraries and logic circuit designs were demonstrated in [4]. Liu et al. presented a design tradeoff study on transistor- and gate-level monolithic 3D design styles [8]. Yet, their design techniques and interconnect options did not resolve the routing congestion problem in transistor-level monolithic 3D designs, which may degrade the design quality much.

18.3 Design Methodologies

In this section, we explain our design methods for MI-T technology in detail. Various practical considerations for high density and high performance MI-T designs are discussed.

18.3.1 Library Construction

The overall library construction steps are summarized in Fig. 18.2. First, we define the two-tier MI-T layer structures and dimensions (see Fig. 18.3b). Based on these structural assumptions, we adopt and modify the original layer structure dimensions and design rules from Nangate 45 nm 2D library [10] accordingly. Then, we build a Virtuoso technology file and a display file, which are required for standard cell layouts in Cadence Virtuoso. Next, we choose the target standard cells and manually draw their monolithic 3D versions in Cadence Virtuoso (see Fig. 18.4). We then generate the physical library using Cadence Abstract Generator. For interconnect RC parasitics, we first write an interconnect technology file using our structural information of MI-T. Lastly, we generate RC parasitic libraries using Cadence Encounter and Techgen.

Our MI-T standard cells are modified from Nangate 45 nm standard cell library as discussed in details in the following Sect. 18.3.2. In [2], it is observed that the transistors on the bottom and the top tier show very similar performance; the performance is not significantly degraded by monolithic 3D integration. Thus, the authors in [4] used the original 2D Nangate timing/power library for 3D cells. We also believe that with the advanced manufacturing technology and standard cell design techniques, the timing/power of monolithic 3D cells will match those of 2D cells. Thus, we use the original Nangate timing/power libraries for our 3D cells.

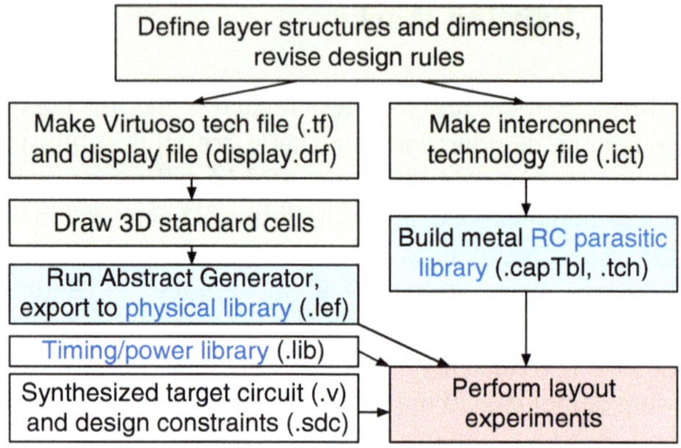

Fig. 18.2 Library construction steps. *Green boxes* indicate manual steps, while *blue* ones are mainly executed by CAD tools. We also show the file extensions for various technology and library related files

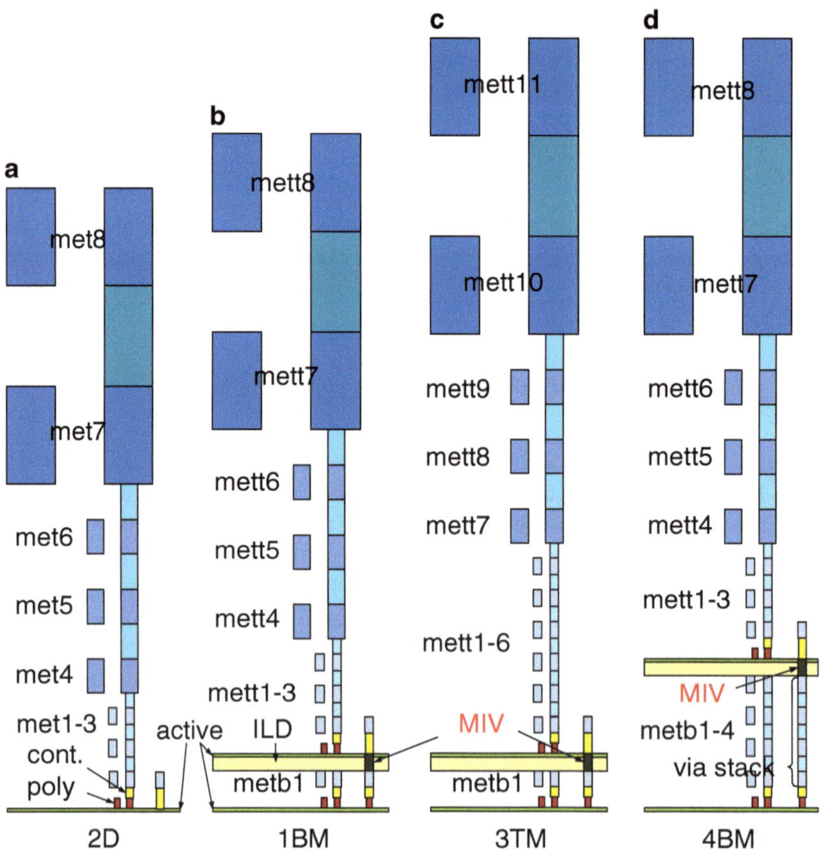

Fig. 18.3 Metal layer stack options. (**a**) 2D, (**b**) baseline MI-T. (**c**) Three local metal layers added to the *top tier*, (**d**) Three local metal layers added to the *bottom tier*. ILD stands for inter-layer dielectric between the *top* and the *bottom tier*. The *bottom tier* substrate and ILD for metal layers are not shown for simplicity. Objects are drawn to scale

Table 18.1 List of monolithic 3D standard cells built in our library

Category	#Cells	Cell names
Logic	23	NAND2_X1, NOR2_X1, OAI22_X1, MUX2_X1, ···
Sequential	4	DFF_X1, DFFR_X1, DFFS_X1, DFFRS_X1
Buffer	12	BUF_X1/2/4/8/16/32, INV_X1/2/4/8/16/32

18.3.2 Standard Cell Design

The MI-T standard cells are completely different from those of traditional 2D ICs. The PMOS and the NMOS transistors are placed on different tiers, and the internal wires connecting the transistors and the power/ground rails need to be redesigned. Carefully designed cells are essential for high logic density and performance. From Nangate 45 nm 2D standard cell library, we chose total 39 standard cells based on their functionality as well as the usage in the synthesized netlist of benchmark circuits. Our monolithic 3D cells are summarized in Table 18.1.

Our standard cell design method differs from Intra-Cell Stacking in [4] for three major reasons:

- We place PMOS transistors on the bottom tier and NMOS transistors on the top. In standard cells, NMOS transistors are smaller than PMOS transistors because of electron/hole mobility mismatch. By placing NMOS transistors on the top tier, the remaining space on the top silicon surface is used for MIVs, reducing wasted space. If PMOS is on the top tier as in [4], we need extra space for MIVs, which increases the cell footprint.
- We apply a folding technique on the original 2D standard cell layout. Compared with the Intra-Cell Stacking technique in [4] which requires a complete redesign of internal connections, our method is straightforward and provides opportunities for reducing internal parasitic resistance.
- We place power and ground rails of standard cells in different metal layers on the bottom side. Compared with the method in [4] which places power/ground rails on the top/bottom of the standard cells, our method further reduces the cell footprint.

Our standard cell design steps are illustrated in Fig. 18.4. First, the original 2D cell is cut to separate NMOS and PMOS parts. Then, the PMOS part is flipped so that the VDD rail is on the bottom side of the cell footprint. The PMOS part is placed on the bottom tier, and the NMOS part is on the top tier. The internal wires are carefully modified to preserve connectivity and keep design rules. The 3D connections are completed with MIVs. The original 2D standard cell height is 1.4 μm, whereas our MI-T standard cell height is 0.84 μm. The cell height is reduced by 40 %, which is more than the reported values in [4] (about 30 %). The MIVs do not require additional silicon space on the bottom tier, thus placing the PMOS part (which is larger than the NMOS part) on the bottom tier is useful to reduce cell

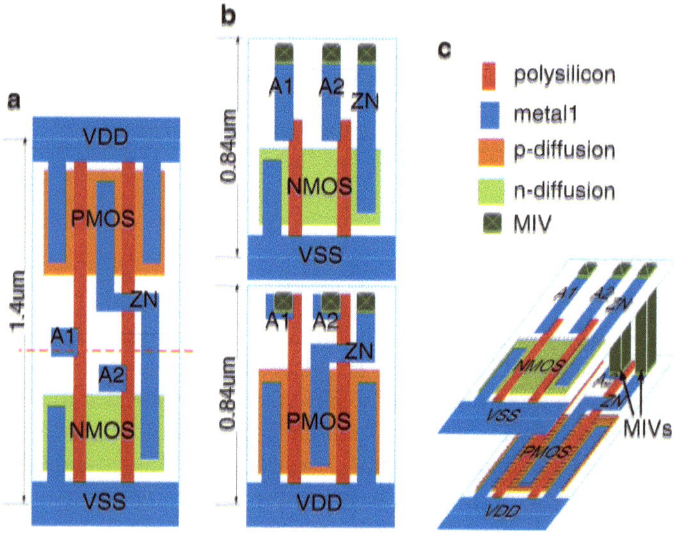

Fig. 18.4 Standard cell design steps for MI-T. (**a**) Original 2D cell with the cut line shown in *red* stippled line, (**b**) after cut, flip, and modification, (**c**) 3D view. The VDD and VSS means power and ground rails. *Blue dotted boxes* denote cell boundary

height. Note that the pins of standard cells are placed in both tiers after the MI-T modification. For example, in Fig. 18.4b, pin A1 is available on both the top and the bottom tier.

Since the polysilicon (or poly) connect PMOS and NMOS devices in both tiers, we attempt to reduce the length of polysilicon. For instance, the length of polysilicon wires in Fig. 18.4a is about 1.3 μm, whereas that in Fig. 18.4b is about 1.2 μm. Since the unit length resistance of polysilicon is quite high, about 156 Ω/μm, the internal wire resistance can be reduced significantly if desired, by carefully determining the locations of MIVs and adjusting poly lines. The parasitic RC of MIVs are small compared with other interconnects inside the cell, thus the performance impact by MIVs is negligible.

As shown in Fig. 18.4c, both the VDD and VSS rails of our MI-T cells are located on the bottom *side* of the standard cell. If the VDD rail is on the top side, the connectivity from VDD through PMOS, MIV, NMOS, to VSS becomes highly complex, and the interconnect routes may need to be modified, especially when the internal connections of the cell is complex (e.g., D flip-flop, full adder cell). It may increase the standard cell height.

Two MI-T standard cells are shown in Fig. 18.5. The red squares on the top side of cells denote MIVs. Depending on its complexity, a cell may have many MIVs. However, since MIVs are very small, they consume a small amount of the cell area. Note that although we placed most of MIVs on the top side, they can be moved down if desired.

Fig. 18.5 Our MI-T standard cell layouts. Per each cell, the *top* shows NMOS part and the *bottom* shows PMOS part. The p/n-well and implant are not shown for simplicity

Table 18.2 Our benchmark circuits

	AES	VGA	DES	JPEG	FFT
#Cells	19,719	68,318	76,088	297,028	582,621
#Nets	20,146	74,696	78,608	381,548	751,399
Average fanout	2.131	2.307	2.034	1.850	2.130
Clock period (ns)	0.5	0.5	0.5	3.0	0.6

18.3.3 Full-Chip Physical Layout

With the libraries built for MI-T, we proceed to full-chip layout experiments. Using Synopsys Design Compiler, we synthesize the benchmark circuits based on our MI-T standard cells and benchmark design constraints. These benchmark circuits are summarized in Table 18.2. Next, we build physical layouts of the circuits using Cadence Encounter. Starting from floorplaning, we perform power delivery network planning, timing-driven placement of cells, clock synthesis, and timing-driven routing. Since a MI-T cell contains both the top and the bottom tier parts and MIVs as a single unit, the placer places the cells in a 2D fashion without any overlap between cells. The MI-T cells have pins on the first metal of *both* the bottom and the top tiers (metb1 and mett1 in Fig. 18.3b).

Unlike the metal layer assumption in [4], we allow our router to use the metal layer on the bottom tier (metb1 in Fig. 18.3b) for local routing as well. In this case, the timing-driven router in Encounter chooses which pin in which layer to connect to, based on routing congestion and timing information. As shown in Fig. 18.6, the router may use metb1 only, mett1 only, or all three: metb1, MIV, and mett1. Note that the router should not place MIVs inside standard cells because these MIVs may touch the internal objects of the cell.

After routing is finished, we perform RC extraction of nets, which is required for timing and power analysis. Once the RC information and the netlist are available, static timing analysis (STA) engine handles the entire top and bottom tiers at once,

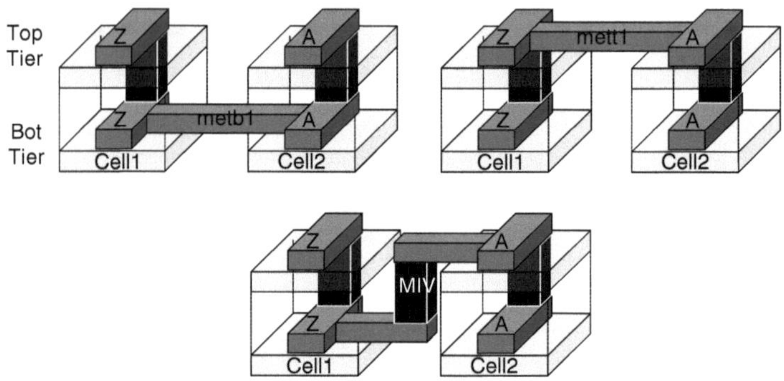

Fig. 18.6 Illustration of net routing cases in MI-T. This net connects pin Z of Cell1 to pin A of Cell2

providing true 3D STA results. For power analysis, we use Synopsys PrimeTime PX. We assume certain switching activity at the primary input pins and the flip-flop outputs (0.2 and 0.1, respectively). Then, the tool propagates switching activity information to the rest of the circuit. Based on the switching activity and library information, power calculation is performed.

Layout snapshots of the benchmark circuit AES is shown in Fig. 18.7. In the zoom-in shots, cells, signal nets, and power rails are shown. For the top tier, only the first two metals (mett1 and mett2) are shown. We observe that Encounter places and routes MI-T cells without any problem. Note that MIVs used in net routing are placed in the white spaces between cells, avoiding any contact. Since we use the state-of-the-art EDA software for layout, the quality of placement and route is very high.

18.4 Routing Congestion Issues in Monolithic 3D ICs

Our study shows that a major problem of MI-T designs is routing congestion. Since our MI-T cells are smaller than the original 2D cells by 40 %, the overall chip footprint is reduced by 40% as well. Yet, the number of cell pins to connect stays the same. As shown in Table 18.3, the *pin density* of MI-T becomes much higher than that of 2D. For instance, the pin density of the MI-T design for AES is 66 % higher than that of the 2D design. The interconnects need to be routed within 40 % smaller footprint, which means increased routing demand. The additional metal layer on the bottom tier of MI-T (metb1) can be used only for local interconnects because the metb1 strips inside cells (internal wires and pins) block cell-to-cell routing. Thus, the routing capacity of MI-T per routing tile (= a tile in $N \times N$ grid for global routing) is almost the same as that of 2D and cannot satisfy the much increased routing demand.

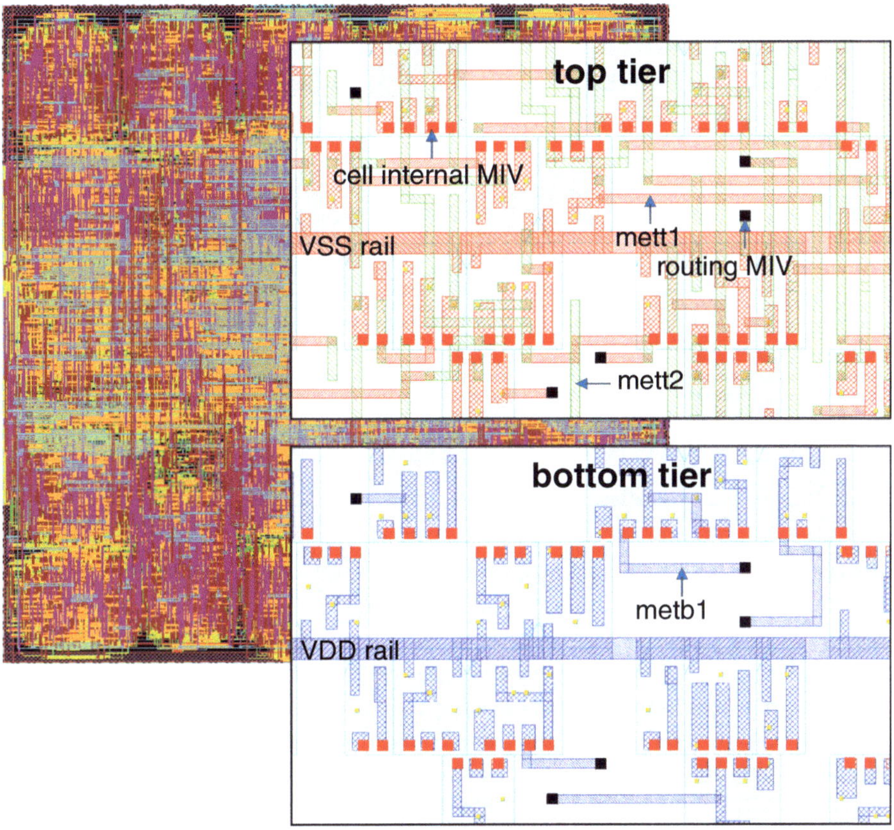

Fig. 18.7 Layout snapshots of the benchmark circuit AES. On the *right*, zoom-in shots of the *top* and the *bottom tier* are shown. *Black* and *red squares* indicate the MIVs used for net routing and cell internal connections, respectively

Table 18.3 Pin density of the benchmark circuits. Cell area and pin density (= #cell pins/cell area) are shown in μm^2 and pins/μm^2, respectively

		AES	VGA	DES	JPEG	FFT
#Cell pins		63,068	247,015	238,488	1,087,390	2,351,692
Cell	2D	20,964	129,977	102,840	639,677	1,357,493
Area	MI-T	12,578	33,728	61,704	383,806	814,496
Pin	2D	3.01	1.90	2.32	1.70	1.73
Density	MI-T	5.01	3.17	3.87	2.83	2.89

Routing congestion maps of the 2D and the MI-T design for a benchmark circuit are shown in Fig. 18.8. It is evident that MI-T (= the 1BM case defined in Sect. 18.5.1) has more severe routing problem than 2D. The overall over-congestion rate (reported by Encounter, calculated from metal layers with maximum shortage) is 0.30 % for 2D case and 4.36 % for MI-T. Because of metal layer changes and

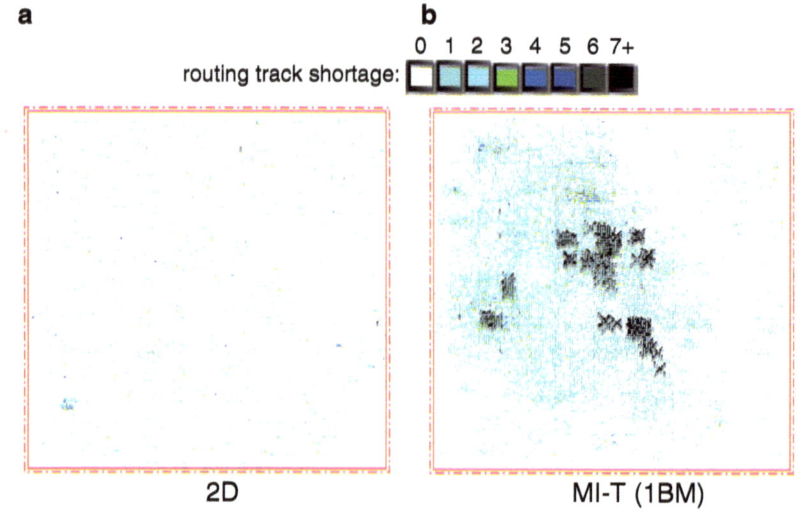

Fig. 18.8 Routing congestion map of VGA with (**a**) 2D and (**b**) MI-T. Black X marks show design rule violations due to routing congestions

detours to deal with routing congestions, the timing and power quality of MI-T is also degraded. In addition, we observe that the routing congestion becomes more severe with circuit optimization because the optimizer inserts buffers and breaks a complex cell into a group of simpler cells to improve timing, which in turn increases pin density considerably.

This routing congestion problem is unique in MI-T technology; it does not happen when the technology node is scaled down, because local metal dimensions and cells shrink at about the same rate. It does not happen for TSV-based 3D ICs either because enough metal layers are available on each tier of TSV-based 3D ICs.

18.5 Impact of Additional Metal Layers

In order for the MI-T technology to be suitable for high density and high performance, the routing congestion problem discussed in Sect. 18.4 needs to be mitigated. In our study, we consider two kinds of metal interconnect modifications: (1) adding more metal layers and (2) reducing metal width and spacing.

18.5.1 Metal Layer Stack Options

Adding more local metal layers is an effective way to increase routing capacity and reduce congestion. We believe that more investment will be made to allow additional metal layers on the top and/or the bottom tier of monolithic 3D ICs if there is a clear

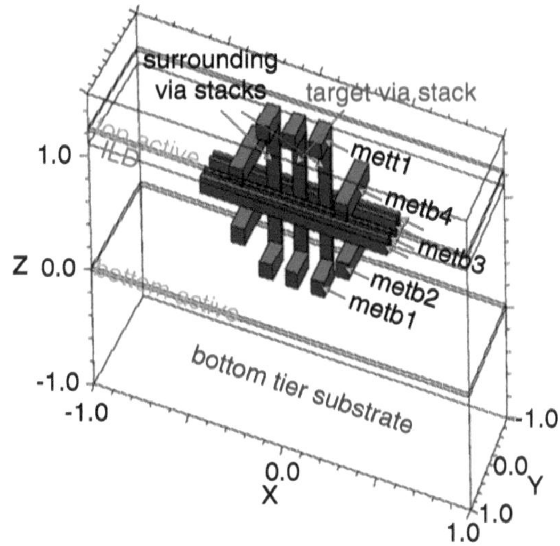

Fig. 18.9 Raphael simulation structure for a via stack and its surrounding objects. The dimensions are shown in μm

evidence that they improve the design quality of MI-T significantly. As shown in Fig. 18.3, we now have three metal layer stack options for MI-T:

- **1BM**: This is our baseline MI-T layer stack with one bottom tier metal layer.
- **3TM**: We add three additional (local) metal layers to the top tier. As a result, we have total six local metal layers on the top tier.
- **4BM**: We add three metal layers to the bottom tier. As a result, we have total four local metal layers on the bottom tier.

18.5.2 RC Modeling of Via Stack in 4BM Case

In 4BM case, as shown in Fig. 18.3d, the connections from a PMOS on the bottom tier to an NMOS on the top tier are made through metal and via layers on the bottom tier (metb1-4, viab1-3) and monolithic inter-tier vias (= MIVs), which we call *via stack* in this chapter. The physical size of a via stack is considerably larger than that of a single MIV. In addition, there could be metal interconnects surrounding a via stack, which may increase its coupling capacitance. Thus, we study the impact of parasitic RC of these via stacks on the timing and power of 4 BM cells.

Using Synopsys Raphael, the capacitance of a via stack is extracted. The structure for our Raphael simulation is shown in Fig. 18.9, where the target via stack is surrounded by neighboring via stacks and metal wires. The capacitance (C_{vs}) of a via stack reported by Raphael is 0.123 fF. The resistance of a via stack (R_{vs}) is dominated by the resistances of local vias (viab1-3) and the MIV. From the values in the technology definition file, the calculated R_{vs} is 20 Ω, which includes contact resistances.

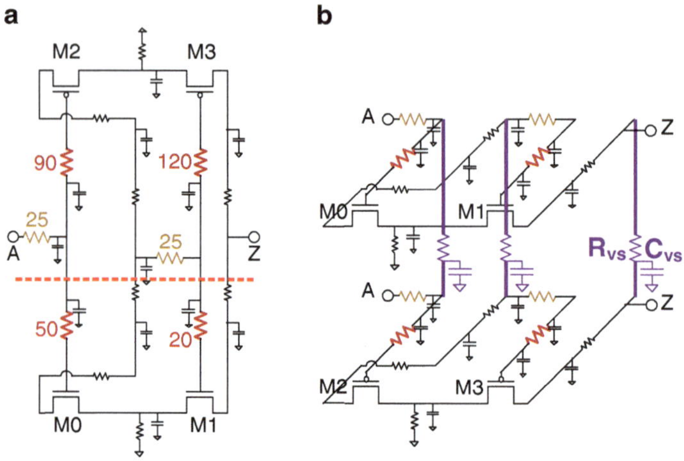

Fig. 18.10 SPICE netlist of a standard cell: (**a**) original netlist, (**b**) with via stack RC. The *dotted line* in (**a**) is the tier boundary, and the values denote internal parasitic resistances in Ω

A lumped RC model of a via stack is incorporated into the SPICE netlist of each standard cell to characterize its timing/power behavior. In Fig. 18.10a, the original SPICE netlist of a buffer cell with internal parasitic RC is shown. The C_{vs} and R_{vs} of via stacks are inserted at the cut locations as shown in Fig. 18.10b. Then, we run Cadence Encounter Library Characterizer to characterize the timing and power of the modified standard cell for the 4 BM case.

In Table 18.4, we compare the timing and power of a buffer cell with or without via stack RC. The delay includes both the cell intrinsic delay and load-dependent delay, and the power is the cell internal power, excluding wire switching and leakage power. In general, when the load capacitance of a cell is small, the impact of via stack RC on timing and power is large; the impact becomes smaller with larger load capacitance. This trend is observed in most of the cells. If a driving net is very short and has a small load capacitance, the timing and power of the driver may degrade by about 10 %. Since the timing and power of the circuit depend on the net delay and net switching power, the overall degradation of timing and power of the entire circuit level is lower – about 2–3 % – which is still significant. Thus, we incorporate via stack RC in all of our 4BM-based designs.

18.5.3 Delay and Power Calculations in MI-T Designs

For a cell driving a net and the sink cells on the net, the delay (D) is:

$$D_{total} = D_{cell} + D_{net} \tag{18.1}$$

Table 18.4 Comparison of timing and power of a cell with and without via stack RC. The values are from the timing/power tables of the characterized libraries

	Delay			Power		
Load cap (fF)	Without RC (ps)	With RC (ps)	Diff. (%)	Without RC (fW)	With RC (fW)	Diff. (%)
0.4	28.4	31.2	9.86	1.15	1.33	15.65
0.8	33.1	35.8	8.16	1.40	1.52	8.57
1.6	42.8	45.4	6.07	1.86	1.98	6.45
3.2	62.4	64.9	4.01	2.81	2.99	6.41
6.4	100.3	103.0	2.69	4.78	4.93	3.14
12.8	175.8	179.9	2.33	8.54	8.74	2.34
25.6	330.0	330.6	0.18	16.17	16.33	0.99

$$D_{cell} = D_{intrinsic} + D_{load-dependent} \tag{18.2}$$

$$D_{load-dependent} = f(C_{load}, input\ slew) \tag{18.3}$$

$$C_{load} = C_{wire} + C_{pin} \tag{18.4}$$

The $D_{intrinsic}$ is the intrinsic delay of the cell. The $D_{load-dependent}$ is a function of C_{load} and the signal slew at the cell input pin. Compared with 2D designs, wires are shorter in MI-T designs, which in turn reduces C_{wire}, C_{load}, and $D_{load-dependent}$. The D_{net} also reduces as wires become shorter. However, the overall delay improvement may not keep up with wirelength reduction. If C_{pin} is larger than C_{wire}, the C_{load} may not decrease significantly because C_{pin} is not reduced. Moreover, $D_{intrinsic}$ also contributes to D_{cell}. Thus, depending on the circuit characteristics and layouts, the delay improvement of MI-T may vary.

Meanwhile, the power consumption (P) of a cell is:

$$P_{total} = P_{internal} + P_{switching} + P_{leakage} \tag{18.5}$$

$$P_{internal} = f(C_{load}, input\ slew) \tag{18.6}$$

$$P_{switching} \propto switching\_activity \times C_{load} \tag{18.7}$$

The $P_{internal}$ is the power consumed for the objects within the cell boundary, which weakly depends on C_{load} and the cell input slew. When the input slew is larger, $P_{internal}$ increases. Usually, $P_{leakage}$ is smaller than $P_{internal}$ and $P_{switching}$. The $P_{switching}$ is proportional to both the switching activity and C_{load}. Assuming that the switching activity is the same for 2D and MI-T designs, the reduction of C_{load} in MI-T designs is the main reason for the total power reduction. Note that if (a) C_{pin} is more dominant than C_{wire}, or (b) $P_{internal}$ is more dominant than $P_{switching}$, the total power reduction of MI-T designs caused by wirelength reduction may not be significant.

18.5.4 Simulation Results and Discussions

The design and analysis results for 2D and MI-T design options are summarized in Tables 18.5 and 18.6. Placement utilization of all designs is 70 %. Compared with 2D designs, the footprints of MI-T designs are 40 % smaller, while the total silicon areas are 20 % larger. Compared with 2D, the total wirelength and clock wirelength of all three MI-T design types are reduced by about 20 %. The total number of MIVs used in routing is about the same for 1BM and 3TM, while 4BM utilizes considerably more MIVs because the bottom tier metals are highly utilized for routing.

To further understand the timing and power improvement, we plot the wirelength distribution of the designs. In Fig. 18.11, the wirelength distribution of 2D, 1BM, 3TM, and 4BM designs for the FFT circuit is shown. We observe that the wirelength reduction on the long nets is the highest for 3TM. For the majority of the nets, the wirelengths are very short ($<10\,\mu m$). For these short nets, the pin cap (C_{pin}) is dominant over the wire cap (C_{wire}). Thus, the timing and the power improvements of these nets are low. For this circuit, the reduction is mostly on long nets. Since long nets tend to be on the critical path, reducing the lengths of long nets improves the critical path delay significantly.

The timing improvement of 3TM is the best among the MI-T design types. For the largest circuit (FFT), the longest path delay improvement of 3TM over 2D is 39.7%. Note that this timing improvement can be used towards power reduction during the timing/power optimization; for the same target clock speed, 3TM may use more power-efficient (slower) cells to reduce power. However, the total power reduction of MI-T designs is less significant than timing improvement. The power reduction of MI-T designs over 2D design is mostly from reduced wire power. However, wire power is only a small fraction of the total power. For instance, the wire power of JPEG for 3TM is 39.2 mW, which is only 13.2 % of the total power. Depending on the quality of Encounter clock tree synthesis (CTS) results, the clock tree power may decrease. We observe that CTS usually produces the best results for 3TM among MI-T designs, because the CTS quality is related to the routing quality. The timing and power of 4BM designs are generally worse than 1BM and 3TM designs mainly because of the RC effect of via stacks inside cells.

18.6 Impact of Reduced Metal Width and Spacing

Another interconnect modification option we study to deal with the routing congestion problem is to reduce the width and spacing of metal layers. The local metal width/spacing is close to the minimum feature size of the technology node. However, if scaling down the metal width/spacing brings large benefits in design quality, process engineers are willing to invest efforts towards it. Thus, the purpose of this metal width/spacing reduction study is to explore the interconnect design space for maximizing the benefit of MI-T; extreme scalings (>20 %) may not be

Table 18.5 Comparison between 2D and monolithic 3D designs. #routing MIVs means the number of MIVs used in net routing, excluding the MIVs used inside the monolithic cells. The WL, LPD, and TNS mean wirelength, longest path delay, and total negative slack, respectively. The values in parentheses show the percentage ratio to the 2D designs

Ckt	Design type	Footprint (μm^2)	Total silicon area (μm^2)	Total WL (m)	Clk WL (mm)	#Routing MIVs	LPD (ns)	TNS (μs)
AES	2D	174 × 172	29,948	0.271 (100)	2.125	0	1.310 (100)	0.226 (100)
	1BM	135 × 134	35,938	0.209 (77.2)	1.819	1,070	1.260 (96.2)	0.202 (89.4)
	3TM	135 × 134	35,938	0.209 (76.9)	1.696	897	1.165 (88.9)	0.190 (83.9)
	4BM	135 × 134	35,938	0.214 (78.8)	1.866	3,266	1.226 (93.6)	0.207 (91.4)
VGA	2D	432 × 430	185,682	1.623 (100)	1.489	0	2.173 (100)	15.29 (100)
	1BM	334 × 333	222,822	1.284 (79.1)	1.236	2,349	1.954 (89.9)	13.01 (85.1)
	3TM	334 × 333	222,822	1.281 (78.9)	1.243	2,357	1.632 (75.1)	10.64 (69.6)
	4BM	334 × 333	222,822	1.363 (84.0)	1.179	18,020	1.843 (84.8)	11.27 (73.7)
DES	2D	384 × 382	146,916	0.849 (100)	32.77	0	1.086 (100)	0.581 (100)
	1BM	297 × 297	176,298	0.659 (77.6)	24.16	3,152	0.968 (89.1)	0.527 (90.8)
	3TM	297 × 297	176,298	0.654 (77.0)	24.54	3,121	0.923 (85.0)	0.503 (86.5)
	4BM	297 × 297	176,298	0.682 (80.3)	25.41	11,300	1.000 (92.1)	0.557 (95.8)
JPEG	2D	957 × 955	913,825	5.148 (100)	163.9	0	6.053 (100)	10.514 (100)
	1BM	741 × 740	1,096,592	4.032 (78.3)	126.4	16,502	5.422 (89.6)	2.999 (28.5)
	3TM	741 × 740	1,096,592	3.997 (77.6)	121.2	17,148	5.096 (84.2)	2.642 (25.1)
	4BM	741 × 740	1,096,592	4.160 (80.8)	127.9	71,944	5.967 (98.6)	4.018 (38.2)
FFT	2D	1,394 × 1,392	1,939,278	12.93 (100)	629.0	0	5.958 (100)	340.0 (100)
	1BM	1,079 × 1,079	2,327,134	10.41 (80.5)	463.6	30,407	4.250 (71.3)	299.0 (87.9)
	3TM	1,079 × 1,079	2,327,134	10.26 (79.4)	462.7	31,478	3.593 (60.3)	250.0 (73.5)
	4BM	1,079 × 1,079	2,327,134	10.75 (83.2)	492.1	163,833	3.810 (63.9)	287.0 (84.4)

Table 18.6 Power comparison between 2D and monolithic 3D designs. The total power includes cell internal, switching, and leakage power. Clock power includes the power of clock buffers and wires. The values in parentheses show the percentage ratio to the 2D designs

Circuit name	Design type	Total power (mW)	Wire power (mW)	Clock power (mW)
AES	2D	13.7 (100)	3.31 (100)	3.89 (100)
	1BM	13.6 (99.3)	2.93 (88.4)	4.26 (109)
	3TM	12.8 (93.4)	2.41 (72.7)	3.72 (95.5)
	4BM	13.7 (100)	2.64 (79.6)	4.20 (108)
VGA	2D	43.5 (100)	13.23 (100)	7.61 (100)
	1BM	41.8 (96.1)	11.59 (87.6)	7.48 (98.4)
	3TM	40.1 (92.2)	9.79 (74.0)	7.33 (96.3)
	4BM	43.8 (101)	11.22 (84.9)	8.24 (108)
DES	2D	134.9 (100)	24.81 (100)	66.1 (100)
	1BM	131.1 (97.2)	20.36 (82.1)	64.0 (96.8)
	3TM	126.1 (93.5)	19.36 (78.1)	60.1 (90.9)
	4BM	130.7 (96.9)	20.30 (81.8)	64.1 (97.0)
JPEG	2D	314.9 (100)	51.31 (100)	53.20 (100)
	1BM	300.6 (95.5)	42.92 (83.6)	46.70 (87.8)
	3TM	296.6 (94.2)	39.20 (76.4)	45.70 (85.9)
	4BM	312.6 (99.3)	41.56 (81.0)	50.10 (94.2)
FFT	2D	1,469.2 (100)	295.9 (100)	1,053.8 (100)
	1BM	1,431.2 (97.4)	248.9 (84.1)	1,025.3 (97.3)
	3TM	1,345.4 (91.6)	226.9 (76.7)	9,48.1 (90.0)
	4BM	1,535.8 (105)	245.2 (82.9)	1,114.4 (106)

Table 18.7 Minimum width/spacing of metal layers with varied reduction ratio. First metal means the lowest metal layer of the top/bottom tier. Unit is nm

Reduction ratio (%)	0	10	20	30	40
First metal	70/65	63/59	56/52	49/46	42/39
Local metals	70/70	63/63	56/56	49/49	42/42
Intermediate metals	140/140	126/126	112/112	98/98	84/84
Global metals	400/400	360/360	320/320	280/280	240/240

manufacturable with the technology node due to lithography limitations or chemical mechanical polishing issues. For all MI-T cases (1BM, 3TM, and 4BM), we reduce the minimum metal width and spacing of all metal layers up to 40 by 10 % step. The diameters of vias and MIVs are also reduced to match the corresponding metal layers. The height of metal layers is not reduced to keep the metal resistance from increasing too much. Table 18.7 summarizes the reduced metal width/spacing. Per each reduced metal dimension setting, the interconnect-related libraries such as capacitance table are rebuilt. Note that we do not modify the cell internal wires.

The unit length resistance and capacitance of local metal layers with reduced metal width/spacing are summarized in Table 18.8. As the metal width (and via diameter) reduces, the unit length resistances of metal layers (and via resistances) increase. Yet, depending on the surrounding metal wires, the unit length capacitance

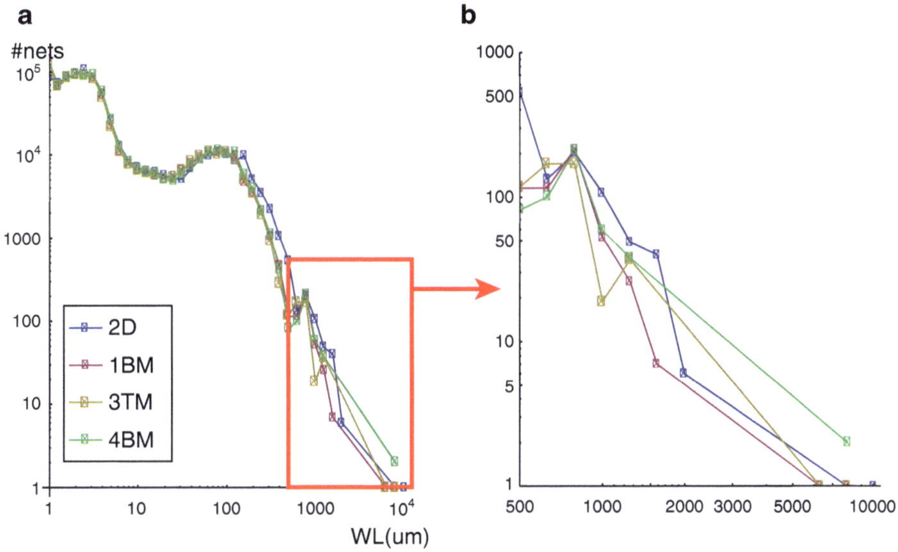

Fig. 18.11 Wirelength distribution of 2D and MI-T designs for FFT circuit: (**a**) full range, (**b**) zoom-in for long nets. Per each wirelength step, a marker is shown only when there are corresponding nets in the design; no marker means there is no net with such a wirelength. The axes are in log scale

Table 18.8 Unit length resistance and capacitance of local metals with varied metal width/spacing reduction ratio. The C_{high} and C_{low} are the max/min total wire capacitance per unit length, depending on the surrounding wires

Reduction ratio (%)	0	10	20	30	40
R (Ω/μm)	3.57	3.97	4.46	5.10	5.95
C_{high} (fF/μm)	0.16	0.18	0.17	0.19	0.22
C_{low} (fF/μm)	0.10	0.10	0.10	0.10	0.09

may increase or decrease as the metal width and spacing reduces. If there are surrounding wires at minimum spacing, the coupling capacitance is higher with reduced metal spacing. On the other hand, if there are not many surrounding wires, the unit length capacitance may decrease because the self capacitance decreases with narrower metal widths. With reduced metal width and spacing, more routing tracks are available. Thus, the router has a better chance for improving timing by carefully routing metal wires to reduce coupling capacitance. However, if the reduction ratio is too high, the metal resistance may increase the net delay and signal slew considerably.

Various design metrics of the JPEG circuit with varied metal width/spacing reduction ratio are shown in Fig. 18.12. The wirelength generally reduces as metal width/spacing reduces, because of less routing congestion and detour.

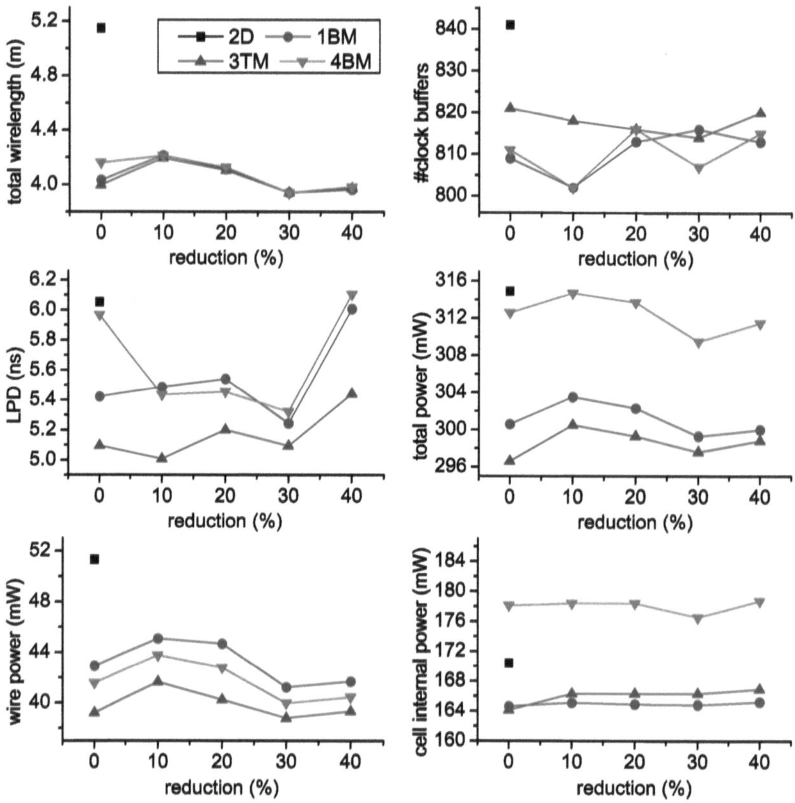

Fig. 18.12 Various results of JPEG with metal width and spacing reduction

The number of clock buffers generally increases when the reduction ratio increases. The reason is that as the metal width/spacing decreases, the metal unit length RC increases, and the clock signal slew degrades. To meet the clock skew/slew specifications, the CTS engine inserts more buffers. For the LPD, the sweet spot of 1BM and 4BM cases is at the 30 % reduction, while that of 3TM is 10 %. Moreover, the LPD improvement of 4BM at the sweet spot over the default setting (=0 % reduction) is larger than 1BM and 3TM cases. We found that the sweet spot changes with respect to the benchmark circuit. The wire power generally decreases with the reduced metal width/spacing. However, we see that the cell internal power increases, which is also related to the signal slew degradation with reduced metal width/spacing. As a result, the total power of 1BM and 4BM is minimum when the reduction ratio is 30 %.

The total wirelength, longest path delay, and total power of the other benchmark circuits are shown in Tables 18.9 and 18.10. For total wirelength, the same trend as with JPEG is observed. The maximum wirelength reduction is 27.8 % for AES with 3TM and 40 % reduced metal width/spacing. However, depending on the circuit characteristics, reducing metal width and spacing may not translate to longest path delay reduction (see VGA and FFT results). In general, 3TM provides the

Table 18.9 Total wirelength and longest path delay comparison with reduced metal width and spacing

		Total WL (m)					LPD (ns)				
Reduction (%)		0	10	20	30	40	0	10	20	30	40
AES	2D	0.271	–	–	–	–	1.310	–	–	–	–
	1BM	0.209	0.214	0.205	0.204	0.200	1.260	1.212	1.216	1.167	1.207
	3TM	0.209	0.208	0.203	0.197	0.196	1.165	1.209	1.153	1.165	1.150
	4BM	0.214	0.213	0.208	0.205	0.200	1.226	1.228	1.256	1.193	1.193
VGA	2D	1.623	–	–	–	–	2.173	–	–	–	–
	1BM	1.284	1.278	1.254	1.256	1.233	1.954	2.113	2.254	2.360	2.421
	3TM	1.281	1.255	1.254	1.242	1.236	1.632	1.940	1.752	2.278	2.194
	4BM	1.363	1.275	1.250	1.251	1.231	1.843	1.949	2.244	2.192	2.224
DES	2D	0.849	–	–	–	–	1.086	–	–	–	–
	1BM	0.659	0.656	0.647	0.644	0.639	0.968	0.932	0.948	0.933	0.960
	3TM	0.654	0.652	0.640	0.638	0.638	0.923	0.921	0.955	0.940	0.935
	4BM	0.682	0.657	0.646	0.638	0.637	1.000	0.969	0.972	1.027	0.961
FFT	2D	12.93	–	–	–	–	5.958	–	–	–	–
	1BM	10.41	10.28	10.14	9.99	10.00	4.250	4.559	4.851	5.082	4.957
	3TM	10.26	10.18	10.07	9.95	9.99	3.593	4.061	4.302	4.569	4.028
	4BM	10.75	10.29	10.16	10.06	10.06	3.810	4.376	4.719	5.000	4.740

Table 18.10 Total power comparison with reduced metal width and spacing

		Total power (mW)				
Reduction (%)		0	10	20	30	40
AES	2D	13.7	–	–	–	–
	1BM	13.6	13.1	13.0	13.0	12.9
	3TM	12.8	12.8	13.0	12.9	13.0
	4BM	13.7	13.5	13.4	13.3	13.3
VGA	2D	43.5	–	–	–	–
	1BM	41.8	42.1	41.3	41.3	41.0
	3TM	40.1	39.6	39.6	39.6	39.6
	4BM	43.8	42.7	42.5	42.5	42.3
DES	2D	134.9	–	–	–	–
	1BM	131.1	128.1	131.0	127.2	127.3
	3TM	126.1	126.2	126.2	126.1	126.0
	4BM	130.7	130.4	126.6	126.7	126.6
FFT	2D	1,469	–	–	–	–
	1BM	1,431	1,465	1,366	1,412	1,354
	3TM	1,345	1,436	1,433	1,429	1,334
	4BM	1,536	1,491	1,528	1,481	1,475

most power improvement over 2D designs. We observe that the maximum power reduction is 9.2 % with 3TM and 40 % reduced metal width/spacing for FFT circuit. Note that depending on the benchmark circuit, the sweet spot changes.

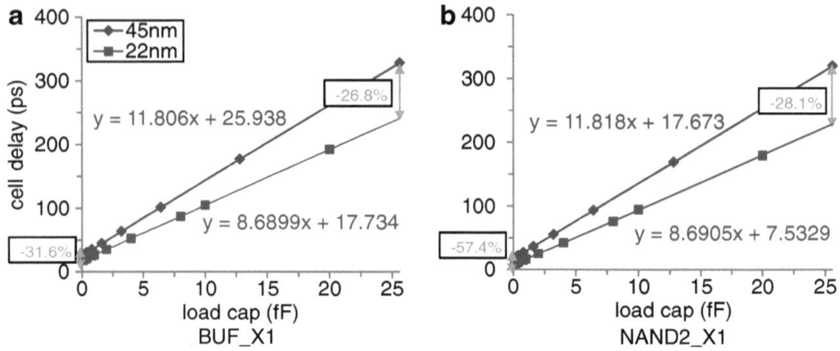

Fig. 18.13 Cell delay comparison between 45 and 22 nm. (**a**) BUF_X1, (**b**) NAND2_X1. The numbers in boxes are improvement rate

18.7 Impact of Device and Interconnect Scaling

We take a step further and predict how the benefits of MI-T change with a future technology node. The motivation behind this study is to see if the benefit would become greater when a better device and interconnect technology is introduced. In general, when a new technology node is introduced, (1) standard cells become faster and more power-efficient, (2) input pin capacitances of cells reduce, and (3) unit length resistance and capacitance of wires increase. Thus, we scale the timing, power, and pin capacitance values in our libraries as well as wire RC in layouts for this study.

To study the performance of future technology node cells, we created several standard cells using the transistor model from the 22 nm predictive technology model [1] and our 22 nm design rules. For the SPICE netlists extracted from the 22 nm cell layouts, the timing and power of cells are characterized by Cadence Encounter Library Characterizer. In Fig. 18.13, the delay of 45 and 22 nm cells are compared. We observe that the ratio of delay improvement from 45 to 22 nm technology node is not uniform but different with respect to the load capacitance; when the load capacitance is smaller, the improvement rate is larger. We apply this uneven scaling ratio to the timing and power library values.

Our 45 vs. 22 nm technology assumptions are summarized in Table 18.11. For cell delay and intrinsic power, the numbers in bracket means [min – max] scaling rate on the minimum/maximum load capacitance points as discussed above (see Fig. 18.13). The modified timing/power library and the wire RC scaling factors are applied to the designs with metal width/spacing reductions per benchmark design for all 2D and MI-T design types.

The longest path delay and total power for the 45 vs. 22 nm technology settings are shown in Table 18.12. Since we are interested in how much timing and power improvement MI-T designs provide over 2D designs, we show the ratio of MI-T design timing/power results over those of the corresponding 2D designs in 45 or

Table 18.11 Summary of the 45 vs. 22 nm technology settings

	45 nm	22 nm
Cell delay	[100–100 %]	[40–80 %]
Cell internal power	[100–100%]	[40–80%]
Cell input pin capacitance	100 %	80 %
Wire unit length resistance	100 %	110 %
Wire unit length capacitance	100 %	105 %

Table 18.12 Longest path delay and total power of target circuits for 2D and MI-T designs. The ratio to the 2D results are shown. In the setting, the suffix 'rXX' means the metal width/spacing reduction ratio. For example, r30 means 30 % reduced metal width/spacing

		LPD (ratio)		Total power (ratio)	
Circuit	Setting	45 nm	22 nm	45 nm	22 nm
AES	1BMr30	0.891	0.894	0.949	0.924
	3TMr20	0.880	0.857	0.949	0.908
	4BMr30	0.911	0.870	0.971	0.929
VGA	1BM	0.899	0.869	0.961	0.941
	3TM	0.751	0.697	0.922	0.882
	4BM	0.848	0.851	1.007	0.964
DES	1BMr30	0.859	0.796	0.943	0.963
	3TMr10	0.848	0.768	0.936	0.922
	4BMr40	0.885	0.798	0.938	0.939
JPEG	1BMr30	0.866	0.722	0.950	0.933
	3TMr10	0.827	0.771	0.954	0.937
	4BMr30	0.879	0.714	0.983	0.951
FFT	1BM	0.713	0.713	0.974	0.952
	3TM	0.603	0.598	0.916	0.884
	4BM	0.639	0.630	1.045	1.002

22 nm. Note that for each circuit, the metal width/spacing options of MI-T designs vary because we picked the best setting. In general, with 22 nm setting, the timing and power improvement of MI-T over 2D becomes larger. This is because with the 22 nm settings, D_{cell} in Eq. (18.2) is reduced. Thus, the reduced D_{net} by the shorter wirelengths in MI-T designs translates to more total delay reduction. In addition, C_{pin} becomes smaller, thus the reduced C_{wire} by the shorter wirelengths in MI-T designs affects C_{load} more. This in turn reduces D_{cell} more. Similarly, in Eq. (18.7), C_{load} reduces more with the 22 nm settings, which reduces $P_{switching}$ more. In summary, with the future technology settings, the timing and power improvement of MI-T over 2D becomes larger.

18.8 Conclusions

In this chapter, we studied the benefits and challenges of monolithic 3D IC technology from designers' perspective. We demonstrated that monolithic 3D technology provides various benefits over traditional 2D technology. We identified routing congestion issues that may hinder the benefit of monolithic 3D technology and studied several options to overcome the problem.

References

1. N. G. at ASU. Predictive Technology Model.
2. P. Batude, M. Vinet, A. Pouydebasque, C. Royer, B. Previtali, C. Tabone, L. Clavelier et al. GeOI and SOI 3D monolithic cell integrations for high density applications, in *Proceedings of the Symposium on VLSI Technology*, 2009, pp. 166–167
3. P. Batude, M. Vinet, A. Pouydebasque, C. L. Royer, B. Previtali, C. Tabone, J.-M. Hartmann, L. Sanchez, L. Baud, V. Carron, A. Toffoli, F. Allain, V. Mazzocchi, D. Lafond, O. Thomas, O. Cueto, N. Bouzaida, D. Fleury, A. Amara, S. Deleonibus, O. Faynot, Advances in 3D CMOS sequential integration, in *Proceedings of the IEEE International Electron Devices Meeting*, 2009, pp. 1–4
4. S. Bobba, A. Chakraborty, O. Thomas, P. Batude, T. Ernst, O. Faynot, D.Z. Pan, G.D. Micheli, CELONCEL: effective design technique for 3-D monolithic integration targeting high performance integrated circuits, in *Proceedings of the Asia and South Pacific Design Automation Conference*, 2011, pp. 336–343
5. N. Golshani, J. Derakhshandeh, R. Ishihara, C. Beenakker, M. Robertson, T. Morrison, Monolithic 3D integration of SRAM and image sensor using two layers of single grain silicon, in *Proceedings of the IEEE International Conferences on 3D System Integration*, 2010, pp. 1–4
6. S.-M. Jung, J. Jang, W. Cho, J. Moon, K. Kwak, B. Choi, B. Hwang, H. Lim, J. Jeong, J. Kim, K. Kim, The revolutionary and truly 3-dimensional $25F^2$ SRAM technology with the smallest S^3 (Stacked Single-crystal Si) Cell, $0.16um^2$, and SSTFT (Stacked Single-crystal Thin Film Transistor) for ultra high density SRAM, in *Proceeding of the Symposium on VLSI Technology*, 2004, pp. 228–229
7. Y.-J. Lee, P. Morrow, S. K. Lim, Ultra high density logic designs using transistor-level monolithic 3D integration, in *Proceedings of the IEEE Internatinal Conference on Computer-Aided Design*, 2012
8. C. Liu, S. K. Lim, A design tradeoff study with monolithic 3D integration, in *Proceedings of the International Symposium on Quality Electronic Design*, 2012, pp. 531–538
9. T. Naito, T. Ishida, T. Onodukal, M. Nishigoori, T. Nakayama, Y. Ueno, Y. Ishimoto, A. Suzuki, W. Chung, R. Madurawe, S. Wu, S. Ikeda, H. Oyamatsu, World's first monolithic 3D-FPGA with TFT SRAM over 90nm 9 layer Cu CMOS, in *Proceedings of the Symposium on VLSI Technology*, 2010, pp. 219–220
10. Nangate, Nangate 45nm Open Cell Library

Chapter 19
Impact of TSV Scaling on 3D IC Design Quality

Abstract TSVs incur two major kinds of overhead in the design of 3D ICs. First, TSVs lead to significant silicon area overhead. In addition, the non-negligible TSV parasitic capacitance causes delay overhead in 3D signal paths. Therefore, the possibility of obtaining all the benefits such as wirelength reduction and better performance from 3D ICs is highly dependent on TSV size and TSV capacitance. Meanwhile, TSVs are getting smaller to minimize their negative effects and sub-micron TSVs are expected to be fabricated in the near future. At the same time, the device size is also being downscaled beyond 32 and 22 nm, so it is highly likely that future 3D ICs are built with sub-micron TSVs and advanced device technologies. In this chapter, we study the impact of sub-micron TSVs on the quality of today and future 3D ICs. For future process technologies, we develop 22 and 16 nm libraries. Using these future process libraries as well as a 45 nm library, we generate 3D IC layouts with different TSV sizes and capacitances and study the impact of sub-micron TSVs thoroughly.

The materials presented in this chapter are based on [7].

19.1 Introduction

Three-dimensional integrated circuits (3D ICs) are expected to offer various benefits such as higher bandwidth, smaller form factor, shorter wirelength, lower power, and better performance than traditional two-dimensional (2D) ICs. These benefits are enabled by die stacking and the use of through-silicon vias (TSVs) for inter-die connections. However, TSVs have two negative effects, occupation of silicon area and non-negligible capacitance, in the design of 3D ICs. The fact that TSVs occupy silicon area has great effects not only on silicon area, but also on wirelength, critical path delay, and power. The reason is as follows. If larger TSVs are inserted in a 3D

S.K. Lim, *Design for High Performance, Low Power, and Reliable 3D Integrated Circuits*, DOI 10.1007/978-1-4419-9542-1_19,

IC layout, footprint area of the design becomes larger, so the average wirelength increases [12]. This wirelength overhead leads to longer critical path delay and higher dynamic power consumption due to increased wire capacitance. In addition, non-negligible TSV capacitance also has a negative effect on critical path delay and dynamic power consumption. One thing to notice is that smaller TSVs do not necessarily have smaller capacitance than larger TSVs. The reason is because TSV capacitance is dependent not only on the TSV diameter and the TSV height, but also on the liner thickness and doping concentration of the substrate [4].

Similarly as devices are scaled, TSVs are also being downscaled [2, 14, 17]. Therefore, negative effects of TSVs will be reduced if smaller TSVs are used.[1] However, this statement is valid only when process technology is fixed and TSV technology advances. In fact, process technology is also advancing, so it is highly likely that future 3D ICs will be fabricated with smaller TSVs and state-of-the-art process technology. In this case, negative effects of TSVs might remain the same or even increase.

In this chapter, we study the impact of sub-micron TSVs on area, wirelength, critical path delay, and power of today and future 3D ICs based on GDSII-level layouts. For future process technologies, we develop 22 and 16 nm process and standard cell libraries. Using these future process technologies as well as an existing 45 nm library, we generate 3D IC layouts with different TSV sizes and capacitances and study the impact of TSVs thoroughly. The following specific topics are covered in this chapter:

- To investigate the impact of sub-micron TSVs on future 3D ICs, we develop a 22 nm and a 16 nm process and standard cell libraries. These libraries enable us to obtain very trustable experimental results.
- We generate layouts with various technology combinations (e.g., 16 nm process with 0.5 μm-diameter TSVs and 0.1 μm-diameter TSVs) and obtain area, wirelength, critical path delay, and power. Therefore, we not only cross-compare 3D ICs built with different process technologies, but also compare 3D ICs built with the same process technology and different TSV sizes and capacitances.
- We also cross-compare 2D designs built with more advanced process technology and 3D designs built with older process technology. Our experimental results show that 3D ICs built with an n-th generation process technology could be beaten by 2D ICs built with an $n+2$-th generation process technology.[2]

[1] If we assume that only the TSV size and the TSV height are downscaled while other design parameters such as the liner thickness and doping concentration are fixed, TSV capacitance decreases as TSVs are downscaled.

[2] This observation is strongly dependent on TSV capacitance used at each process node.

19.2 Preliminaries

19.2.1 Design Overhead of TSVs

The use of TSVs in 3D ICs have two negative effects on the quality of 3D ICs: area and delay overhead. According to recent research on TSV area overhead [9], silicon area occupied by TSVs is quite significant, which in turn reduces the wirelength benefit of 3D ICs. In addition, according to recent research on TSV capacitance overhead [5], TSV capacitance is a significant source of delay on 3D signal paths. Although buffer insertion can reduce delay overhead caused by TSV capacitance, buffer insertion itself also causes another problem: additional silicon area for buffer insertion and additional power consumption.

The degree of negative effects of TSVs on 3D ICs is dependent on various technology and design parameters. For example, if we use 5 μm TSVs[3] with state-of-the-art process technology such as 32 nm technology in 3D IC designs, these TSVs may cause a huge area overhead. On the other hand, if we use 5 μm TSVs with relatively old technology such as 0.18 μm technology, these TSVs may not cause any area overhead because the ratio between area occupied by a TSV and average gate area of the old technology is smaller than the ratio of the advanced technology. Similarly, small TSVs (e.g., 1 μm TSVs) can have huge capacitance depending on the liner thickness and doping concentration of the substrate. In this case, small TSVs may not cause serious area overhead, but they will cause serious delay overhead.

19.2.2 Motivation

Downscaling of devices reached 32 nm node [19] in 2009, and 22 and 16 nm technologies are currently under development. As devices are downscaled as process technology advances, TSVs are also being downscaled as TSV manufacturing technology advances. Recently, it was demonstrated that 0.7 μm-diameter TSVs could also be fabricated reliably [14]. In addition, according to the ITRS prediction on TSV diameter and TSV aspect ratio, TSV diameter will continue to decrease while TSV aspect ratio will increase. Therefore, we expect that sub-micron TSVs will be developed and be ready for use within the next few years.

However, all of the existing work on the impact of TSVs on the quality of 3D IC designs focus on using micron-size TSVs and current (32 or 45 nm) or even old (90 and 130 nm) process technology. For example, a 45 nm technology and 1.67 μm TSVs are used in [8] and a 45 nm technology and TSVs whose width is

---

[3] A "X μm TSV" in this chapter denotes a TSV whose width (= for square-shaped TSVs) or diameter (= for cylindrical-type TSVs) is X μm.

approximately 4 μm are used in [20]. However, none of them discuss what will occur if smaller TSVs are used in 45 nm technology or what will occur if the same-size TSVs are used with different process technologies (e.g., a 90, 32, or 22 nm). However, it is crucial to accurately predict the impact of new TSV technology on the design quality of 3D ICs in order to refine the technology or justify the investment and cost. Our goal in this chapter is to study the impact of sub-micron TSVs on the area, wirelength, critical path delay, and power of today and future 3D IC designs. For our future process technology, we develop a 22 nm and a 16 nm process and standard cell libraries. We also use various sets of TSV-related dimensions in our GDSII-level 3D IC layouts. Lastly, we present a thorough study on the impact of sub-micron TSVs on the design quality of today and future 3D ICs.

19.3 Library Development Flow

In this section, we demonstrate the development flow of our 22 and 16 nm process and standard cell libraries. For 22 and 16 nm transistor models, we use the high-performance transistor model of the predictive technology model (22 nm PTM HP model V2.1 and 16 nm PTM HP model V2.1) [21]. The supply voltages of the 22 nm and the 16 nm models are 0.8 and 0.7 V, respectively.

19.3.1 Overall Development Flow

For the development of a 22 nm and a 16 nm process and standard cell libraries, we follow a typical library development flow illustrated in Fig. 19.1. We first define device and interconnect layers. From the defined device and interconnect layers, we create a tech file (.tf), a display resource file (.drf), an interconnect technology file (.ict), a design rule file, a layout-versus-schematic (LVS) rule file, and an RC parasitic extraction rule file. With the tech file and the display resource file, we draw standard cell layouts. After the layout generation, we perform abstraction on these layouts to create a library exchange format file (.LEF), and run RC extraction and create SPICE netlists (post_xRC.cdl). With these SPICE netlists and the PTM transistor models, we perform library characterization to create timing and power libraries (.lib and .db). We also generate a capacitance table and a .tch file for sign-off RC extraction and timing analysis.

19.3.2 Interconnect Layers

We create interconnect layers of our 22 and 16 nm technology based on ITRS interconnect prediction [3], downscaling trends of other standard cell libraries, and

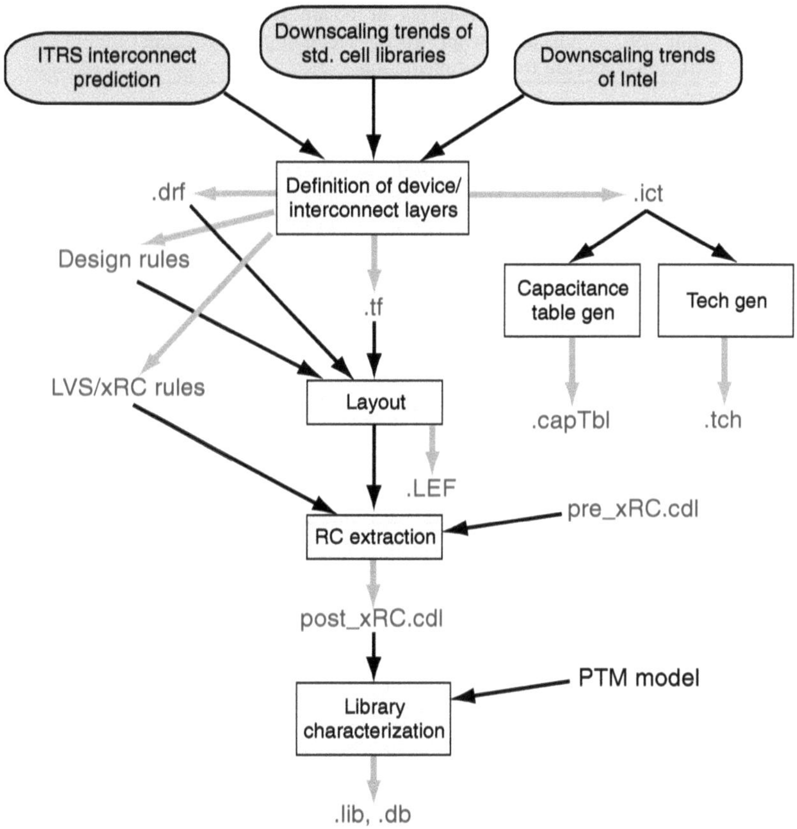

Fig. 19.1 Development flow of our 22 and 16 nm process and standard cell libraries

the downscaling trends of Intel process technology [1, 16, 19]. According to ITRS prediction on interconnect layers, for example, the pitch of the metal 1 wire at 22 nm is about 72 nm and that at 16 nm is about 48 nm, and the pitch of a semi-global wire at 22 nm is about 160 nm and that at 16 nm is about 130 nm. From these values as well as extrapolation of interconnect layers of Intel process technology and other standard cell libraries, we predict interconnect layers at 22 and 16 nm. Table 19.1 shows the contacted gate pitch and the pitches of metals 1–8 layers at each process node. Table 19.2 shows widths and thicknesses of all metal layers of our 22 and 16 nm process libraries. Notice that the 22 nm and the 16 nm libraries have the same width in metals 9–12 layers. Since these metal layers are sometimes used for special purposes such as power/ground lines and clock distribution, we did not scale them down. The aspect ratio of the 22 nm library is set to 1.8 and that of the 16 nm library is set to 1.9. Since we assume the use of low-k inter-layer insulator material, we use 1.9 for the dielectric constant of the inter-layer dielectric material and 3.8 for the dielectric constant of the barrier material for both the 22 nm and the 16 nm libraries.

Table 19.1 Interconnect layers of 65 nm [1], 45 nm [16], 32 nm [19], 22 nm, and 16 nm process technology. The 22 and the 16 nm layers are from our prediction

	Pitch (nm)				
Layer	65 nm	45 nm	32 nm	22 nm	16 nm
Contacted gate	220	160	112.5	86	62
Metal 1	210	160	112.5	76	46
Metal 2	210	160	112.5	76	46
Metal 3	220	160	112.5	76	46
Metal 4	280	240	168.8	130	72
Metal 5	330	280	225.0	206	98
Metal 6	480	360	337.6	206	146
Metal 7	720	560	450.1	390	240
Metal 8	1,080	810	566.5	390	240

Table 19.2 Width (w) and thickness (t) of metal layers used in our 22 and 16 nm process libraries. The aspect ratio for the 22 nm library is 1.8 and that for the 16 nm library is 1.9

	22 nm		16 nm	
Layer	w (nm)	t (nm)	w (nm)	t (nm)
Metal 1, 2, 3	36	64.8	22	41.8
Metal 4	60	108	32	60.8
Metal 5	96	172.8	44	83.6
Metal 6	96	172.8	66	125.4
Metal 7, 8	180	324	110	209
Metal 9, 10	400	720	400	760
Metal 11, 12	800	1,440	800	1,520

19.3.3 Standard Cell Library

We first create a tech file defining device and interconnect layers and a set of design rules such as minimum poly-to-contact spacing, minimum metal-to-metal spacing, and so on. Then, we draw standard cell layouts with this tech file and the design rules.[4] We created about 90 cells and Table 19.3 shows the list of these standard cells except antenna and filler cells. The placement site width and height of our 22 nm standard cell library are 0.1 and 0.9 μm, respectively, and those of our 16 nm library are 0.06 and 0.6 μm, respectively. Figure 19.2 shows the smallest (1×) two-input NAND gates of the 45 nm, our 22 and 16 nm standard cell libraries. After creating the standard cell layouts, we perform DRC and LVS of each layout and extract parasitic RC of each standard cell. We also characterize all the standard cells to create timing and power libraries.

[4] We referred to the standard cell layouts of the Nangate 45 nm standard cell library [18].

Table 19.3 Standard cells in our 22 and 16 nm standard cell libraries

Type	Available sizes
AND2/3/4, AOI21/211/221	1×, 2×, 4×
BUF, INV	1×, 2×, 4×, 8×, 16×, 32×
LOGIC 0, LOGIC 1	1×
MUX2	1×, 2×
NAND2/3/4/, NOR2/3/4	1×, 2×, 4×
OAI21/22/211/221/222	1×, 2×, 4×
OAI33	1×
OR2/3/4	1×, 2×, 4×
XNOR2, XOR2	1×, 2×
DFF	1×, 2×
FA, HA	1×

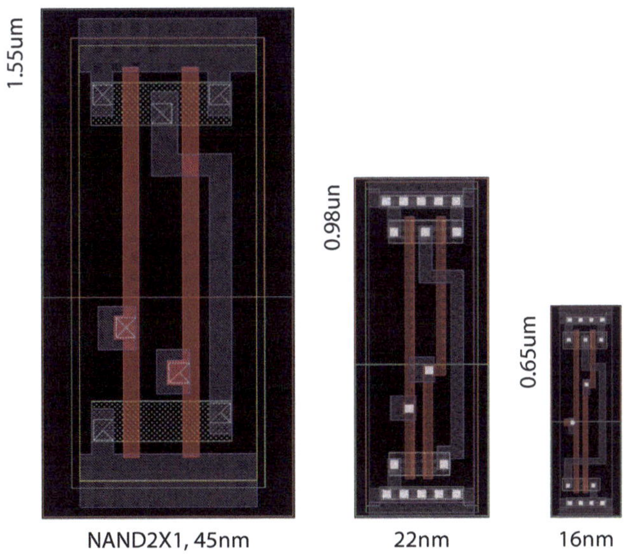

Fig. 19.2 The smallest (1×) two-input NAND gates of the 45 nm [18], and our 22 and 16 nm libraries (drawn to scale)

19.4 Comparison of 45, 22, and 16 nm Libraries

Before we proceed to the comparison of 2D and 3D ICs in Sect. 19.6, we compare the Nangate 45 nm, our 22 nm, and our 16 nm standard cell libraries and transistor characteristics.

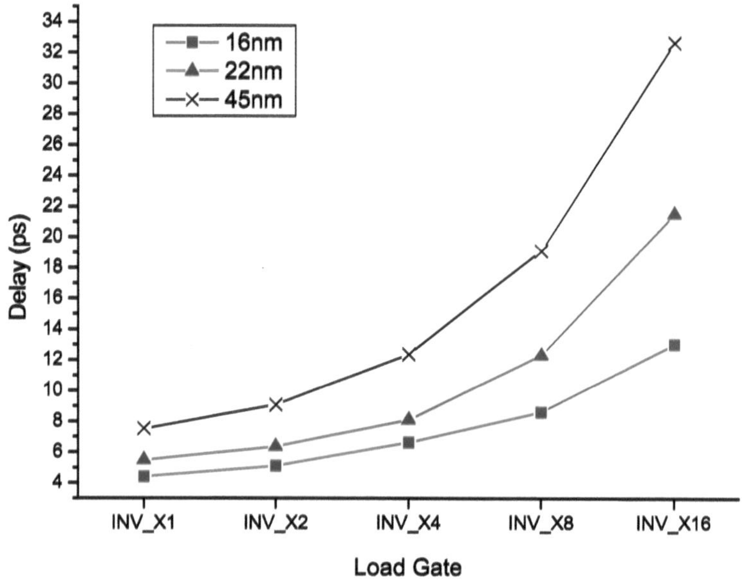

Fig. 19.3 Delay of a minimum-size inverter driving an N× inverter (N = 1, 2, 4, 8, 16), where both inverters are in the same process. RC parasitics are included

19.4.1 Gate Delay and Input Capacitance

Gate delay and drive strength are determined by transistor characteristics and the gate size. Therefore, our first experiment is to compare the transistor characteristics. In this experiment, a minimum-size inverter in a process library drives another minimum-size inverter, which drives an $N\times$ inverter in the same library. We obtain the delay of the second minimum-size inverter (driving the $N\times$ inverter) by SPICE simulation. Figure 19.3 shows the delay. We observe that the 16 nm inverter has the shortest delay and the 45 nm inverter has the longest delay. Quantitatively, we observe approximately 30 % improvement when the process moves from 45 to 22 nm and about 20 % improvement when the process moves from 22 to 16 nm. Notice that this SPICE simulation does not consider interconnect parasitic resistance and capacitance. Table 19.4 also shows the FO4 delay at each process technology.

Since gate input capacitance is also an important factor determining delay and power, we show input capacitances of 45, 22, and 16 nm standard cells in Table 19.5. As shown in the table, the average input capacitance of the 22 nm standard cells is approximately 48 % of the average input capacitance of the 45 nm standard cells. On the other hand, the average input capacitance of the 16 nm standard cells is approximately 83 % of the average input capacitance of the 22 nm standard cells.

Table 19.4 FO4 delay, standard cell heights, wire sheet resistance, and unit wire capacitance (fF/μm)

	45 nm	22 nm	16 nm
FO4 delay (ps)	15.15	13.63	12.28
Std. cell. height (μm)	1.4	0.9	0.6
Wire sheet resistance (Metal 1)	0.38	0.26	0.40
(Metal 4)	0.21	0.16	0.28
(Metal 7)	0.08	0.05	0.08
Unit wire capacitance (Metal 1)	0.20	0.15	0.16
(Metal 4)	0.20	0.15	0.13
(Metal 7)	0.20	0.14	0.14

Table 19.5 Input capacitance of selected standard cells in the 45 nm, the 22 nm, and the 16 nm libraries

	Cap (fF)		
Cell	45 nm	22 nm	16 nm
AND2 1×	0.54 (1.00)	0.25 (0.46)	0.22 (0.41)
AOI211 1×	0.64 (1.00)	0.30 (0.47)	0.25 (0.39)
AOI21 1×	0.55 (1.00)	0.23 (0.42)	0.20 (0.36)
BUF 4×	0.47 (1.00)	0.28 (0.60)	0.29 (0.62)
DFF 1×	0.90 (1.00)	0.41 (0.46)	0.26 (0.29)
FA 1×	2.46 (1.00)	1.31 (0.53)	1.36 (0.55)
INV 4×	1.45 (1.00)	0.69 (0.48)	0.56 (0.39)
MUX2 1×	0.95 (1.00)	0.42 (0.44)	0.34 (0.36)
NAND2 1×	0.50 (1.00)	0.24 (0.48)	0.22 (0.44)
OAI21 1×	0.53 (1.00)	0.25 (0.47)	0.20 (0.38)
OR2 1×	0.60 (1.00)	0.26 (0.43)	0.20 (0.33)
XOR2 1×	1.08 (1.00)	0.55 (0.51)	0.45 (0.42)
Average	(1.00)	(0.48)	(0.40)

19.4.2 Interconnect Layers

Characteristics of interconnect layers also have a big effect on the performance of a library, so we show wire sheet resistance and unit wire capacitance of short, semi-global, and global metal layers in Table 19.4. The resistivity of the 45 nm technology is about 5.0×10^{-8}, so the sheet resistance of the library is relatively high compared to the 22 nm library. On the other hand, the resistivity of the 22 nm and the 45 nm technology is 1.7×10^{-8}, which is the resistivity of copper. This is why the sheet resistances of the 22 nm metal layers are lower than those of the 45 nm metal layers although the thickness of the 45 nm metal layers is larger than that of the 22 nm metal layers. On the other hand, as the technology moves from 22 to 16 nm, the sheet resistance goes up because both of them use the same resistivity, but the metal layer thickness of the 16 nm library is smaller than that of the 22 nm library.

The unit wire capacitance of the 45 nm library is also slightly higher than that of the 22 nm library. This is because the dielectric constant used for the 45 nm library is 2.5 while the 22 nm library uses 1.9 for its dielectric constant. If the same dielectric material ($\varepsilon_r = 1.9$) is used for the 45 nm library, the unit wire capacitance becomes 0.15, which is close to the unit wire capacitance of the 22 nm library.

Table 19.6 Benchmark circuits

			Total cell area		
Circuit	# Gates	# Nets	45 nm	22 nm	16 nm
BM1	352 K	372 K	0.632	0.218	0.098
BM2	518 K	680 K	1.288	0.437	0.198

Table 19.7 Comparison of 2D layouts

	BM1			BM2		
	45 nm	22 nm	16 nm	45 nm	22 nm	16 nm
Area (mm^2)	1.00	0.36	0.17	2.56	0.81	0.42
Wirelength (m)	10.65	4.22	2.75	15.17	8.90	6.19
Delay (ns)	3.19	2.61	2.38	6.51	4.10	3.93
Power (W)	0.352	0.0684	0.068	0.521	0.154	0.133

19.4.3 Full-Chip 2D Design

In this experiment, we design 2D circuits using the three standard cell libraries and compare area, wirelength, critical path delay, and power. The experimental flow is as follows. We prepare two benchmark circuits shown in Table 19.6, synthesize, design, and optimize them using each standard cell library and commercial tools. For all the libraries, we use the same area utilization (60 %) for fair comparison and find the fastest operation frequency for each library.

Table 19.7 shows the comparison results for the 2D designs. The chip area of the 45 nm designs is about three times larger than that of the 22 nm designs on average, and the chip area of the 22 nm designs is approximately two times larger than that of the 16 nm designs on average. In addition, the total wirelength of the 16 nm designs is approximately 1.48× shorter than that of the 22 nm designs, and 3.08× shorter than that of the 45 nm designs. Regarding the critical path delay, the 16 nm designs are 1.49× faster than the 45 nm designs on average and 1.07× faster than the 22 nm designs on average. Power consumption of the 16 nm designs is approximately 4.5× smaller than that of the 45 nm designs and 1.1× smaller than that of the 22 nm designs. Overall, the delay and power enhancement coming from 22-to-16 nm transition is not as significant as the enhancement coming from 45-to-22 nm transition because 45 and 22 nm technologies are two generations apart while 22 and 16 nm technologies are only one generation apart, and the quality (sheet resistance and unit wire capacitance) of the interconnect layers of the 45 nm library is worse than that of the 22 nm library.

19.5 Full-Chip 3D IC Design and Analysis Methodology

To generate 3D IC layouts, we use the 3D RTL-to-GDSII tool obtained from [20]. This tool works as follows: For a given 2D gate-level (flattened) netlist, this tool partitions gates in the x-, y-, and z-directions iteratively to globally place gates in grids in 3D. After the global placement, it constructs a 3D Steiner tree for each net and inserts TSVs into each placement grid based on the locations of vertical edges in the 3D Steiner tree. Then, it runs detailed placement in each placement grid using Cadence Encounter. Routing for each die is also performed by Encounter. The output of the tool includes a Verilog netlist and a design exchange format (DEF) file containing TSV locations for each die, a top-level Verilog netlist containing die-to-die connections, and a top-level standard parasitic exchange format (SPEF) file. One thing to notice is that the minimum number of TSVs to be inserted in the 3D design is dependent on the cut sequence, which is the order of x-, y-, and z-direction partitioning we apply for global placement. For example, if we apply z-direction partitioning in early steps, we are likely to obtain fewer inter-die connections. On the other hand, if we apply z-direction partitioning later, we are likely to obtain more inter-die connections [20]. This variation of the number of TSVs enables us to produce different global placement solutions with different TSV counts.

After generating 3D IC layouts, we perform 3D timing optimization. We first perform initial timing optimization for each die. Then, we feed all the layouts, timing analysis results, and the target clock frequency into the 3D timing optimization tool obtained from [15]. This 3D timing optimization tool iterates the following steps: (a) it performs RC extraction and obtains an SPEF file for each die; (b) it performs 3D timing analysis using the SPEF files and the top-level SPEF file using Synopsys PrimeTime; (c) based on the timing analysis result and the target clock frequency, the tool scales the target delay of each 3D path and creates a timing constraint file for each die; (d) since each die has its own netlist and timing constraint file, we perform timing optimization for each die separately. We iterate this timing optimization process several times until the overall timing improvement saturates.

3D power analysis needs (a) a top-level netlist as well as a netlist for each die, (b) a top-level SPEF file as well as a SPEF file for each die, and (c) switching activities of cells and nets. To obtain switching activities of cells and nets, we load Verilog netlists generated by the 3D RTL-to-GDSII tool obtained from [20] into Encounter and run power analysis. This power analysis internally generates and stores switching activities of cells and nets, so we dump this information into an output file after the power analysis. Then, we load all the netlists, SPEF files, and the switching activity files into PrimeTime and run power analysis. This power analysis method produces true full-chip 3D power analysis results.

19.6 Experimental Results

19.6.1 Simulation Settings

We use two benchmark circuits, BM1 and BM2, as shown in Table 19.6. For the 45 nm process node, we use the Nangate 45 nm standard cells library [18]. We also

Table 19.8 TSV-related dimensions, design rules, and TSV capacitance

Dimensions	TSV-5	TSV-1	TSV-0.5	TSV-0.1
Width (μm)	5	1	0.5	0.1
Height (μm)	25	5	8	5
Aspect ratio	5	5	16	50
Liner thickness (nm)	100	30	20	10
Barrier thickness (nm)	50	15	10	5
Landing pad width (μm)	6	1.6	1	0.18
TSV-to-TSV spacing (μm)	2	0.8	0.6	0.1
TSV-to-device spacing (μm)	1	0.4	0.3	0.1
TSV capacitance (fF)	20	2.67	3.2	0.8

use four sets of TSV-related dimensions shown in Table 19.8. In our simulations, we use 5 and 0.5 μm TSVs with the 45 nm technology, 1 and 0.1 μm TSVs with the 22 nm technology, and 0.5 and 0.1 μm TSVs with the 16 nm technology. Since the standard cell height of the 45 nm library is 1.4 μm, a 5 μm TSV including its keep-out zone occupies five standard cell rows while a 0.5 μm TSV including its keep-out zone occupies one standard cell row, as shown in Table 19.8. Similarly, a 1 μm TSV and a 0.1 μm TSV occupy three standard cell rows and 0.26 standard row, respectively, when they are used with our 22 nm standard cell library. If 0.5 and 0.1 μm TSVs are used for our 16 nm standard cell library, a 0.5 μm TSV occupies 1.33 standard cell rows and a 0.1 μm TSV occupies 0.5 standard cell row. Figure 19.4 shows the four different TSVs in a top-down view and a side view and Fig. 19.5 shows GDSII images of TSVs and standard cells at 45, 22, and 16 nm technology. We also assume that face-to-back bonding and via-first TSVs are used for 3D integration. Die thickness that we use for each TSV dimension set is same as the TSV height, which ranges from 5 to 25 μm. Although 5 μm thickness is extremely thin, it is practical [11, 13].

19.6.2 Impact on Silicon Area

Figure 19.6 shows footprint area of 2D designs and two-die 3D BM1 designs at each technology node. If the TSV size is zero, the footprint area of a two-die 3D design should be approximately half of its 2D counterpart. Since the TSV size is not zero, however, the footprint area of a two-die 3D design is usually greater than half of its 2D counterpart. For example, the area of the 45 nm 2D design is 1.0 mm^2, but the area of the 45 nm 3D design using 5 μm TSVs is about 0.85 mm^2, which is 85 % of the 2D design. Similarly, the area of the 45 nm 3D design using 0.5 μm TSVs is about 0.63 mm^2, which is 63 % of the 2D design. The same trend is found in the 22 nm and the 16 nm designs. However, if the TSV size is 0.1 μm, the footprint area of a two-die 3D design becomes almost half of its 2D counterpart. We find similar trends in BM2 designs as shown in Fig. 19.7.

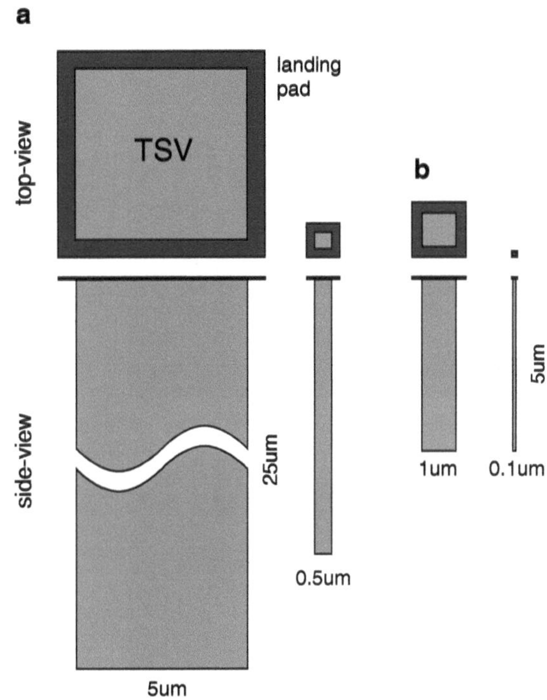

Fig. 19.4 Size comparison of the four TSVs used in our study: (**a**) 5 and 0.5 μm width used for 45 nm technology, (**b**) 1 and 0.1 μm width used for 22 nm technology

All these trends depend on TSV size and the number of TSVs used in the designs. Of course, using smaller TSVs helps achieve smaller footprint area, which can reduce chip cost. However, smaller TSVs could be more expensive due to manufacturing difficulties, so the use of smaller TSVs does not necessarily lead to lower chip cost. Using fewer TSVs also helps achieve smaller footprint area. However, several studies show that using more TSVs than the minimum number of TSVs helps reduce wirelength and achieve better performance [8, 10, 20]. Thus, there exist tradeoffs among TSV size, the number of TSVs used in the design, and chip cost.

19.6.3 Impact on Wirelength

Figure 19.6 shows wirelength of BM1 designs. When 5 μm TSVs are used with the 45 nm technology, 3D designs have longer wirelength than 2D designs. However, when 0.5 μm TSVs are used with 45 nm technology, the wirelength of the 3D design is about 10 % shorter than that of the 2D design. When 1 and 0.1 μm TSVs are used with the 22 nm technology, however, we do not observe large wirelength reduction. On the other hand, when 0.5 and 0.1 μm TSVs are used with the 16 nm technology, we find 15 % wirelength reduction.

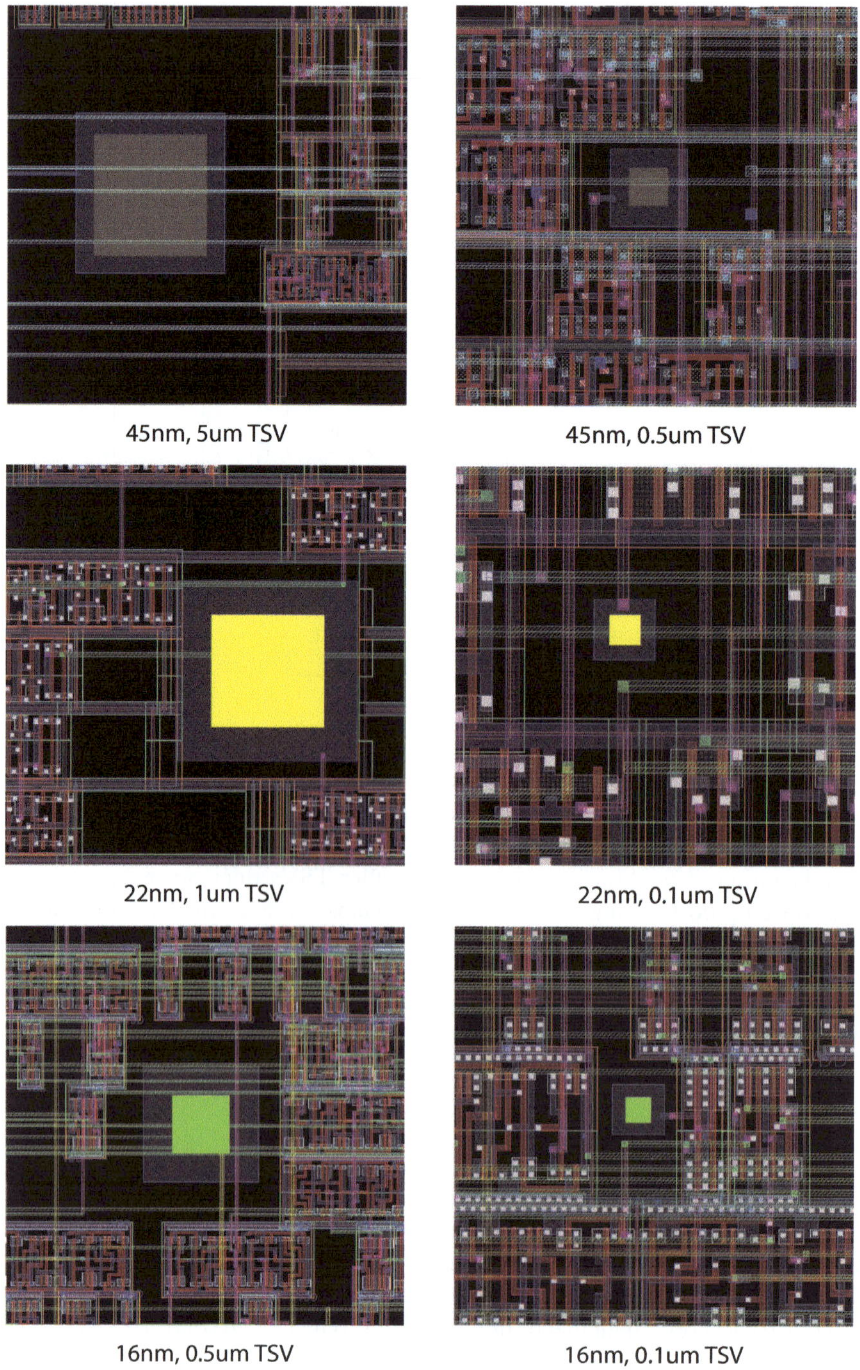

Fig. 19.5 Zoom-in GDSII layouts of the six types of designs studied in this chapter. Each TSV is surrounded by its keep-out-zone

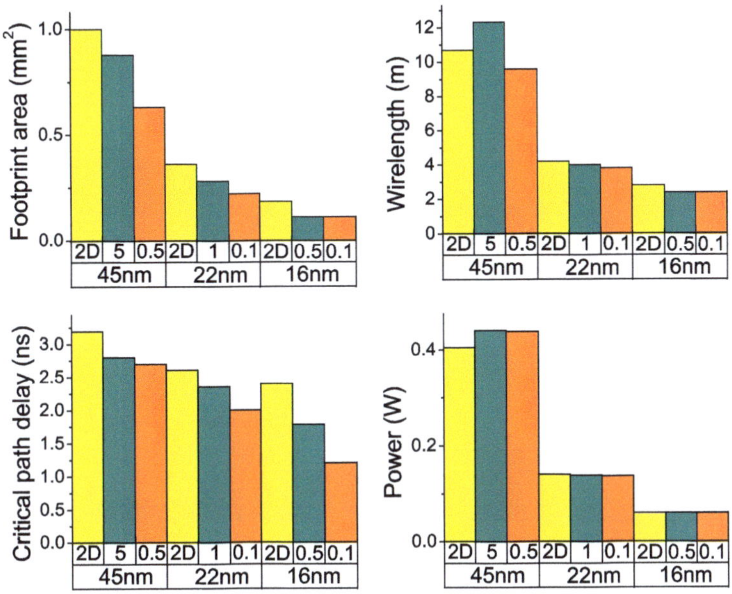

Fig. 19.6 Comparison of the optimized 2D designs and two-die 3D designs (BM1) in 45, 22, and 16 nm technology. The x-axis shows the technology combination (the first row shows TSV diameter in μm)

We find similar trends in BM2 designs too as shown in Fig. 19.7. Above all, 45 nm 3D designs have longer wirelength than 2D designs. However, when 1 and 0.1 μm TSVs are used with the 22 nm technology, we observe 9 and 13% wirelength reduction, respectively. Similarly, when 0.5 and 0.1 μm TSVs are used with the 16 nm technology, we observe 12 and 15 % wirelength reduction, respectively.

Wirelength reduction that we can obtain by moving from 2D ICs to 3D ICs comes mainly from smaller footprint area. However, wirelength reduction is also dependent on the quality of the 3D global placement algorithm, the TSV insertion (3D routing) algorithm, TSV size, and characteristics of benchmarks circuits. Therefore, it is possible to obtain higher wirelength reduction ratio or vice versa depending on those factors. If, however, all other factors such as the placement algorithm and the TSV insertion algorithm except the TSV size do not change, the TSV size is the main factor affecting the wirelength. For instance, in Fig. 19.6, by shrinking the TSV size from 5 to 0.5 μm in the 45 nm 3D designs, we obtain 22 % wirelength reduction. However, when we shrink the TSV size from 0.5 to 0.1 μm in the 16 nm 3D designs, we obtain almost no wirelength reduction. This is because 0.5 μm TSVs are already sufficiently small, so shrinking the TSV size does not lead to further wirelength reduction.

One thing to note is that 3D designs at the n-th generation process node has longer wirelength than 2D designs at the $n+1$-th generation process node.

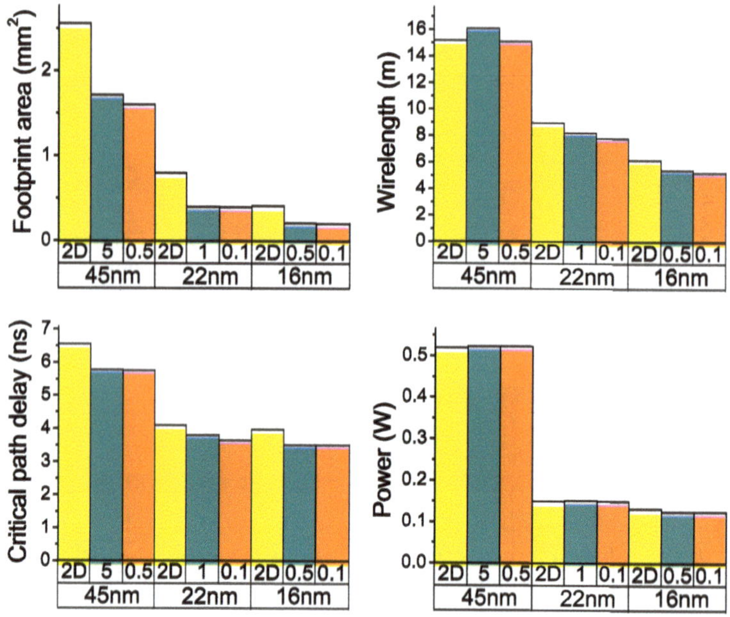

Fig. 19.7 Comparison of the optimized 2D designs and two-die 3D designs (BM2) in 45, 22, and 16 nm technology. The x-axis shows the technology combination (the first row shows TSV diameter in μm)

Therefore, shrinking the TSV size is important to reduce the wirelength, but moving to the advanced process node is also important for wirelength reduction. This also coincides with the prediction result presented in [6].

19.6.4 Impact on Performance

Figure 19.6 shows the critical path delay of 2D and 3D designs for the BM1 benchmark circuit. As seen in the figure, the critical path delay of a 3D design having longer wirelength than (or similar wirelength to) its 2D counterpart can be smaller than that of the 2D design. For example, the wirelength of the 3D design built with 5 μm TSVs and the 45 nm technology is 15% longer than that of the 2D design, but the critical path delay of the 3D design is 12% smaller than that of the 2D design. Similar trends are also found in the BM2 benchmark circuit as shown in Fig. 19.7.

One important observation is that the critical path delay of 3D designs built with the n-th generation process node could be smaller than the critical path delay of 2D designs built with the $n+1$-th generation process node. For example, the BM1 3D design built with 0.1 μm TSVs with the 22 nm technology has approximately 20%

Table 19.9 Additional TSV-related statistics. "c.p." denotes critical path

	BM1					
	45 nm		22 nm		16 nm	
TSV diameter (μm)	5	0.5	1	0.1	0.5	0.1
# TSVs in c.p.	1	0	3	4	2	4
	BM2					
	45 nm		22 nm		16 nm	
TSV diameter (μm)	5	0.5	1	0.1	0.5	0.1
# TSVs in c.p.	0	0	1	2	1	2

smaller delay than the 2D design built with the 16 nm technology. Similarly, the BM2 3D design built with 0.1 μm TSVs with the 22 nm technology has about 9 % smaller delay than the 2D design built with the 16 nm technology.

For more in-depth analysis, we show the number of TSVs used in the critical paths in Table 19.9. If the TSV count is zero, the critical path is a 2D path existing in a single die. If the TSV count is three, the critical path alternates three times (e.g., die0 – die1 – die0 – die1) between two dies since all the layouts are two-die designs. Especially, if the TSV count is zero and the critical path delay is shorter than the critical path delay of its 2D counterpart design, the shorter critical path delay of the 3D design is primarily due to the shorter wirelength achieved by the smaller footprint area. On the other hand, if the TSV count is non-zero, the critical path delay comes from both the smaller footprint area and the shorter wirelength.

19.6.5 Impact on Power

Figures 19.6 and 19.7 show power consumption for BM1 and BM2 benchmark circuits, respectively. As seen in the figures, moving from 2D ICs to 3D ICs does not necessarily lead to power reduction even if 3D designs have shorter wirelength than 2D designs. The reason is as follows. Reduction in power consumption by building 3D ICs comes from smaller dynamic power consumption due to shorter wirelength.[5] However, TSV capacitance can essentially be thought of as wire capacitance. Therefore, the total capacitance is the sum of the total TSV capacitance and the total wire capacitance. This means that the total TSV capacitance should be less than the reduced wire capacitance to achieve power reduction.[6] In other words, achievement of power reduction needs smaller TSV capacitance, use of fewer TSVs, and wirelength reduction. However, there again exist tradeoffs among the number of TSVs, the amount of wirelength reduction, and power consumption. Inserting fewer

[5]There exist many kinds of 3D integration and some of them (e.g., core-DRAM stacking) provide a huge amount of power saving by removing long chip-to-chip connections.

[6]Note that this is a simplified analysis. In reality, the total power should be computed in a more sophisticated fashion taking switching activities of nets and gates into account.

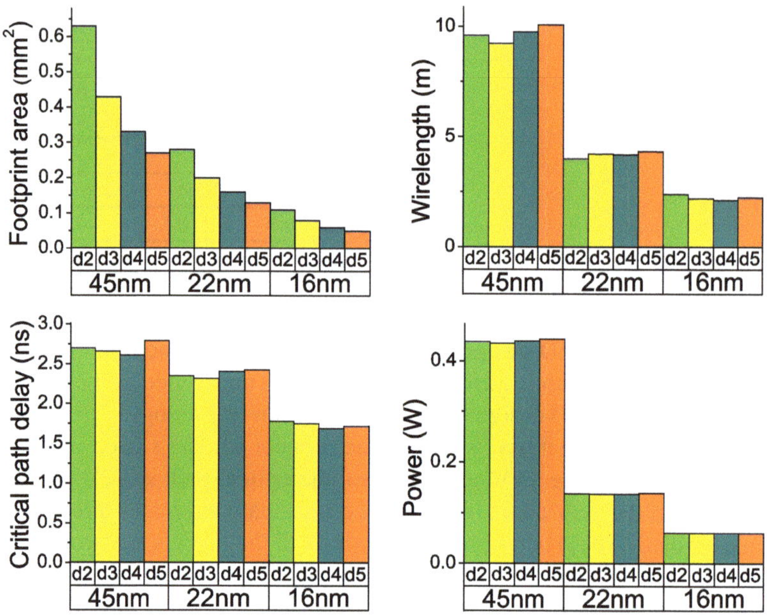

Fig. 19.8 Comparison of optimized 3D designs (BM1) implemented in multiple dies. "d*n*" denotes *n*-die implementation. We use 0.5, 1, and 0.5 μm TSVs for the 45, 22, and 16 nm technologies, respectively

TSVs may not reduce the total wirelength as much as expected. Similarly, the use of fewer TSVs may not reduce the dynamic power consumption. Inserting more TSVs, however, may reduce the total wirelength more than 10–20 % [8], but then the total TSV capacitance also increases, so the total capacitance could be larger than the total capacitance of 2D designs.

19.6.6 Impact of the Number of Dies

In the above sections, we build 3D designs in two dies. However, the number of dies also has an impact on area, wirelength, critical path delay, and power [8]. In this section, therefore, we vary the number of dies and study the impact of TSVs on the four metrics. Figures 19.8 and 19.9 show footprint area, wirelength, critical path delay, and power for BM1 and BM2 benchmarks when the number of dies varies from two (d2 cases in the figure) to five (d5 cases in the figure). To limit the simulation space size, we use 0.5 μm TSVs for the 45 nm technology, 1 μm TSVs for the 22 nm technology, and 0.5 μm TSVs for the 16 nm technology.

As the number of dies increases, the footprint area decreases as expected. Assuming that the TSV diameter is zero and the same utilization is used for all layouts, the footprint area of an *n*-die design of a circuit is approximately A_{2D}/n where A_{2D} is the area of the 2D design of the circuit. However, the TSV size is not

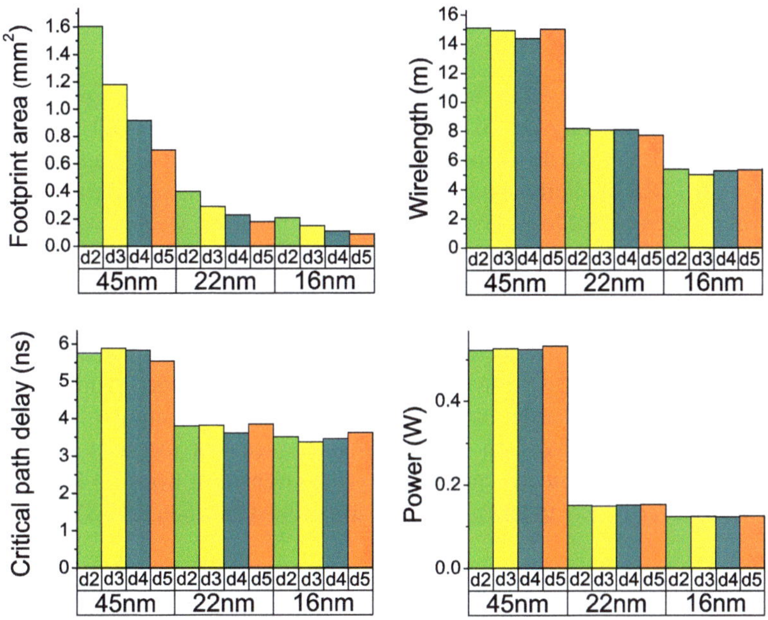

Fig. 19.9 Comparison of optimized 3D designs (BM2) implemented in multiple dies. "dn" denotes n-die implementation. We use 0.5, 1, and 0.5 μm TSVs for the 45, 22, and 16 nm technologies, respectively

negligible and more TSVs are inserted as more dies are stacked, so the footprint area of a circuit designed in n dies is slightly larger than A_{2D}/n.

On the other hand, stacking more dies does not necessarily result in shorter wirelength although stacking more than two dies helps reduce the wirelength in most cases of our simulation. The largest wirelength reduction ratio between more-than-two-die designs and two-die designs is about 11 % in our simulation (16 nm two-die implementation vs. 16 nm four-die implementation of BM1). In addition, in many cases of our simulation, stacking five dies does not produce shorter wirelength than two- to four-die designs. The main reason is because stacking more dies needs more TSVs in general, and inserting more TSVs causes wirelength overhead because of area overhead.

Regarding the critical path delay, stacking three or four dies reduces the critical path delay compared to two-die designs. The largest critical path delay ratio between more-than-two-die designs and two-die designs is about 5 % in our simulation (16 nm four-die implementation of BM1). However, stacking more than four dies does not reduce the critical path delay effectively. On the other hand, power consumption varies in a very small range. The reason is because gate internal power is dominant, so the combination of reducing wirelength (a positive effect) and inserting more TSVs (a negative effect due to TSV capacitance) leads to the very small change in total power consumption.

19.7 Conclusions

In this chapter, we studied the impact of sub-micron TSVs on the quality of today and future 3D ICs using on GDSII-level layouts. To generate 3D IC layouts of future 3D IC layouts, we developed 22 and 16 nm process and standard cell libraries based on the ITRS prediction and downscaling trends of other standard cell libraries and Intel process technology. With these realistic libraries, we generated today and future 3D IC layouts and compared footprint area, wirelength, critical path delay, and power consumption. The simulation results showed that (1) footprint area is strongly dependent on the TSV size, so the use of sub-micron TSVs is the most important factor for area reduction; (2) wirelength is also dependent on the TSV size, but if the TSV size is sufficiently small (0.5 μm TSVs for 16 nm technology), shrinking the TSV size further does not help wirelength reduction; (3) critical path delay is strongly dependent on the TSV capacitance, but footprint area also has a non-negligible effect on critical path delay; (4) transition from 2D ICs to 3D ICs does not necessarily lead to less power consumption even when the TSV capacitance is small.

References

1. P. Bai et al., A 65 nm logic technology featuring 35 nm gate lengths, enhanced channel strain, 8 Cu interconnect layers, low-k ILD and 0.57 μm^2 SRAM cell, in *Proceedings of the IEEE International Electron Devices Meeting*, IEEE, Piscataway, 2004
2. R.E. Farhane, M. Assous, P. Leduc, A. Thuaire, D. Bouchu, H. Feldis, N. Sillon, A successful implementation of dual damascene architecture to copper TSV for 3D high density, in *Proceedings of the IEEE International 3D Systems Integration Conference*, IEEE, Piscataway, 2010
3. ITRS. International Technology Roadmap for Semiconductors 2007 Edition Interconnect. http://www.itrs.net
4. G. Katti, M. Stucchi, K.D. Meyer, W. Dehaene, Electrical modeling and characterization of through silicon via for three-dimensional ICs. IEEE Trans. Electron Devices, IEEE, Piscataway, **57**(1), 256–262 (2010)
5. D.H. Kim, S.K. Lim, Through-silicon-via-aware delay and power prediction model for buffered interconnects in 3D ICs, in *Proceedings of the ACM/IEEE International Workshop on System Level Interconnect Prediction*, ACM, New York, 2010, pp. 25–31
6. D.H. Kim, S.K. Lim, Impact of through-silicon-via scaling on the wirelength distribution of current and future 3D ICs, in *Proceedings of the IEEE International Interconnect Technology Conference*, IEEE, Piscataway, 2011
7. D.H. Kim, S.K. Lim, Design quality trade-off studies for 3D ICs built with sub-micron TSVs and future devices. IEEE J. Emerg. Sel. Top. Circuits Syst. IEEE, Piscataway, **2**(2), 240–248 (2012)
8. D. H. Kim, K. Athikulwongse, S.K. Lim, A study of through-silicon-via impact on the 3D stacked IC layout, in *Proceedings of the IEEE International Conference on Computer-Aided Design*, IEEE, Piscataway, 2009, pp. 674–680
9. D.H. Kim, S. Mukhopadhyay, S.K. Lim, Through-silicon-via aware interconnect prediction and optimization for 3D stacked ICs, in *Proceedings of the ACM/IEEE International Workshop on System Level Interconnect Prediction*, ACM, New York, 2009, pp. 85–92

10. D.H. Kim, S. Mukhopadhyay, S.K. Lim, TSV-aware interconnect length and power prediction for 3D stacked ICs, in *Proceedings of the IEEE International Interconnect Technology Conference*, IEEE, Piscataway, 2009, pp. 26–28
11. Y.S. Kim, A. Tsukune, N. Maeda, H. Kitada, A. Kawai, K. Arai, K. Fujimoto, K. Suzuki, Y. Mizushima, T. Nakamura, T. Ohba, T. Futatsugi, M. Miyajima, Ultra thinning 300-mm wafer down to 7-um for 3D wafer integration on 45-nm node CMOS using strained silicon and Cu/low-k interconnects, in *Proceedings of the IEEE International Electron Devices Meeting*, IEEE, Piscataway, 2009, pp. 14.6.1–14.6.4
12. D.H. Kim, S. Kim, S.K. Lim, Impact of sub-micron through-silicon vias on the quality of today and future 3D IC Designs, in *Proceedings of the ACM/IEEE International Workshop on System Level Interconnect Prediction*, ACM, New York, 2011
13. D.H. Kim, K. Athikulwongse, M.B. Healy, M.M. Hossain, M. Jung, I. Khorosh, G. Kumar, Y.-J. Lee, D.L. Lewis, T.-W. Lin, C. Liu, S. Panth, M. Pathak, M. Ren, G. Shen, T. Song, D.H. Woo, X. Zhao, J. Kim, H. Choi, G.H. Loh, H.-H.S. Lee, S.K. Lim, 3D-MAPS: 3D massively parallel processor with stacked memory, in *Proceedings of the IEEE International Solid-State Circuits Conference*, IEEE, Piscataway, 2012, pp. 188–190
14. M. Koyanagi, T. Fukushima, T. Tanaka, High-density through silicon vias for 3-D LSIs. Proc. IEEE, IEEE, Piscataway, **97**(1), 49–59 (2009)
15. Y.-J. Lee, S.K. Lim, Timing analysis and optimization for 3D stacked multi-core microprocessors, in *Proceedings of the International 3D System Integration Conference*, IEEE, Piscataway, 2010
16. K. Mistry et al., A 45 nm logic technology with high-k+metal gate transistors, strained silicon, 9 Cu interconnect layers, 193 nm dry patterning, and 100 % Pb-free packaging, in *Proceedings of the IEEE International Electron Devices Meeting*, IEEE, Piscataway, 2007
17. M. Motoyoshi, Through-silicon via (TSV). Proc. IEEE, IEEE, Piscataway, **97**(1), 43–48 (2009)
18. Nangate, Nangate FreePDK45 open cell library. http://www.nangate.com
19. P. Packan et al., High performance 32 nm logic technology featuring 2nd generation high-k + metal gate transistors, in *Proceedings of the IEEE International Electron Devices Meeting*, IEEE, Piscataway, 2009
20. M. Pathak, Y.-J. Lee, T. Moon, S.K. Lim, Through-silicon-via management during 3D physical design: when to add and how many? in *Proceedings of the IEEE International Conference on Computer-Aided Design*, IEEE, Piscataway, 2010, pp. 387–394
21. PTM, Predictive technology model. http://ptm.asu.edu

Chapter 20
3D-MAPS: 3D Massively Parallel Processor with Stacked Memory

Abstract This chapter describes the architecture, design, analysis, and simulation and measurement results of the 3D-MAPS (3D massively parallel processor with stacked memory) chip built with a 1.5 V, 130 nm process technology, and a two-tier 3D stacking technology using 1.2 um-diameter, 6 um-height through-silicon vias (TSVs) and 3.4 um-diameter face-to-face bonding pads. 3D-MAPS consists of a core tier containing 64 cores and a memory tier containing 256 KB SRAM. Each core communicates with its dedicated 4 KB SRAM using face-to-face bonding pads. The maximum feasible memory bandwidth of 3D-MAPS is extremely wide (70.9 GB/s at 277 MHz) while the data transfer delay between the core tier and the memory tier is negligible. The maximum operating frequency is 277 MHz, the peak measured memory bandwidth usage is 63.8 GB/s, and the peak measured power is approximately 4 W based on eight parallel benchmarks.

The materials presented in this chapter are based on [7].

20.1 Introduction

Three-dimensional integrated circuits (3D ICs) are expected to provide numerous benefits. If a traditional two-dimensional (2D) IC is redesigned in multiple tiers vertically stacked in a 3D IC, the footprint area of 3D IC reduces significantly. This smaller form factor is a very important design consideration for some applications such as bio-chips. However, 3D IC also reduces the total wirelength from short vertical interconnections enabled by through-silicon vias (TSVs) and/or face-to-face (F2F) bond pads. Shorter interconnection improves the performance of the chip and reduces the dynamic power consumption. In addition, if the area of the 2D IC is very large, the fabrication cost of the 3D version could reduce [3, 16]. 3D ICs also enable heterogeneous integration by which circuit components built using different technologies are integrated into a single 3D chip.

S.K. Lim, *Design for High Performance, Low Power, and Reliable 3D Integrated Circuits*, DOI 10.1007/978-1-4419-9542-1_20,

Although 3D ICs built with TSVs or F2F bond pads provide very high inter-tier communication bandwidth, the development of architectures and applications that can fully exploit the wide bandwidth is still under research. A recently popular application expected to exploit the wide bandwidth is so called *wide-I/O* memory. The communication bandwidth between a processor chip and a memory chip on a PCB is usually limited by the maximum number of pins of the chips. If processor and memory chips are stacked and connected through TSVs or F2F bond pads, however, the memory bandwidth can increase dramatically. Since more and more applications such as image/video processing and large database management are demanding wider memory bandwidth, wide-I/O memory can fulfill the memory bandwidth requirement [10, 11, 13, 14].

There exist three main challenges to fully exploit the extreme memory bandwidth that 3D IC provides. First, the memory itself should be able to support wide-I/O capability. In other words, the memory should have a well-organized and well-parallelized memory hierarchy to fully utilize the wide memory bandwidth and/or operate at a very high frequency so that the memory can effectively process data request. Second, processors (cores and processing elements) should also be able to effectively process data that they receive from the memory chip. If the processing time is too long, the 3D IC cannot fully benefit from the wide memory bandwidth. Third, applications should be developed in such a way to exploit the wide memory bandwidth. Overall, the memory architecture, the processor architecture, and the application should all be optimized to fully benefit from the wide-I/O memory.

Several existing works have demonstrated the benefits of 3D ICs [2, 5, 6, 8, 9, 13], but none of them involves a fully functioning multi-core processor, memory stacking, and parallelized benchmarks. In this chapter, we present the details of our 3D-MAPS (3D massively parallel processor with stacked memory) chip built with GlobalFoundries 130 nm process technology and Tezzaron two-tier 3D stacking technology that provides 1.2 um-diameter, 6 um-height TSVs for chip-to-package connections and 3.4 um-diameter face-to-face bond pads for inter-tier communication. 3D-MAPS consists of two tiers, one (core tier) for processors and the other for memory (memory tier). The core tier contains 64 cores in 8×8 array, and the memory tier contains 64 memory tiles in the same 8×8 array. The memory operates at the same clock frequency (277 MHz) as the processors. We design the processor so that we can run memory read/write operations *every clock cycle* to fully exploit the wide memory bandwidth. We also execute eight parallel benchmark applications to fully utilize the 64 cores and benefit from 3D stacking.

This chapter is organized as follows. In Sect. 20.2, we present the architecture design of 3D-MAPS. Section 20.3 describes benchmark applications used to evaluate 3D-MAPS. In Sect. 20.4, we explain the 3D bonding and fabrication technologies used to manufacture 3D-MAPS. Section 20.5 demonstrates the physical design flow and methodology used to design 3D-MAPS. In Sect. 20.6, we present analysis methodologies and tools used to analyze 3D-MAPS layouts. Section 20.7 briefly explains the package and board design of 3D-MAPS. In Sect. 20.8, we present measurement results. Finally, Sect. 20.9 summarizes this chapter.

20.2 Architecture Design

In this section, we present the instruction set architecture, the single- and multi-core architectures, and the off-chip interface of 3D-MAPS.

20.2.1 *Instruction Set Architecture*

Figure 20.1 shows an overview of the 3D-MAPS architecture. The instruction set architecture (ISA) of 3D-MAPS supports 32-bit, two-way, very long instruction word (VLIW) instructions. The pipeline depth is 5, and we do not support out-of-order execution to keep the core architecture simple. The issue width is 2, so each instruction bundle is 64-bit wide (two instructions, one arithmetic and one memory). By reserving one execution path for each operation type (arithmetic/memory), we are able to maximize the memory bandwidth utilization. The 3D-MAPS ISA, which is similar to that of MIPS ISA, consists of arithmetic, logical, branch, move, and memory operations. The memory instructions handle both word and byte operands. Each core owns a 4 KB scratchpad SRAM memory tile that is stacked below it and connected using F2F bonding. Since there are 64 cores and 64 memory ties, the total memory tier capacity is 256 KB.

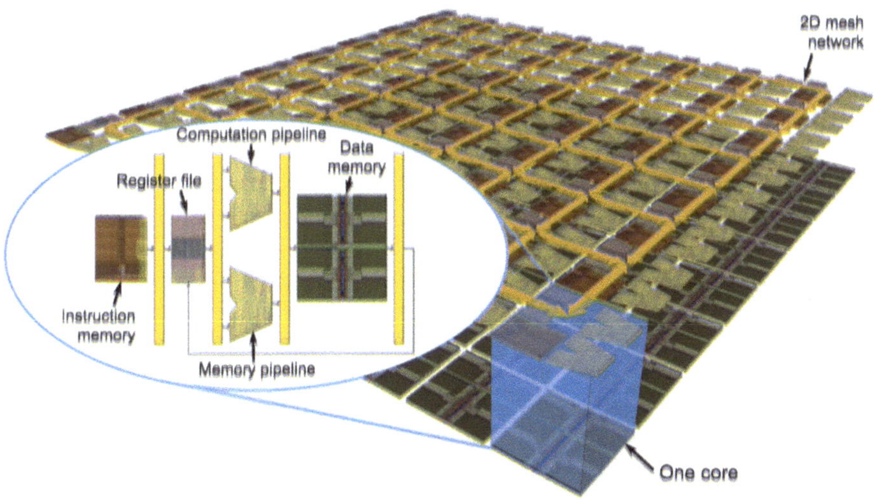

Fig. 20.1 3D-MAPS architecture

20.2.2 Single-Core Architecture

The goal of 3D-MAPS is to fully utilize the extremely high memory bandwidth enabled by vertical connections. Since the area assigned to a single core is very tight (560 um by 560 um),[1] we explored various design options such as pipeline depth, register file size, issue width, and supporting arithmetic functions. Because of the tight area budget, we excluded components requiring large silicon area such as complex decoders, dynamic instruction schedulers, reorder buffers, branch predictors, floating-point units, and integer dividers. Instead, we implemented these functionalities in software. For example, when we program a benchmark, we manually optimize the assembly code so that in-order and out-of-order executions of the program have similar performance. In addition, we have a five-stage pipeline depth, two issue width, a 64-bit 1.5 KB instruction memory, and a 32-bit dual-pump register file having 32 registers, three input (write) ports, and four output (read) ports.

Although we assign an instruction issue slot to memory instructions, our ISA also runs certain commonly used non-memory instructions in the memory pipeline when the memory instruction is absent. In addition, the ISA also supports auto-increment to further increase memory bandwidth utilization by improving the ratio between the memory instruction count and the non-memory instruction count.

20.2.3 Multi-core Architecture

In order to reduce routing complexity, simplify the multi-core network, and minimize power consumed by inter-core interconnections, we employ a point-to-point 2D mesh communication network controlled by explicit communication and synchronization instructions in the 3D-MAPS multi-core architecture. In particular, we choose the 2D mesh network instead of other alternatives such as 2D torus and 2D-folded torus network topologies for the following reasons. First, the 2D mesh network does not require long wires connecting two cores at both ends of a row or a column in the 2D torus network. Second, the total wirelength of 2D mesh structure is a half of that of 2D-folded torus network. Third, routing resources available for core-to-core communication in the core tier are very limited because the space between two adjacent cores is small and over-the-core routing is not allowed. To support our 2D mesh network topology, each core has buffers for sending (or receiving) data to (or from) its north, south, east, and west neighbors. To synchronize cores, we use an H-tree-shaped global barrier instruction.

[1]The footprint area given to us as part of 2009 DARPA/Tezzaron 3D IC multi-project wafer run was 5×5 mm, and the size of core was determined so that we can fit the maximum number of cores.

20.2.4 Off-Chip Interface

The primary design goal for the off-chip interface is to minimize the signal pin count. We model the interface using the IEEE 1149.1 standard with two key deviations. First, we design a custom test control state machine (TCSM) that controls the 3D-MAPS chip, manages functional test, initializes memory, and executes programs. Second, we have four pairs of test data in (TDI) and test data out (TDO) ports. The 64 cores in the core tier are grouped into four sectors of 16 cores each, and TDI k and TDO k are in charge of Sector k. We isolate test structures of each sector from other sectors so that defective sectors do not affect non-defective sectors during testing.

20.3 Benchmark Applications

As mentioned in Sect. 20.1, exploiting wide memory bandwidth requires parallelization of application softwares. To maximize the use of wide memory bandwidth, we design and parallelize the following eight benchmarks: AES encryption, edge detection, histogram, K-means clustering, matrix multiplication, median filter, motion estimation, and string search.

- AES Encryption: Advanced Encryption Standard (AES) [12] is a highly standardized kernel widely used for secure communication of confidential data. Specifically, AES is an iterative, symmetric-key block cipher that can use keys of 128, 192, and 256 bits, and encrypts and decrypts data in blocks of 16 bytes. Iterative ciphers use a loop structure that repeatedly performs permutations and substitutions of input data. Our implementation of AES encryption algorithm uses a seed key size of 256 bits that offers maximum security to the target data by 13 rounds of permutation and substitution. The kernel is parallelized by distributing the input plain text equally to the 64 cores of 3D-MAPS. As the computation of the kernel occurs only in a local data block, this benchmark does not include core-to-core communication.
- Edge Detection: Edge detection for an image is to find points at which the image has discontinuities of brightness. We use the sobel operator in image processing for edge detection. Each pixel has two 3×3 convolution matrices, one for horizontal kernel and the other for vertical kernel. Since we can compute each pixel individually, the edge detection algorithm can be highly parallelized. To enable each core to process a part of a given test image, we split the test image into an 8×8 array and assign each image tile in the array to its corresponding core. We also avoid boundary processing problems (edges exist on the boundaries of the split images) in the parallelization by assigning to each core an image tile slightly larger than the original 8×8 image tile.
- Histogram: Histogram requires many memory-intensive operations in various applications such as sorting a set of characters in the order of their occurrence

frequencies and processing images to approximate the gray-level co-occurrence matrix among neighboring pixels. The characteristics of the histogram problem make it suitable for exploiting the wide memory bandwidth. At each step in the kernel, one arithmetic operation (an increment) exists along with two load and one store operations. Our implementation uses the first 1 KB of the data memory in the memory tier as a counter for the 128 ASCII characters. To obtain the final histogram, we parallelize the kernel by distributing input data evenly to the 64 cores and accumulate locally computed histogram results across the cores.

- K-means clustering: K-means clustering is an NP-hard problem in which a given set of n-dimensional vectors is partitioned into k sets (clusters) minimizing the within-cluster sum of squares. For parallelization, we use the algorithm presented in [4]. In the beginning, each data node is pre-assigned to a particular group. We calculate the distance between each node and the center of its group in parallel. Then, we reassign each node to the group closest to it. We calculate the new center point of each group. We repeat this process until every node converges.
- Matrix Multiplication: Matrix Multiplication is extensively used in almost all applications. To leverage the 4 KB SRAM capacity in the current 3D-MAPS design, we use Cannon's algorithms [1] for distributed matrix multiplication. Cannon's algorithm requires constant memory space to process each matrix element, and the computation is easily distributable over a square mesh of 8 by 8 cores. Each of the input matrices is of 8×8 size and each core works on a single element of the two source matrices at a time. After each element-wise multiplication, the matrix elements are systematically rotated among the neighboring cores so that every core gets a fresh pair of elements after each rotation. The partial computation of element-wise multiplication is accumulated, and finally the product is resulted after a series of row-wise and column-wise rotation.
- Median filter: Median filter is a well-known kernel that is heavily used for removing noise from images. In our implementation, we apply 3×3 filter operations across a two-dimensional image similar to [15]. We parallelize this kernel in a two-dimensional manner so that each of our 8×8 cores is responsible for its evenly assigned region of the input image. We also heavily optimize this kernel by unrolling loop iterations and exploiting software pipelining techniques to schedule instructions more efficiently.
- Motion estimation: Motion estimation is a popular kernel that is used to detect a motion vector that can minimize the difference between two 2D images, typically two successive frames in a video sequence. In our implementation, we use 16×16 macro blocks and find a motion vector that can minimize the difference by shifting these macro blocks in a two-dimensional image space. We parallelize this kernel in a two-dimensional manner similar to Median filter. We also optimize this kernel by unrolling the innermost loop and carefully scheduling instructions using software pipelining techniques.
- String search: String Search is a popular kernel for analyzing text content. Our benchmark implements a naive string matching algorithm for searching a target pattern in the input text. We parallelize the kernel by distributing the input text

Table 20.1 3D technology specifications for 3D-MAPS

2D	
Process technology	Global foundries 130 nm
# metal layers	6
Supply voltage	1.5 V
3D	
Die-to-die bonding	Face-to-face
# dies	2
TSV shape	Octagon
TSV type	Via-first
TSV diameter	1.2 um
TSV height	6 um
Minimum TSV landing pad width	2.2 um
Minimum keep-out zone spacing	0.5 um
Minimum TSV-to-TSV pitch	2.5 um
TSV resistance (R)	0.6 Ω
TSV capacitance (C)	15 fF
Face-to-face connection pad shape	Octagon
Face-to-face connection pad diameter	3.4 um
Face-to-face connection pad pitch	5 um
Face-to-face connection pad R & C	Negligible

equally to the 64 cores. Each core first performs a search on the corresponding segment of the data allocated to its SRAM. Upon completion of the search on the local data, neighboring cores share their data to search the pattern in the overlapped region of the input text. Our implementation assumes a target pattern of maximum length 21 characters; and hence, 20 characters are transferred from the SRAM of a core to its preceding core in the grid layout of 8×8 cores. After communication of the overlapped data, a final search is performed on the shared text.

20.4 TSV and Stacking Technology

Table 20.1 shows the profile of the 3D technologies used to build 3D-MAPS. The device technology is based on 1P6M 130 nm process provided by GlobalFoundries. The supply voltage is 1.5 V. Tezzaron 3D technology stacks two dies using face-to-face (F2F) bonding. The thickness of the bottom die is 765 um, and the top die is thinned after bonding as illustrated in Fig. 20.2. The thickness of the top die is 12 um (6 um for the metal layers and 6 um for the silicon substrate). TSV resistance is very small ($=0.6\,\Omega$) compared with wire resistance ($0.7\,\Omega/\square$), but TSV capacitance is 15 fF, which is not negligible.

Via-first TSVs are inserted into the thin die. TSV diameter is 1.2 um, and the height is 6 um. The minimum Metal 1 TSV landing pad width is 2.2 um.

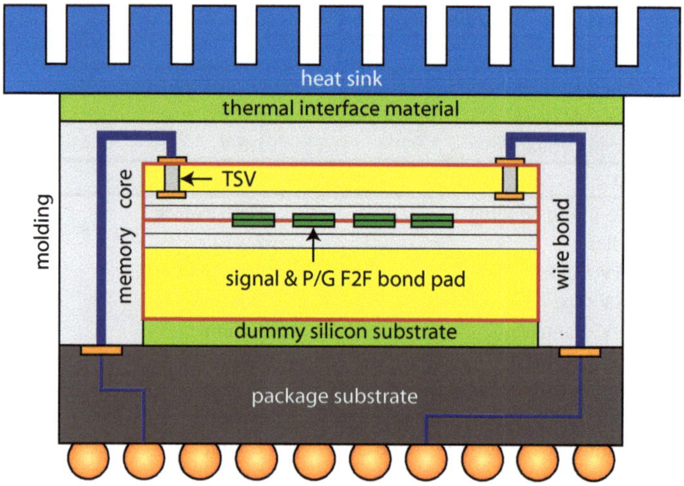

Fig. 20.2 3D stacking, TSVs, face-to-face bond pads, and chip-to-package connections

The minimum keep-out zone (KOZ) spacing that refers to the distance between the vertical surface of a TSV and a device is set to 0.5 um. The minimum TSV-to-TSV pitch is 5 um. The backside of the bottom (= thick) die is not used at all (it is attached to a dummy silicon substrate for packaging as illustrated in Fig. 20.2). The backside of the top die is used for wire bonding. Most of the TSVs in the top die are used for I/O, while others are used for dummy TSVs to satisfy the minimum TSV density rule.

Tezzaron provides F2F bond pads for the die-to-die communication. The Metal 6 layer is dedicated to F2F connections, so the actual number of metal layers that can be used for signal and power/ground routing is 5. The width of a F2F bond pad is 3.4 um, and the pitch between two adjacent F2F bond pads is 5 um. Since the electrical characteristics of the bond pads are very similar to a typical via, the resistance and capacitance of F2F bond pad is negligibly small. The location of all F2F bond pads is aligned to a grid structure with 5 um pitch, and approximately one million (= 1,000 by 1,000 grid) F2F bond pad locations are available between the two dies. This allows very high degree of freedom in choosing F2F bond pads for signal and P/G routing.

20.5 Physical Design of 3D-MAPS

20.5.1 Overview of 3D-MAPS Layout

The die footprint is 5 × 5 mm, and the area of a single core is 560 × 560 um. The core-to-core spacing is 10 um. Between a core and its memory tile, we have 668

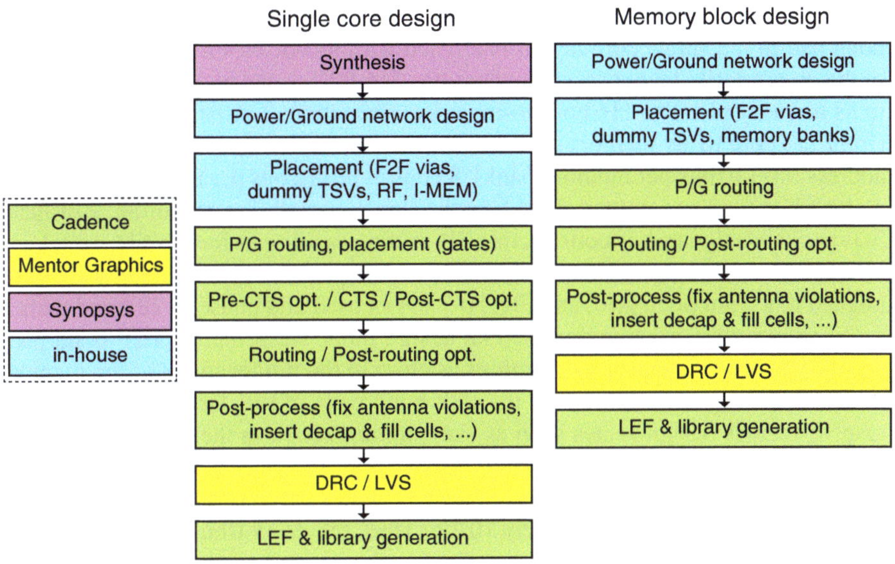

Fig. 20.3 3D-MAPS physical design flow

F2F connections for power and ground (P/G). Therefore, the total number of F2F bond pads used for power delivery is 42,752 (=64 × 668). Since the resistance of a single F2F bond pad is approximately 3 mΩ, ignoring the contact resistance, the resistance through P/G F2F bond pads is almost negligible. We also have 116 F2F connections for signal between a core and its memory tile that includes 32-bit data in/out, memory address, clock, and control signals. Therefore, the total number of F2F bond pads used for signal is 7,424 (=64 × 116). The longest path delay in the design is 3.61 ns at 1.5 V, which gives the maximum frequency of 277 MHz. 3D-MAPS contains about 50 K TSVs and 50 K F2F bond pads.

20.5.2 Single Core and Memory Tile Design

Figure 20.3 shows our single core and memory tile design flow. Our RTL-to-GDSII tool chain is based on commercial tools from Cadence, Synopsys, and Mentor Graphics and enhanced with our in-house tools to handle TSVs and 3D stacking. With initial design constraints, the entire 3D netlist is synthesized by Design Compiler. The layout of each die is designed separately in Encounter. The power distribution network is designed by our in-house tools. In addition, all the signal F2F pins are placed by our in-house tool, with proper alignments to the grid structure of Metal 6 bond pads. Clock tree synthesis is performed with proper boundary conditions at the clock F2F pins to memory die, followed by signal routing.

Figure 20.4 shows the single core and single memory tile layouts. Each F2F bond pad in the single core layout is aligned with its corresponding F2F bond pad in the memory tile layout. A single core has 116 F2F signal and clock connections to its memory tile. The 116 core-to-memory connections consist of 32 data-out bits (from the core to its memory), 32 data-in bits (from the memory to the core), 40 address bits (10 bits per memory bank), 8 control bits (2 bits per memory bank), and 4 clock bits (one per memory bank). In the single core design, therefore, a primary 3D specific design task is constructing 3D routing topologies for the 116 signal and clock core-to-memory connections. The F2F bond pads located on Metal 6 are to be aligned to a 5 um pitch grid, and Fig. 20.5 shows used and unused F2F bond pads.

To fully utilize existing commercial design tools, we pre-place F2F bond pads for all the core-to-memory connections as close to the pin locations of the memory blocks as possible. Placement of the F2F bond pads in this fashion enables us to resolve various design issues such as skew minimization in the 3D clock tree and timing optimization of 3D signal paths. For example, the longest wirelength between a signal or clock F2F bond pad pin (called P1) and its corresponding pin (called P2) in the memory tile is approximately 10 um. Thus, the delay from P1 to P2 (or from P2 to P1 depending on the signal direction) is almost negligible. Therefore, we can perform clock skew minimization and timing optimization in the core tier only.

The next steps are similar to the traditional IC design flows: power/ground distribution network design, placement, pre-clock tree synthesis (CTS) optimization, CTS, post-CTS optimization, routing, and post-routing optimization. For CTS, the F2F bond pad pins are defined as clock sinks. We also insert dummy TSVs in the single core design before standard cell placement to satisfy the TSV density design rule. Power delivery to the logic cells in the core is done with a P/G ring along the core boundary and horizontal P/G stripes. Power delivery to the memory tier is done using P/G F2F pads placed on the top and the bottom of each core/tile as shown in Fig. 20.4.

20.5.3 Top-Level Design and Power Delivery Network

Figure 20.6 shows the layout of the entire core and memory tiers. The design procedure for the core and memory tier is straightforward. We first place I/O cells on the periphery of the core tier. In each I/O cell, we utilize 204 *redundant* TSVs as shown in Fig. 20.7. This number is not chosen based on the current requirement, but mainly based on the area available. Next, all 64 cores and 64 memory tiles form a 8×8 array in each tier. Since the F2F bond pad grid is pre-fixed, we shift each core and memory tile to align the F2F bond pads to the grid structure in Metal 6 layer.

The P/G ring in each core are connected with additional P/G wires that run in between the cores. These wires are then connected to the P/G I/O cells. Power delivery to the memory tier is done using P/G F2F pads in each core. Clock routing is done at the full-chip level to connect the clock driver in clock I/O cell to the clock entry point in each core with minimum skew.

single core layout

single memory tile layout

Fig. 20.4 Single core and memory tile layout. *Red squares* in the *middle* are signal F2F pads, and the *red/green squares* on the *top* and the *bottom* are power/ground F2F pads

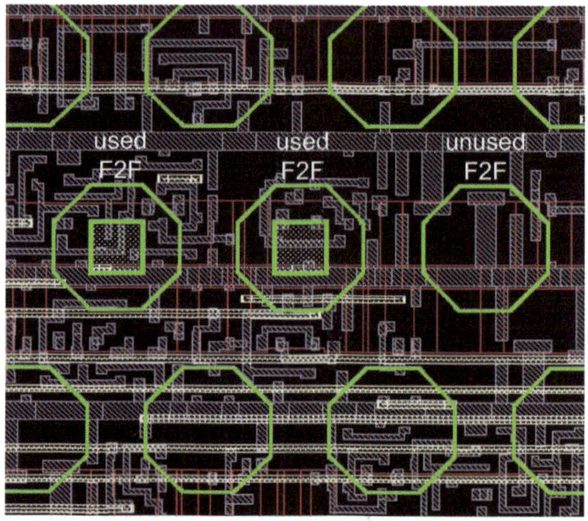

Fig. 20.5 Used and unused F2F bond pads. The pitch between adjacent F2F pad is 5 um

20.6 Design Evaluation and Verification of 3D-MAPS

Existing commercial tools built to simulate and analyze 2D ICs do not naturally handle 3D ICs. In this section, we describe our strategy to extend existing commercial tools to handle 3D IC design and analysis.[2]

20.6.1 Timing and Signal Integrity Analysis

Figure 20.8 shows our flow for 3D static timing and signal integrity (SI) analysis flow. 3D SI-aware timing and SI analysis requires a netlist and a standard parasitic exchange format (SPEF) file for each tier. Therefore, we obtain Verilog netlists of the core and the memory tiers from Encounter and SPEF files of the tiers using Cadence QRC extraction. We also need top-level netlist and SPEF files. The top-level netlist has two modules, one for the core tier and the other for the memory tier. We represent the face-to-face connections between the two tiers as wires connecting the two modules, and the TSVs between the core tier and the back side metal landing pads as wires connecting the core-tier module and the primary I/Os. The top-level SPEF file contains TSVs represented by the PI-model. Lastly, we feed all the netlist and SPEF files to PrimeTime to perform SI-aware timing analysis. Similarly, we feed the same files to Cadence CeltIC to perform SI analysis.

[2]We do not perform on-chip thermal analysis mainly because our processor is low power and consumes up to 4 W as shown in Sect. 20.8. We observe that our package-level solutions are enough to keep the processor temperature low using a simple air-cooled heatsink as shown in Fig. 20.13.

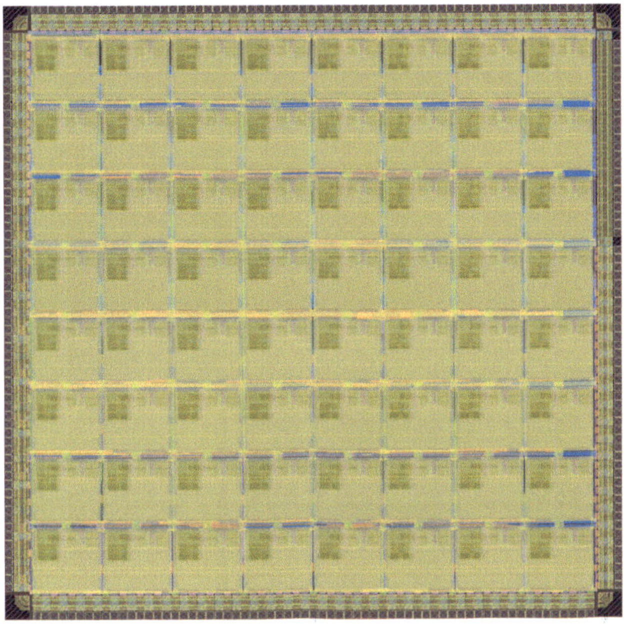

64 core tier layout

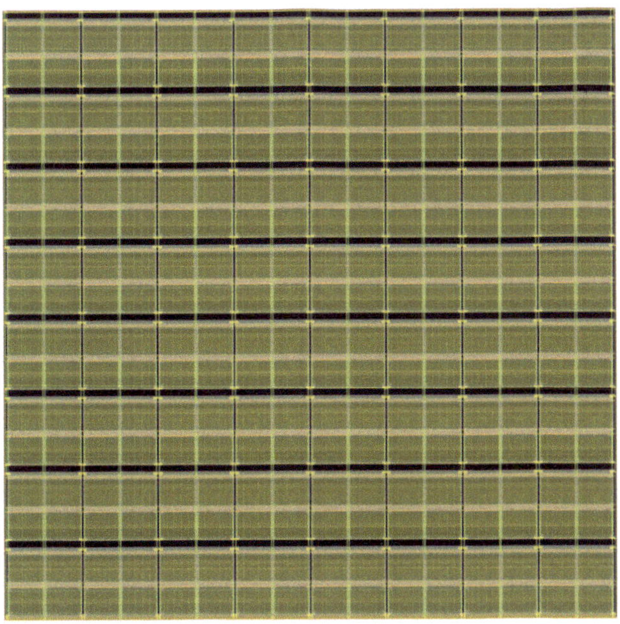

256KB SRAM tier layout

Fig. 20.6 Core die (= 64 cores) and memory die (= 64 SRAM tiles) layouts

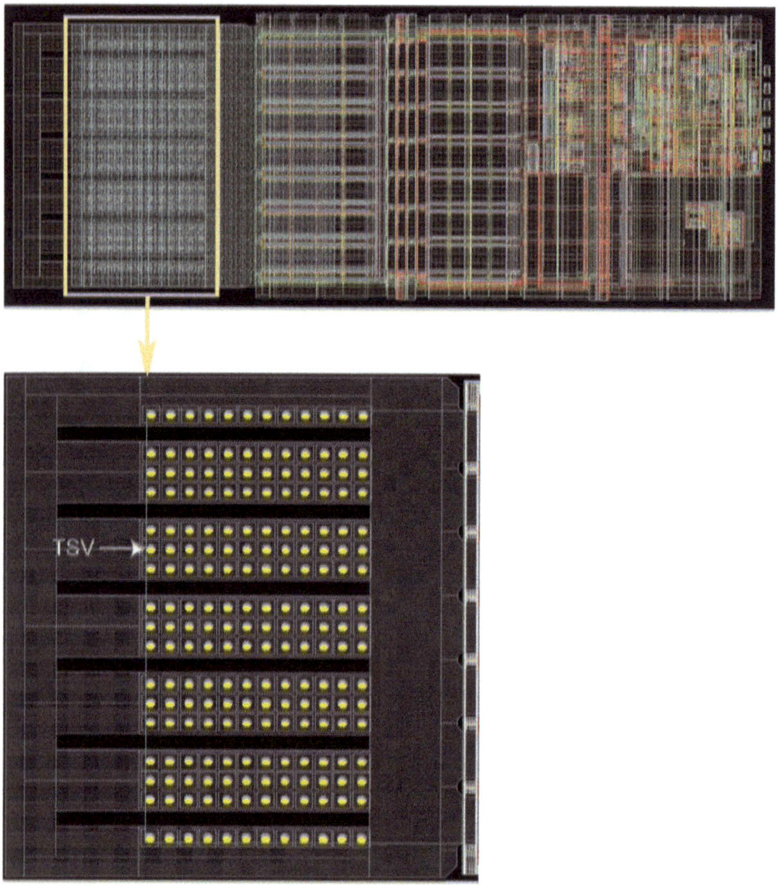

Fig. 20.7 Redundant TSVs inserted in an I/O pad

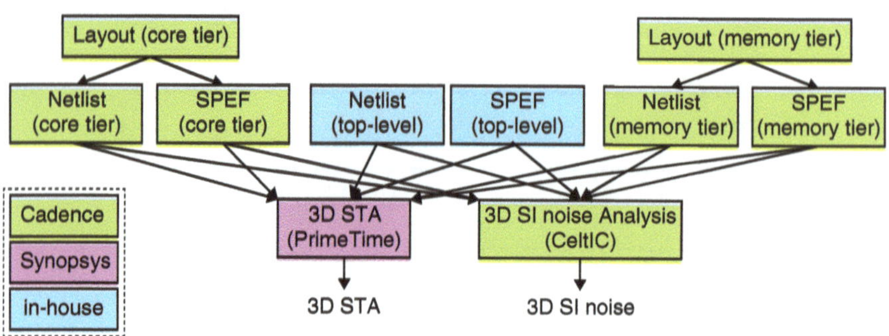

Fig. 20.8 Our 3D static timing and signal integrity analysis flow

Table 20.2 Timing-critical path of our single core. PS means pipeline stage. Net delays are negligibly small and thus not shown

Block	Type	Delay (ns)
PS 2–3	gate (SDFFHQX2)	0.253
MUX	gate (AO22X1)	0.228
	gate (AOI22X1)	0.272
	gate (NAND2X2)	0.168
adder	gate (NOR2X2)	0.172
	gate (NOR2X1)	0.113
	gate (AOI21X1)	0.176
	gate (OAI21X1)	0.119
	gate (AOI21X1)	0.324
	gate (BUFX3)	0.196
	gate (XNOR2X1)	0.135
	gate (XOR2X1)	0.142
	gate (AO22X4)	0.230
PS 3–4	gate-setup (SDFFQX1)	0.487
	Slack	+0.019

Table 20.2 shows the timing-critical path of our single core, based on our 3D static timing analysis (STA) engine described above. The path is from a scan-enabled F/F of the pipeline stage 2–3, through a MUX and an adder, to a scan-enabled F/F of the pipeline stage 3–4. In the slack calculation, the STA engine uses clock arrival time at the begin point (1.270 ns) and the end point (0.802 ns) of the path. The path delay is 2.625 ns, excluding the setup time of the end point F/F. The maximum crosstalk noise value on the worst net is 674 mV, which is close to the noise limit of 750 mV.

In Tezzaron process technology, two dies are bonded by face-to-face connections, so die-to-die capacitive coupling exists at the bonding interface. Therefore, capacitance extraction tools should consider the top metal layers in both dies in face-to-face bonding. However, we extract parasitic RC in each die separately and ignore this die-to-die capacitive coupling. The main reason is that currently there is no commercial tool that can handle the extraction of this kind of parasitics. However, we believe that this is acceptable because we do not use Metal 5 in the memory tier, so the distance between the top surface of the Metal 5 layer in the core tier and that of the Metal 4 layer in the memory tier is approximately ten times greater than the distance between two adjacent metal layers.

20.6.2 Power and Power Supply Noise Analysis

Figure 20.9 shows our 3D power analysis flow. We conduct 3D power consumption analysis as follows. We prepare netlists and standard delay format (SDF) files for both the core and the memory tiers. We also prepare bitstreams for each benchmark. With all these files, we run Mentor Graphics ModelSim to obtain switching activities of gates and nets. Finally, we perform power simulation using Encounter with the netlist, SPEF, and switching activity files.

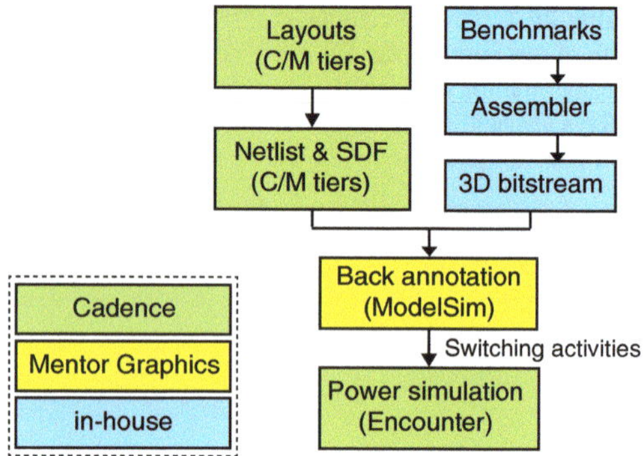

Fig. 20.9 Our 3D power analysis flow

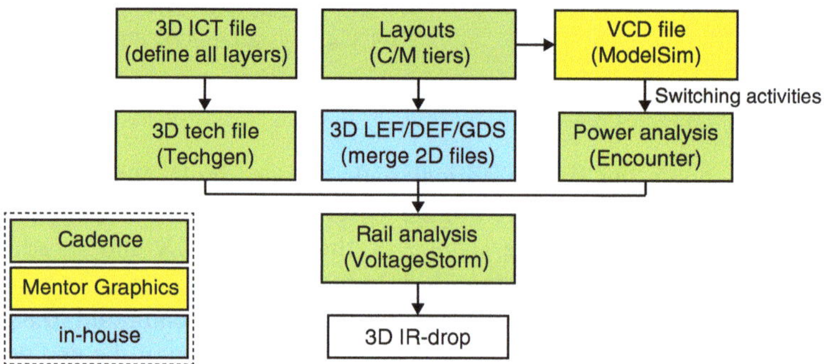

Fig. 20.10 Our 3D IR-drop analysis flow

Figure 20.10 shows our 3D IR-drop analysis flow. We perform 3D power noise analysis using Cadence VoltageStorm. The stand-alone VoltageStorm takes in a DEF file, technology files, and power dissipation files to generate both peak and average power noise values. For our 3D-MAPS design, we perform true 3D power noise analysis with VoltageStorm as follows. We first create a 3D interconnect technology file (ICT) containing all the layers in the top and the bottom tiers. Then we run Cadence Techgen to compile it. We also construct a new LEF file containing instances specific to each tier. For instance, we represent an inverter cell in the core tier as INVC, whereas we represent an inverter cell in the memory tier as INVM. Similarly, we define INVC using Metal 1C wires, which is the Metal 1 layer in the core tier, whereas we define INVM using Metal 1 M wires, which is the Metal 1 layer in the memory tier. Then, we construct a new DEF file containing both the top and the bottom tiers using the new LEF file. Finally, we obtain 3D power noise values using VoltageStorm with the new technology, LEF, and DEF files as well as the power dissipation file obtained from our 3D power analysis.

Figure 20.11 shows our 3D IR-drop analysis results, where we show the IR-drop map of a single core. The maximum IR drop happens with the clock buffers, and the level is 60 mV. We also performed the full 64-core IR-drop analysis. Because the power/ground I/O cells are located along the periphery, the cores in the middle

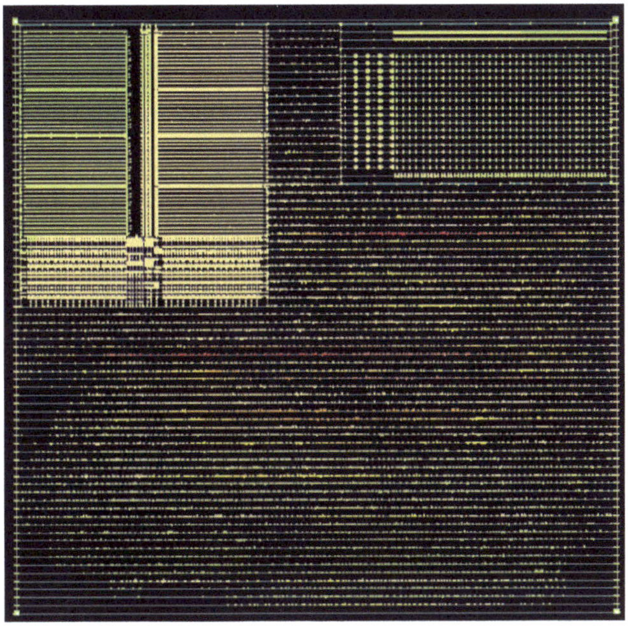

Fig. 20.11 IR-drop map of single core, where the maximum drop is 60 mV

experience the worst IR-drop of 78 mV. But, this is under our threshold, which is set to 150 mV. We found that adding thick P/G rings for the cores and thick P/G wires in between the cores help reduce IR-drop significantly.

20.6.3 DRC and LVS

Design rule checking (DRC) of 3D ICs consists of 2D and 3D DRC. 3D design rules are the minimum spacing between two adjacent TSVs, the minimum overlap between a TSV and its Metal 1 landing pad, the minimum TSV-to-device spacing, the minimum TSV density, and so on. Since there exists no inter-tier design rules, we run 2D and 3D DRC for each tier separately.

On the other hand, we perform 3D layout-versus-schematic (LVS) for both core and memory tiers simultaneously. Especially, the netlist extraction tool should be able to understand multi-tier layouts. For 3D LVS, we assign a unique number to each layer of the layouts. For example, we assign 32 and 42 to the poly layer and the active layer in the core tier, respectively, and 132 and 142 to the poly layer and the active layer in the memory tier, respectively. We also modify the extraction rule files so that the same layers in the core tier and the memory tier are treated individually. For example, the extraction rule includes both "a crossing of 32 and 42 forms a transistor" and "a crossing of 132 and 142 forms a transistor". In this case, even if the x- and the y-coordinates of a transistor in the core tier overlap with those in the memory tier, the transistors are extracted correctly because their layer numbers are different. Once a netlist is extracted from the given layouts, comparison between the extracted netlist and the given schematic netlist is performed by a commercial LVS tools.

20.7 Package and Board Design

3D-MAPS is wire bonded to a four-layer, 0.8 mm-pitch land grid array (LGA) package as shown in Fig. 20.12. The LGA contains 324 land pads out of which 294 pads are used for power and ground to supply high current demand (~2.7A) and 30 pads are used for clock and signal. The package is designed to accommodate high temperature (~90 °C from our simulation) caused by high power density (~16 W/cm^2). A dummy silicon substrate is inserted between the memory tier of 3D-MAPS and the package substrate to increase thermal conductivity from 3D-MAPS to the package. In addition, the center region of the LGA is implemented as a single large copper pad and dedicated to ground to decrease thermal resistance from the ground plane to the outside.

To verify the functionality of 3D-MAPS, we design a four-layer PCB test board having additional features. For example, to vary the power supply voltage from 0.9 to 1.9 V, we add an additional power circuitry on the PCB. I/Os are connected to an FPGA test board (Xilinx Vertex-6) for verification. The average parasitic values of our signal package routes are $R = 377.5\,m\Omega$, $L = 4.1\,nH$, and $C = 1\,pF$. Figure 20.12 shows the bare die of 3D-MAPS and its package.

To load data, execute benchmarks, read final data out, and verify the functionality of 3D-MAPS, we use scan chains and the test structure mentioned in Sect. 20.2. Figure 20.13 shows testing boards for 3D-MAPS. The scan cells are ordered into several scan chains that can be bypassed at the chain and core granularity. The scan chains are inserted and optimized with Synopsys Design Compiler. The test architecture is composed of four independent test sectors that enable coarse-grained fault isolation; each sector can be tested and operated completely independently, up to and including the scan-in and scan-out package pins. A custom test controller is used to manage the configuration, test, and operation of the chip. 3D-MAPS contains 49,408 scan FFs (772 per core) and 16 chains (4 per sector). As there is no traditional off-chip memory interface, the test system serves as the only access mechanism for this chip. An FPGA development kit serves as the external test driver, delivering the driving bit-streams, managing the chip operation, and observing the results.

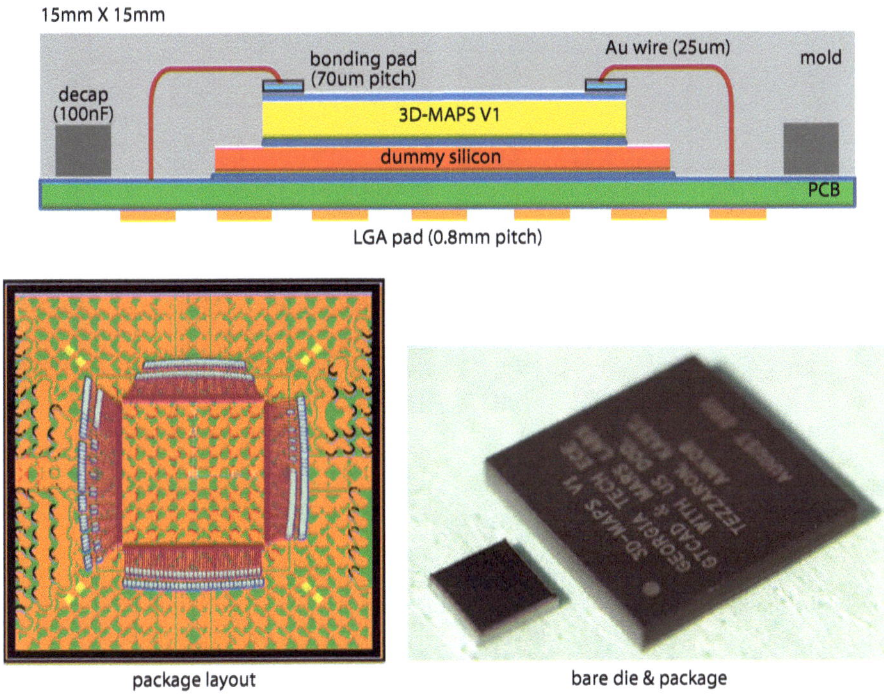

Fig. 20.12 Package design for 3D-MAPS

Fig. 20.13 Test board for 3D-MAPS

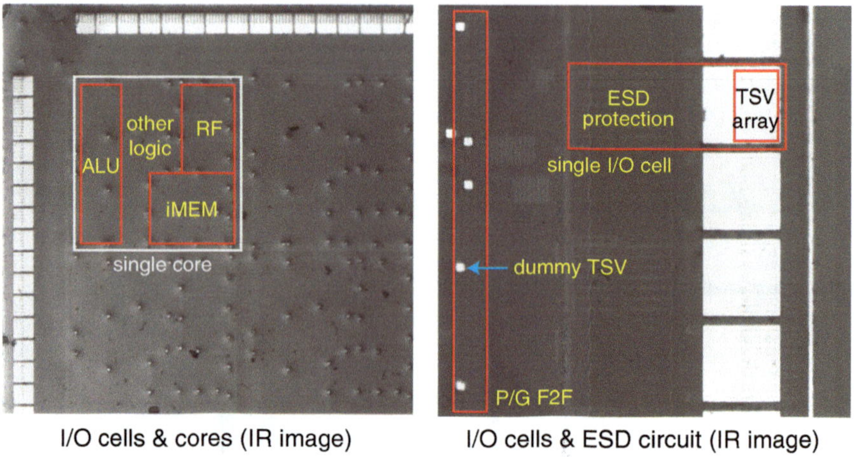

Fig. 20.14 Die photographs

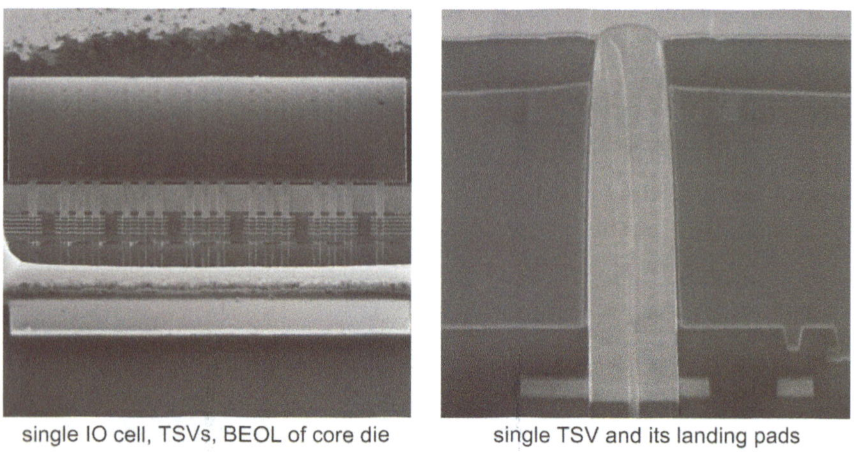

Fig. 20.15 SEM image of Tezzaron TSVs and face-to-face bond pads

20.8 Die Shots and Measurement Results

Figure 20.14 shows die photographs that contain logic blocks, TSVs, I/O cells, and F2F pads. Figure 20.15 shows details of TSVs and F2F bond pads between the two dies. Figure 20.16 shows SEM image of TSVs in an I/O cell, front-side and back-side metal I/O pads, and P/G face-to-face bond pads.

Table 20.3 lists the peak memory bandwidth we measured in gigabytes per second (GB/s), instructions per cycle (IPC) per core, and measured power consumption of 3D-MAPS for each benchmark. The maximum memory bandwidth that we achieved is 63.8 GB/s (Median filter). The theoretical maximum memory bandwidth that 3D-MAPS can achieve is:

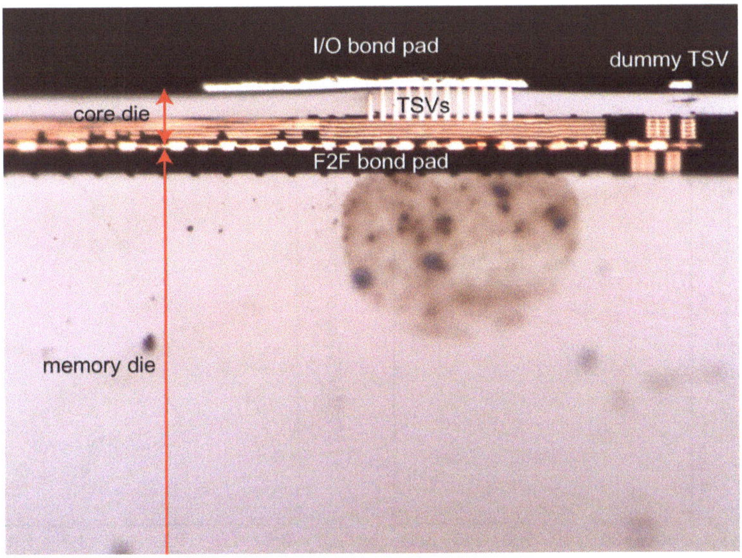

Fig. 20.16 SEM image of TSVs in an I/O cell, *front-side* and *back-side* metal I/O pads, and P/G face-to-face bond pads

Table 20.3 Measured memory bandwidth, IPC per core, and measured power consumption results

Benchmark	Bandwidth (GB/s)	IPC/core	Power (W)
AES encryption	49.5	0.97	4.032
Edge detection	15.6	0.95	3.768
Histogram	30.3	0.90	3.588
K-Means clustering	40.6	0.94	4.014
Matrix multiplication	13.8	0.32	3.789
Median filter	63.8	1.62	4.007
Motion estimation	24.1	1.20	3.830
String search	8.9	0.65	3.876

$$64 \times 4 \times (277 \times 10^6) = 70.912\,\text{GB/s} \tag{20.1}$$

where 64 is the total number of cores, four the word size (4 bytes), and 277 MHz the operating frequency. Therefore, the median filter benchmark uses up to 90 % of the theoretical maximum memory bandwidth, whereas the string search benchmark uses 13 %. The theoretical maximum memory bandwidth is higher than that of modern processors (e.g., Intel Core i7) whose maximum memory bandwidth is approximately 64 GB/s when DDR3-1333 memory is used. If we simply scale the clock frequency of 3D-MAPS up to 1,333 MHz, the maximum memory bandwidth becomes 341 GB/s, which is much higher than the modern memory bandwidth.

We measure power consumption of 3D-MAPS using a Watts-Up Pro power meter. The peak power consumption ranges from 3.5 to 4.0 W as listed in Table 20.3.

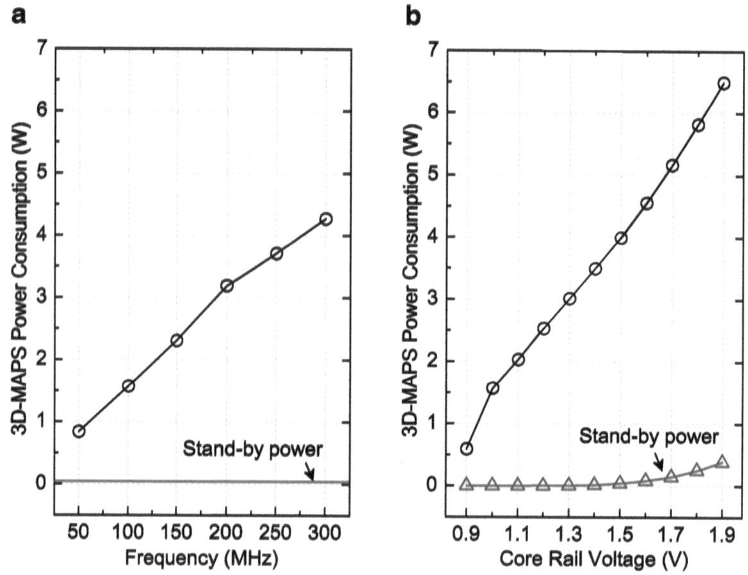

Fig. 20.17 (**a**) Frequency vs. power (at 1.5 V), (**b**) Voltage vs. power (at 250 MHz) for the AES encryption benchmark

Although the median filter benchmark shows the highest memory bandwidth usage, the AES encryption benchmark has the highest peak power consumption because arithmetic and logic operations consume more power than memory operations. We also measure power consumption when the clock frequency and the core rail voltage vary. Figure 20.17 shows power consumption of 3D-MAPS for the AES encryption benchmark when the clock frequency varies from 50 to 277 MHz (at 1.5 V) and when the core rail voltage varies from 0.9 to 1.9 V (at 250 MHz), respectively. The power consumption increases almost linearly with the frequency while the stand-by power is negligible. The core rail voltage vs. power graph also shows that the power consumption is almost linearly dependent on the supply voltage while the stand-by power consumption increases slightly. The only exception is at 0.9 V where the chip begins to suffer from near-threshold effects since the threshold voltage of our 130 nm PMOS is approximately 0.85 V.

20.9 Conclusions

We presented the architecture, design flow and methodologies, analysis, and measurements of 3D-MAPS, 3D massively parallel processor with stacked memory, designed to exploit extremely high memory bandwidth obtained from core-memory stacking. The 3D stacking technology that we use supports numerous inter-die connections through face-to-face bonding pads, thereby enabling the very high memory bandwidth. To maximize the utilization of the wide memory bandwidth, we designed the architecture so we can run memory read/write operations every clock cycle. We also built a custom RTL-to-GDSII toolset for 3D ID designs based on 2D IC commercial tools and our add-ins. We selected and parallelized the eight benchmarks to fully utilize the 8×8 2D mesh multi-core architecture. 3D-MAPS runs at 277 MHz, and the maximum peak memory bandwidth utilization is 63.8 GB/s while consuming 4 W power.

References

1. L.E. Cannon, A cellular computer to implement the Kalman Filter algorithm. PhD thesis, Montana State University, 1969
2. G.V. der Plas, P. Limaye, I. Loi, A. Mercha, H. Oprins, C. Torregiani, S. Thijs, D. Linten, M. Stucchi, G. Katti, D. Velenis, V. Cherman, B. Vandevelde, V. Simons, I.D. Wolf, R. Labie, D. Perry, S. Bronckers, N. Minas, M. Cupac, W. Ruythooren, J.V. Olmen, A. Phommahaxay, M. de Potter de ten Broeck, A. Opdebeeck, M. Rakowski, B.D. Wachter, M. Dehan, M. Nelis, R. Agarwal, A. Pullini, F. Angiolini, L. Benini, W. Dehaene, Y. Travaly, E. Beyne, P. Marchal, Design issues and considerations for low-cost 3D TSV IC technology. IEEE J. Solid-State Circuits **46**(1), 293–307 (2011)
3. X. Dong, Y. Xie, System-level cost analysis and design exploration for three-dimensional integrated circuits (3D ICs), in *Proceedings of Asia and South Pacific Design Automation Conference*, Pacifico Yokohama, Jan 2009, pp. 234–241
4. B. Hohlt, Pthread parallel K-means, 2001
5. U. Kang, H.-J. Chung, S. Heo, S.-H. Ahn, H. Lee, S.-H. Cha, J. Ahn, D. Kwon, J.H. Kim, J.-W. Lee, H.-S. Joo, W.-S. Kim, H.-K. Kim, E.-M. Lee, S.-R. Kim, K.-H. Ma, D.-H. Jang, N.-S. Kim, M.-S. Choi, S.-J. Oh, J.-B. Lee, T.-K. Jung, J.-H. Yoo, C. Kim, 8Gb 3D DDR3 DRAM using through-silicon-via technology, in *Proceedings of IEEE International Solid-State Circuits Conference*, San Francisco, Feb 2009, pp. 130–131
6. Y. Kikuchi, M. Takahashi, T. Maeda, H. Hara, H. Arakida, H. Yamamoto, Y. Hagiwara, T. Fujita, M. Watanabe, T. Shimazawa, Y. Ohara, T. Miyamori, M. Hamada, M. Takahashi, Y. Oowaki, A 222mW H.264 full-HD decoding application processor with x512b stacked DRAM in 40nm, in *Proceedings of IEEE International Solid-State Circuits Conference*, San Francisco, Feb 2010, pp. 326–327
7. D.H. Kim, K. Athikulwongse, M.B. Healy, M.M. Hossain, M. Jung, I. Khorosh, G. Kumar, Y.-J. Lee, D.L. Lewis, T.-W. Lin, C. Liu, S. Panth, M. Pathak, M. Ren, G. Shen, T. Song, D.H. Woo, X. Zhao, J. Kim, H. Choi, G.H. Loh, H.-H. S. Lee, S.K. Lim, 3D-MAPS: 3D massively parallel processor with stacked memory, in *Proceedings of IEEE International Solid-State Circuits Conference*, San Francisco, 2012
8. J.-S. Kim, C.S. Oh, H. Lee, D. Lee, H.R. Hwang, S. Hwang, B. Na, J. Moon, J.-G. Kim, H. Park, J.-W. Ryu, K. Park, S.K. Kang, S.-Y. Kim, H. Kim, J.-M. Bang, H. Cho, M. Jang, C. Han, J.-B. Lee, J.S. Choi, and Y.-H. Jun. A 1.2 V 12.8 GB/s 2 Gb mobile wide-I/O DRAM with 4×128 I/Os using TSV based stacking. IEEE J. Solid-State Circuits **47**(1), 107–116 (2012)
9. M. Koyanagi, Y. Nakagawa, K.-W. Lee, T. Nakamura, Y. Yamada, K. Inamura, K. tae Park, H. Kurino, Neuromorphic vision chip fabricated using three-dimensional integration technology, in *Proceedings of IEEE International Solid-State Circuits Conference*, San Francisco, Feb 2001, pp. 270–271

10. C.C. Liu, I. Ganusov, M. Burtscher, S. Tiwari, Bridging the processor-memory performance gap with 3D IC technology. IEEE Des. Test Comput. **22**(6), 556–564 (2005)
11. G.H. Loh, 3D-Stacked memory architectures for multi-core processors, in *Proceedings of IEEE International Symposium on Computer Architecture*, Beijing, June 2008, pp. 453–464
12. National Institute of Standards and Technology, Advanced encryption standard (AES), http://csrc.nist.gov/publications/fips/fips197/fips-197.pdf
13. H. Saito, M. Nakajima, T. Okamoto, Y. Yamada, A. Ohuchi, N. Iguchi, T. Sakamoto, K. Yamaguchi, M. Mizuno, A chip-stacked memory for on-chip SRAM-rich SoCs and processors. IEEE J. Solid-State Circuits **45**(1), 15–22 (2010)
14. D.H. Woo, N.H. Seong, D.L. Lewis, H.-H.S. Lee, An optimized 3D-stacked memory architecture by exploiting excessive, high-density TSV bandwidth, in *Proceedings of IEEE International Symposium on High-Performance Computer Architecture*, Bangalore, Jan 2010
15. Xilinx, Implementing median filters in XC4000E FPGAs
16. J. Zhao, X. Dong, Y. Xie. Cost-aware three-dimensional (3D) many-core multiprocessor design, in *Proceedings of ACM Design Automation Conference*, Anaheim, June 2010, pp. 126–131

MIX
Papier aus verantwortungsvollen Quellen
Paper from responsible sources
FSC® C105338

If you have any concerns about our products,
you can contact us on
ProductSafety@springernature.com

In case Publisher is established outside the EU,
the EU authorized representative is:
Springer Nature Customer Service Center GmbH
Europaplatz 3, 69115 Heidelberg, Germany

Printed by Libri Plureos GmbH
in Hamburg, Germany